WEARABLE SENSORS

WEARABLE SENSORS

Fundamentals, Implementation and Applications

Edited by

EDWARD SAZONOV

MICHAEL R. NEUMAN

AMSTERDAM • BOSTON • HEIDELBERG • LONDON
NEW YORK • OXFORD • PARIS • SAN DIEGO
SAN FRANCISCO • SINGAPORE • SYDNEY • TOKYO

Academic Press is an imprint of Elsevier

Academic Press is an imprint of Elsevier
525 B Street, Suite 1800, San Diego, CA 92101-4495, USA
225 Wyman Street, Waltham, MA 02451, USA
The Boulevard, Langford Lane, Kidlington, Oxford OX5 1GB, UK

British Library Cataloguing-in-Publication Data
A catalogue record for this book is available from the British Library

Library of Congress Cataloging-in-Publication Data
A catalog record for this book is available from the Library of Congress

ISBN : 978-0-12-418662-0

For information on all Academic Press publications
visit our website at http://store.elsevier.com/

Typeset by MPS Limited, Chennai, India
www.adi-mps.com

Contents

List of Contributors

Ruzena Bajcsy Electrical Engineering and Computer Science, University of California, Berkeley, California, USA

Paolo Barsocchi ISTI-CNR, Pisa, Italy

Fernando Benito-Lopez CIC microGUNE, Arrasate-Mondragón, Spain; INSIGHT: Centre for Data Analytics, National Centre for Sensor Research, Dublin City University, Dublin, Ireland

Mattia Bertschi Tampere University of Technology, Tampere, Finland

Laura Caldani Smartex s.r.l., Pisa, Italy

Alexander J. Casson Imperial College, London, UK

Gert Cauwenberghs Department of Bioengineering at University of California, San Diego, La Jolla, CA, USA

Guangwei Chen Imperial College, London, UK

Yu M. Chi Cognionics, Inc., San Diego, CA, USA

Kyunghee Chung Georgia Institute of Technology, School of Materials Science & Engineering, Atlanta, Georgia, USA

Shirley Coyle INSIGHT: Centre for Data Analytics, National Centre for Sensor Research, Dublin City University, Dublin, Ireland

Vincenzo F. Curto INSIGHT: Centre for Data Analytics, National Centre for Sensor Research, Dublin City University, Dublin, Ireland

Dermot Diamond INSIGHT: Centre for Data Analytics, National Centre for Sensor Research, Dublin City University, Dublin, Ireland

Xiaorong Ding Department of Electronic Engineering, The Chinese University of Hong Kong, Hong Kong SAR, China

Tobias Dräger Fraunhofer Institute for Integrated Circuits IIS, Nuremberg, Germany

Lucy Dunne University of Minnesota, Minneapolis, MN, USA

John Dykstra MetaLogics Inc., Minneapolis, Minnesota, USA

Larisa Florea INSIGHT: Centre for Data Analytics, National Centre for Sensor Research, Dublin City University, Dublin, Ireland

Juan M. Fontana The University of Alabama, Tuscaloosa, Alabama, USA

Maysam Ghovanloo GT-Bionics Lab, School of Electrical and Computer Engineering, Georgia Institute of Technology, Atlanta, Georgia, USA

Sohmyung Ha Department of Bioengineering at University of California, San Diego, La Jolla, CA, USA

Xueliang Huo Interactive Entertainment Business, Microsoft, Redmond, Washington, USA

Jung-Hwan Hwang Electronics and Telecommunications Research, Daejeon, Korea

Chang-Hee Hyoung Electronics and Telecommunications Research, Daejeon, Korea

Roozbeh Jafari University of Texas at Dallas, Richardson, Texas, USA

Sundaresan Jayaraman Georgia Institute of Technology, School of Materials Science & Engineering, Atlanta, Georgia, USA

Chul Kim Department of Bioengineering at University of California, San Diego, La Jolla, CA, USA

Ilkka Korhonen Tampere University of Technology, Tampere, Finland

Yuichi Kurita Institute of Engineering, Hiroshima University, Hiroshima, Japan

Tomohiro Kuroda Kyoto University Hospital, Kyoto, Japan

Mathieu Lemay Tampere University of Technology, Tampere, Finland

Ming Li Utah State University, Logan, Utah, USA

Vitali Loseu MPI Lab, Samsung Research America

Ningqi Luo Department of Electronic Engineering, The Chinese University of Hong Kong, Hong Kong SAR, China

I. Maglogiannis Department of Digital Systems, University of Piraeus, Piraeus, Greece

Masaaki Makikawa Ritsumeikan University, Kyoto, Japan

A.L.R. Mansano Delft University of Technology, Delft, The Netherlands

Atsuji Masuda Industrial Technology Center of Fukui Prefecture, Fukui, Japan

Loreto Mateu Fraunhofer Institute for Integrated Circuits IIS, Nuremberg, Germany

Iker Mayordomo Fraunhofer Institute for Integrated Circuits IIS, Nuremberg, Germany

Ed Melanson University of Colorado Anschutz Medical Campus, Denver, Colorado, USA

Shima Okada Kinki University, Osaka, Japan

Maria Pacelli Smartex s.r.l., Pisa, Italy

Rita Paradiso Smartex s.r.l., Pisa, Italy

Jakub Parak Tampere University of Technology, Tampere, Finland

Sungmee Park Georgia Institute of Technology, School of Materials Science & Engineering, Atlanta, Georgia, USA

Markus Pollak Fraunhofer Institute for Integrated Circuits IIS, Nuremberg, Germany

Carmen C.Y. Poon Department of Electronic Engineering and Department of Surgery, The Chinese University of Hong Kong, Hong Kong SAR, China

Francesco Potortì ISTI-CNR, Pisa, Italy

Halley Profita University of Minnesota, Minneapolis, MN, USA

Philippe Renevey Tampere University of Technology, Tampere, Finland

Esther Rodriguez-Villegas Imperial College, London, UK

Giovanni Salvatore ETH Zürich, Zürich, Switzerland

H.G. Sandalidis Department of Computer Science and Biomedical Informatics, University of Thessaly, Lamia, Greece

Edward Sazonov The University of Alabama, Tuscaloosa, Alabama, USA

W.A. Serdijn Delft University of Technology, Delft, The Netherlands

Edmund Seto School of Public Health, University of Washington, Seattle, Washington, USA

Lu Shi University of Arkansas at Little Rock, Little Rock, Arkansas, USA

Naruhiro Shiozawa Ritsumeikan University, Kyoto, Japan

Josep Sola Tampere University of Technology, Tampere, Finland

M. Stoopman Delft University of Technology, Delft, The Netherlands

Neil Szuminsky MetaLogics Inc., Minneapolis, Minnesota, USA

Hideya Takahashi Osaka City University, Osaka, Japan

Toshiyo Tamura Osaka Electro-Communication University, Neyagawa, Japan

Gerhard Tröster ETH Zürich, Zürich, Switzerland

Jian Wu University of Texas at Dallas, Richardson, Texas, USA

Shucheng Yu University of Arkansas at Little Rock, Little Rock, Arkansas, USA

Clint Zeagler University of Minnesota, Minneapolis, MN, USA

Yuan Ting Zhang Department of Electronic Engineering, The Chinese University of Hong Kong, Hong Kong SAR, China; Key Laboratory for Health Informatics of Chinese Academy of Science (HICAS), Shenzhen, China

Yali Zheng Department of Electronic Engineering, The Chinese University of Hong Kong, Hong Kong SAR, China

Introduction

Recent years witnessed an explosive growth in wearable technology. A wearable device is essentially a tiny computer with sensing, processing, storage and communications capabilities. May wearable devices also include interfaces and actuation capabilities that provide feedback to the user. The concept of a wearable device is not new, but the area is experiencing a rapid growth in popularity due to several factors that we try to explore below.

Historically, the concept of a wearable device goes back for centuries, with the pocket and wrist watches being the best know examples of a widely popular device that is still in use today. First watches designed to be worn appeared in the sixteenth century and were serving more as a decoration rather than a practical device for keeping time. These wearable mechanical devices evolved from rather cumbersome pieces worn as attachments to clothing or on a chain around the neck to a pocket and then to a wrist worn device. The evolution of the mechanical watch technology involved invention of self-winding watches, or early form of what today is known as energy harvesting. The technology also migrated from purely mechanical device to electromechanical and then to electronic devices that provide better accuracy and better functionality at a much lower cost. However, this evolution took almost 500 years. The pace of progress has increased dramatically in recent times and now the one of the oldest wearable technologies is being morphed with a relative newcomer that is less than 50 years old.

Handheld mobile phones are extremely popular descendants of mobile phone technology originally developed for use in vehicles. The handheld mobile phone began its history in the 1970s, with the first commercial services starting in the 1980s, and they took the world by storm. The capability of the new device to provide instant communication made the technology widely accepted and extremely successful. The co-evolution of the mobile phones and electronic technology lead to continuous miniaturization, making the mobile phone a truly wearable device found in billions of pockets on every single day. The merging of computer capabilities and instant communication created the modern generation of the smartphones that provide capabilities well beyond of those expected from a phone in its classical definition. These devices connect us to the world, organize our daily life, capture the memories though images and video, provide entertainment and serve in many other roles. Here is where the old meets the new in the emergence of "smart watches" that combine a century-old concept of a wrist worn device with the modern technological advances that pack the power of a smartphone in a watch form factor. The boundary between the devices is becoming increasingly blurry, with many people delegating time keeping to their phone or having phone functionality in their watches.

The acceptability of watches and mobile phones across different cultures, generations and societies demonstrates that wearable technology is a major part of our life and will be becoming increasingly involved in every aspect. Wearable devices provide utility and convenience not available through any other means and extend our capabilities as humans. Wearable sensors are extremely popular as personal tracking devices that today are used primarily as monitoring devices, but the age of wearable is yet to come with multiple applications in personalized health, wellness/sports/fitness, rehabilitation, personal entertainment, social communications and lifestyle computing. Whether being able to tell time, connect to almost anyone almost anywhere, track one's exercise or monitor a patient on a hospital, wearable technology makes us more efficient by relying on its multiple capabilities.

Modern technology is the true enabler for the current and future generations of wearable devices. The roots of the rapid growth can be tracked to a few fundamental advances:

— Sensors: Advances in Micro-Elecro-Mechanical Systems (MEMS) and Nano technologies enable creation of miniature and inexpensive sensing solutions for a variety of electrical, mechanical, optical, chemical and other variables of interest that can be captured from or around the human body.
— Storage: The data captured by the sensors often need to be accumulated and stored for processing, especially in situations where a communication link may not be readily available. The recent advances in the solid state storage opens the door for collection of large datasets using wearable devices.
— Computation: Dramatic increase in the processing capabilities, reduction in power consumption and decrease in size of embedded processors facilitates real-time signal processing and pattern recognition in battery-powered wearable devices. The techniques of machine learning and computational intelligence enable a new, never before seen capability to provide real-time pattern recognition in the sensor data which may identify events of interest (for example, an elderly person experiencing a fall) and acting upon or providing real-time feedback on such events.
— Communication: Rapidly evolving wireless communication techniques permit instantaneous information delivery virtually anywhere in world and processing of the data in the "cloud". The data from a wearable device can be almost instantaneously delivered anywhere in the world. The communications channels can carry sensor data or messages and alerts generated by the device.
— Interface: Modern interfaces go well beyond LEDs and flatscreen technology, delivering results of the sensor data processing in a compelling, easy to understand manner accessible by an individual without advanced technological knowledge though tactile, audible, visual or haptic interfaces.

These fundamental advances are supplemented by complementary technologies that enable operation of body-worn devices: ongoing miniaturization of electronic devices, advances in materials science leading to creation of smart textiles, and flexible electronics, advances in battery technology, energy harvesting and so on.

Overall, wearable technology epitomizes the interaction of humans and technology and as such, covers a very broad area that requires a mixture of expertise quite often hidden in a disjointed array of academic publications in fields of science and technology that normally may not interact with each other. How many electrical engineers took classes in knitting technology? How many behavioral scientists studied sensors? The list can continue. This book attempts to present a holistic view of the various aspects of wearable technology, originating from different fields of study. While by no means all-inclusive, the book is organized into several sections that act as an umbrella over a few key aspects of wearable technology. The chapters in each section are written by the world's best experts that present their unique view on the field.

Section 1, "Wearable design issues and user interfaces" provides an introduction to the world of wearables, talks about key applications that drive development of wearable systems, discusses issues of social acceptance, practicality and convenience and provides an example of a novel user interface.

Section 2, "Fundamentals of sensors in wearable devices" focuses on various aspects of sensor technology, ranging from physiological monitoring, inertial, bio and chemical sensors, to optical and heat flow sensors. The sensor technologies described in this section provide an overview of the fundamental sensor modalities used in wearable technology.

Section 3, "Smart fabrics and flexible electronics" describes two key techniques used in production of smart textiles: knitting and weaving, and also provides several examples of practical applications of smart textile technology. The final chapter of this section portrays in depth the emerging field of flexible electronics technology that finds one its primary applications in the production of smart textiles.

Section 4, "Energy harvesting from human body" looks at the possible ways to perform battery charging or battery-free operation from the energy generated by the human body. This section discusses a variety of techniques and issues associated with energy harvesting.

Section 5, "Analog and digital signal processing, pattern recognition and data analysis" looks at techniques for processing of sensor data. Being severely energy-constrained, wearable sensors need to rely on a combination of energy-efficient analog and digital signal processing and pattern recognition techniques. Wearable sensors may also generate massive amounts of data that can be viewed as a part of the "Big Data" challenge and need advanced algorithms for data mining. This section describes techniques for efficient on-body processing as well as off-line mining of sensor data.

Section 6, "On-body communications and body area networks" deals with issues of communication around or through the human body. The chapters in this section describe design issues in wireless networks, channel propagation models, security and trust in wireless communications, and localization of sensors.

Section 7, "Applications" covers several examples of wearable sensor technology used in less conventional applications such as use of wearables as assistive technology for wheelchair-bound individuals, use of sensors worn inside of the body for gastric disease monitoring, and detection and characterization of diet, food intake and ingestive behavior in community-dwelling individuals.

In summary, the book describes the fundamentals, practical implementation and applications of wearable technology. Written by the leading experts in the field, the book also provides insights into future directions of the technology. We hope that this volume will be useful for researchers in academy and industry, students and practitioners who would like to obtain a comprehensive overview of the fascinating world of wearables.

Wearables: Fundamentals, Advancements, and a Roadmap for the Future

Sungmee Park[1], Kyunghee Chung[2] and Sundaresan Jayaraman[3]

Georgia Institute of Technology, School of Materials Science & Engineering, Atlanta, Georgia, USA

1. WORLD OF WEARABLES (WOW)

In today's digital world the term "wearable" has a new meaning! It no longer conjures up images of clothing such as an elegant evening dress or a heated Sherpa jacket worn by a mountaineer at a base camp on Mount Everest. Rather, today it brings up images of accessories such as a smart watch on a business executive's wrist, a head-mounted display worn by an immersive gamer, a tiny sensor on a cyclist's helmet, or a smart garment a runner uses to track and monitor his steps. In recent years, the dimensions of fashion and protection typically associated with the traditional wearability of clothing have expanded to include "functionality" on the go. This functionality can essentially be characterized as mobile information processing — whether it is the executive checking e-mail, the gamer shooting at a target that is also being simultaneously chased by a fellow gamer on the other side of the world, the cyclist's trainer ensuring that the rider is maintaining proper posture on the curve, or the runner tracking his workout for the day. Just as clothing can be personalized and customized for each person (depending on the physical dimensions,

[1] On leave of absence from Georgia Tech to Kolon Corporation, Seoul, South Korea

[2] Kolon Corporation, Seoul, South Korea

[3] To whom correspondence should be addressed (sundaresan.jayaraman@gatech.edu)

1

taste, and style preferences) and/or occasion (business, evening, casual, home, and hiking), the new wearable too can be configured for personalized mobile information processing for specific applications such as immersive gaming, fitness, public safety, entertainment, healthcare, etc. In short, the world of wearables (WOW) is transforming our lives.

Figure 1 shows a snapshot of people interacting with their personalized wearables. Today's avid gamers want total immersion and expect the gaming experience to be "natural." They do not want to be constrained by traditional interfaces (e.g., joysticks, keyboards, mice, etc.), but prefer games that let them perform body movements that are realistic [1]. For example, when hitting a ball, players prefer swinging their arm or leg, rather than sliding a mouse or pressing a button. Moreover, wearables are enabling immersive multi-player games with tangible and physical interaction not ever experienced by anyone [2]. As a result, the videogame market is growing and the revenue, including mobile games on smartphones and tablets, was $66 billion in 2013 (up from $63 billion in previous year) and is expected to grow to $78 billion in 2017 [3].

Wearables are not just for fun though. They are also used to keep first responders safe and alive by monitoring their physical conditions (e.g., vital signs) and the ambient environment for the presence of dangerous gases and hazardous materials. Without these, the casualties amongst the world's 25.3 million first responders — 2.3 million of them on the frontlines and others supporting them — would be significantly higher [4].

Wearables are also used to monitor racecar drivers. The racecar driver is experiencing 4 + G-force while traveling at over 190 miles per hour, all the time losing water, which can

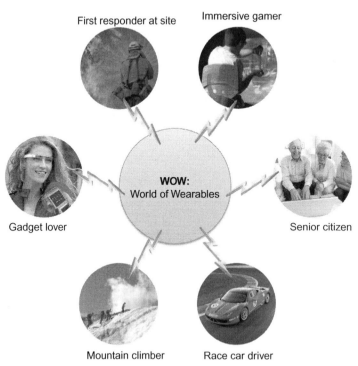

First responder at site   Immersive gamer

Gadget lover

WOW: World of Wearables

Senior citizen

Mountain climber   Race car driver

FIGURE 1   WOW: The world of wearables is enabling digital lives.

be up to 10 pounds over a three-hour period. While this feeling can be exhilarating for the driver, there is also potential risk to the driver's health. Using wearables, the driver's pit crew and manager can mine real-time data to track his physical condition and decide whether he is at risk. At the same time, the video stream from the driver's wearable (camera) can provide a unique "on-the-track" experience for fans.

Likewise, as shown in the figure, wearables can deliver unique value to users and to those accessing the data being collected by the wearables. For example, wearables can help the "sandwich generation" caring for elderly parents monitor their health and well-being and increase their independence. Wearables have also been used to help parents care for young children.

1.1 The Role of Wearables

Fundamentally, wearables can perform the following basic functions or unit operations in each of the scenarios shown in Figure 1:

- Sense
- Process (Analyze)
- Store
- Transmit
- Apply (Utilize)

Of course, the specifics of each function will depend on the application domain and the wearer, and all the processing may occur actually on the individual or at a remote location (e.g., command and control center for first responders, fans watching the race, or viewers enjoying the mountaineer's view from the Mount Everest base camp).

Figure 2 is a schematic representation of the unit operations associated with obtaining and processing situational data using wearables. For example, if dangerous gases are detected by a wearable on a first responder, the data can be processed in the wearable and an alert issued. Simultaneously, it may be transmitted to a remote location for confirmatory testing and the results — along with any appropriate response (i.e., put on a gas mask) — can be communicated to the user in real-time to potentially save a life [5]. This

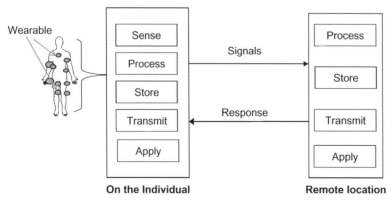

FIGURE 2 Unit operations in obtaining situational awareness: role of wearables.

same philosophy can also be used by an avid gamer who might change his strategy depending on what "weapons" are available to him and how his opponents are performing. Each of these scenarios requires *personalized mobile information processing*, which can transform the sensory *data* to *information* and then to *knowledge* that will be of *value* to the individual responding to the situation.

While wearables are being used in many fields, as discussed, this chapter will focus primarily on wearables in the healthcare domain. Inferring the potential of wearables in other application domains should be straightforward and can be accomplished by instantiating the fundamental principles and concepts presented here.

1.2 Data-Information-Knowledge-Value Paradigm

Figure 3 shows the data-value transformation paradigm [6]. Let's consider a patient visiting a physician. In triage, the nurse documents the vital signs gathered using instruments (e.g., thermometer, blood pressure monitor, electrocardiogram, or EKG machine) that convert the raw signals (the *data*) from the body into meaningful *information* (temperature, heart rate, diastolic/systolic pressure) and thus add value as shown in the figure. When the physician processes this information, he or she gains insight into the potential condition of the patient. The physician adds value by drawing upon the *knowledge* — expertise and experience accumulated over time — to come up with a diagnosis and a plan of action or treatment. This course of treatment — in the form of medication and other interventions — is the *value* derived by (or delivered to) the patient resulting in the curing of the illness. Thus, the raw data gathered by the instruments is valuable only when it is properly transformed and harnessed to benefit the individual. For this transformation to occur seamlessly there is a need for an information/knowledge processing ecosystem.

1.2.1 The Emerging Concept of Big Data

Park and Jayaraman discussed the role of wearables in relationship to "big data" [7]. Big data refers to large amounts and varieties of fast-moving data from individuals and groups that can be processed, analyzed, and integrated over time to create significant value by revealing insights into human behavior and activities. According to McKinsey, if the US healthcare system could use "big data creatively and effectively to drive efficiency and quality," the potential value from data in the sector could be more than $300 billion

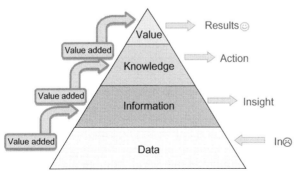

FIGURE 3   Data-information-knowledge-value transformation paradigm.

every year, two-thirds of which would be in the form of reducing national healthcare expenditures by about 8% [8]. Likewise, in the private sector, McKinsey estimates that a retailer using big data to the fullest extent has the potential to increase its operating margin by more than 60%. McKinsey also projects a 50% decrease in manufacturing and assembly costs. Thus, it is clear that there is value in harnessing big data in a wide spectrum of activities and industries. Let's consider one such example in healthcare, an application domain in which wearables are being increasingly deployed.

1.2.2 *Medical Loss Ratio and Wearables*

The Patient Protection and Affordable Care Act of 2010 requires health insurance companies to spend at least 80 to 85% of premiums collected on providing medical care [9]. Known as the Medical Loss Ratio (MLR), the objective behind this provision is to bring down the overhead costs of providing medical care and limit it to 20% for individual and small-group coverage and to 15% for large-group coverage. With the increasing shift from volume-based to value-based reimbursement for services rendered, healthcare providers are incentivized to provide holistic care to patients by closely monitoring them to ensure compliance with medication and promoting healthier lifestyles.

Wearables enable this remote health monitoring of patients. The health data can be wirelessly sent to the physician's office by the wearable, negating the need for office visits. Consequently, the cost of care decreases. Moreover, the ability to continuously track patients' health can help identify any potential problems through preventive interventions and thus enhance the quality of care while eliminating unnecessary procedures since the cost of prevention is significantly less than the cost of treatment. The resulting higher quality of care at lower costs would also contribute to better operating efficiencies and lower overhead costs for insurance companies since their resources can be better spent on actually providing care and not on measures to ensure that a high quality of care is being provided.

Thus, at the heart of the concept of "big data" is the individual who is simultaneously the source of the data and the recipient of the resulting "value" after the processing/harnessing of the data. This is where wearables have a critical role to play in creating and serving as the core of an ecosystem essential for facilitating the seamless transformation of data to deliver value.

1.3 The Ecosystem Enabling Digital Life

The advancements in, and convergence of, microelectronics, materials, optics, and bio-technologies, coupled with miniaturization, have led to the development of small, cost-effective intelligent sensors for a wide variety of applications. These sensors are now so intimately interwoven into the fabric of our lives that they are not only pervasive, but are also operationally "invisible" to end-users. The user interface is so simple that with the touch of a few buttons a different "programming" sequence can be launched by anyone — from a young kid to a senior citizen — for a wide variety of tasks, e.g., from monitoring vital signs to controlling the ambience in the room. Thus, the ease of the user interface coupled with the invisibility of the "embedded" technology in the various devices and systems has contributed to the proliferation of these sensors in various applications such as those represented in Figure 1. By effectively taking advantage of these technological advancements, it is possible to create an ecosystem that facilitates the harnessing of large amounts of situational awareness data.

1.3.1 Smart Mobile Communication Devices

A key component of the ecosystem is the smart mobile communications device — smartphone and/or tablet — that provides a platform for "information processing on the go" for anyone, anytime, and anywhere. According to *The Economist*, both the number of individuals with access to and connecting to the Internet is increasing as are the number of places providing connectivity at ever increasing speeds [10]. Citing Ericsson, the telecoms-network provider, *The Economist* states that the volume of mobile data traffic in 2017 is expected to be 21 times greater than it was in 2011, while the number of mobile-broadband subscriptions (mostly for smartphones) will jump from 900 million to 5 billion.

1.3.2 Social Media Tools

Easy-to-use social media tools complete the ecosystem that is digitizing, connecting, and continuously transforming our lives. Indeed, virtually everything is being captured and is being reduced to a sequence of 0 s and 1 s inside the hardware, but with significant value to the user/viewer on the outside!

Now that we have defined a wearable, established the important role of wearables, and have defined the components of an ecosystem to enable digital life with wearables at its core, we will discuss the salient attributes of wearables, develop the taxonomy, and discuss the advancements in the field.

HUMAN SKIN AS THE ULTIMATE SENSOR

While the different types of sensors and wearables are relatively *new* in the timeline of civilization, there has been one piece of "sensing" technology that has been there since the dawn of civilization: human skin. It is the *ultimate* sensor. As the largest organ of the human body, it not only provides a physical barrier that protects a human's insides from the outside elements, it also senses, adapts, and responds based on both external and internal stimuli such as heat, cold, fear, pleasure, and pain. In fact, it has the intrinsic and rather unique ability to respond to all the five senses of touch, sight, sound, smell, and taste. Physically, it is soft, smooth, flexible, strong, and evolves over time to meet the changing needs of the individual, including physical needs. When injured or damaged, it heals, and, in most instances, returns to its "original" state with very little, if any, residual impact of the injury.

Interestingly, in the computing paradigm, the skin is an input/output (I/O) device that senses and passes the stimulus (input) to the brain (the CPU), which draws upon its knowledge (the processing power of the CPU) to come up with the interpretation and action that is eventually reflected in the skin's response (the output).

Thus, human skin is a powerful and versatile sensor that nature has designed and is akin to an I/O device in a computing system. The Holy Grail in sensor or wearables design is to create one that has all the desired attributes of human skin and performs as well as it does!

---

2. ATTRIBUTES OF WEARABLES

A sensor is defined as "a device used to detect, locate, or quantify energy or matter, giving a signal for the detection of a physical or chemical property to which the device responds" [11]. Not all sensors are necessarily wearable, but all wearables, as discussed earlier and shown in Figure 2, must have sensing capabilities. The key attributes required of an ideal wearable are shown in Figure 4.

From a physical standpoint, the wearable must be lightweight and the form factor should be variable to suit the wearer. For instance, if the form factor of the wearable to monitor the vital signs of an infant prone to sudden infant death syndrome prevents the infant from (physically) lying down properly, it could have significant negative implications. The same would apply to an avid gamer — if the form factor interferes with her ability to play "naturally," the less likely that she would be to adopt or use the technology. Aesthetics also plays a key role in the acceptance and use of any device or technology. This is especially important when the device is also seen by others (the essence of fashion). Therefore, if the wearable on a user is likely to be visible to others, it should be aesthetically pleasing and, optionally, even make a fashion statement while meeting its functionality. In fact, with wearables increasingly becoming an integral part of everyday lives, the sociological facets of the acceptance of wearables opens up exciting avenues for research. Ideally, a wearable should become such an integral part of the wearer's clothing or accessories that it becomes a "natural" extension of the individual and "disappears" for all intents and purposes. It must have the flexibility to be shape-conformable to suit the desired end use; in short, it should behave like the human skin.

The wearable must also have multi-functional capability and be easily configurable for the desired end-use application. Wearables with single functionality (e.g., measuring just the heart rate) are useful, but in practical applications, more than one parameter is typically monitored; and, having multiple wearables — one for each function or data stream — would make the individual look like a *cyborg* and deter their use even if the multiple data

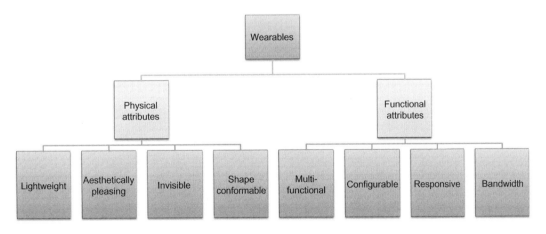

FIGURE 4   Key attributes of wearables.

streams could be effectively managed. The wearable's responsiveness is critical, especially when used for real-time data acquisition and control (e.g., monitoring a first responder in a smoke-filled scene). Therefore, it must be "always on." Finally, it must have sufficient data bandwidth to enable the degree of interactivity, which is key to its successful use.

Thus, the design of wearables must be driven by these attributes.

2.1 Taxonomy for Wearables

Figure 5 shows the proposed taxonomy for wearables. To begin with, they can be classified as single function or multi-functional. They can also be classified as invasive or noninvasive. Invasive wearables (sensors) can be further classified as minimally invasive, those that penetrate the skin (subcutaneous) to obtain the signals, or as an implantable, such as a pacemaker. Implantable sensors require a hospital procedure to be put into place inside the body. Non-invasive wearables may or may not be in physical contact with the body; the ones not in contact could either be monitoring the individual or the ambient environment (e.g., a camera for capturing the scene around the wearer or a gas sensor for detecting harmful gases in the area). Non-invasive sensors are typically used in systems for continuous monitoring because their use does not require extensive intervention from a healthcare professional.

Wearables can also be classified as active or passive depending upon whether or not they need power to operate; pulse oximetry sensors fall into the former, while a temperature probe is an example of a passive wearable that does not require its own power to operate. Yet another view of wearables is the mode in which the signals are transmitted for processing — wired or wireless. In the former, the signals are transmitted over a physical data bus to a processor; in the wireless class of wearables, the communications capability is built into it, which transmits the signals wirelessly to a monitoring unit. Sensors can be for one-time use or they can be reusable. Finally, wearables can be classified based on their field of application, which can range from health and wellness monitoring to position tracking as shown in the figure. "Information processing" is listed as one of the application areas because many of these traditional functions such as processing e-mail can now

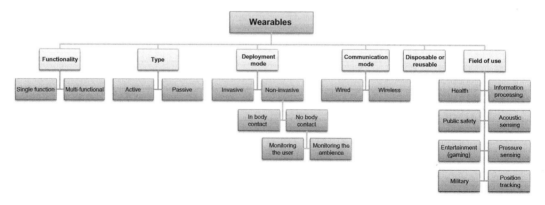

FIGURE 5   The taxonomy for wearables.

be done on a wearable in the form of a wristwatch. It is important to note that not all the classes are mutually exclusive. For instance, a wearable can be multi-functional, active, non-invasive, and be reusable for health monitoring.

The proposed taxonomy serves two key functions: first, it helps in classifying the currently available wearables so that the appropriate ones can be selected depending upon the operating constraints; second, it helps in identifying opportunities for the design and development of newer wearables with performance attributes for specific areas that need to be addressed.

2.2 Advancements in Wearables

Today's "wearables" can be traced back to the concept of wearable computers. Jackson and Polisky provide an excellent account of the development of wearable computers going back to the early 1960 s with the work of Thorp and Shannon to predict the performance of roulette wheels [12]. In the 1980s, Mann defined the following attributes for wearable computers: Constant, unrestrictive to the user, unmonopolizing of the user's attention, observable by the user, controllable by the user, attentive to the environment, and personal [13]. Mann's criteria for wearable computers include it being eudaemonic, existential, and in constant operation and interaction [14]. Weiser proposed the concept of ubiquitous computing in which the computers themselves "vanish into the background" [15]. In 2002, Xybernaut introduced its Poma Wearable PC, but it was not a commercial success. One of the reasons this paradigm of "wearable computers" did not catch on was because they were technology-driven; they only focused on making the bulky computer "wearable" and did not attempt to rethink the information-processing paradigm itself to address the usability of the technology. Moreover, the resulting systems (e.g., Xybernaut) were far from aesthetically pleasing, which further hindered their acceptance.

2.2.1 The Wearable Motherboard – a User-Centric Approach to the Design of Wearables

Beginning in late 1996, Jayaraman and co-workers took a fundamentally different approach to the field of wearables and developed the concept of a wearable motherboard. Driven by the needs of soldiers – the end user – to be monitored in real-time in the battlefield so that they would receive medical care in the event of being shot, they developed fabric-based wearable technology to monitor the vital signs of soldiers in an unobtrusive manner and also to detect any shrapnel penetration when shot [13–17]. This concept was called the wearable motherboard as it is conceptually analogous to a computer motherboard.

The computer motherboard provides a physical information infrastructure with data paths into which chips (memory, microprocessor, graphics, etc.) can be plugged in to meet performance requirements for specific end uses such as gaming, image processing, high-performance computing, etc. Likewise, the wearable motherboard – in the form of a fabric or a piece of clothing such as an undershirt – provides an information infrastructure into which the wearer can plug in sensors and devices to achieve the desired functionality, say, for example, vital signs monitoring. Thus, it fulfills the twin roles of being: (i) a flexible information infrastructure to facilitate the paradigm of ubiquitous computing, and (ii) a platform for monitoring the vital signs of individuals in an efficient and cost-effective manner with a "universal" interface of clothing. This development essentially led to the birth of the field of smart textiles. According to Park and Jayaraman, "clothing can indeed have the third dimension of 'intelligence' embedded into it and spawn the growth of *individual networks* or *personal networks* where each garment has its own IN (individual network) address much like today's IP (Internet protocol) address for information-processing devices [18]. When such IN garments become the *in* thing, personalized mobile information processing would indeed have become a reality for all of us!" Looking back (twelve years later), this prediction has turned out to be true with today's "Internet of Things" paradigm.

Following the promise of the technology spawned by the Smart Shirt (a more common name for the wearable motherboard), a considerable amount of research has been going on in this field, judging by the number of books, special issues of journals, number of papers, and the establishment of the IEEE Technical Committee on wearable biomedical systems [19–28].

2.2.2 Research in Flexible Electronics

Another class of wearables – known as flexible electronics – is focused on printing electronics (thin-film transistors, thin-metal films, nanomaterials, and carbon nanotubes, among others) onto elastomeric substrates resulting in "electronic skins" with pressure and temperature sensing capabilities, among others; these can be directly applied to the human body [29].

2.2.3 The Latest Trends in Commercial Wearables

The newest generation of wearables is shown in Figure 6 [30–33]. These typically have only some of the attributes of wearables discussed earlier and shown in Figure 4. For instance, most of these perform a *single* function (e.g., measuring heart rate during a workout) and so their application domain is limited. The recent arrival of Google Glass® appears to be the tipping point in terms of accelerating the wearables movement into the

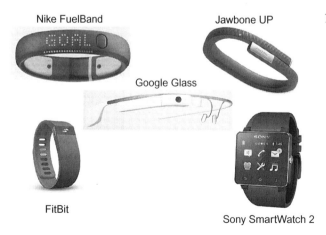

Nike FuelBand

Jawbone UP

FIGURE 6   The emerging set of wearables.

Google Glass

FitBit

Sony SmartWatch 2

mainstream since Google is an entity with sizable technical and financial resources. Yet another form of wearable is smartwatches, like those from from Sony [34] and Samsung [35]. These have the ability to check e-mail and surf the Internet.

3. TEXTILES AND CLOTHING: THE META-WEARABLE

A critical need for extensive deployment of wearables for personalized mobile information processing is that they should not impose any additional social, psychological, or ergonomic burden on the individual. For instance, Google Glass significantly impacts social dynamics since the ones without this wearable device are not sure of what the wearer is doing with the device while being part of the conversation. What is therefore needed is an infrastructure or *platform* that will be unobtrusive, natural and pervasive, and not adversely impact social interaction.

Moreover, for many real-world applications, some of which are shown in Figure 5, multiple parameters must be simultaneously acquired, processed, and used to develop an effective response. This leads to the following requirements for creating and developing a useful wearable sensor system [36]:

- Different *types* of sensors will be needed for various parameters to be monitored *simultaneously*; for instance, sensors to monitor the various vital signs (e.g., heart rate, body temperature, pulse oximetry, blood glucose level) are of different types. Likewise, for monitoring hazardous gases, another class of sensors (e.g., carbon monoxide detection) will be required. Accelerometers will be required to continuously monitor the posture of the gamer or an elderly person to detect falls.
- Different *numbers* of sensors may be needed to obtain the signals to compute a single parameter (e.g., at least three sensors are required to compute the electrocardiogram or EKG).
- Sensors need to be positioned in *different* locations on the body to acquire the necessary signals (e.g., sensors for EKG go in three different locations on the body, whereas pulse-ox sensors and accelerometers go in other locations on the body).
- Different *subsets* of sensors and devices may be used at different times, necessitating their easy attachment and removal, or *plug and play*. For instance, the gamer may want to record how his body feels and reacts while being immersed in the game and, at other times, may also want to record his experience.
- The signals from the various sensors and in different physical locations (such as first responders responding to a disaster scene) have to be *sensed, collected, processed, stored,* and *transmitted* to the remote control and coordination location.
- Signals from different types of sensors (e.g., body temperature, EKG, accelerometers) have to be processed in parallel to evaluate the various parameters in real-time.
- Since a large number of sensors is usually required, these sensors would have to be low cost and hence would likely have minimal built-in (on-board) processing capabilities.
- The sensors should be power-aware (i.e., have low power requirements).
- Power must be supplied (distributed) to the various sensors and processors.

Thus, there is a need for a platform that has both a physical form factor and an integral information infrastructure. In addition to serving as a wearable in its own right, the platform must be able to *host* or hold other "wearables" or sensors in place and provide *data buses* or pathways to carry the signals (and power) between sensors and the information-processing components in the wearable network [37–38]. Simply attaching different types of sensors and processors to different parts of the body is not the ideal solution. What is needed is a *meta-wearable* [7]. That *meta-wearable* is textiles.

3.1 Attributes of the Textile Meta-Wearable

A textile is a *meta-wearable* because it meets all the attributes of the wearables in Figure 4. For instance, textile yarns, which are an integral part of the fabric, can serve as *data buses* or communication pathways for sensors and processors and can provide the necessary bandwidth required for interactivity. The topology, or structure of placement of these data buses, can be engineered to suit the desired sensor surface distribution profile, making it a versatile technology platform for wearables. In addition, textiles and clothing have the following key attributes [17,32–33]:

- Humans are used to wearing clothes so, in general, no special "training" is required to wear them, i.e., to use the interface. In fact, it is probably the most *universal* of human—computer interfaces and is one that humans need, use, have familiarity with, and which can be easily customized. Often termed the "second skin," it is the next best wearable (other than a smile).
- Humans enjoy clothing and this universal interface of clothing can be "tailored" to fit individual preferences, needs, and tastes, including body dimensions, budgets, occasions, and moods in which the wearables will be used.
- Textiles are flexible, strong, lightweight, and generally withstand different types of operational (stress/strain) and harsh environmental (biohazards and climatic) conditions.
- Textiles, unlike other engineering structures such as buildings, are unique in combining strength and flexibility in the same structure, and so they conform to the desired shape when bent but retain their strength.
- Textiles can be made in different form factors including desired dimensions of length, width, and thickness, and hence "variable" surface areas that may be needed for "hosting" varying numbers of sensors and processors for the desired application can be created.
- Textiles provide the ultimate flexibility in system design by virtue of the broad range of fibers, yarns, fabrics, and manufacturing techniques (e.g., weaving, knitting, non-wovens, and printing) that can be deployed to create products with engineered performance characteristics for desired end-use applications.
- Textiles are easy to manufacture in a relatively cost-effective (inexpensive) manner — roll-to-roll — compared to traditional printed circuit boards.
- Textiles obviate issues associated with entanglement and snags when using the system since the data buses or communication pathways are an *integral* part of the fabric.

- Textiles can easily accommodate "redundancies" in the system by providing multiple communication pathways in the network.
- Textile structures enable easy power distribution from one or more sources through the textile yarns integrated into the fabric, thus minimizing the need for on-board power for the sensors.

Therefore, from a technical performance perspective, a textile fabric (or clothing) is a true meta-wearable, making it an excellent platform for the incorporation of sensors and processors to harness situational awareness data while retaining its aesthetic and comfort attributes, among many other textile-unique properties.

3.2 Realization of the Meta-Wearable: The Wearable Motherboard

The Wearable Motherboard or Smart Shirt briefly mentioned earlier is the first such meta-wearable that has been successfully developed [39]. It has since paved the way for today's wearables revolution. The comfort or base fabric provides the necessary physical infrastructure for the wearable motherboard shown in Figure 7. The base fabric is made from typical textile fibers (e.g., cotton, polyester) where the choice of fibers is dictated by the intended application. The conducting yarns integrated into the fabric serve as data buses and constitute the information infrastructure. An interconnection technology has been developed and used to route the information (signals) through desired paths in the fabric, thereby creating a motherboard that serves as a flexible and wearable framework into which sensors and devices can be plugged.

For instance, when sensors for vital signs such as heart rate, electrocardiogram, and body temperature are plugged in, the wearer's physical condition is monitored.

3.2.1 Wearable Motherboard Architecture

The wearable motherboard architecture is shown in Figure 8. The signals from the sensors flow through the flexible data *bus* integrated into the structure to the multi-function

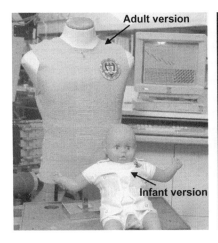
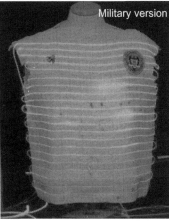

FIGURE 7 The wearable motherboard: adult, baby, and military versions.

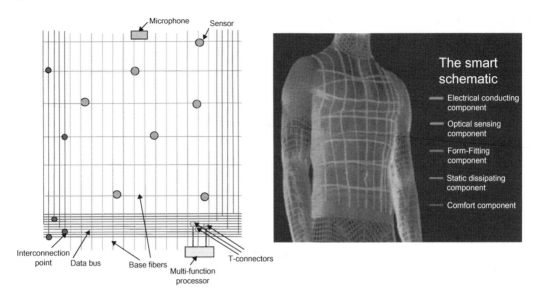

FIGURE 8   Wearable motherboard architecture.

processor/controller. This controller, in turn, processes the signals and transmits them wirelessly (using the appropriate communications protocol) to desired locations (e.g., doctor's office, hospital, battlefield triage station). The bus also serves to transmit information *to* the sensors (and hence, the wearer) from external sources, thus making the Smart Shirt a valuable bi-directional information infrastructure. The controller provides the required power (energy) to the wearable motherboard. With the advent of the smartphone, all the processing and communication can be shifted to it, thereby obviating the need for the controller.

The advantage of the motherboard architecture is that the *same garment* can be quickly reconfigured for a different application by changing the suite of sensors. For example, to detect carbon monoxide or hazardous gases in a disaster zone, special-purpose gas sensors can be plugged into the same garment and these parameters in the ambient environment can be monitored along with the first responder's vital signs. Similarly, by plugging in a microphone into the Smart Shirt, voice can be recorded. Optionally, the conducting fibers in the wearable motherboard can themselves act as "sensors" to capture the wearer's heart rate and EKG (electrocardiogram) [40]. Likewise, the military version of the Smart Shirt shown in Figure 7 uses optical fibers to detect bullet wounds in addition to monitoring the vital signs of the soldier during combat conditions. The wearable motherboard can be tailored to be a head cap so that the gamer's brain activity can be tracked by recording the electroencephalogram (EEG). Thus, the wearable motherboard is an effective meta-wearable and the structure has the *look* and *feel* of traditional textiles with the fabric serving as a comfortable information infrastructure.

3.2.2 Convergence and Interactive Textiles

The wearable motherboard is a platform that enables true convergence between electronics and textiles. Due to the modularity of the design architecture, the extent and duration of convergence can be controlled by the user. For example, as long as the sensors and processors are plugged into the wearable motherboard, there is true convergence and the resulting wearable (in the form of clothing) is smart and can perform its intended function, e.g., monitor the wearer's vital signs or other situational awareness data. When this task is completed, the sensors and processors can be unplugged and the garment laundered like other clothes. Thus, the usually *passive* textile structure is temporally transformed into a *smart interactive* structure and embodies the new paradigm that clothing is an *information processing structure* that *also* protects the individual while making him/her fashionable.

3.3 Applications of Wearables

Figure 9 is an artist's rendering of the role of wearables during the day in the life of a typical family. It is clear from the illustration that the number of applications is only limited by the imagination. They range from monitoring babies to senior citizens, i.e.,

FIGURE 9 Wearables in the twin continuum of life and activities.

TABLE 1   Applications of Wearables

Sports Application: Coyle et al. have developed a wearable sensing system that integrates a textile-based fluid handling system for sample collection and transport with a number of sensors including sodium, conductivity, and pH sensors [41]. Together with sensors for sweat rate, EKG, respiration, and blood oxygenation, they were able to successfully monitor a number of physiological parameters together with sweat composition in real-time.

Public Safety (Protection) Application: The ProeTEX project, funded by the European Commission, has developed a wearable system for the protection of first responders [42]. It monitors the health of the users (heart rate, breathing rate, body temperature, blood oxygen saturation, position, activity, and posture) and environmental variables (external temperature, presence of toxic gases, and heat flux passing through the garments) and transmits the information to the coordination center to ensure the safety of the personnel.

Entertainment Application: The Philips Lighting project for the Black Eyed Peas group created a wearable system for the singers with organic light-emitting diodes (OLED) and LEDs that light up during the performance and provide a new experience for the audience and the entertainers [43].

the continuum of life, and span the continuum of activities in which the individuals are engaged.

Table 1 provides a summary of the various fields of application along with the typical parameters monitored for that application; the target population is also shown. In each application example, the wearable system is responsible for sensing, processing, analyzing, and transmitting the results to the user. Illustrative examples for three of the major classes of applications are presented here.

Despite the promise of wearables in the various fields of application and the projected market sizes, they have not become an integral part of many users' "must-have" accessories or technologies. We will now discuss the challenges encountered by wearables and the opportunities to successfully transition the technology to the marketplace.

4. CHALLENGES AND OPPORTUNITIES

The success of any innovative product in the marketplace depends on:

- Its effectiveness in successfully understanding the user's needs and meeting them
- Its compatibility with or similarity to existing products or solutions
- The extent of behavioral change needed to use the new product
- The reduction in the cost of current solutions or technologies it aims to supplant
- The improvement in the quality of service (or performance)
- The enhancement of the user's convenience

The innovation should provide a tangible advantage to the user and it should be consistent and compatible with the user's values, beliefs, and needs. Many an innovative technology has not been a marketplace success for one or more of these aforementioned reasons. For example, Apple's Newton, the first handheld device did not make it in the market, but spawned the highly successful Palm Pilot and generations of personal digital assistants because the latter addressed many of the issues that plagued the Newton. In the process, they spawned the ongoing innovation in tablets. Thus, factors related to the diffusion of innovation must be considered in addition to the technical and business challenges to ensure the successful transition of wearables from the laboratory to the highly competitive marketplace. A roadmap analyzing the technical, business, and public policy issues, including the need for a "killer app" to influence the adoption and acceptance of wearables, has been proposed [44].

4.1 Technical Challenges

The key technical challenges in the adoption of wearables are as follows:

- The success of wearables depends on the ability to connect them seamlessly in a body-worn network. This means the meta-wearable framework must have the ability to route the signals and power between desired points in the structure (Figure 8). The interconnection process for creating such junctions in textile materials has been manual to-date. The concept of textillography to automate interconnections during the fabric manufacturing process has been proposed [45]. An automated process that can provide precise, rugged, and flexible interconnections will help facilitate mass production and also lower the costs associated with wearables.
- In the event of damage to the data buses in the meta-wearable framework, the "failure" in the network must be recognized and alternate "data paths" must be established in the fabric to maintain the integrity of the network by taking advantage of the redundant data buses in the fabric. Preliminary work on the concept of "soft" interconnects has resulted in a programmable network in a fabric that enables real-time routing that can be configured on the fly [46].
- Currently, the so-called "t-connectors" and button snaps are being used for connecting sensors and processors to the meta-wearable. There is therefore the need for a common interface similar to the RJ-11 jack for telephones for connecting these sensors and processors to the meta-wearable so that general-purpose sensors and devices can be developed, thereby reducing their cost.
- Many of the wearables, especially those used for health monitoring applications and immersive gaming, are prone to motion artifacts, which can potentially affect the integrity of the results. There is therefore the need for in-depth studies to develop robust signal processing algorithms and systems to ensure the quality of the data generated by the wearables.
- While currently available conductive fibers can fulfill the basic requirements for the first generation of textile-based wearables, it is important to develop new materials that will have the conductivity of copper and the properties of textile fibers such as cotton, polyester, or nylon, and be available in commercial quantities. Research is needed to develop fibers that can also retain their conducting properties after repeated laundering.
- Today's wearables are powered by lithium-ion rechargeable batteries, which is another limiting factor in the adoption of the technologies due to the rigidity of the battery in relation to the flexible nature of the wearables, a key desired attribute of wearables

shown in Figure 4. This bottleneck is being addressed by research on two fronts, piezoelectric-based energy-harvesting systems and flexible textile battery, respectively. A textile battery, developed using a woven polyester fabric as a substrate, has exhibited comparable electrochemical performance to those of conventional metal foil-based cells even under severe folding–unfolding motions simulating actual wearing conditions [47]. The 13 mAh battery retained 91.8% of its original capacity after 5,500 deep folding–unfolding cycles. The researchers also successfully integrated the flexible textile battery with lightweight solar cells on the battery pouch to enable convenient solar-charging capabilities.

- The seamless integration of wearables in healthcare settings and for remote monitoring faces the challenge of ensuring compatibility with existing wireless technologies and established operational protocols in those settings [48]. Strategies and solutions must be developed to address this important aspect to help the adoption of wearables for remote monitoring.
- The challenges associated with protection of individual privacy, data security, and other social aspects of the acceptance of wearables must be addressed because the wearables are collecting personal information. The electronics and communications industry in collaboration with privacy protection organizations must develop appropriate protocols that will identify proper technology and public policy solutions to further the free acceptance and use of wearables.
- The supply chains for textiles/clothing and electronics are significantly different. Apparel manufacturing is a labor-intensive operation whereas electronics manufacturing is highly automated. Consequently, the production rates are much higher in electronics manufacturing. The apparel industry is not as precise in terms of topology and interfaces between the different components when compared to the electronics industry whose operating paradigm is precision. Thus, the differences between these manufacturing paradigms must be addressed for the widespread adoption of textile-based meta-wearables for the various applications listed earlier in Table 1.
- Finally, the same wearable may be used in a range of environmental conditions – indoors to outdoors – which may include disaster zones involving high temperatures (e.g., fire) and hazardous materials. Therefore, they should be designed to function effectively and seamlessly in a wide range of ambient environments.

4.2 Making a Business Case

The litmus test for wearables lies in demonstrating their value to the end user and those involved in paying for the technology. The key activities for transitioning the technology from the laboratory to the market are shown in Figure 10. It begins with articulating the need for the technology in a chosen domain and demonstrating its effectiveness through the metrics of cost, quality, and convenience. The various stakeholders responsible for effecting the transition are also shown in the figure. The end user – a patient in the case of a wearable for the healthcare market or a gamer in the gaming market, and so on – must experience the value of the technology which will motivate the user or the payer (the healthcare insurance company in the case of healthcare or the individual gamer) to pay for

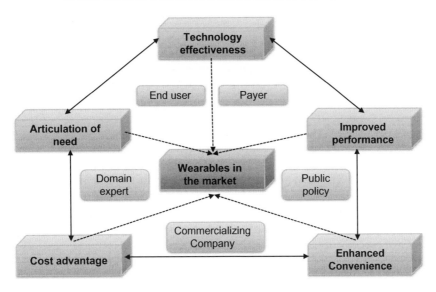

FIGURE 10   Making a business case: stakeholders and metrics.

it as it would benefit the payer in the long-run. Public policy comes into play in the adoption of wearables because of the associated privacy and data access issues. Once the needs are articulated well by the domain expert and the benefits are corroborated by the end user, the commercializing company will be incentivized to proceed with developing and marketing the wearable technology. Thus, all the stakeholders are critical for the success of wearables in the marketplace and a total lifecycle approach that goes beyond just the basic cost of the technology must be adopted in developing the market for wearables.

We will now attempt to gaze into the future of wearables and define a research roadmap to realize it.

5. THE FUTURE OF WEARABLES: DEFINING THE RESEARCH ROADMAP

The paradigm of "Information Anywhere, Anytime, Anyone" is a reality today. For instance, a racing car enthusiast in Cupertino, California can see — on his mobile device — the driver's view of the track as he negotiates the Daytona Speedway. He can instantaneously access all the "stats" associated with the lap, the race, the standing, the history, and so on, thanks to the convergence of high-performance computing, communications, video, and data fusion technologies.

5.1 Imagine the Future

What if the driver's racing suit changes color as the G forces acting on different parts of his body change during the course of the race [7]? What if his suit also captures biometrics such as heart rate, electrocardiogram, body temperature, water loss, and calories burned, and displays these parameters on the fan's mobile device? What if the pit crew can use this real-time data and integrate that with the archival data to decide on *when* to take the next pit stop and *what* actions to take during the stop? Imagine further if the fan in California could *physically* "experience" the G forces acting on the driver during the race with varying degrees of compression on his body?

The meta-wearable of clothing with its integrated sensors and devices can make this possible. The driver's biometric and contextual/experiential data can be captured through the driver's smart clothing — a meta-wearable. This information can be wirelessly transmitted to the fan; the fan's meta-wearable — the smart clothing called *ExpWear* for Experience Wear — can, in turn, transform the data and recreate the remote ambient environment so that the fan's clothing lights up the same way as the driver's and the fan also experiences the G forces experienced by the driver through a suite of sensors, actuators, and other devices integrated into the garment. What if the *ExpWear* also displays the fan's biometrics on the left sleeve and the driver's on the right sleeve? In other words, imagine a world in which the fan can recreate and experience in Cupertino the remote ambience in Daytona through his meta-wearable *ExpWear*. This is the world of *sportatainment* that represents the integration and transformation of sports actions into entertainment using the meta-wearable of textiles and clothing.

Another example: It is the Super Bowl 2014 and Peyton Manning's Smart Jersey — the meta-wearable — is monitoring him and his heart rate is displayed on it (Figure 11). With just 45 seconds left in the game, he is tackled; the force he experiences is displayed on his Jersey. Immediately, on another continent, a football fan watching the game in his *ExpWear* experiences the *pain* of the tackle! Indeed, he feels like he is "in the game" thanks to the meta-wearable of textiles and clothing. While the sports domain has been chosen as an example, it is easy to visualize transformations in other areas and to see the potential for wearables in the dynamic world of Internet of Things.

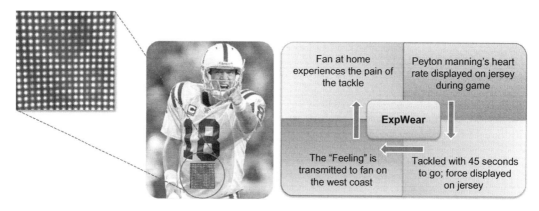

FIGURE 11   Sportatainment: Enabled by the meta-wearable of clothing.

5.2 The Research Roadmap: A Transdisciplinary Approach to Realizing the Future

There is a need for a transdisciplinary approach to realize the future of wearables, which means that it should be pursued as a *new* field of endeavor that brings together knowledge (both foundational and technological advancements) from other established fields such as materials/textile science and engineering, electronics, manufacturing and systems engineering, computing and communications, industrial design, and social sciences [49].

Figure 12 attempts to capture this transdisciplinary approach to wearables research. The major building blocks of wearables, viz., sensors, actuators, processors, energy sources, and interconnections are shown in the figure; the standards governing the design and use of wearables, which must be developed, are also shown in the figure. The materials and manufacturing methods that are integral to the realization of wearables are shown in the left and right panels to signify their key roles in bringing the building blocks together and making the wearable a reality. A change to any of the building blocks will affect the others and, in turn, influence the wearable that is shown in the center of the figure. It is therefore important to view this as a unified ecosystem rather than as a collection of individual pieces. For this key reason, the transdisciplinary paradigm should be adopted to drive the advancements in the field of wearables. Such an approach will bring an innovative perspective leading to revolutionary advancements. This is because a transdisciplinary inquiry focuses on *the* issue, viz., the wearable, rather than what each of the disciplines can *individually* bring to the table and "contribute" to it (the interdisciplinary mode of inquiry and research).

In closing, wearables are increasingly becoming an integral part of our digital lives and the potential application areas are only limited by our imagination. Indeed, it is hard to fathom life without a wearable! A transdisciplinary approach will indeed move us rapidly forward on this exciting journey towards the Holy Grail of wearables and, in the process, enable us to "do well by doing good."

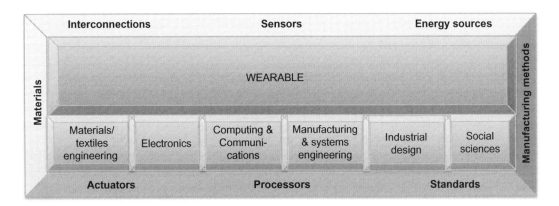

FIGURE 12   Research roadmap for wearables: Need for a transdisciplinary approach.

References

[1] M. Rocetti, G. Marfia, A. Semeraro, Playing into the wild: A gesture-based interface for gaming in public spaces, J. Vis. Commun. Image Representation 23 (3) (2012) 426−440.

[2] J. Tedjokusumo, S. Zhou, S. Winkler, Immersive multiplayer games with tangible and physical interaction, IEEE Trans. Syst. Man Cybern. A Syst. Hum. 40 (1) (2010) 147−157.

[3] M. Nayak, FACTBOX - A look at the $66 billion video-games industry, <http://in.reuters.com/article/2013/06/10/gameshow-e-idINDEE9590DW20130610>, (Last Accessed: 18.11.13).

[4] T.A. Cellucci, Commercialization: The First Responders' Best Friend, US Department of Homeland Security: Science and Technology, Washington, DC, 2009.

[5] S. Park, S. Jayaraman, Wearable sensor network: a framework for harnessing ambient intelligence, J. Ambient Intell. Smart Environ. 1 (2) (2009) 117−128.

[6] S. Jayaraman, Reinventing Clothing: Clothing of Nations to the Health of Nations, The Second TIIC International Symposium, Seoul, South Korea, 2012.

[7] S. Park, S. Jayaraman, The Wearables Revolution and Big Data: The Textile Lineage, Proceedings of the 1st International Conference on Digital Technologies for the Textile Industries, Manchester, UK, 2013.

[8] McKinsey Global Institute, Big Data: The Next Frontier for Innovation, Competition and Productivity, 2011.

[9] <http://www.cms.gov/CCIIO/Programs-and-Initiatives/Health-Insurance-Market-Reforms/Medical-Loss-Ratio.html>, (Last Accessed: 29.11.13).

[10] P. Lane, A Sense of Place, Special Report, The Economist, October 27, 2012, <http://www.economist.com/news/special-report/21565007-geography-matters-much-ever-despite-digital-revolution-says-patrick-lane>, (Last Accessed: 02.11.13).

[11] E. Kress-Rogers, Biosensors and electronic noses for practical applications, in: E. Kress-Rogers (Ed.), Handbook of Biosensors and Electronic Noses: Medicine, Food, and the Environment, CRC Press, New York, NY, 1997, pp. 3−39.

[12] K.L. Jackson, L.E. Polisky, Wearable computers: information tool for the twentyfirst century, Virtual Real., 3, 147−156.

[13] S. Mann, On the bandwagon or beyond wearable computing? Pers. Technol., 1:203−207.

[14] S. Mann, Smart clothing: the wearable computer and WearCam. Pers. Technol., 1:21−27.

[15] M. Weiser, The Computer for the 21st Century, Sci. Am. (1991) 94−104.

[16] S. Park, C. Gopalsamy, R. Rajamanickam, S. Jayaraman, The Wearable motherboard™: an information infrastructure or sensate liner for medical applications, pp. 252−258. Studies in Health Technology and Informatics, vol. 62, IOS Press, 1999.

[17] C. Gopalsamy, S. Park, R. Rajamanickam, S. Jayaraman, The wearable motherboard™: the first generation of adaptive and responsive textile structures (ARTS) for medical applications, Virtual Real. 4 (1999) 152−168.

[18] S. Park, S. Jayaraman, Adaptive and responsive textile structures, in: X. Tao (Ed.), Smart Fibers, Fabrics and Clothing: Fundamentals and Applications, Woodhead Publishing Limited, Cambridge, UK, 2001, pp. 226−245.

[19] A. Lymberis, D. DeRossi, Wearable eHealth Systems for Personalised Health Management, IOS Press, Amsterdam, The Netherlands, 2004.

[20] D. DeRossi, A. Lymberis, New generation of smart wearable health systems and applications, IEEE Trans. Inf. Technol. Biomed. 9 (3) (2005) 293−294.

[21] A.V. Halteren, R. Bults, K. Wac, N. Dokovsky, G. Koprinkov, I. Widya, et al., Wireless body area networks for healthcare: the MobiHealth project, Stud. Health Technol. 108 (2004) 181−193.

[22] A. Lymberis, D. Dittmar, Advanced wearable health systems and applications: Research and development efforts in the European union, IEEE Eng. Med. Biol. 23 (3) (2007) 29−33.

[23] R. Paradiso, A. Gemignani, E.P. Scilingo, D. DeRossi, Knitted bioclothes for cardiopulmonary monitoring, Proc. 25th Ann. Int. Conf. IEEE EMBS 4 (2003) 3720−3723.

[24] M.D. Rienzo, F. Rizzo, G. Parati, G. Brambilla, M. Ferratini, P. Castiglioni, MagIC system: a new textile-based wearable device for biological signal monitoring. Applicability in daily life and clinical setting, Proc. 27th Ann. Int. Conf. IEEE EMBS (2005) 7167−7169, Shanghai, China.

[25] P. Bonato, Wearable sensors/systems and their impact on biomedical engineering, IEEE Eng. Med. Biol. 22 (3) (2003) 18−20.

[26] M.A. Hanson, H.C. Powell, A.T. Barth, K. Ringgenberg, B.H. Calhoun, J.H. Aylor, et al., Body area sensor networks: challenges and opportunities, Computer 42 (2009) 58–65.

[27] K. Cherenack, L. von Pieterson, Smart textiles: challenges and opportunities, J. Appl. Phys. 112 (2012) 091301–091314.

[28] S. Patel, H. Park, P. Bonato, L. Chan, M. Rodgers, A Review of wearable sensors and systems with application in rehabilitation, J. Neuroeng. Rehabil. 9 (2012) 21.

[29] S. Bauer, Flexible electronics: sophisticated skin, Nat. Mater. 12 (2013) 871–872.

[30] Fitbit®, <http://www.fitbit.com/>, (Last Accessed: 02.11.13).

[31] Jawbone®, <https://jawbone.com/up>, (Last Accessed: 30.08.13).

[32] Nike FuelBand®, <http://store.nike.com/us/en_us/pd/fuelband-se/pid-924485/pgid-924484?cp = usns_kw_AL! 1778!3!30651044462!e!!g!nike%20fuelband>, (Last Accessed: 15.11.13).

[33] Welcome to a World Through Glass, <http://www.google.com/glass/start/what-it-does/>, (Last Accessed: 30.08.13).

[34] <http://store.sony.com/smartwatch-2-zid27-SW2ACT/cat-27-catid-Smart-Watch>, (Last Accessed: 15.11.13).

[35] <http://www.samsung.com/us/mobile/wearable-tech/SM-V7000ZGAXAR>, (Last Accessed: 15.11.13).

[36] S. Park, S. Jayaraman, Sensor networks and the i-textiles paradigm, Proceedings of the Next Generation PC 2005 International Conference, COEX, Seoul, Korea, 2005, pp. 163–167.

[37] S. Jayaraman, Fabric is the Computer: Fact or Fiction? Keynote Talk at Workshop on Modeling, Analysis and Middleware Support for Electronic Textiles (MAMSET) at ASPLOS-X (Tenth International Conference on Architectural Support for Programming Languages and Operating Systems), San Jose, CA, October 6, 2002.

[38] S. Park, S. Jayaraman, Smart textiles: wearable electronic systems, MRS Bull. (2003) 586–591.

[39] R. Rajamanickam, S. Park, S. Jayaraman, A structured methodology for the design and development of textile structures in a concurrent engineering environment, J. Text. Inst. 89 (Part 3) (1998) 44–62.

[40] S. Jayaraman, S. Park, A novel fabric-based sensor for monitoring vital signs, US Patent No 6,970,731, November 29, 2005.

[41] S. Coyle, King-Tong Lau, N. Moyna, D. O'Gorman, D. Diamond, F. Di Francesco, et al., BIOTEX—biosensing textiles for personalised healthcare management, IEEE Trans. Inf. Technol. Biomed. 14 (2) (2010) 364–370.

[42] D. Curone, E.L. Secco, A. Tognetti, G. Loriga, G. Dudnik, M. Risatti, et al., Smart garments for emergency operators: the proetex project, IEEE Trans. Inf. Technol. Biomed. 14 (3) (2010) 694–701.

[43] See What Light Can Do — Black Eyed Peas Project, <http://www.lighting.philips.com/main/connect/ lighting_university/led-videos/black-eyed-peas-project.wpd>, (Last Accessed: 17.11.13).

[44] S. Park, S. Jayaraman, Smart textile-based wearable biomedical systems: a transition plan for research to reality, IEEE Trans. Inf. Technol. Biomed. 14 (1) (2010) 86–92.

[45] S. Jayaraman, S. Park, Method and apparatus to create electrical junctions for information routing in textile structures, US Patent 7, 299,964, November 27, 2007.

[46] S. Park, K. Mackenzie, S. Jayaraman, The Wearable Motherboard: A Framework for Personalized Mobile Information Processing (PMIP), Proceedings of DAC 2002, New Orleans, Louisiana, USA, 2002, pp. 170–174.

[47] Y. Lee, J. Kim, J. Noh, I. Lee, H. Kim, S. Choi, et al., Wearable textile battery rechargeable by solar energy, Nano. Lett. 13 (2013) 5753–5761.

[48] H.S. Ng, M.L. Sim, C.M. Tan, C.C. Wong, Wireless technologies for telemedicine, BT Technol. J. 24 (2) (2006) 130–137.

[49] S. Park, S. Jayaraman, The engineering design of intelligent protective textiles and clothing, in: P. Kiekens, S. Jayaraman (Eds.), Intelligent Textiles and Clothing for Ballistic and NBC Protection: Technology at the Cutting Edge, NATO Science for Peace and Security Series B: Physics and Biophysics, Springer, The Netherlands, 2012, pp. 1–27.

1.2

Social Aspects of Wearability and Interaction

Lucy Dunne, Halley Profita, and Clint Zeagler

University of Minnesota, Minneapolis, MN, USA

1. INTRODUCTION

The concept of "wearability" in wearable systems is typically understood as relating to either the physical ability to mount a device on the body (e.g., to be "wearable" or "not wearable"), or to the physical and perhaps mental comfort of the wearer (e.g., to be comfortable physically and to easily interact with the device). However, since wearable systems are by definition worn on the body surface, they can be subject to the social perceptions and norms established by the clothing, accessories, worn artifacts, and body modifications that make up "dress" (as defined by Roach-Higgins and Eicher, [1]). Here, we consider the social facet of "wearability," the variables and factors that influence how socially comfortable an individual feels while wearing a piece of technology. This often-overlooked aspect of wearability we find to be crucial to user adoption of a technology: if a user refuses to adopt the technology because of social factors, the functional benefit is entirely lost.

Wearable systems are almost exclusively discussed with a functional focus, i.e., in terms of what they do and how well they do it. By contrast, the key functions of dress consist of not only what dress "does" for the wearer, but also what dress communicates to others about the wearer through aesthetics and expressive elements. The latter has been argued as the more important "function" of dress — and in our climate-controlled modern world, there is certainly a less-obvious need for some central functions of apparel, such as thermal protection. Given the capabilities of modern technology, it is more likely that clothing is worn because of social conventions than because of the protection it affords. Certainly, the sheer variety of redundant garments owned and worn by an individual is evidence that functionality is not the only requirement (e.g., see [2], where we studied the size and use of individual wardrobes). Indeed, even the history of clothing reflects this: the earliest forms of apparel evolved in climates where protection from the elements was not a central, everyday concern, and in many early societies protective garments like shoes were not worn even in climates that modern humans would consider quite painful. It is believed that early apparel was most likely developed for spiritual (to ward off evil spirits) and communication (to signify status or group membership, to make visual reference to powerful or beautiful things) purposes [3].

The development of next-generation smart clothing and wearable systems brings a new facet to our understanding of apparel and dress. Similarly, though, approaching wearability of smart clothing through theories of dress leads us to emphasize the social wearability of wearable systems equally with their physical wearability.

In this chapter we address two key facets of social wearability: the "static" visual perception of a wearable system and the visual perception of dynamic interactions with a wearable system.

2. SOCIAL INTERPRETATION OF AESTHETICS

Clothing and wearable devices are primarily perceived through visual and somatosensory (sensations of the body) processes. Somatosensory perception is more pertinent to the wearer's own experience of wearability, and more particularly to the experience of physical wearability and body comfort. The communication functions of dress — the ways in which dress helps to define the individual and group identity of the wearer and the wearer's context — are therefore achieved mostly through visual communication. The degree to which perceptions of one's visual appearance are comfortable for the wearer can be interpreted as the social wearability of clothing and worn artifacts.

2.1 Visual Processing of Aesthetics

DeLong [4] describes the clothed or adorned body as the "apparel-body-construct," the combined influence of the visual properties of the body and the visual properties of worn artifacts. The body alone and a garment alone may have discrete properties, but both are modified as they are brought together. The body and the garment each have visual properties which encode meaning — sometimes literally, in the sense of a graphic t-shirt with a text-based message, and sometimes in a more abstract manner, in the sense of the social judgments that accompany wearing a fashion trend, or having a certain body shape. The intersection of the properties of the garment and the body also contributes to the viewer's perception of meaning in "decoding" the identity and context of the individual. For example, a too-tight waistband may constrict the wearer's waistline, producing folds above and below the waist, and distorting the original shape of the garment. This visual cue may be read in any number of ways, depending on other contributing visual factors, and attributed to elements of identity (too lazy to buy new clothes, too vain to admit a size change) or of context (temporarily wearing ill-fitting clothes, following a fashion for cinched waist-lines), or both.

A wearable device worn under clothing may produce a bump or distortion to the body shape. This distortion may or may not be obviously attributable to something being worn under the clothes. Figure 1 shows three body-worn volumes (a rectangular shape with square corners, a curved shape, and a bracelet) worn under and on top of clothing. While curved shapes may be perceived as more "ergonomic" or comfortable, when worn under clothing they may be more easily read as a protrusion of the body's surface rather than a concealed technology.

DeLong identifies two types of characteristics of the apparel-body-construct: expressive characteristics and referential characteristics (see the left side of Figure 2). Expressive characteristics are direct characteristics of the form itself (visual elements of shape, color, texture, etc.). Referential characteristics are interpreted by the viewer; they are characteristics of the form the viewer understands as related to something outside of the form (such as a brand logo, a visual reference to another time period or a symbol of an occupational role like a badge or white coat). In some ways expressive characteristics are less open to interpretation by the viewer. They tend to play on innate responses and associations (such as bolder colors being perceived as more aggressive, flowing shapes being perceived as softer and more gentle), whereas referential characteristics depend more on the experiences and prior knowledge of the viewer. For example, it is common for teen fashion trends to reference various decades of the past (e.g., 1970s fashion references in the late 1990s and early 2000s; 1980s references in the 2010s). However, while these references may evoke direct memories in older viewers, many of the adopters of these trends have no memory of the time periods they are referencing (and may not even be aware that there is a reference), and may therefore find a different kind of meaning in the aesthetic. The same holds true for wearable systems: while the aesthetic effect of the system may hold one meaning for the designer of the technology, to an observer this meaning may be completely lost, or interpreted as something else entirely. Starner et al. [5] found that the expressive characteristics of a wearable computer were often (at the time) interpreted as being those of a medical device, the nearest mental model that most viewers could compare the wearable to. They found that altering the color of the head-mounted display quickly translated the device into a new referential association: white or light-colored devices were more often interpreted as medical devices, but grey or black devices were more often interpreted as consumer products. This division in device color specifically has blurred to some extent since 1996, but other expressive characteristics of devices still display trends that afford referential grouping by viewers.

Meanings are often defined and agreed upon by groups and sub-groups within a society (Figure 2, right side). Bell described one facet of this definition process as "sartorial morality" — the codes and mores established by a society that govern "appropriateness" of dress. These codes change with context, such that a form of dress that may be appropriate on the beach is rejected in an office setting, or clothing that is appropriate for a younger person may be inappropriate for an older person [6]. While some codes may be explicitly enforced, more often they simply carry undesired social weight: social repercussions in the form of unwanted attention or negative responses that an individual may receive when "inappropriately" dressed [7].

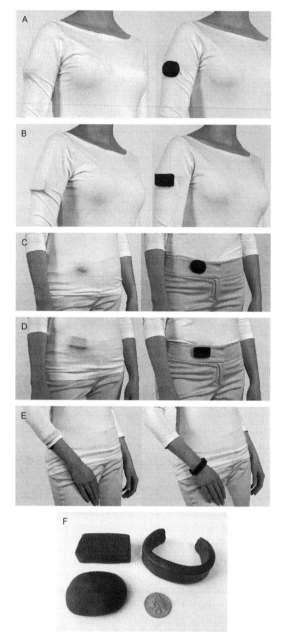

FIGURE 1   Three styles of wearable volume in three body locations, worn under and on top of clothing.

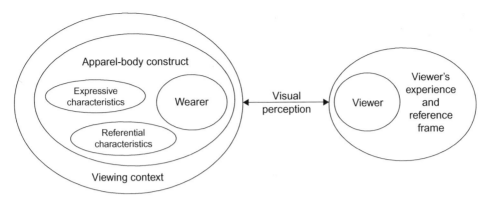

FIGURE 2 Visual perception of aesthetics.

2.2 Visual Expression of Individual and Group Identity

One of the most important ways in which the aesthetics of dress are interpreted is in understanding and assigning group identity. In some ways, the individual can be perceived as the intersection of the many groups of which she is a member: explicit groups such as her occupation, age, and socio-economic status, as well as less explicit groups like the kind of trends she adopts, products she buys, and the way in which she wears her clothes. In wearable sensing, a device may communicate group affinities like "handicapped," "sporty," or "high-tech," depending on its aesthetics and context. For example, the wearable volumes shown in Figure 1 may evoke very different group affinities depending on the shape, body location, and visibility of the "device." By assembling the many referential characteristics of her appearance, the observer forms an aggregate impression of this individual's identity. This impression is also heavily influenced by the observer's experience: for example, a young man wearing very tight jeans may be displaying allegiance to a certain trend, communicating to others that he is style-conscious enough to know that this is a trend and confident enough to pull it off. An older observer, however, may not be aware of this trend or that there exists such a group of younger people who all agree that this is a stylish way to dress, and may interpret the tight jeans by drawing on other previous experiences in which men were observed wearing tight jeans. If no suitably similar previous experiences are found, the viewer may perhaps interpret the visual appearance in another context, such as evidence of rebellion, lack of social awareness, or any number of other reasons.

Because wearable systems have not yet achieved widespread adoption, it may be difficult for observers to identify the group representation afforded by a visible device. As such, the aesthetics of the system may be grouped with a nearest-neighbor reference point, by interpreting expressive characteristics that fit a known category (such as the aforementioned influence of color).

In addition to communicating group identity to others, the way we dress also communicates some element of identity to ourselves. The "role" theory of social organization posits that an individual's understanding of how he is perceived by others mediates the roles he believes he is able to perform, and can actually affect his abilities and skills. A 2012 study by Adam and Galinsky [8] found effects in cognitive performance between groups of participants wearing identical white lab coats. The group told that their coat was a doctor's coat performed significantly better than the group told that their coat was a painter's coat. In wearable systems, similar effects can be found. Wearing a medical device may result in the wearer adopting a "sick person" or "patient" role, causing them to reduce their physical activity and restrict their movements in some way. For example, Costa et al. [9] found that patients wearing an ambulatory blood-pressure monitor were significantly less active when wearing the device.

3. ADOPTION OF INNOVATION AND AESTHETIC CHANGE

Wearable systems are at the same time required to be worn on the body, and not part of the current set of things typically worn on the body. Pervasive wearable systems rely on users wearing devices long-term, in their everyday contexts, which is generally a use pattern that relies heavily on social wearability. In turn, social wearability often depends on the normalcy of the wearer's appearance. Further, attempts to circumvent the requirement that worn devices subscribe to the aesthetic limitation of current forms of body adornment (devices that attempt to impose an entirely different aesthetic, such as the aesthetics of carried devices or small electronics) have to date not been successful in widespread adoption. It seems that attention to the structure and mechanisms of aesthetic changes in body adornment may benefit the development of wearable sensing devices. The following section will present theories of trend propagation and adoption of innovation in wearable products and technology. The case study at the end of this chapter applies these theories to a specific wearable technology product.

3.1 The Fashion Cycle: Aesthetic Change in Fashion

While the adoption of wearable devices has not been studied in depth, adoption of apparel fashion trends is well-researched. The term "fashion" as pertains to aesthetic changes in dress can be described as a gradual evolutionary process, inspired by changes in cultural identity and values, and established through a process of social consensus. A trend is not a trend until it is agreed upon by enough people, and innovation most often happens gradually, couched in the context of current norms [10]. (An aesthetic that is truly entirely new, with very few references to current norms, is more likely to look crazy or silly than to look cool.) Successful innovations often must resonate with the cultural "zeitgeist" [11] — the moods, values, and focus of a group or sub-group of a given society at a given point in time. This process is fuzzy and difficult to define (if it were straightforward, we'd all be rich!) and in practice relies heavily on the intuition and subconscious processing of members of the design community (i.e., designers, stylists, trend forecasters, journalists).

While the origin of trends and direction of trend propagation varies widely, the adoption and progress of a trend generally follows a similar structure. Innovators and early adopters are the first to adopt a new concept. (For fashion, early adopters are those invested in aesthetics and aesthetic change, who devote considerable time and attention to perceiving and developing trends, and who are usually more interested in differentiating themselves through aesthetics rather than conforming.) As the concept progresses and gains traction (is adopted by more and more people), certain factors are at play. Rogers [10] identifies the following factors that contribute to an individual's decision to adopt or reject an innovation: relative advantage, compatibility, complexity, observability, and trialability. Many of these factors relate to questions like: Do I understand it? Do I like it? Have I seen anyone else doing it? (And if so, who? Were they people like me?) Can I try it out before committing? For wearable devices, the relative advantage of the innovation is usually the key focus of the designer (making the device do something that the designer perceives as useful). However, if the remaining factors are not also taken into account, relative advantage may not be enough for widespread adoption.

If the innovation is adopted by early adopters and opinion leaders (Figure 3), and also meets the needs of the mass-market consumer, it will continue to pick up speed. Fashion consumers in the middle of the consumer spectrum are generally less sensitive to noticing trends, and are more interested in conformity in aesthetics ("fitting in") than in differentiating themselves ("standing out") [12]. Therefore, a mass-market consumer or late adopter must see the device adopted by far more people than a fashion innovator before they perceive it as something they could adopt.

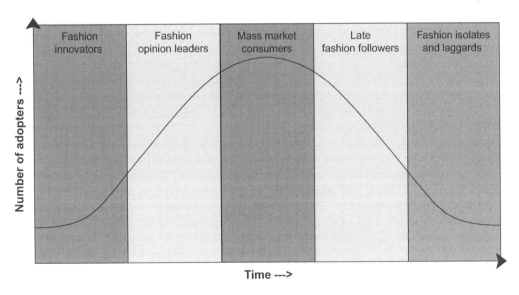

FIGURE 3 The consumer spectrum for fashion trends. *Adapted from Rogers [10].*

Finally, as innovations are adopted by the most-resistant adopters (fashion laggards), they can either become staple or non-fashion items (if they are still in use by fashion innovators and opinion leaders), or they can become symbols of insensitivity to fashion trends (e.g., become "uncool," if they have passed out of use for fashion innovators and opinion leaders). Laggards have also been characterized as particularly insensitive to aesthetic trends: for technology, this user group may actually be more willing to adopt a technological innovation that fashion innovators might perceive as socially unwearable, precisely because they are less sensitive to the emotional and identity-based factors of social wearability.

3.2 Social Leadership in Fashion

Many researchers and theorists have discussed possible influences of social roles on diffusion of fashion ideas. One of the first of these is the "trickle-down" theory, reinforced by Veblen's theory of "conspicuous consumption" [13]. In this model, the wealthy upper classes consume new fashions as a means of displaying wealth and influence. These status-symbol fashions are then disseminated through lower classes. However, with the advent of mass manufacturing and mass communications, innovations became more accessible at all social strata. King [14] discussed the "horizontal flow" theory, wherein innovations "trickle across" populations rather than down from higher social classes. Finally, "trickle-up" effects, where innovations are generated by youth and social minorities and subsequently adopted by a broader section of the population, have become more prominent in the last few decades (an example of this would be the rise of "street fashion" trends).

While these theories differ in their perspective on the origin and direction in which an idea spreads, the mechanisms in each theory are similar. The influences outlined by Rogers hold true regardless of the social status of individuals in each adopter group.

In the "fashion" of behavior (including adoption of technology), there are similarly several theories that have been used to explain the manner in which new technologies are explored and adopted by user groups. Ajzen's [15] theory of planned behavior describes the influences on an individual's consciously intended behavior as being influenced by both the person's attitude toward the behavior, the person's perception of his ability to perform the behavior, and the person's normative beliefs about the behavior (perception of what others believe he should do).

Davis's Technology Acceptance Model [16] reduces influences on a user's decision to adopt a technology to two major facets: the perceived usefulness and perceived ease-of-use of the technology. Malhotra and Galletta [17] expand this model to incorporate the social aspect of "psychological attachment," the user's perception of how well the technology fits her value system.

Wearable systems can be seen as the intersection between the processes that influence technology adoption, the processes that influence behavior change and the processes that influence fashion adoption: the usability and utility influence of technological systems, the social influence on behavior, and the emotional identity influence of fashion and worn artifacts.

4. ON-BODY INTERACTION: SOCIAL ACCEPTANCE OF GESTURE

Many wearable systems require some kind of interaction between the user and the device. Through interaction, the visual aesthetics of the device take on a new dimension: a dynamic dimension influenced by the interaction being performed. With interactive wearable systems, the user is visually evaluated not just by how they look, but by what they're doing. From the technology designer's standpoint, there are two major types of interaction that can take place with a wearable system: passive interactions and active interactions.

Passive interactions are sensed by the device without any explicit intent by the user. That is to say, passive interactions are activities, actions, and body states the device senses continuously. A wearable system may use this kind of information to infer context or to respond to a specific set of events.

Active interactions are actions performed by the user specifically for the purpose of interacting with the device. These activities can be seen as separate from (although often in parallel with) "natural" actions that are happening for a purpose other than interacting with the device.

Because passive interactions rely on detecting natural activities that have been defined within contexts outside of a wearable device, they are generally easy to socially understand (as they tend to be commonly-observed activities like sitting, standing, eating, etc.). However, the design and placement of the device itself will still affect the static aesthetics of the system (the aesthetics of the system "at rest"), and the relevant impact must be taken into account. By contrast, many active interactions introduce entirely new "vocabulary" into the visual vernacular in addition to the aesthetics of the device.

Here, we concentrate on active interactions that are performed with the hands or body. As such, we describe all active interactions as being "gestural" in nature (as in, requiring a specific movement of a part of the body). However, it is important to note that active interactions can be performed in other modalities as well — most significantly, through vocal interactions. The social perceptions of gesture have an effect similar to that of static aesthetics on the social wearability of a sensing device. Similarly, though not perceived in the visual domain, verbal interactions are also subject to social constraints on "aesthetic" and contextual appropriateness. While the following discussion concentrates on variables pertinent to physical interactions, a parallel can be drawn to similar variables pertinent to verbal and auditory interactions (voice commands and audio output).

4.1 Conspicuity and Social Weight

While active gestures are inherently more conspicuous than passive gestures, even passive gestures can be conspicuous when performed in an inappropriate context. For instance, a device that uses eye movements as inputs to perform specific commands (such as seen in [18]) may rely on movements that are often performed in daily life, such as blinking. However, if such a movement is performed when there is no readily observable reason for it to occur, it can become conspicuous as an active command (i.e., excessive blinking or long, drawn-out blinking).

Similarly, not all active gestures have the same conspicuity. Toney et al. [7] discuss the impact of device interactions on social interactions as "social weight": the degree to which interacting with a device has a negative impact on a simultaneous social interaction. This is an effect familiar to most users of technology: for example, interacting with a smartphone while mid-conversation can have a distinct social effect. If an interaction can be disguised as a "natural" gesture (in the case of Toney et al., gestures like touching buttons in a pocket or hem of a suit jacket), it may decrease the perceived social weight of the interaction.

4.2 Impact of Body Location and Handedness

Conspicuity of gesture is tightly coupled to two other design variables: body location and handedness. In previous work [19], we have explored the effect of body location on perceived acceptability of on-body interactions. When viewers observe users interacting with devices placed in various body locations, the perceived acceptability shows some interesting trends. For example, interactions that occurred in body areas where existing technology is already heavily worn and interacted with (the wrist and the forearm) were much more preferred to socially sensitive areas on the body core (e.g., the torso, which is proximal to a woman's breasts, or the pocket area, which is proximal to a male's genitalia).

Gestures that access readily reachable body locations (such as the forearm, as seen in Figure 4, or torso) may be more "naturally" performed than gestures that require contortion of the body to reach (such as the ankle or back). It is also clear that the more intensive the interaction with respect to body resources it occupies, the more conspicuous it may become. For example, a gesture that needs only one hand is likely less to be conspicuous than a gesture that needs both hands (even more so when one or more hands may be occupied prior to initiating the gesture). Furthermore, a gesture that also demands visual attention becomes even more conspicuous.

FIGURE 4 Distance (left) and close-up (right) views of interaction with the Jogwheel on female forearm (USA).

Humans are sensitive to the direction of gaze in others — for example, McAtamney and Parker [20] found that a head-mounted visual display re-directed gaze sufficiently to interrupt the flow of a conversation. To mitigate the effect of re-directed gaze, in circumstances where interface visibility is somewhat or entirely obscured, device surface topography can be employed to help tactilely guide the user toward the appropriate interaction (see an example in Figure 5). In previous work we employed design techniques to improve the "gropability" of a wearable display to minimize the demands for visual attention in interaction [21,22]. Tangible protrusions such as buttons, beveled or chamfered edges, or discernible apertures can help to orient the user for interface operation and reduce fumbling or accidental triggering.

4.3 Impact of Cultural Norms

From our previous research (Figure 6) it has also become clear that visual perceptions are strongly dependent on cultural norms and background. As these perceptions are largely judged relative to previous experience, the source of this influence is clear. The physical appearance and fashionable properties of an item may largely be driven by culture and context [23]. Additionally, perceptions of technology use may vary as normative behavior is not identical in all cultures [24].

In our work, which looked at societal attitudes toward textile-based on-body technology placement and interactions [19], we found that Americans and South Koreans reflected a similar affinity for on-body technology placement and interactions that occurred on the wrist and forearm, and a similar distaste for their least preferred on-body locations and location-based interactions: the torso and collarbone. However, in America, there were

FIGURE 5   E-textile interface with discernible surface topography to aid touch (non-visual) interaction.

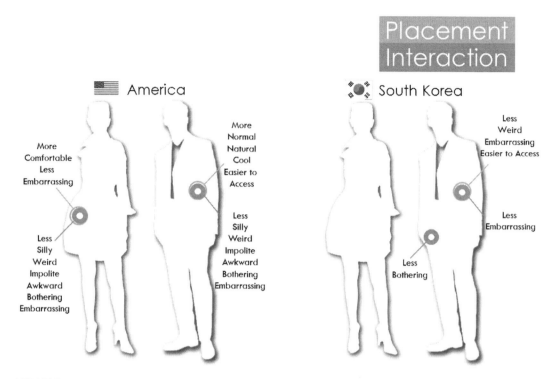

FIGURE 6 Cultural and gender comparison of on-body device placement and interactions. These were some of the least preferred locations, and despite this there were still significant attitudes toward gender preference.

some distinctions between attitudes toward on-body interactions. While, overall, the torso was one of the least preferred areas for device placement, interactions occurring at the torso were still significantly less acceptable on a female actor versus a male actor. Opposingly, interactions occurring at the pocket area were significantly less acceptable on a male actor versus a female actor. In South Korea, by contrast, this gender distinction was minimally present. Despite overall positive attitudes toward a particular placement or wearable, it was almost always less acceptable for the interactions to be conducted by a female actor versus a male actor, especially at areas on the upper body core.

4.4 The "Vocabulary" of Gesture

Social acceptability involves the manner in which one presents oneself so as to act comfortably within society [25], thus, societally prescribed conventions will ultimately play a role in how the physicality of interaction methods are perceived — i.e., whether they are deemed acceptable when performed within a particular context. Furthermore, the previously discussed facets of wearable technology (interaction technique, on-body placement, appearance, and aesthetics) and their perceived level of social acceptability will have a strong influence on their overall adoption.

Given these circumstances, how should gestural interactions be "designed"? Several precedents may lend insight. Many gestural interactions employ techniques common to hand-held and stationary devices, such as button-presses, taps, and touches. Newer gestural norms have emerged with hand-held touch screens, often based on real-world interactions, such as "swipe" (to move) and "pinch" (to zoom in/out) gestures. However, not all gestural inputs based on real-world interactions have taken hold: the "pinch" and "swipe" may be nearly ubiquitous now, but the "shake" gesture (to erase or undo) is not as well-entrenched.

Ashbrook and Starner [26] used a gestural toolkit to gather gesture inputs for a music player as designed by naïve consumers. While participants were relatively successful at creating distinct gestures for the audio player operations, a few participants expressed having quite a bit of difficulty devising gesture-based interactions that did not overlap with "everyday" gestures. Based on this study, a number of design strategies, such as "gestures should be easy to remember" and "gestures should be socially acceptable," were developed to help guide future gesture design.

Many investigations of gestural input have focused on gestures made by the hands, perhaps because affixing sensor devices to phones, watches, and other devices mounted on or held in the hands is a relatively accessible application. Other natural gestures like foot tapping, shrugging, head tilting, etc., have been less fully explored, perhaps because they require sensing more "distant" body parts. An electronic-textile interface that is seamlessly integrated into clothing might benefit from input methods that exploit current clothing interactions, such as pushing long sleeves up one's arm, or adjusting one's tie. Capitalizing on current clothing-based interactions may help minimize any cognitive dissonance experienced by a third-party viewer.

4.5 Differentiating Passive and Active Gestures

In the space between conspicuous and inconspicuous active gestures is the ambiguity of differentiating between active and passive gestures. Gestures that appear more "normal" often begin to approximate passive gestures, and can therefore begin to be confused with actual passive gestures (risking false positive triggering of interactions with the device). Large sweeping gestures may be easily and reliably detected as inputs to a device, but may also bring unwanted attention to a user, not to mention physical fatigue or obstruction of the use of other devices. Therefore, gestural interactions should be designed so that operations delineate intent from passive actions [7]. Ashbrook and Starner [26] describe the "push-to-gesture" interaction, where an "intent" button is activated before commencing the input gesture. Our "gropable" interface used a thumb anchor-pad to both tangibly orient the hand relative to the interface, and to enable touch-based activation of interface inputs [21]. Similar techniques include micro-gestures that frame an interaction to trigger the "input" phase: an audio analog is the "OK Glass" verbal activation command interaction with Google Glass.

There is also a crucial interaction between the static aesthetics of the device and the dynamic aesthetics of the interaction. Interaction with a device that is visible to third parties may self-evidently communicate intent; however, a conspicuous interaction with an inconspicuous device may obscure intent and induce confusion for observers. A solution is to pair an inconspicuous device with an inconspicuous interaction, as seen in Karrer et al. [27] in their Pinstripe e-textile interface, developed for eyes-free continuous input. Pinstripe can be embedded on the underside of one's shirt sleeve or pant pocket, and is manipulated by pinching the fabric between one's fingers. User preference was found in matching the inconspicuity of the device with a subtle interaction (pinching the inner and outer side of one's front pant pocket), thus not calling attention to the placement of the device.

5. CASE STUDY: GOOGLE GLASS

Because the spread of an innovation is so socially driven, and the genesis of a "good" idea so contextually sensitive, it can be difficult to force an idea to take root in a society. In technology, the concept of a "killer app" supposes that a very compelling function can overcome many of these social limitations to the spread of an idea. Similarly, in fashion there is a not-uncommon idea that a supremely influential innovator (individual, brand) can dictate a change that others will follow without question. In practice, both of these are difficult to achieve and prove more sensitive to subtleties of cultural context than expected. Gregory Abowd is credited with the observation that "It's not about a killer application with wearables; it's about a killer existence!" [5].

The Sony Walkman is an example of a technology that required a significant and abrupt aesthetic change (asking users to wear a conspicuous device on their heads and bodies, for the first time), yet also proposed a significant technological benefit (pervasive access to personal music). Sony's approach to encouraging the spread of their innovation was to hire young, attractive models who might be considered fashion innovators. These models were depicted in advertisements donning the device (a conspicuous trans-formation) and were also hired to wear the device in everyday situations, in major cities [28]. Through this strategy, Sony increased the observability of the device and to some extent, the trialability, lowering the barrier to adoption for fashion opinion leaders. Because the device already had a clear relative advantage, and, in Sony's opinion, an acceptable level of complexity and compatibility with existing systems (e.g., cassette tapes), the observability/trialability barriers were the key to facilitating adoption.

This model seems compelling: with the right application and enough marketing clout, it might be possible to push an aesthetic transformation into the mainstream. Google is cur-rently in the middle of just such an effort, through the Explorer program for Google Glass. Google Glass is a head-mounted display, designed for micro-interactions with information technology. In Figure 7, the display itself is the blue component on the wearer's right side. In current models, the display is suspended from a metal band worn across the forehead, supported on the nose and ears as eyeglasses would be. (The device model shown in Figure 7 is custom-built onto a pair of eyeglasses. Consumer models would be worn on top of existing eyewear and would not contain lenses.)

The Explorer program is the mechanism by which Google has distributed early models of the device to users outside of Google employees. Glass Explorers were selected by the Glass team, and by a highly publicized open application process through which everyday

FIGURE 7 Google Glass, worn by project technical lead Thad Starner.

users could pitch Google on why they should be selected to purchase one of the first Glass devices. This strategy implements something of a trickle-down approach (by putting Glass in the hands of celebrities and other thought-leaders) as well manipulating the observability and trialability of the device through a wide variety of "everyday" users.

In parallel, putting the device into the world expands what Thad Starner (Technical Lead for Glass) calls the "living laboratory" in which the device can be further developed.

"Most of our experience comes from the team wearing the devices ourselves. With enough people wearing the device in normal life, a "living laboratory" is formed where a lot of these issues become self-evident. The main reason for the announcement of Glass as a Google Project in 2012 was to enable Googlers to wear Glass in public, expanding the living laboratory to everyday life on the street." [29]

This approach both increases the exposure of the device, allowing potential adopters to see it on everyday users (and expanding the likelihood that a potential user will see the device in use by someone similar to themselves), and helps illuminate some of the usability and social acceptability issues of the device as it is used "in the wild."

The design of Glass has inherent expressive characteristics but is intended to integrate with existing eyewear in a modular fashion, to allow users their own influence on the overall aesthetic. Glass designer Isabelle Olsson says:

> "Glass is different from eyewear that you are used to seeing both in its form and function. We took a reductionist approach to the design and removed everything that wasn't essential to create a light, minimalistic and transformable design. The key challenge for the design team was how can we create a product that people can make their own, you walk into a glasses store and there are thousands of styles to choose from. With this in mind we made Glass modular, so people can choose their eyewear whether it be sunglasses, prescription glasses or the band that's free from lenses. Glass at its core is ever evolving." [29]

As such, the design team took the approach of designing a sleek, modern device that adopts the future-facing aesthetics of other current consumer products such as mobile phones, rather than the nostalgic aesthetics that are currently popular in eyewear. Although the intent of such a modular design was to allow the device to be paired with other eyewear, the contrast in aesthetic between Glass and current trends in eyeglasses (many of which are nostalgic and reference vintage styles) could be abrupt. However, the designer's challenge is clear: the amount of aesthetic variability possible in traditional eyewear is significantly larger than the amount of variability that is feasible in the first consumer version of an emerging technology. If enabling the user's current norms in terms of choice and aesthetics is not possible, the alternative approach is to create enough demand around one single (novel) aesthetic that it overcomes the demand for variability in aesthetic. Demonstrating Glass in the context of fashion opinion leaders increases the chance of success for a transformative aesthetic that transcends current norms. Indeed, historically many significant aesthetic changes are attributed to either periods of dramatic social change, or to the invention of new technology.

In addition to the impact of static aesthetics on the social acceptability of Glass, the device's interactive aesthetics are also an important influence. Glass is designed to be interacted with either through verbal commands, or through a few gestures. Gestural interactions include looking up (to activate the device), swiping the touch-pad in front of the ear or pressing a button on the top of the right-hand temple. All of these gestures ostensibly fall into the category of "natural" gestures (or can be disguised as natural gestures, such as adjusting one's glasses), but as previously discussed, may be distinguished as contextually inappropriate depending on the context of interaction. Among the gestural interactions, head-tilt is the gesture most likely to suffer from confusability with natural movements (and, in anecdotal experience, it often does): because the gesture depends on angle of the head and can be calibrated to the user's specifications, the user must decide whether to rely on a potentially conspicuous head gesture (lifting the head to a more dramatic angle) or to risk accidental activation (when the head is lifted to a "natural" level).

By contrast, the verbal activation command ("OK, Glass") is explicitly distinct from phrases that would likely come up in conversation. In order to distinguish activation commands from everyday speech, it was necessary to design a unique verbal command. An alternative approach is what Lyons et al. [30] identify as "dual-purpose speech," natural words and phrases used in everyday conversation that can also serve as input to a wearable system. They use examples like appointment scheduling, where the user activates a "push-to-talk" button to activate the speech interface, then proceeds with natural conversational structures like "I'll pencil you in on Thursday at 1 pm" to deliver commands to the system. Starner describes "OK, Glass" as an explicit signal to a conversational partner that the user is now interacting with the device. He says:

"One thing was to make the use of the device observable. The display is mounted above the user's right eye. Using the device is a social gesture. When I say "let me look that up," I literally look up into the display. When I do a Google search, I say "OK Glass, Google..." which implicitly tells bystanders that I'm doing a web search. Taking a picture is "OK Glass, take a picture" or pressing a button on top of the device close to the camera, similar to the gesture of triggering the shutter on a SLR or a point and shoot camera. All of these interactions are social gestures, just like looking at the time on a wristwatch. They cue your conversational partners to the interaction and help include them in it. Having a transparent display, where your conversation partners can actually see the interface as it is being used or at least get a sense of it, further includes them in the interaction. In essence, one is making the interface not just for the user, but for others in the social context of the user." [29]

The link between commands/gestures and visual feedback helps others to build a more accurate mental model of the functionality of Glass than a system like the Private Eye (discussed in [5]) allowed, which may help build more accurate first impressions of the device.

The combination of the viewer's mental model of what the device is and what it does and the viewer's social impressions of the identity conferred by the device on the wearer mediates the decision of whether or not to adopt a new device like Glass. Starner goes on to say that "First impressions are important with these devices. Unless potential users are willing to at least put the device on their head to try it, it does not matter how good the interaction is nor how powerful the interface. When I taught with my old systems, only 50% of the class would be willing to try it on. With Glass, that number is 99%. That is a major success for the project."

Sony's experience with the Walkman is one of the few examples of a top-down, trickle-down approach to the design of a conspicuous wearable device being commercially successful. However, the Walkman is closer to a "carried" device than a "worn" device, an important emotional distinction (carried technology has less influence on identity and sense of self than "worn" technology — a carried device is "something I have," where a worn device becomes "something I am"). Like Sony, however, Google has many of the important social influencers in place: a brand name with considerable caché, enough clout to put the device visibly in use by opinion leaders, and (for some, arguably) a compelling function.

By recruiting celebrities and innovators in many demographics and social strata, the device will gain targeted observability in everyday life. Assuming the relative advantage is significant enough and compatibility and complexity are in-line with user expectations, in theory the device will gain wider adoption. However, in the case that this approach is not successful, it will be less straightforward to determine which obstacle prevented user adoption. Glass as a case study for trickle-down adoption is better positioned than most other technologies for such an approach to be successful, and to date has seen acceptance/adoption by some high-profile thought leaders (for example, *Vogue* Magazine independently decided to feature Glass as an accessory in a fashion spread in their iconic September 2013 issue.) As yet, it is too early to tell whether or not it will gain mainstream traction, but this "living laboratory" may serve as a long-term example of how such an approach plays out in the marketplace.

6. CONCLUSION

Fully functional and high-performing devices have oftentimes been abandoned because aesthetics were not considered. This has been exceptionally true with respect to assistive devices [31]. Physical properties of the system will also dictate how readily a wearable device is received and adopted. Weight, volume, and placement [32] should be taken into account so as to avoid situating the device in areas that can hinder device use, obstruct other activities, or garner unwanted attention from outside observers.

As discussed in this chapter, the question of social wearability is complex, ever-evolving, and influenced by human psychology and sociology in many ways. Because of the inherent complexity of social dynamics, it is not feasible to prescribe a "how-to" or comprehensive rubric for designing effectively for social wearability. However, the following questions may help guide the design process:

1. For how long and in what contexts will the device be worn/interaction take place? How frequently will the system be accessed?
2. What resources does the device/interaction require? Which body parts, and which sensory systems? How much attention?
3. Where is the device located in relation to how it will be accessed? Is it visible?
4. Is the device/interaction similar to or different from "natural" interactions or existing aesthetics? In what ways?
5. How widely acceptable is the device/interaction to different user groups? What mechanisms are available to mediate that acceptance process?

Although it can be tempting to assume that a novel functionality may "trump" the influence of social wearability in mediating the adoption decision-making process, in most cases social factors have an unexpectedly strong influence on adoption decisions. Most importantly, the nature and influence of social factors for wearable products is distinctly different (and higher-impact) than the nature and influence of social factors for other types of personal devices (including handheld and mobile devices). The transition from mobile to wearable also includes a transition of the device into the wearer's intimate, emotional identity space. The impact of this transition on decisions around adoption and use of a device is paramount for wearable devices and must be creatively addressed to produce a successful product.

References

[1] M.E. Roach-Higgins, J.B. Eicher, Dress and Identity, Cloth Text Res J vol. 10 (no. 4) (1992) 1–8.
[2] LE. Dunne, V Zhang, and L Terveen, "An Investigation of Contents and Use of the Home Wardrobe," in Proceedings of the ACM Conference on Ubiquitous Computing, Pittsburgh, PA, 2012.
[3] B. Payne, G. Winakor, J. Farrell-Beck, The history of costume: from ancient Mesopotamia through the twentieth century, HarperCollins, New York, NY, 1992.
[4] M.R. DeLong, The way we look: dress and aesthetics, Fairchild Publications, New York, 1998.
[5] T Starner, B Rhodes, and J Weaver, "Everyday-use Wearable Computers," Georgia Tech Technical Report Georgia Tech Technical Report, 1999.
[6] Q. Bell, On Human Finery, Allison & Busby, 1992.
[7] A. Toney, B. Mulley, B.H. Thomas, W. Piekarski, Social weight: designing to minimise the social consequences arising from technology use by the mobile professional, Pers Ubiquitous Comput vol. 7 (no. 5) (2003) 309–320.
[8] H. Adam, A.D. Galinsky, Enclothed cognition, J Exp Soc Psychol vol. 48 (no. 4) (2012) 918–925.
[9] M. Costa, M. Cropley, J. Griffith, A. Steptoe, Ambulatory Blood Pressure Monitoring Is Associated With Reduced Physical Activity During Everyday Life, Psychosom Med vol. 61 (no. 6) (1999) 806–811.
[10] E.M. Rogers, Diffusion of Innovations, fifth ed., Free Press, 2003.
[11] B. Vinken, Fashion zeitgeist: trends and cycles in the fashion system, Berg, Oxford; New York, 2005.
[12] A. Cholachatpinyo, I. Padgett, M. Crocker, B. Fletcher, A conceptual model of the fashion process – part 2: An empirical investigation of the micro-subjective level, J Fash Mark Manag vol. 6 (no. 1) (2002) 24–34.
[13] T. Veblen, The Theory of the Leisure Class: An Economic Study of Institutions, The Macmillan Company, New York, 1899.
[14] C. King, Fashion Adoption: A Rebuttal to the "Trickle Down" Theory, in: S. Greyser (Ed.), Toward Scientific Marketing, American Marketing Association, Chicago, 1963.

[15] I. Ajzen, The theory of planned behavior, Organ Behav Hum Decis Process vol. 50 (no. 2) (1991) 179–211.

[16] F.D. Davis, "A technology acceptance model for empirically testing new end-user information systems: theory and results," Thesis, Massachusetts Institute of Technology, 1985.

[17] Y. Malhotra and D.F. Galletta, "Extending the technology acceptance model to account for social influence: theoretical bases and empirical validation," in Proceedings of the 32nd Annual Hawaii International Conference on Systems Sciences, 1999. HICSS-32, 1999, vol. Track1, p. 14.

[18] H. Manabe, M. Fukumoto, and T. Yagi, "Conductive rubber electrodes for earphone-based eye gesture input interface," in Proceedings of the 17th annual international symposium on International symposium on wearable computers, New York, NY, USA, 2013, pp. 33–40.

[19] H.P. Profita, J. Clawson, S. Gilliland, C. Zeagler, T. Starner, J. Budd, and E.Y.L Do," Don't mind me touching my wrist: a case study of interacting with on-body technology in public," in Proceedings of the 17th annual international symposium on International symposium on wearable computers, New York, NY, USA, 2013, pp. 89–96.

[20] G. McAtamney and C. Parker, "An examination of the effects of a wearable display on informal face-to-face communication," in Proceedings of the SIGCHI Conference on Human Factors in Computing Systems, New York, NY, USA, 2006, pp. 45–54.

[21] N. Komor, S. Gilliland, J. Clawson, M. Bhardwaj, M. Garg, C. Zeagler, and T. Starner, "Is It Gropable?-Assessing the Impact of Mobility on Textile Interfaces," in Proceedings of the 2009 International Symposium on Wearable Computers, Washington, DC, USA, 2009, pp. 71–74.

[22] S. Gilliland, C. Zeagler, H. Profita, and T. Starner, "Textile Interfaces: Embroidered Jog-Wheel, Beaded Tilt Sensor, Twisted Pair Ribbon, and Sound Sequins," in 2012 16th International Symposium on Wearable Computers, Los Alamitos, CA, USA, 2012, vol. 0, pp. 60–63.

[23] J. Craik, The face of fashion: cultural studies in fashion, Routledge, London; New York, 1994.

[24] S.W. Campbell, Perceptions of Mobile Phone Use in Public Settings: A Cross-Cultural Comparison, Int J Commun vol. 1 (no. 1) (2007) 20.

[25] E. Goffman, The presentation of self in everyday life, Peter Smith Publisher, Inc., 1999.

[26] D. Ashbrook and T. Starner," MAGIC: a motion gesture design tool," in Proceedings of the SIGCHI Conference on Human Factors in Computing Systems, New York, NY, USA, 2010, pp. 2159–2168.

[27] T. Karrer, M. Wittenhagen, F. Heller, and J. Borchers, "Pinstripe: eyes-free continuous input anywhere on interactive clothing," in Adjunct proceedings of the 23nd annual ACM symposium on User interface software and technology, New York, NY, USA, 2010, pp. 429–430.

[28] P. Du Gay, Doing cultural studies: the story of the Sony Walkman, Sage, in association with The Open University, London; Thousand Oaks [Calif], 1997.

[29] T. Starner, I. Olsson, Design and development of Google Glass (2013).

[30] K. Lyons, C. Skeels, T. Starner, C. M. Snoeck, B. A. Wong, and D. Ashbrook, "Augmenting conversations using dual-purpose speech," in Proceedings of the 17th annual ACM symposium on User interface software and technology, New York, NY, USA, 2004, pp. 237–246.

[31] A. Kintsch and R. Depaula, "A framework for the adoption of Assistive Technology," in ASSETS 2002, 2002, pp. 1–10.

[32] F. Gemperle, C. Kasabach, J. Stivoric, M. Bauer, and R. Martin, "Design for wearability," in Proceedings of the Second International Symposium on Wearable Computers, Pittsburgh, PA, USA, 1998, pp. 116–122.

Wearable Haptics

Yuichi Kurita

Institute of Engineering, Hiroshima University, Hiroshima, Japan

1. INTRODUCTION

The field of haptics has grown in recent decades, and haptic display is now viewed as a highly promising approach to interfaces in human-computer interaction because it provides users with nuanced information in virtual environments. In applications ranging from physical rehabilitation to tele-operation of remote robotic systems, haptic feedback can provide intuitive and valuable information, thereby facilitating the user's understanding of force and motion.

The word *haptic* comes from the Greek word *haptikos*, which relates to the sense of touch — also known as the tactile sense [1]. Humans rely on the sense of touch in real environments. This sense allows us to understand the nature of objects. We also use haptic information to hold objects and manipulate them. The goal of a haptic feedback system is to help users feel physical characteristics, interact with virtual or distant environments, and intuitively manipulate remote objects.

Humans utilize tactile and haptic information to determine object properties using thousands of mechanoreceptors located in the hands [2]. Researchers have determined that different mechanoreceptors work collectively to provide different cutaneous sensations [3]. Various haptic illusions can be created by applying vibration and/or deformation to tactile mechanoreceptors, which exhibit either rapidly adapting (RA) or slowly adapting (SA) response characteristics. RA and SA receptors each have two sub-classes: RA I (Meissner corpuscles) and RA II (Pacinian corpuscles), and SA I (Merkel cells) and SA II (Ruffini endings). These four major mechanoreceptors provide information for detecting touch, pressure, frequency, slip, and the texture of objects.

2. THE NEED FOR WEARABLE HAPTIC DEVICES

Tactile and haptic information is essential for completing tasks and control operations. Haptic feedback systems can be classified into three types according to their mechanical grounding configuration [4]: the grounded type, the non-grounded type, and the wearable type. Grounded haptic devices readily provide information on weight sensation and three-dimensional forces. PHANToM is one of the most widely used commercially available haptic devices [5]. It has a serial link mechanism with a counterbalance weight and a back-drivable arm, and can exert forces along multiple axes and torques at the device's pen tip. Sato et al. developed a series of wire-driven haptic devices (the SPIDAR series) [6–8] such as SPIDAR-G, which has seven degrees of freedom (DOF) for force feedback: three for translation, three for rotation, and one for grasping. The device provides smooth force feedback with minimal inertia, no backlash, and superior levels of scalability and safety. Arata et al. developed a haptic device that can exert three-axis force control (DELTA-4) [9]. Its parallel link mechanism contributes to the high rigidity of the device. In the robotics field, various haptic feedback methods have been developed for a tele-operation system [10–13]. These devices provide powerful haptic feedback with a simplified design. Their weight is of little significance because they are usually fixed to a static object or to the ground. However, haptic interaction with virtual environments is limited to small workspaces. Non-grounded haptic devices employ a counteractive mechanism that creates linear or angular momentum [14–16] and does not require ground connection, thus allowing for the device to be mobile or wearable. However, such devices cannot generate large forces. Exerting force in multiple directions and maintaining continuous intensity are also problematic.

Wearable haptic devices allow for greater freedom and a larger workspace than grounded systems. As the computing environment moves from desktop to mobile platforms, new computer interfaces and human-computer interaction tools for mobile systems are required. Wearable sensor systems have been developed to allow monitoring of people anywhere and at any time [17–19] and wearable haptics can be utilized as a part of the feedback mechanism. Because tactile stimulation can be perceived through the skin all over the body, utilizing the sense of touch is key in maximizing attentional resources [20]. A device that can be strapped to the user is desirable for the provision of more intuitive sensation because placement near the user's skin and/or joints enables more direct haptic feedback.

Technologies used in the development of wearable haptic and tactile display are divided into three categories: force feedback, vibro-tactile feedback, and electronic feedback. Force feedback provides stimuli for the kinematic sense. Vibro-tactile feedback is aimed more at the tactile senses. Electronic feedback leverages the electrical conductivity of human skin to evoke tactile sensations.

3. CATEGORIES OF WEARABLE HAPTIC AND TACTILE DISPLAY

3.1 Force Feedback Devices

There are two types of haptic feedback systems with force feedback capability that can be considered as wearable devices: exoskeletons and fingertip-mounted devices. Exoskeletons allow users to manipulate virtual objects based on various types of haptic feedback. A glove-type exoskeleton is provided as a simple and practical design for fingers and hands.

The CyberGrasp [21] from Immersion Corporation is a commercially available exoskeleton system. It has four fingers and allows force feedback on different phalanxes with a cable-driven exoskeleton structure attached to the back of the user's hand. SPIDAR-8 [22] is a two-handed multi-finger haptic interface device. It has eight fingertip attachment devices connected with strings to enable calculation of the 3D position of each fingertip. Haptic force is displayed by controlling the tension of the strings. HapticGEAR also employs a wire tension mechanism to display force, but the user feels force via the tip of a pen-type grip with a backpack-type mechanism to minimize fatigue for the wearer [23]. The Rutgers Master II [24] is a light and compact system composed of four small air cylinders placed inside the palm and attached to the fingertips. Its configuration does not depend on finger size, and its calibration is simplified. The telexistence cockpit system [25] represents the force of gravity by displaying multi-DOF haptic force at the operator's wrist. Yang et al. proposed a seven-DOF haptic device that can be worn on a human arm. Its operation involves the use of mechanical links/joints and human bones/joints [26]. Kobayashi et al. developed a two-fingered body-grounded haptic device called ExoPhalanx, which provides force to the distal segment of the human operator's thumb and middle finger and to the basipodite of the middle finger [27]. They provide both tactile and visual feedback using servo motor actuators and tendons attached to each of the user's fingers.

The primary problem with exoskeletons is the need for user-specific calibration. Considering the strength-mass and power-mass ratios of existing materials and actuators, fingertip devices are an interesting solution compared to exoskeletons [28]. A variety of fingertip-mounted devices have been developed for finger-based interaction [29–31]. For example, Gosselin et al. proposed a wearable haptic interface developed for precise finger interaction within virtual reality applications in large environments [32]. Aoki et al. developed a fingertip-mounted haptic device to provide contact for cutaneous sensation using wires for mixed-reality environments [33]. Minamizawa et al. integrated an underactuated mechanism with one-point kinesthetic feedback from the arm with multipoint tactile feedback [34]. Kawasaki et al. developed a hand haptic interface for use in palpation training in virtual reality environments. The device consists of finger-pad force display devices and a 3D fingertip haptic display device [35]. Chinello et al. developed a small-scale lightweight wearable haptic display that allows fingertip stimulation with a wide range of contact forces [36]. Ando et al. developed a fingernail-mountable tactile display that allows the user to feel various textures when touching smooth objects [37].

The complexity, size, and weight of fingertip-mounted devices is lower than that of exoskeletons, but they still allow most interactions including kinematic and tactile senses. However, only precision grasping can be efficiently simulated because no force is fed back to the phalanxes.

3.2 Vibro-Tactile Feedback Devices

Vibration is widely used as a haptic technology for cutaneous feedback to stimulate the tactile senses. Haptic display based on vibro-tactile feedback is easy to implement and provides robust sensation. Such feedback is a useful tool for guiding users with limited visual cues. Konyo et al. proposed a wearable stimulation device that can produce various distributed stimuli on human skin in response to hand movements using ICPF (Ionic Conducting Polymer gel Film) actuators [38]. Kevin et al. developed a wearable wireless haptic piano instruction system composed of five small vibration motors (one for each finger) inside a glove [39]. Kapur et al. developed a rehabilitation system that employs vibro-tactile feedback for stroke victims [40]. Phong and Chellali proposed two forms of vibro-tactile information known as continuous and frequency-modulation-based vibro-tactile (FMBV) feedback [41]. Garcia-Hernandez et al. evaluated subjects' tactile ability to discriminate small virtual ridge patterns using a portable-wearable tactile device [42]. Ding et al. developed the wearable InterfaceSuit, which enables human motion replication and learning in co-space, and also proposed a human-to-human motion replication methodology with multi-modal feedback mechanisms [43].

The use of vibro-tactile feedback has expanded to many different applications. Haptic illusion displays have been developed in consideration of vibration sensitivity differences by applying carefully designed vibration patterns to cutaneous mechanoreceptors [44–46]. However, tactile sensation of motion from vibrating motors can be masked by kinesthetic movement of the human body in mobile conditions [47].

3.3 Electro-Tactile Feedback Devices

Electro-tactile stimulation evokes tactile sensation on the skin by passing a current through surface electrodes. Although electro-tactile display requires direct contact between the skin and the surface electrode, small flexible electrodes are now commercially available.

Kajimoto et al. developed various electro-tactile display approaches, including a haptic interface with electrotactile-kinesthetic integration for dexterous manipulation [48], an electro-tactile display with a repeated electrode structure for enlarged display [49], and an electro-tactile display with real-time impedance feedback using pulse width modulation [50]. Lee and Starner proposed a design for wearable textile-based electro-tactile display in a wristband that can be integrated with current mobile phones and wearable computers [51]. Tamaki et al. developed PossessedHand, which controls the user's fingers by applying electrical stimulus to the muscles around the forearm [52].

The sensation delivered through surface electrodes sometimes causes pain to users. In this context, controlling sensation to a comfortable level is challenging in the design of electro-tactile displays [53].

4. DISPLAY OF FRICTION AND WEIGHT ILLUSIONS BASED ON FINGERTIP MANIPULATION

4.1 Creation of Haptic Sensation via Finger Pulp Manipulation

Other studies have investigated the creation of haptic sensation based on finger pulp manipulation to stimulate cutaneous mechanoreceptors by producing slip/shear force on the skin. Lee et al. developed a type of display that pushes pins up to apply pressure to the user's fingertips [54]. The device was found to be capable of forming shapes and patterns for subjects to identify. Webster et al. proposed a tactile haptic slip display involving the combination of a slip display with a force feedback haptic device [55]. Chen et al. developed a shear/slip feedback device that stimulates different texture sensations by rotating at various speeds [56]. Minamizawa et al. proposed a wearable ungrounded haptic display that creates realistic gravity sensation for virtual objects via fingerpad manipulation [57]. Yokota et al. produced the sensation of softness using an electrostatic device with a slider film that controls spacing between the user's finger and the device [58]. Bark et al. developed a wearable haptic feedback device that imparts rotational stretch to hairy skin for motion display [59]. Tsukada and Yasumura developed a wearable interface called ActiveBelt that enables users to obtain multiple directional information via the tactile sense [60]. Damian et al. developed a wearable single-actuator haptic device that relays multimodal haptic information on variables such as grip force and slip speed [61].

Although the skin stretch method is considered to generate more direct sensation than vibro- and electro-feedback because the skin is physically deformed, these devices can display relative haptic sensation (i.e., levels heavier/lighter or rougher/smoother than the default value) based on an ad hoc control strategy. A sophisticated control strategy is needed to display target object parameters such as weight and friction. Kurita et al. explored the challenge of displaying friction and weight illusions based on fingertip deformation [62] when lifting a weight with two fingers. The proposed device has three significant characteristics. The first is its wearability, which ensures that it occupies only a small space. Users pinch the device with the thumb and index finger. The device is fixed to the finger phalanges, and the contact area at the fingertip is manipulated via the sliding of a plate at the part where the user holds it. The second characteristic is the device's absence of a force sensor in favor of a small built-in camera that allows estimation of manipulation at the fingertip by visualizing the contact area between a transparent plate and the fingertip (Figure 1).

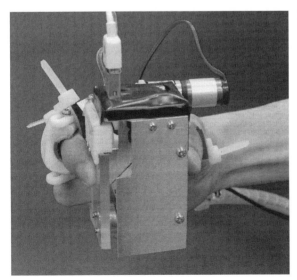

FIGURE 1   Wearable weight and friction display.

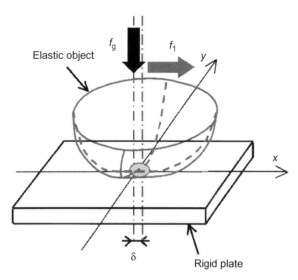

FIGURE 2 Deformation of an elastic body during contact.

The third characteristic is the device's capability to display arbitrary weight and friction values using *eccentricity* control. Eccentricity is used as a quantitative index of deformation at the contact area. Its mathematical relationship with skin deformation can be determined based on known material parameters, normal and tangential forces applied to the contact area, and the friction coefficient [63].

4.2 Deformation of the Contact Area

Contact between an elastic sphere and a rigid plate is called Hertzian contact. Figure 2 shows a schematic image of the contact occurring when a normal (vertical) force f_g and a tangential (lateral) force f_l are applied. Here, relative displacement (i.e., δ in Figure 2) occurs due to the deformation of the elastic object. The analytic solution of δ is given by:

$$\delta = \frac{3\mu f_g}{16a}\left(\frac{2-v}{G}\right)\left\{1-\left(1-\frac{f_l}{\mu f_g}\right)^{\frac{2}{3}}\right\}\cdots \tag{1}$$

where the radius of the contact area $a = \left(\frac{3f_g R}{2E}\right)^{\frac{1}{3}}$, $G = \frac{E}{2(1+v)}$, R is the radius of the elastic object, μ is the friction coefficient of the contact area, and E, v are the equivalent Young's modulus and Poisson's ratio of the elastic object.

Variables f_l and μ correspond to the weight of the object and the static friction of the contact surface when the object is held. This implies that haptic illusions in weight and friction sensation can be created by applying the appropriate relative displacement δ, which can be determined by calculating the distance between the contact center at the default position and that observed when the fingertip is deformed. However, it is difficult to measure relative displacement precisely with a low-resolution camera because displacement relating to fingertip deformation is very small. In this study, eccentricity [63,64] was used as an index of elastic object deformation. It is easier to determine eccentricity than to

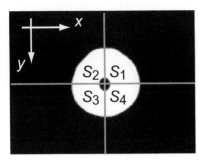

FIGURE 3 Separation of the contact area.

directly measure displacement because its calculation is based on the contact area between an elastic object and a rigid plate.

The contact area can be separated into four parts as shown in Figure 3 based on its center. The deformation of this area along the y axis S_y is calculated as follows:

$$S_y = (S_3 + S_4) - (S_1 + S_2) \cdots$$ (2)

where S_1, S_2, S_3 and S_4 represent the respective areas of each part. The eccentricity e_y along the y axis is defined as follows:

$$e_y = \frac{S_{ty}}{S_t} - \frac{S_{sy}}{S_s} \cdots$$ (3)

where S_s and S_t represent the whole contact area, and S_{sy} and S_{ty} are calculated for the neutral position and that after the application of traction force, respectively. Eccentricity is a dimensionless value normalized by the contact area.

When $\frac{\delta}{a} \ll 1$, the following approximation is obtained:

$$e_y \approx \frac{4\delta}{\pi a} \cdots$$ (4)

Eccentricity along the y axis is calculated by substituting Eq. (1) into Eq. (4):

$$e_y = K\mu f_g^{\frac{1}{3}} \left\{ 1 - \left(1 - \frac{f_l}{\mu f_g}\right)^{\frac{2}{3}} \right\} \cdots$$ (5)

$$K = \frac{1}{\pi}(2 - \nu)(1 + \nu)\left(\frac{3}{2ER^2}\right)^{\frac{1}{3}} \cdots$$ (6)

When an object is lifted up, f_l and μ correspond to its weight and friction. This implies that weight and friction sensation can be represented by measuring and controlling e_y rather than δ. Figure 4 shows the eccentricity calculated using Eq. (6) when the object weight changes from 50 to 500 g in (a), and the friction changes from 0.4 to 1.8 in (b). The influence of weight on eccentricity is relatively large; the difference is observed at the beginning $(t < 0.1)$. The influence of friction is smaller than that of weight; the difference appears late (around $t = 0.3$) and becomes larger as the friction coefficient decreases.

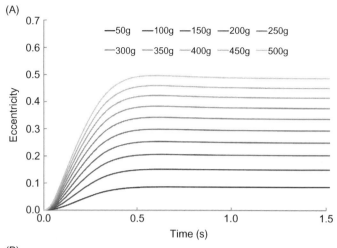

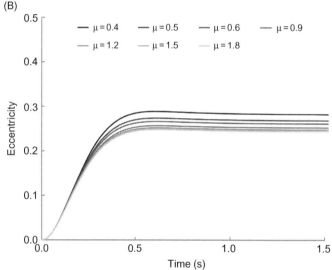

FIGURE 4   (a) Eccentricity when the weight changes. (b) Eccentricity when the friction changes.

4.3 Weight and Friction Illusion Display

A prototype display device was developed to produce haptic illusions of weight and friction via eccentricity control for the contact surface of a human fingertip. The device consists of a transparent acrylic plate, a motor to actuate the plate, and a camera to capture images of the contact surface.

An overview of the device and a schematic image of its operating principle are shown in Figure 1 and Figure 5. The total weight of the device is 210 g. The user inserts the thumb and index finger into the rings attached to the device and pinches it so that the index finger is on the center of the transparent plate. Here, the user's hand was fixed to the body of the device at the middle phalanx of the thumb and at the middle and proximal phalanges of the index finger using plastic belts. The camera captures images of the contact surface at

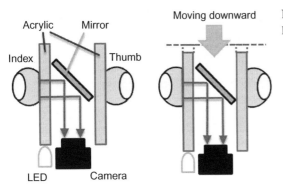

Acrylic   Mirror

Moving downward

Index   Thumb

LED   Camera

FIGURE 5 Schematic image of the operating principle.

 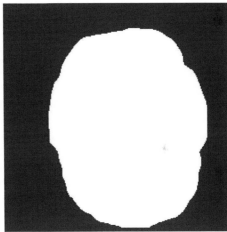

FIGURE 6   Captured image by the camera (left) and the resultant contact area (right).

60 fps. Here, morphological image processing for opening and smoothing was applied to the images captured, and binarized contact images as shown in Figure 6 were obtained. The contours of the contact images were used to calculate eccentricity.

The actuator controls motor torque to achieve the desired eccentricity. As shown in Figure 7, the desired and measured eccentricity with the target weight and friction were set as $(m, \mu) = (100, 1.0), (200, 1.0)$, and $(300, 1.0)$. Highly favorable eccentricity display performance was confirmed.

The efficacy and usefulness of the prototype were investigated in three evaluation experiments involving weight display and friction display with 10 healthy subjects aged from 22 to 24 years old. The weight-friction illusion device was attached to the dominant hand of the subjects, who were instructed to place the same hand on a table and to relax their torso and non-dominant hand. They were then asked to lift the device to a height of 200 mm in 2 seconds. Eccentricity feedback control was started when the device left the table. After lifting and deposition were complete, the experimenter detached the device from the subject's hand. The subjects were then asked to lift a test object up and down freely.

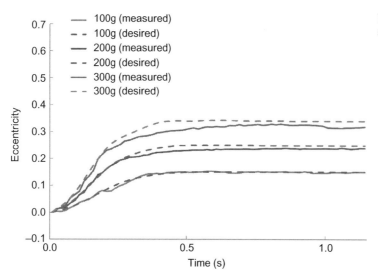

FIGURE 7 Desired and measured eccentricity.

In the weight display test, the displayed weights were $m = 100, 200$, and 300 g. When the target weight is heavier than the weight of the device itself, the actuator moves the plate downward where the subjects pinched; when the target weight is lighter than the device, the actuator moves the plate upward. The texture of the contact surface between the display device and the fingertip was fixed using oil to have a value of approximately $\mu = 0.9$. The subjects could change the weight of the test object to 50, 100, 150, 200, 250, 300, 350, and 400g by placing weights on the test object container. The surface texture of the test object was the same as that of the display device. The subjects were asked to select the one they judged to have the same weight as the displayed value. In the friction display test, the displayed friction values were $\mu = 0.6, 0.9$, and 1.2. The weight of the display device was fixed at 200 g. The weight of the test object was the same as that of the display device. The subjects could change the friction coefficient of the test object to approximately $\mu = 0.4, 0.6, 0.9, 1.2$, and 1.7 by replacing the contact plate. The subjects were asked to select the one they judged to have the same friction as the displayed value. Figure 8 and Figure 9 show the subjects' answers in the weight and friction display test, respectively. The results of the experiments indicate that the prototype successfully displayed weight and friction illusions to the subjects. However, gaps were also observed between the target sensations displayed by the device and the subjective sensations actually felt by the subjects. Measurement of forces during the task and utilization of data to modulate the desired eccentricity are expected to improve the range of weight and friction values that can be presented to users using more sophisticated models and devices.

5. A WEARABLE SENSORIMOTOR ENHANCER

5.1 Improvement of Haptic Sensory Capability for Enhanced Motor Performance

Improving haptic and tactile sensory capability helps to enhance motor performance. In this context, a fingertip-wearable device that improves motor performance could be used

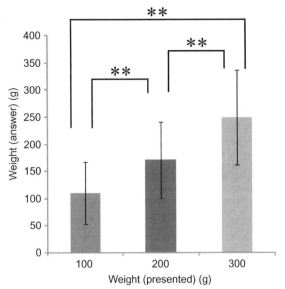

FIGURE 8 Result of the weight presentation experiment. The results of Steel-Dwass testing revealed significant differences among all the cases (**: $p < 0.01$).

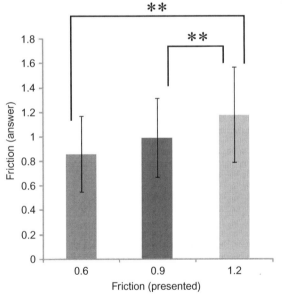

FIGURE 9 Result of the friction presentation experiment. The results of Steel-Dwass testing revealed significant differences between $\mu = 0.6$ and 1.2, and between $\mu = 0.9$ and 1.2 (**: $p < 0.01$).

to assist workers engaged in tasks requiring high-precision manual dexterity. A variety of such devices have been developed in the past.

Mascaro and Asada developed a photoplethysmograph fingernail sensor that measures two-dimensional variations in blood volume beneath the fingernail for estimation of normal force, lateral shear force, longitudinal shear force and bending angle based on readings from the sensor [65]. Provancher and Sylvester revealed that the perceived magnitude of friction rendered by conventional force feedback can be increased with the addition of fingertip skin stretch [66]. Tanaka et al. developed a tactile device called a tactile nail chip that can be mounted on the fingernail and deforms it to change tactile perceptivity [67]. Jeong et al. reported that vibration sensitivity changes as a result of tangential vibration application to the skin surface [68]. Romano et al. developed the SlipGlove, which provides tactile cues associated with slippage between the glove and a contact surface [69].

Some researchers have also reported that Stochastic Resonance (SR) improves tactile and haptic sensitivity in the feet, hands, and fingers [70–73]. Improving tactile sensitivity is known to influence the sensorimotor functions of humans [74,75]. Kurita et al. reported the results of a study on a wearable device called a sensorimotor enhancer, which is considered to improve the tactile sensitivity of fingertips. The proposed device has two important qualities [76]. The first is its utilization of a piezoelectric stack actuator featuring high-frequency vibration generation capacity in a compact body. The second is its design, which promotes dexterity by leaving the palmar region of the fingertips free.

5.2 A Wearable Sensorimotor Enhancer Based on the Stochastic Resonance Effect

Figure 10 shows a fingercap-type wearable sensorimotor enhancer prototype. In the device, a compact lead zirconate titanate (PZT) piezoelectric stack actuator is attached to the fingercap. The actuator is used to generate white-noise vibration, which is transmitted to tactile receptors around the finger pulp.

Four tests were conducted with 11 healthy subjects aged from 24 to 38. The amplitude threshold of vibration perception for each subject was determined prior to the experiment. Each subject was asked to report the perception of vibration at the contact point with the actuator when the signal amplitude changed. The maximum amplitude that the subject could not feel was recorded using a limit-based method. In the subsequent experiments, random order was used to apply conditions of no vibration and five different amplitudes (with 50, 75, 100, 125, and 150% of the perception threshold, denoted as 0.5T, 0.75T, 1.0T, 1.25T, and 1.5T, respectively). The sensorimotor enhancer was attached to the subject's index finger.

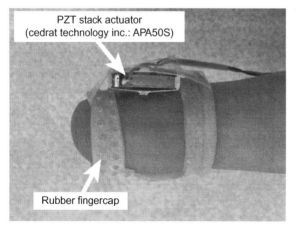

FIGURE 10 Fingercap-type wearable sensorimotor enhancer.

5.2.1 Two-Point Discrimination Test

The subjects were instructed to place the hand with the sensorimotor enhancer attached on a table and close their eyes. The experimenter gently pressed two sharp points of a measuring compass against the palmar surface of each subject's fingertip as shown in Figure 11. The subjects were asked to report when they felt definite contact from the two points. Two series of ascending and descending distances between the points (i.e., a total of four) were tested for each vibration condition, and the average was recorded as the result of each trial. The distance interval was 0.5 mm. The measured distance in the controlled case was normalized against the mean distance measured in the no-vibration case. Figure 12 shows the results of the experiments. The data show that the mean distances in

FIGURE 11 Two-point discrimination test: a measuring compass was pressed against the palmar side of the fingertip.

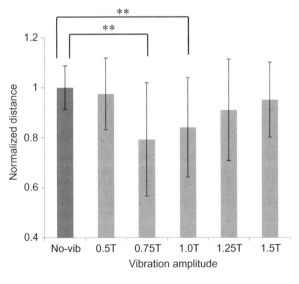

FIGURE 12 Results of two-point test: discriminable distances in all five controlled cases were smaller than that in the no-vibration case.

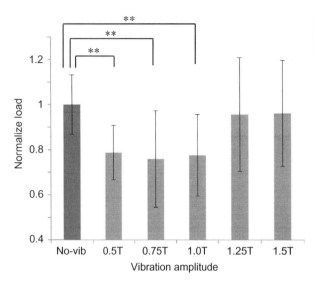

FIGURE 13 Results of one-point test: perceivable forces in all five controlled cases were smaller than that in the no-vibration case.

all five controlled cases were smaller than that in the no-vibration case. The results of post hoc Dunnett testing also revealed significant differences against the no-vibration case for the 0.75T (p < 0.01) and 1.0T (p < 0.01) cases.

5.2.2 One-Point Touch Test

The subjects were instructed to place the hand with the sensorimotor enhancer attached on a table and close their eyes. The experimenter pressed a monofilament against the palmar surface of each subject's fingertip until buckling occurred, held it, and then removed it. The subjects were asked to report when they felt contact from the filament. Two series of ascending and descending loads (i.e., a total of four) were tested for each vibration condition, and the average was recorded as the result of each trial. A total of five Semmes-Weinstein monofilaments (Touch-Test Sensory Evaluator; 0.008, 0.02, 0.04, 0.07, and 0.1 g) were used as stimuli. The measured load in the controlled case was normalized against the mean load measured in the no-vibration case. Figure 13 shows the results of the experiment. Smaller values of mean force equate to better tactile sensitivity. These outcomes show that the forces in all five controlled cases were smaller than that in the non-vibration case. The results of post hoc Dunnett testing also revealed significant differences against the non-vibration case for the 0.5T (p < 0.01), 0.75T (p < 0.01), and 1.0T (p < 0.01) cases.

5.2.3 Active Sensory Test — Texture Discrimination

The subjects were asked to touch nine pieces of sandpaper with CAMI grit sizes of #40, #80, #120, #150, #180, #220, #240, #280, and #320 and select the one they judged to have the same texture as the piece on the other side. All the pieces were glued to one side of a plastic board provided to the subjects, who were not allowed to see the sandpaper but were permitted to touch and feel it. Attached to the other side of the board was a test piece of sandpaper with a grit size matching one of the nine sandpaper types. Figure 14 shows the results of the experiment. Higher ratios equate to better tactile sensitivity. As the

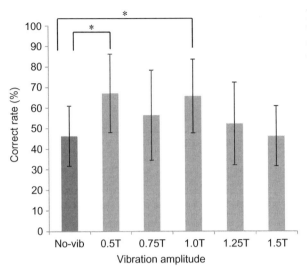

FIGURE 14 Results of texture discrimination test: mean correct ratios in all five controlled cases were higher than that in the no-vibration case.

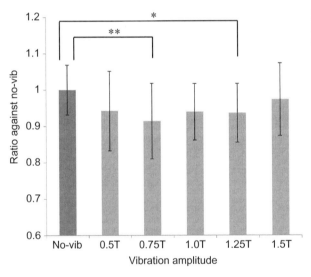

FIGURE 15 Results of grasping test: grasping forces for all of the controlled cases were smaller than that of the no-vibration case.

figure indicates, the mean correct ratios for all the controlled cases tended to be higher than that of the no-vibration case. The results of post hoc Dunnett testing also revealed significant differences against the non-vibration case for the 0.5T ($p < 0.05$) and 1.0T ($p < 0.05$) cases.

5.2.4 Motor Skill Test — Minimal-Force Grasping

The subject was asked to pinch and hold an object with a weight of 140 g for 3 seconds with as little force as possible without allowing it to slip. The force measured in the controlled case was normalized against the mean force measured in the no-vibration case. Figure 15 shows the results of the experiment. Smaller forces equate to better motor performance in terms of pinch grasping. Improvements in motor performance were observed for all the controlled cases. The results of post hoc Dunnett testing also revealed significant differences against the no-vibration case for the 0.75T ($p < 0.01$) and 1.25T ($p < 0.05$) cases.

6. CONCLUSIONS

Haptic display represents a promising interface for human-computer interaction tools that provide users with nuanced information in a virtual environment. Humans utilize tactile and haptic information to determine object properties using thousands of mechanoreceptors located in the hands. Various haptic illusions can also be created by applying vibration and/or deformation to tactile mechanoreceptors. Improving haptic and tactile sensory capability helps to enhance motor performance. Such devices can be used to assist workers engaged in tasks requiring high-precision manual dexterity.

This chapter describes state-of-the-art wearable haptic display devices, and also introduces a variety of previously developed wearable devices that improve motor performance. In this chapter, two research studies are presented in detail. The first explored the challenge of displaying friction and weight illusions based on fingertip deformation. In this study, only phenomena occurring at the fingertips were considered. However, weight is sensed using not only tactile receptors in fingertips, but also proprioceptive information from muscles and tendons. Further investigation is necessary to understand human perception characteristics of weight and friction, and clarify the effects of tactile and haptic sensitivity on other types of motor performance. The second study investigated a wearable sensorimotor enhancer that improves tactile sensitivity in human fingertips. The development of a wearable device for fingertips is expected to assist individuals in workplaces, including those engaging in laboratory work with palpation for tumors, handling of very small objects, texture design for products, and high-precision manual assembly. Continued research in this area may also lead to the development of a novel device that helps individuals with incomplete peripheral neuropathy to use their hands more reliably in daily activities and at work.

The author hopes this chapter will be useful to anybody interested in wearable haptic and tactile devices.

References

[1] M.C. Lin, M.A. Otaduy, Sensation-preserving haptic rendering, IEEE. Comput. Graph. Appl. 25 (4) (2005) 8–11.
[2] R.S. Johansson, A.B. Vallbo, Spatial properties of the population of mechanoreceptive units in the glabrous skin of the human hand, Brain. Res. 184 (2) (1980) 353–366.
[3] S.A. Wall, S. Brewster, Sensory substitution using tactile pin arrays: human factors, technology and applications, Signal Processing 86 (12) (2006) 3674–3695.
[4] S.J. Biggs, M.A. Srinivasan, Haptic interfaces, in: K.M. Stanney (Ed.), Handbook of Virtual Environments, Lawrence Erlbaum Associates, 2002, pp. 93–115.
[5] T.H. Massie, J.K. Salisbury, The phantom haptic interface: a device for probing virtual objects, Symp. Haptic Interfaces Virtual Environ. Teleoperator Syst. 55 (1) (1994) 295–301.

[6] M. Sato, Y. Hirata, H. Kawarada, Rotating shape modeling with SPIDAR, J. Rob Mechatron. 4 (1) (1992) 31–38.

[7] S. Kim, S. Hasegawa, Y. Koike, M. Sato, Tension based 7-dof force feedback device: Spidar-G, IEEE Virtual Real. Conf. (2002) 283–284.

[8] J. Murayama, Y. Luo, K. Akahane, S. Hasegawa, M. Sato, 2004. A haptic interface for two-handed 6DOF manipulation-SPIDAR G&G system. IEICE Transactions on Information and Systems E87-D(6), 1415–1421.

[9] J. Arata, H. Kondo, M. Sagaguchi, H. Fujimoto, Development of a haptic device delta-4 using parallel link mechanism, IEEE Int. Conf. Rob. Autom. (2009) 294–300.

[10] Z. Zhang, T. Chen, Study on the control of 6-DOF manipulators system with force feedback, Int. Conf. Intell. Comput. Technol. Autom. 1 (2008) 498–502.

[11] T. Endo, H. Kawasaki, T. Mouri, Five-fingered haptic interface robot: HIRO III, EuroHaptics Conf. Symp. Haptic Interfaces Virtual Environ. Teleoperator Syst. (2009) 458–463.

[12] P. Berkelman, M. Dzadovsky, Extending the motion ranges of magnetic levitation for haptic interaction, EuroHaptics Conf. Symp. Haptic Interfaces Virtual Environ. Teleoperator Syst. (2009) 517–522.

[13] T. Endo, T. Kanno, M. Kobayashi, H. Kawasaki, Human perception test of discontinuous force and a trial of skill transfer using a five-fingered haptic interface, J. Rob. (2010) 542360.

[14] H. Gurocak, S. Jayaram, B. Parrish, U. Jayaram, Weight sensation in virtual environments using a haptic device with air jets, J. Comput. Inf Sci. Eng. 3 (2) (2003) 130–135.

[15] A. Chang, C. O'Sullivan, Audio-haptic feedback in mobile phones, Int. Conf. Hum. Comput. Interact. (2005) 1264–1267.

[16] K.N. Winfree, J. Gewirtz, T. Mather, J. Fiene, K.J. Kuchenbecker, A high fidelity ungrounded torque feedback device: the iTorqU 2.0, World Haptics (2009) 261–266.

[17] P. Bonato, Wearable sensors and systems, IEEE Eng Med Biol Mag (2010) 25–36.

[18] A. Bonfiglio, D. De Rossi, Wearable Monitoring Systems, Springer, New York, 2011.

[19] R. Paradiso, G. Loriga, N. Taccini, A wearable health care system based on knitted integrated sensors, IEEE Trans Inf Technol Biomed 9 (3) (2005) 337–344.

[20] A. Gallace, H.Z. Tan, C. Spence, The body surface as a communication system: The state of the art after 50 years, Presence: Teleoperators Virtual Environ. 16 (6) (2007) 655–676.

[21] F. Gosselin, C. Andriot, P. Fuchs, 2006. Les dispositifs matriels des interfaces retour deffort. In Le Trait de la Ralit Virtuelle. 3rd ed, pp 135–202, Les Presses de l'Ecole des Mines.

[22] S. Walairacht, M. Ishii, Y. Koike, M. Sato, Two-handed Multi-fingers String-based Haptic Interface Device, IEICE Trans. Inf. Syst. E84D (3) (2001) 365–373.

[23] M. Hirose, K. Hirota, T. Ogi, Y. Hiroaki, N. Kakehi, et al., HapticGEAR: the development of a wearable force display system for immersive projection displays, IEEE Virtual Real. (2001) 123–129.

[24] M. Bouzit, G. Burdea, G. Popescu, R. Boian, The Rutgers Master II - new design force feedback glove, IEEE/ ASME Trans. Mechatron. 7 (2) (2002) 256–263.

[25] S. Tachi, K. Komoriya, K. Sawada, T. Nishiyama, T. Itoko, et al., Telexistence cockpit for humanoid robot control, Adv. Rob. 17 (3) (2003) 199–217.

[26] G. Yang, H.H. Leong, W. Chen, W. Lin, S.H. Yeo, M.S. Kurbanhusen, A haptic device wearable on a human arm, IEEE Conf. Rob., Autom. Mechatron. 1 (2004) 243–247.

[27] F. Kobayashi, G. Ikai, W. Fukui, F. Kojima, Two-fingered haptic device for robot hand teleoperation, J. Rob. (2011) 419465.

[28] F. Gosselin, Guidelines for the design of multi-finger haptic interfaces for the hand, Romansy 19 - Robot Design, Dyn. Control CISM Int. Centre Mech. Sci. 544 (2013) 167–174.

[29] A. Frisoli, F. Simoncini, M. Bergamasco, F. Salsedo, Kinematic design of a two contact points haptic interface for the thumb and index fingers of the hand, J. Mech. Des. 129 (2007) 520–529.

[30] M. Fontana, A. Dettori, F. Salsedo, M. Bergamasco, Mechanical design of a novel hand exoskeleton for accurate force displaying, Int. Conf. Rob. Autom. (2009) 1704–1709.

[31] C. Giachritsis, J. Barrio, M. Ferre, A. Wing, J. Ortego, Evaluation of weight perception during unimanual and bimanual manipulation of virtual objects, Int. Conf. World Haptic. (2009) 629–634.

[32] F. Gosselin, T. Jouan, J. Brisset, C. Andriot, Design of a wearable haptic interface for precise finger interactions in large virtual environments, World Haptic. (2005) 202–207.

[33] T. Aoki, H. Mitake, D. Keoki, S. Hasegawa, M. Sato, Int. Conf. Adv. Comput. Entertain. Technol., Wearable haptic device to present contact sensation based on cutaneous sensation using thin wire (2009) 115—122.

[34] K. Minamizawa, D. Prattichizzo, S. Tachi, Simplified design of haptic display by extending one-point kinesthetic feedback to multipoint tactile feedback, IEEE Haptic. Symp. (2010) 257—260.

[35] H. Kawasaki, Y. Doi, S. Koide, T. Endo, T. Mouri, Hand haptic interface incorporating 1D finger pad and 3D fingertip force display devices, IEEE Int. Symp. Ind. Electron. (2010) 1869—1874.

[36] F. Chinello, M. Malvezzi, C. Pacchierotti, D. Prattichizz, A three DoFs wearable tactile display for exploration and manipulation of virtual objects, IEEE Haptic. Symp. (2012) 71—76.

[37] H. Ando, T. Miki, M. Inami, T. Maeda, SmartFinger: nail-mounted tactile display, ACM SIGGRAPH (2002) 78.

[38] M. Konyo, K. Akazawa, S. Tadokoro, T. Takamori, Wearable haptic interface using ICPF actuators for tactile feel display in response to hand movements, J. Rob. Mechatron. 15 (2) (2003) 219—226.

[39] K. Huang, E.Y. Do, T. Starner, Pianotouch: a wearable haptic piano instruction system for passive learning of piano skills, IEEE Int. Symp. Wearable Comput. (2008) 41—44.

[40] P. Kapur, S. Premakumar, S.A. Jax, L.J. Buxbaum, A.M. Dawson, K.J. Kuchenbecker, Vibrotactile feedback system for intuitive upper-limb rehabilitation, IEEE World Haptic. Conf. (2009) 621—622.

[41] H. Phong Pham, R. Chellali, Frequency modulation based vibrotactile device for teleoperation, IEEE Int. Conf. Space Mission Chall. Inf. Technol. (2009) 98—105.

[42] N. Garcia-Hernandez, N.G. Tsagarakis, D.G. Caldwell, Human tactile ability to discriminate variations in small ridge patterns through a portable-wearable tactile display, Int. Conf. Adv. Comp. Hum. Interact. (2010) 38—43.

[43] Z.Q. Ding, I.M. Chen, S.H. Yeo, The development of a real-time wearable motion replication platform with spatial sensing and tactile feedback, IEEE/RSJ Int. Conf. Intell. Rob. Syst. (2010) 3919—3924.

[44] K. Kyung, S. Kim, D. Kwon, M.A. Srinivasan, Texture display mouse kat: vibrotactile pattern and roughness display, IEEE Int. Conf. Intell. Rob. Syst. (2006) 478—483.

[45] T. Ahmaniemi, J. Marila, V. Lantz, Design of dynamic vibrotactile textures, IEEE Trans. Haptic. 3 (4) (2010) 245—256.

[46] S. Okamoto, Y. Yamada, Lossy data compression of vibrotactile material-like textures, IEEE Trans. Haptic. 6 (1) (2013) 69—80.

[47] I. Oakley, J. Park, Did you feel something? Distracter tasks and the recognition of vibrotactile cues, Interact. Comput. 20 (3) (2008) 354—363.

[48] K. Sato, H. Kajimoto, N. Kawakami, S. Tachi, Electrotactile display for integration with kinesthetic display, IEEE Int. Symp. Rob. Hum. Interact. Commun. (2007) 3—8.

[49] H. Kajimoto, Enlarged electro-tactile display with repeated structure, IEEE World Haptic. Conf. (2011) 575—579.

[50] H. Kajimoto, Electrotactile display with real-time impedance feedback using pulse width modulation, IEEE Trans. Haptic. 5 (2) (2012) 184—188.

[51] S.C. Lee, T. Starner, Stop burdening your eyes: a wearable electro-tactile display, IEEE Int. Symp. Wearable Comput. (2008) 115—116.

[52] E. Tamaki, T. Miyaki, J. Rekimoto, PossessedHand: techniques for controlling human hands using electrical muscles stimuli, SIGCHI Conf. Hum. Factors Comput Syst. (2011) 543—552.

[53] K.A. Kaczmarek, J.G. Webster, P. Bach-Y-Rita, W.J. Tompkins, Electrotactile and vibrotactile displays for sensory substitution systems, IEEE Trans. Biomed. Eng. 38 (1991) 1—16.

[54] J.M. Lee, C.R. Wagner, S.J. Lederman, R.D. Howe, Spatial low pass filters for pin actuated tactile displays, Symp. Haptic Interfaces Virtual Environ. Teleoperator Syst. (2003) 57—62.

[55] R.J. Webster, T.E. Murphy, L.N. Verner, A.M. Okamura, A novel two-dimensional tactile slip display: Design, kinematics and perceptual experiments, ACM. Trans. Appl. Percept. 2 (2) (2005) 150—165.

[56] X. Chen, A.G. Song, J.Q. Li, A new design of texture haptic display system, IEEE Int. Conf. Inform. Acquis. (2006) 1122—1126.

[57] K. Minamizawa, H. Kajimoto, N. Kawakami, S. Tachi, A wearable haptic display to present the gravity sensation - preliminary observations and device design, World Haptic. (2007) 133—138.

[58] H. Yokota, A. Yamamoto, H. Yamamoto, T. Higuchi, Producing softness sensation on an electrostatic texture display for rendering diverse tactile feelings, World Haptic. (2007) 584—585.

[59] K. Bark, J. Wheeler, P. Shull, J. Savall, M. Cutkosky, Rotational skin stretch feedback: a wearable haptic display for motion, IEEE Trans. Haptic. 3 (3) (2010) 166–176.

[60] K. Tsukada, M. Yasumura, Tactile feel display for virtual active touch. Ubiquitous computing, Lect. Notes. Comput. Sci. 205 (2004) 384–399.

[61] D.D. Damian, M. Ludersdorfer, Y. Kim, A.H. Arieta, R. Pfeifer, A.M. Okamura, Wearable haptic device for cutaneous force and slip speed display, IEEE Int. Conf. Rob. Automat. (2012) 1038–1043.

[62] Y. Kurita, S. Yonezawa, A. Ikeda, T. Ogasawara, Weight and friction display device by controlling the slip condition of a fingertip, IEEE/RSJ Int. Conf. Intell. Rob. Syst. (2011) 2127–2132.

[63] J. Ueda, A. Ikeda, T. Ogasawara, Grip-force control of an elastic object by vision-based slip margin feedback during the incipient slip, IEEE Trans. Rob. 21 (6) (2005) 1139–1147.

[64] Y. Kurita, A. Ikeda, J. Ueda, T. Ogasawara, A fingerprint pointing device utilizing the deformation of the fingertip during the incipient slip, IEEE Trans. Rob. 21 (5) (2005) 801–811.

[65] S.A. Mascaro, H.H. Asada, Measurement of finger posture and three-axis fingertip touch force using fingernail sensors, IEEE Trans. Rob. Automat. 20 (1) (2004) 26–35.

[66] W. Provancher, N. Sylvester, Fingerpad skin stretch increases the perception of virtual friction, IEEE Trans. Haptic. 2 (4) (2009) 212–222.

[67] Y. Tanaka, A. Sano, M. Ito, H. Fujimoto, A novel tactile device considering nail function for changing capability of tactile perception. haptics: perception, devices and scenarios, Lect. Notes. Comput. Sci. 5024 (2008) 543–548.

[68] H. Jeong, M. Higashimori, M. Kaneko, Improvement of vibration sensitivity by tangential vibration, J. Rob. Mechatron. 21 (4) (2009) 554–562.

[69] J.M. Romano, S.R. Gray, N.T. Jacobs, K.J. Kuchenbecker, Toward tactilely transparent gloves: Collocated slip sensing and vibrotactile actuation, World Haptic. Conf. (2009) 279–284.

[70] N.T. Dhruv, J.B. Niemi, J.D. Harry, L.A. Lipsitz, J.J. Collins, Enhancing tactile sensation in older adults with electrical noise stimulation, Neuroreport 13 (5) (2002) 597–600.

[71] L. Khaodhiar, J.B. Niemi, R. Earnest, C. Lima, J.D. Harry, A. Veves, Enhancing sensation in diabetic neuropathic foot with mechanical noise, Diabetes Care 26 (12) (2003) 3280–3283.

[72] N. Harada, M.J. Griffin, Factors influencing vibration sense thresholds used to assess occupational exposures to hand transmitted vibration, Br. J. Ind. Med. 48 (1991) 185–192.

[73] G.A. Gescheider, S.J. Bolanowski, J.V. Pope, R.T. Verrillo, A four-channel analysis of the tactile sensitivity of the fingertip: frequency selectivity, spatial summation, and temporal summation, Somatosens. Mot. Res. 19 (2) (2002) 114–124.

[74] J.J. Collins, T.T. Imhoff, P. Grigg, Noise-enhanced tactile sensation, Nat. Lond. 383 (1996) 770.

[75] J.J. Collins, T.T. Imhoff, P. Grigg, Noise-mediated enhancements and decrements in human tactile sensation, Phys. Rev. E 56 (1) (1997) 923–926.

[76] Y. Kurita, M. Shinohara, J. Ueda, Wearable sensorimotor enhancer for fingertip using stochastic resonance effect, IEEE Trans. Hum. Mach. Syst. 43 (3) (2013) 333–337.

Wearable Bio and Chemical Sensors

*Shirley Coyle[1], Vincenzo F. Curto[1], Fernando Benito-Lopez[1,2],
Larisa Florea[1], and Dermot Diamond[1]*

[1]INSIGHT: Centre for Data Analytics, National Centre for Sensor Research, Dublin City
University, Dublin, Ireland, [2]CIC microGUNE, Arrasate-Mondragón, Spain

1. INTRODUCTION

Chemical and biochemical sensors have experienced tremendous growth in the past decade due to advances in material chemistry combined with the emergence of digital communication technologies and wireless sensor networks (WSNs) [1]. The emergence of wearable chemical and biochemical sensors is a relatively new concept that poses unique challenges to the field of wearable sensing. This is because chemical sensors have a more complex mode of operation, compared to physical transducers, in that they must interact in some manner with specific molecular targets in the sample medium. To understand the challenges in developing wearable chemical and biochemical sensors the traits of these devices will be discussed in this introductory section. Following this the potential parameters of interest are presented and examples of wearable systems are discussed. A range of sampling techniques and methods of chemical sensing are presented along with integration issues and design challenges. Finally, some of the main application areas of this novel technology are discussed.

1.1 Chemical and Biochemical Sensors

Typically, a chemical or biochemical sensor consists of a recognition element (receptor) coupled with a transduction element (Figure 1). In the receptor part, the chemical information, e.g., the concentration of a given compound, pH, etc., is converted into a form of energy that can be measured by a transducer. The function of the receptor is fulfilled in many cases by a thin layer that is able to interact with the analyte molecules, catalyze a reaction selectively, or participate in a chemical equilibrium together with the analyte [2].

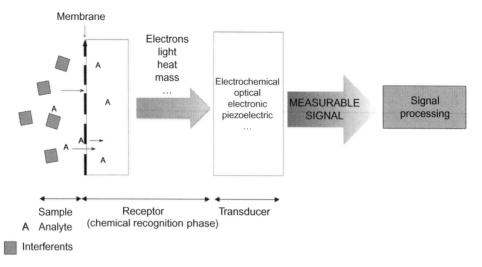

FIGURE 1   Schematic representation of the working principle of a chemical (biochemical) sensor. Some details of a real sensor are indicated. As an example, a filter or membrane may be used to separate the analyte to be detected from the interfering particles.

The transducer translates the chemical information about the sample into a useful analytical signal. The transduction element may be electrochemical, optical, electronic, or piezoelectric in nature. In recent years, novel chemical and biochemical sensors have been developed by scientific groups all over the world. This is evident by the growing number of international scientific articles in this area.

In spite of the clear advances in sensing technologies and high number of laboratory prototype sensors that appear every year, only a small fraction of these sensors reach commercialization. Apart from the complexity of the receptor design, the main reason for this is the incompatibility of the sensor platform with the real-world application. Sensors exposed to real-life scenarios are subjected to many environmental effects that can affect stability, reproducibility, and sensitivity. Effective sampling methods are crucial to avoid contamination and ensure controlled delivery to the active sensor surface. For example, a chemical sensor analyzing a body fluid such as sweat, blood, or tears must collect and deliver a sample to the sensor's active surface, whereupon a binding event happens and a signal is generated [3]. Other issues relate to system integration, sensor miniaturization, and low-power sensor interface circuitry design. One possible route toward the development of new and improved chemical sensing technologies is the emergence of nanoscience, related new nanostructured materials and multi-functional polymers. It is therefore not surprising that the last decade has witnessed rapid development in the field of chemical and biochemical sensors in parallel with the development of new materials [4].

Several chemical and biochemical wearable sensors have been developed in recent years, and they include pH sensors for sweat, sensors for several electrolytes (sodium, chloride, potassium) in sweat, the oximeter sensor, and others; however, most of them require further optimization and assessment in clinical trials before exploitation and routine use can occur.

1.2 Parameters of Interest

The gold-standard fluid for diagnostics is blood, however, blood analysis requires invasive sampling, which is undesirable for long-term continuous use. In recent years, significant effort has been focused on sampling and analysis of alternative body fluids, such as interstitial fluid [5], tears [6], saliva [7], and sweat [8]. The transition from blood to other body fluids provides a less invasive means of sampling, which in turn may provide a route to facilitate longer-term continuous monitoring.

Sweat is probably the most accessible fluid to be collected by a garment [9]. Many relevant physiological analytes such as sodium, chloride, potassium, calcium, ammonia, glucose, and lactate are normally present in sweat. The sweat test of sodium and chloride concentrations is the gold-standard technique for the diagnosis of cystic fibrosis [10]. Analysis of sweat loss and sweat composition can also offer valuable information regarding hydration status and electrolyte balance, which can be especially important during sports activities.

Saliva is another large accessible bodily fluid present in the oral cavity, in which it is possible to get access to a large number of analytes and biomarkers. Saliva contains ions such as sodium, potassium, chloride, bicarbonate, nitrates, urea, uric acid, creatinine, and more than 400 types of proteins [11]. However, the use of saliva can be quite difficult due to the presence of mucus and residual food and blood, which can interfere with sensor operation.

Breath analysis can reveal a diverse signature of physiological processes that occur in the body. Conventional methods of breath analysis typically involve gas chromatography coupled with a form of mass spectroscopy [12]. These techniques require complex and expensive equipment not conducive to wearable formats. Chemical sensor arrays or electronic noses (e-noses) are an alternative cheaper, portable, and faster approach that have shown promising results in the fields of renal disease, lung cancer, and diabetes [13]. With these methods breath analysis still has challenges regarding standardization of methods of collection, treatment, and conditioning of samples.

Tears are another body fluid in which many proteins and electrolytes are normally present [14]. Glucose is one of the most important biomarkers present in tears, and may be useful in personal monitoring of diabetes. Recently, glucose in interstitial fluids also gained much attention for the realization of wearable sensors, and in particular through the use of microneedles, as discussed in more detail in section 2.12.

Wound exudate may be analyzed to monitor the healing process by monitoring different cell types, growth factors, and nutritional factors. Temperature and pH of the wound are also recognized to be critical for healing. For instance, Voirin et al. [15] realized an optical-integrated wound sensor using pH responsive hydrogels to provide information on the wound-healing process.

In addition to monitoring samples from the body, wearable chemical sensors may be used to sense substances in the environment around the wearer and thereby warn of any hazards that may be present. This has great implications for those working in challenging or extreme conditions such as first responders and military personnel. Detection of gases that are dangerous to health such as carbon monoxide or potential explosive agents can obviously help to ensure safety.

2. SYSTEM DESIGN

This section focuses on the components of a wearable chemical or biochemical sensor system. A sample, either a body fluid or a gas, must be collected and delivered in the first instance. Then a suitable sensing mechanism must be integrated within the sensing platform also incorporating associated electronics and power supply. This overall system must be designed so that it can be placed on the body in a suitable location for long-term use in a reliable and comfortable manner.

2.1 Sample Handling

Micro-fluidics have significantly contributed to the development of personalized healthcare and point-of-care (POC) diagnostic devices. Micro-fluidic systems can deal with micro-volumes of the target sample and generate exactly the same information as standard analytic techniques [16]. This smaller sample volume of body fluid reduces the size and improves the reliability of the fluid-handling system. Another advantage is the possibility of integrating the active sensing area inside the micro-fluidic channel to minimize sample dead-volume, and consequently also the delay between sampling and analysis. For a wearable configuration, textile compatible methods are preferable. One approach is to use low or zero-power passive pumps based on capillary force driven micro-fluidics. This is discussed in the following section.

2.1.1 Transport of Fluids in a Textile

The wetting and wicking properties of natural and synthetic fibers can provide efficient means for controlling fluid flows without the need for external pumps. A recent example of a textile-based interfacial micro-fluidic platform was reported by Xing et al. [17], in which a 3D micro-fluidic network was fabricated by "stereo-stitching" of cotton yarn on a superhydrophobic textile. Taking advantage of the difference in wettability between the cotton yarn (contact angle = 0°) and the superhydrophobic textile (contact angle = 140°), a surface tension-driven micro-fluidic system was created in which the flow was inversely proportional to the flow resistance of the hydrophilic yarn. Under these conditions, the interfacial micro-fluidic can operate in two different regimes, the discrete transport mode and the continuous transport mode (Figure 2).

Other interesting examples of fabric-based micro-fluidics presented by Nilghaz et al. [18] were both two-dimensional and three-dimensional cloth-based devices achieved through the deposition of hydrophobic wax, which confined the flow within the wax border in a similar manner to paper-based micro-fluidic technology. The patterned fabric is capable of retaining its flexibility and mechanical strength, which in turn makes it suitable for future use in wearable technologies. Other approaches using hydrophobic and hydrophilic silk yarns have been proposed by Bhandari et al. [19]. They demonstrated the use of a handloom machine, but with the potential to be scaled up for mass production as it uses a well-established manufacturing technique within the textile industry.

A textile-based fluidic system was developed under the EU BIOTEX project to collect and analyze sweat on the body in real-time [20]. The system employed a passive pumping

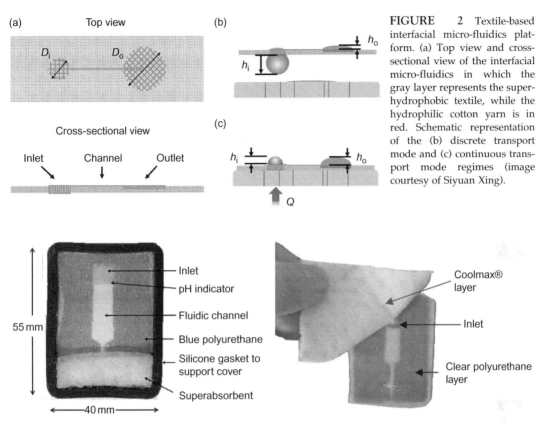

FIGURE 2 Textile-based interfacial micro-fluidics platform. (a) Top view and cross-sectional view of the interfacial micro-fluidics in which the gray layer represents the superhydrophobic textile, while the hydrophilic cotton yarn is in red. Schematic representation of the (b) discrete transport mode and (c) continuous transport mode regimes (image courtesy of Siyuan Xing).

FIGURE 3   Textile-based platform to collect and analyze sweat developed during EU BIOTEX project.

mechanism based on capillary action using a combination of moisture-wicking fabric and a highly absorbent material. A fabric fluidic channel was formed by screenprinting a hydrophobic material (polyurethane) on either side of a polyester/Lycra® blend. An acquisition layer of fabric (Coolmax®/polyester/polyester) was stitched at this inlet to maximize sweat collection over a larger area (Figure 3). Sensing elements were placed in contact with the sample in the channel, including a dye for colorimetric analysis of pH and electrochemical sensors printed onto a Kapton film [21]. It was found that a priming time of 10 to 20 minutes was needed for the pH sensor, and around 35 minutes for the conductivity/sodium sensor due to the dead volume of the platform. A key goal in wearable chemical sensor platform design should therefore be to minimize dead volume and thereby reduce the delay between sampling and analysis.

Further to the development of the BIOTEX platform, Curto et al. [22] developed a miniaturized flexible plastic micro-fluidic system based on the same principle. The micro-fluidic device was capable of drawing fresh sweat from the skin surface to the detection area using a cotton thread and superabsorbent inside a microfluidic channel. The authors also demonstrated the use of this device coupled with a wireless platform for sweat pH

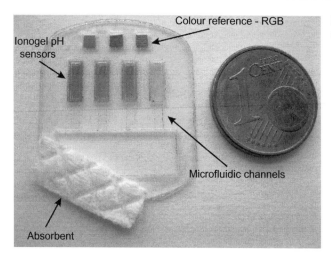

FIGURE 4 Wearable micro-fluidic barcode for pH analysis of sweat.

monitoring over a period of 50 minutes during cycling activity. Parallel to this, an alternative approach was developed using an electronic-free visual device [23]. The device presents four different sensing areas made of an ionogel (ionic liquids encapsulated in a cross-linked polymeric gel matrix) doped with four different pH dyes, forming a barcode-type sensor array. Three colored reference patches (Red, Green, and Blue — RGB) were incorporated at the top of the device to allow for automatic correction of ambient light (Figure 4). The color change and pH can be estimated by eye, or more accurately, by using a digital image or smartphone application.

These wearable sensors were mainly developed for single-use applications since integration of textile bio-chemical sensors into clothes that can be used multiple times and for long periods of time is still very challenging. These sensors are subject to the same problems of other common physical sensors like environmental noise and motion artifacts, but are also subject to additional constraints that come from the nature of the sensing mechanism itself (chemical or biochemical reaction, changes in the sensor surface, contact of sensor surface with analyte, etc.). In addition, for everyday use of the textile sensors, external factors like sensor-surface contamination, which decreases the accuracy of the sensor, need to be considered. Moreover, since the sensors are ideally integrated into clothes, the possibility of washing the sensors needs to be taken into account; therefore the sensing area needs to be somehow protected or the sensor has to be washing-compatible.

2.1.2 Microneedle Technology

Microneedle arrays offer a minimally invasive means of biosensing through highly integrated biocompatible devices. There is the possibility of making them wearable through small devices and patches directly in contact with the skin. The trend is to make these arrays easy to fabricate on an industrial scale and at low cost.

The principal applications of these microneedle arrays involve fluid sampling [24] and extraction [25] since they provide the opportunity to overcome the skin barrier and thereby reach dermal biofluids, which are more reflective of systemic levels of key analytes. This novel technology can be used as a therapeutic tool for transdermal drug delivery, including insulin, and more recently as a diagnostic tool for the analysis of the biofluid contents. In this regard, glucose is being intensely investigated using microneedles in interstitial fluid. For instance, Sakaguchi et al. [26] developed a sweat-monitoring patch for measuring glucose using a minimally invasive interstitial fluid extraction technology based on microneedles. The advantage of this technique is that sweat contamination during interstitial fluid glucose extraction is avoided. Good correlations were found between interstitial fluid and reference plasma glucose levels.

Other analytes measured using microneedle sampling were hydrogen peroxide and ascorbic acid in which sampling was integrated with sensing via chemically modified carbon fiber bundles [27], and lactate, using carbon paste microneedle arrays. In the latter case, the need for integrated microchannels was avoided and extraction of the interstitial fluid was avoided, as the microneedles themselves were provided with inherent sensing capabilities [28]. Highly linear lactate detection was achieved over the entire physiological range, along with high selectivity, sensitivity, and stability of the carbon paste microneedle array. Recently, Miller et al. [29] reported the use of similar microneedle configurations for the simultaneous detection of multiple analytes in physiologically relevant tissue environments. The microneedles selectively detected changes in pH, lactate, and glucose, showing their potential use for applications in sports science. Biopotential measurements were demonstrated by O'Mahony et al. [30] using microneedles as dry electrodes for detecting electrocardiography (ECG) and electromyography (EMG). The microneedle-based dry electrodes compared to conventional wet electrodes are shown in Figure 5.

The fabrication of functional wearable devices that incorporate microneedle technology is very challenging. Nowadays, micro-fluidics are becoming the most reliable approach to control fluid transport in these sensors systems, where in many cases, only small sample volumes may be available (e.g., interstitial fluid flow is lower than $10\,\mu L\,h^{-1}$). In this regard, Strambani et al. [31] developed a silicon microchip for transdermal injection/sampling applications. The microneedles were connected with a number of independent reservoirs integrated in the back side of a silicon die. Flow rate through the needle array as a

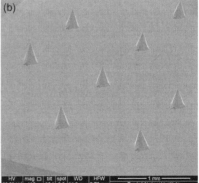

FIGURE 5 (a) Traditional wet electrode (top left), assembled prototype electrodes (bottom left), and microneedle arrays (right). (b) Microneedle-based dry electrode, illustrating an array of microneedles [30].

function of the pressure drop applied to the chip for injection/drawing purposes was investigated. Using an array with 38,000 active needles, the flow rate can be finely controlled from a few mL min^{-1} up to tens of mL min^{-1}. Other considerations that need to be taken into account when using microneedle arrays for *in vivo* sensing are the clogging of the microneedles and the potential of structural deformation upon their insertion into the skin, which may change the dynamics of the sampling and thereby introduce an unpredictable delay from time of sampling to detection.

2.1.3 Sampling Gases

Gas sensors may be employed to monitor the air surrounding the wearer (e.g., to detect dangerous levels of toxic gases). Delivery of the gas sample to the sensing surface can be controlled by choosing a fabric of suitable permeability to encase the sensor. Such a textile layer would also serve to protect the sensing surface against pollution and mechanical damage. In addition to the choice of fabric, the location of wearable gas sensors on the body needs careful consideration (e.g., depending on the density of the gas being detected compared to air).

Another possible wearable gas sensor is a breath sensor. However, collection of breath samples in a wearable system is challenging to do in an unobtrusive yet accurate manner. An approach taken by the portable Lapka™ breath alcohol monitor [32] is to use a small cylindrical device to collect the sample with the hand closed around it as a mouthpiece. The device is a ceramic cylinder (57 mm x 23 mm) with embedded sensors, which are wirelessly connected to a mobile device with an associated app. Perhaps in the future a headset-based design with an app, similar to the one created by BreathResearch [33] for measuring breath acoustics, capable of measuring chemical and biochemical parameters in breath, would be feasible.

2.2 Types of Sensors

There are various sensing techniques that can be employed for producing wearable chemical and biochemical sensors, and this influences the platform design and thus the materials and electronics required for integration into a textile or wearable device. Colorimetric and electrochemical sensing have both been demonstrated for wearable chemical and biochemical sensing, and examples of each will be discussed in this section.

2.2.1 Wearable Colorimetric Sensing Platforms

Colorimetric sensors involve a color change at the active sensor surface, which can then be measured using optical techniques. One way of combining sensing with a textile fluidic system is to use colorimetric methods. Nilghaz et al. [34] have summarized recent progress in the development of micro-fluidic devices based on multifilament threads and textiles for clinical diagnostic and environmental sensing.

A colorimetric approach to pH measurement in sweat using pH sensitive dyes and surface mount LEDs and photodiodes has been developed by the Adaptive Sensors Group at Dublin City University, as discussed in section 2.1.1. Li et al. [35] first demonstrated the application of thread-based micro-fluidic devices for the development of low-cost colorimetric diagnostic sensors. The capillary wicking of a fabric remarkably improves fluid transport characteristics, and therefore such devices can be readily integrated into clothing like T-shirts through the knitting of chemical-sensitive yarns within the fabric. In fact, Reches et al. [36] further explored the concept of thread micro-fluidics, showing the possibility of performing multiplexed colorimetric analysis through the use of a single thread sewn through a wearable adhesive plaster. This has the potential for sensing a wide range of analytes, such as proteins, ketones, and nitrates.

A wearable colorimetric biosensor within a contact lens has been developed to detect glucose levels in tears. Badugu et al. [37] demonstrated the use of disposable contact lenses embedded with boronic acid-based fluorophores for the colorimetric detection of glucose. The contact lens changes color according to the amount of glucose in tears, and is monitored by the wearer by simply looking into a mirror and comparing the color to a pre-calibrated color strip (Figure 6).

2.2.2 Electrochemical

Electrochemical sensors involve using electrodes to measure the electrochemical changes that occur when chemicals interact with a sensing surface. Recently, a great deal of effort has been applied to the development of wearable electrochemical sensors through the employment of novel materials, such as conductive polymers and carbon nanomaterials. For instance, carbon nanomaterials present excellent electrical and chemical properties, and they have been extensively explored for the realization of biosensors.

Guinovart et al. [38] developed potentiometric sensors for the detection of pH, potassium, and ammonium levels in sweat using cotton yarns coated with single-walled carbon nanotubes (SWCNTs). The sensors were realized by dip-coating the yarns inside a SWCNTs aqueous solution used as ink, followed by further integration of the ion-selective membrane. These devices exhibited reproducible responses over a period of two months, being used once a week. The integration of the sensor into a band-aid for real-time detection of potassium on a manikin was demonstrated, although, as the authors well highlighted, the incorporation of the sensor into a truly wearable platform would require further investigation. Also using a potentiometric sensor, Schazmann et al. [8] developed a wearable platform to perform real-time measurements of sodium concentration in sweat. A liquid contact ion selective electrode (ISE) was integrated within a sodium sensor belt (SSB), which was then applied to the comparison of sodium concentration in sweat emitted by healthy and cystic fibrosis-positive people during exercise. However, for wearable potentiometric ISE sensors, solid contact ISE (scISE) are preferable to classic liquid contact ISE. In fact, scISEs can be easily fabricated on

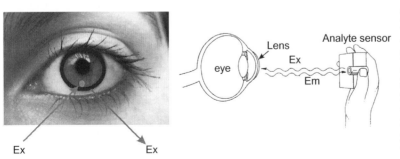

FIGURE 6 Continuous tear glucose monitoring using bored doped contact lenses (left). The hand held device works by flashing a light into the eye (Ex) and measuring the emission (Em) intensity. Schematic representation of the tear glucose-sensing device (right). *Adapted from [37].*

all-plastic flexible substrates, which in turn makes them more compatible with wearable technology, as they can be fabricated in a more conformal fashion. Following this approach, Bandodkar et al. [39] demonstrated real-time epidermal pH monitoring via a tattoo-based scISE. The sensor was fabricated using screen printing technology in the form of a "smiley face," in which the eyes are, respectively, the reference electrode and the scISE, the latter incorporating a pH-sensitive layer made of polyaniline-conductive polymer. Interestingly, the same concept can be readily expanded for the development of similar epidermal ISE tattoo-sensors for the detection of other ionic species present in sweat, such as sodium, potassium, chloride, and ammonium, through the use of appropriate selective membranes for the targeted ionic analyte. Using a similar approach, Jia et al. [40] fabricated a flexible printed temporary-transfer tattoo electrochemical biosensor that could be used for the real-time detection of lactate in human perspiration. The biosensor was tested when applied to the deltoid of an athlete for real-time lactate monitoring during cycling, and showed good correlation with parallel *in vitro* evaluation of sweat samples collected during the exercise. Khodagholy et al. [41] also demonstrated a conformal organic electrochemical transistor (OECT) incorporating an ionogel for the electrochemical determination of lactate in sweat. The use of OECTs for lactate sensing represents an interesting and novel strategy for future integration of this biosensor into a wearable platform. There are several advantages to the use of OECTs compared to more conventional sensor platforms, such as easy integration into electronic circuits and operation at low applied bias compared to traditional field-effect transistors (FET). In addition to this, OECTs can be entirely made of Poly(3,4-ethylenedioxythiophene) Polystyrene sulfonate (PEDOT:PSS) conductive polymer, which makes them very low cost and compatible with industrial stamping processes, such as roll-to-roll technology.

Yang et al. [42] demonstrated that electrochemical sensors could be printed onto clothing for the realization of wearable biosensors. Carbon strips were printed directly on the elastic waistband of underwear, offering a tight and direct contact with the skin. The sensors were capable of maintaining electrochemical behavior after several cycles of folding or stretching of the supporting textile, indicating the potential application of this sensor configuration for wearable electrochemical chemo-/biosensors on clothes.

Claramunt et al. [43] have developed a flexible sensor array using bare and metal-decorated carbon nanofibers as an active layer for on-body gas sensing. Inkjet printing was used to deposit four interdigited silver electrodes onto a Kapton film, and a heater was printed onto the reverse side of the film. Inkjet printing has the advantage of being non-contact and is suitable for many substrates. Metal-decorated carbon nanofibers were spray coated onto the electrodes and the sensor response was characterized for ammonia and nitrogen dioxide. Carbon nanotubes were also demonstrated for sensing gases and biological molecules [44,45].

3. CHALLENGES IN CHEMICAL BIOCHEMICAL SENSING

3.1 Sensor Stability

Ideally, a wearable sensor should be autonomous and reliable over time. However, long-term stability of chemical sensors and biosensors in autonomous operation remains a major issue. These sensors typically need an active (responsive) surface to operate, and changes to this surface due to fouling or leaching of reagents can lead to baseline drift and variations in sensitivity. There may also be a cross-response to interferents that may be present in the sample. Therefore, there is a need for chemical and biochemical sensors to be frequently calibrated, which requires liquid handling for sampling, reagents, and waste. Consequently, systems must include pumps, valves, and liquid storage, which are clearly impractical in terms of current technologies for autonomous wearable devices. This is one of the main reasons that there are few practical systems at present. One approach is to develop low-cost disposable sensors. In this way, part of the system may be reusable, e.g., control electronics for measurement, detection, and wireless capability, but the sensor surface may be inserted/removed easily. For example, the removable/disposable element may be a small micro-fluidic chip. This approach also provides an attractive cost model in terms of commercialization of devices.

3.2 Interface with the Body

A recent survey on wearable biosensor systems for health showed that wearability was a major issue because sensor, battery, and on-body hardware size tend to be too bulky. Sensor systems rapidly become redundant if patients or clinicians do not want to work with them. A review of patients' and clinicians' preferences for non-invasive, body-worn sensor systems found that a body-worn sensor system should be compact, embedded, simple to operate, and should not affect daily behavior nor seek to directly replace a healthcare professional [46]. Therefore, wearable chemical sensors should be seen as an additional tool to assist in healthcare delivery and to provide better continuity of care.

Ideally, the sensor should be flexible, comfortable to wear, and lightweight. Miniaturization of sensors can help to reduce power requirements, and also makes the device less obtrusive for the wearer. Through device miniaturization, we also reduce the size of the sample needed for the sensor to operate.

In the case of sweat analysis, samples from a single site on the body may not give an accurate representation of the total body sweat loss as there are regional variations in sweat composition and volume. Therefore multiple sampling sites may be needed, and a study by Patterson et al. [47] has identified four sampling regions (the chest, scapula, forearm, and thigh) to estimate mean whole-body concentrations of electrolytes in sweat.

Technology is becoming more and more portable. Sensors, including imaging cameras, GPS, and accelerometers, are being increasingly integrated into mobile phones. This is stimulating the emergence of new apps for monitoring personal health. For example, pulse oximetry apps are now available based on wearable Bluetooth-enabled optical sensors on mobile phones, and research into colorimetric sensing is increasing. The smartphone may become part of a wearable sensor system serving as the detection element, and also for data management and communications. An example of the smartphone being used for colorimetric analysis is presented by Oncescu et al., in which sweat and saliva pH measurements are taken using pH-sensitive paper inserted into a smartphone accessory [48]. An example of the data management capability of smartphones is the iBGStar [49] by Sanofi Diabetes, which allows users to track and manage their blood glucose levels by linking to an iPhone. Through mobile communications, this information can be shared by email, websites, or text messages to facilitate a movement toward remote management of personal health.

Increasingly, the smart watch is being considered as an interface to the mobile phone, and is facilitating integration of other wearable sensors. The great advantage of the watch form is that it is something that people are familiar with and used to wearing. An example of a wrist-worn device to monitor chemical parameters is a device for transdermal alcohol sensing by Giner, Inc. The device known as WrisTAS™ (Wrist Transdermal Alcohol Sensor) is a wrist-worn prototype to objectively track patterns of human alcohol consumption [50].

Newly developed electronic "tattoo" technology discussed previously in section 2.2.2 is a non-invasive means of producing conformal and mechanically stable sensors that can be applied directly to the skin. These devices may be easily fabricated using well-established printing technologies, and have great potential for providing access to multi-parameter sensing platforms. For example, MC10 [51] is developing body-conformable electronics that are stretchable and bendable. Figure 7 shows the biostamp technology that is capable of measuring physiological signals such as body temperature and heart rate.

3.3 Textile Integration

Section 2.1.1 presented various approaches to handling body fluids using textile yarns and fabrics. However, for textile-based chemical sensors to become a practical reality, their incorporation must be compatible with traditional manufacturing techniques such as weaving and lamination. For analysis of body fluids, the fluid-handling and sensing elements must both be compatible in order to create a real system that can be used on the body and yield valid results for real-time measurements at an acceptable cost.

Two groups in Switzerland, Ecole Polytechnique Fédérale de Lausanne (EPFL) and Eidgenoessiche Technische Hochschule Zurich (ETHZ), have demonstrated the

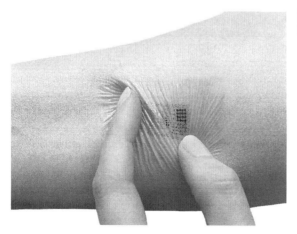

FIGURE 7 Biostamp technology developed by MC10 Inc. (*photo courtesy of mc10 Inc.*)

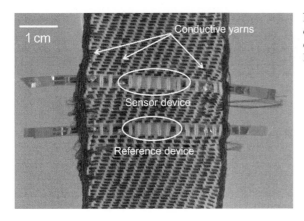

FIGURE 8 Sensing components on thin strips of Kapton film woven into fabric (photo courtesy of Ecole Polytechnique Fédérale de Lausanne, EPFL).

integration of gas sensors into a woven textile. Sensing components on thin strips (25 mm) of Kapton film have been integrated into the weaving process as shown in Figure 8. The sensing regions were well encapsulated and found to retain functionality after weaving and repeated stress/strain testing [52]. The sensing element is an interdigited thin-film capacitor coated with a humidity-sensitive polymer (cellulose acetate butyrate). Changes in ambient humidity caused changes in the capacitance of the sensing layer. In principle, this humidity-sensitive polymer layer could be replaced by a gas-sensitive polymer to target other analytes. Troester's group in ETHZ has demonstrated the use of carbon black/polymer sensors in a similar woven Kapton film configuration to detect vapor such as acetone [53].

3.4 Power Requirements

Limited battery life, size, and weight are the most significant inconveniences for most wearable electronic devices. Low-power device performance should be the first priority through smart integration and miniaturization of the electronic components. To power these devices using lightweight, comfortable components, flexible batteries, energy-harvesting methods, or remote charging may be considered.

Practical miniature devices that can harvest sufficient kinetic energy from the human body to power a wireless bio-sensor are still in their infancy. Energy-harvesting feasibility depends mainly on four factors: the typical power consumption of the device; the use pattern; the device size (and thus the acceptable harvester size); and the motion to which the device is subjected. All these factors need to be considered in order to generate an autonomous, self-powering, wearable device for monitoring body parameters.

Chemical sensors based on RFID tags have been demonstrated as passive devices requiring no power supply to be connected to the sensor. An RFID tag typically consists of a microchip attached to a radio antenna. An RFID reader transmits signals to the RFID tag and reads the signal sent back from the tag. By creating a sensing film on the antenna of an RFID tag, the analyte-induced changes can affect the impedance of the antenna and hence the resonant frequencies. GE Global Research has demonstrated this principle and performed RFID sensing in water to investigate the levels of ions and organic solvents, and in gases to detect toluene and ammonia [54]. More recent work has involved the development of self-correction methods against fluctuations in ambient temperature [55]. McAlpine's group at Princeton University used an external radio transmitter to power their tooth "tattoo" sensor used to monitor respiration and bacteria in saliva [56]. However, this sensing technology is still in its early stages as adhesion to the tooth, along with sampling issues, still need to be investigated further.

4. APPLICATION AREAS

4.1 Personal Health

Wearable sensors are applicable for all stages of treatment in healthcare, including prevention, immediate care, rehabilitation, and long-term support. If it is possible to find a link between non-invasive or minimally invasive chemical/biochemical detection methods and well-established but invasive blood testing, then a new range of tools may be created for the healthcare industry.

Diabetes is a chronic condition that involves close monitoring of glucose levels, typically using finger-prick samples. Menarini Diagnostics have developed a continuous monitor for blood glucose measurement called the GlucoDay® S [57]. The device uses a microdialysis probe inserted into the abdomen for continuous sampling of interstitial fluid. The GlucoDay® S is specifically for clinical use and is worn by the subject for a 48-hour period. Such continuous monitoring can show general trends that may be missed by single-point measurements. The data generated by the GlucoDay® S is the equivalent of approximately 1,000 finger-prick glucose values. The main goal for the device is to chart an individual's daily glucose activity, especially during sleeping hours. With this information the healthcare team and the patient can modify the insulin regimen in order to achieve better control patterns.

Wearable chemical and biochemical sensors have the potential to improve the quality of life in the healthcare arena through real-time assessment. In terms of market statistics, the biggest market is for sensors that can "sense" and "monitor" key diagnostic and therapeutic markers. In the U.S. alone, the Sensors Market in Healthcare Applications is expected to Reach $13.11 billion by 2017 [58].

4.2 Sports Performance

Technology has a huge role to play in improving sports performance, from textiles to simulate a shark's skin gliding through the water for swimwear to moisture-wicking fabrics keeping athletes comfortable and cool. Physiological monitoring has helped our understanding of the body and its response to exercise, and through physiological measurement individual training regimes can be planned and optimized. During prolonged periods of training, diverse physiological conditions can be easily reached. The most commonly monitored parameter in sports science is blood lactate, and while there are portable systems available for this measurement, it is still an invasive approach, giving point measurements from blood samples. Continuous monitoring via a non-invasive approach would be far more favorable. However, it will be necessary to investigate the correlations between the levels of critical parameters in blood compared to other body fluids if this is to provide useful information. In a review of several studies, the conclusion was that the relationship between blood lactate and lactate in other body fluids is unclear [59].

Real-time sweat analysis during exercise could give valuable information on dehydration and changes in the amount of important biomolecules and ions. Dehydration increases perceived effort, so if exercise feels hard, then people may be less inclined to partake in it. In the case of elite athletes, dehydration can greatly impact performance. Sweat electrolyte concentration varies greatly between individuals, being affected by genetics, sweat rate, type of training, degree of hydration, and the state of heat acclimatization. The opportunity of gathering important information on the physical condition of athletes during sport activities could potentially provide huge benefits, arising from personalized training regimes and optimized rehydration/nutrition strategies.

4.3 Safety and Security

There are many scenarios in which personal safety could be supported through the use of wearable sensors for continuous monitoring of external hazards. Some high-risk professions include first responders, construction workers, police, security, and military personnel. Wearable sensors for these applications must of course be lightweight, ergonomic, robust, and accurate. Working in a demanding environment, they need to be easy to operate and ideally part of a wider sensor network. In addition to personal alerts, communication between team members is needed for management and localization of overall team safety.

An EU project, ProeTEX, is focused on the safety of firefighters through the integration of toxic gas sensors among other sensing capabilities into protective clothing [60]. Carbon monoxide (CO) and carbon dioxide (CO_2) sensors were integrated into the jacket and boots of a firefighter's uniform. Figure 9 shows a CO_2 sensor integrated into the boot of a firefighter's uniform and a CO sensor integrated close to the jacket collar. The choice of placement was based on the densities of each gas. The CO and CO_2 gas sensors already have an outer protective membrane built in to filter out small particles (dust, soot, etc.). On the garment, the gas-sensing devices are further protected by full encapsulation into a purposely designed pocket with a heat resistance, waterproof, air-permeable membrane, hence protecting them from potential damage from heat, water, and airborne particles.

Researchers at Arizona State University have developed a wearable sensor to monitor air quality. The device communicates with a smartphone for user feedback and is suitable for indoor and outdoor environmental personal exposure studies. In a recent study [60] the device was used to test air quality in three scenarios: indoors after room remodeling and painting, outdoors near a busy highway and, finally, outdoors in areas affected by an oil spill. Such a sensor has the potential to greatly improve our knowledge of personal exposures and to help protect human health as well as the environment.

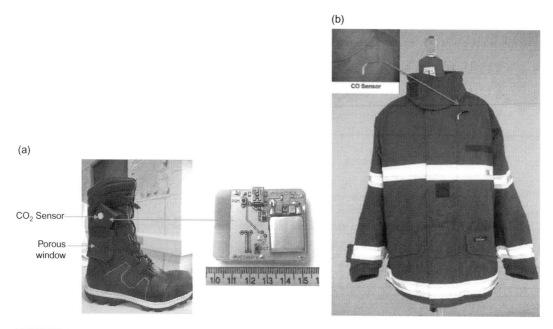

FIGURE 9 Wearable chemical sensing in ProeTEX project (a) Carbon dioxide sensor integrated into a firefighter's boot and (b) carbon monoxide sensor integrated into a firefighter's jacket.

5. CONCLUSIONS

Smart fabrics and interactive textiles are a relatively new area of research with many potential applications in the field of biomedical engineering. The ability of smart textiles to interact with the body provides a novel means to sense the wearer's physiology and respond to the needs of the wearer. The advantage of this technology relies on the integration of sensors in clothes that are worn on a daily basis, providing the capacity to continuously monitor the wearer and his/her environment. To date, most of the research in the field of wearable sensing has focused on physical sensors that respond to changes in their immediate environmental proximity (e.g., electric fields, heat, and movement). Relatively little has been done on the development of similar wearable sensors for real-time monitoring of chemical and biological parameters of interest. The main reasons for this are due to the fact that these sensors require an active surface at which a chemical reaction occurs, which generates the observable signal. However, this surface is subjected to samples, reagents, waste products, and contaminants from the environment that can affect its functionality. Sample handling requires additional hardware such as fluidic systems with valves and pumps, and for this requirement micro-fluidics could play an important role. Furthermore, wearability is probably the most important requirement that these sensors must fulfill without losing functionality and, above all, compromising the wearer's comfort and/or safety.

It is clear that wearable chemical sensors have the potential to provide a means to continuously monitor the physiology of the wearer over extended periods of time, in an innocuous manner that does not interfere with the daily routines of wearers. Through this technology, the wearer can be kept informed of his/her well-being in a dynamic manner, and individuals can become actively involved in the management of their personal health, making pHealth (pervasive health) a feasible goal. There is also great potential in the field of sports science, particularly in the real-time analysis of sweat during exercise. The other main application area of wearable chemical and biochemical sensors is in personal safety in warning of hazardous chemicals in the environment, which is of particular importance to first responders in disaster situations, and those involved in security and safety. This chapter has detailed a summary of the state of the art while also highlighting the current issues and challenges for the future of this exciting field of research.

Acknowledgment

This review work was supported by the Science Foundation Ireland under Insight award SFI/12/RC/2289 and the Gobierno Vasco, Dpto. Industria, Innovación, Comercio y Turismo under ETORTEK 2012 with Grant No. IE12-328, ETORTEK 2013 with Grant No. IE13-360. FBL thanks the Ramón y Cajal programme (Ministerio de Economía y Competitividad), Spain.

References

[1] R. Byrne, F. Benito-Lopez, D. Diamond, Materials science and the sensor revolution, Mater. Today vol. 13 (2010) 9—16.

[2] A. Lobnik, M. Turel, Š. Korent Urek, Optical Chemical Sensors:Design and Applications, Advances in Chemical Sensors, InTech, 2012.

[3] S. Coyle, Y. Wu, K.-T. Lau, D. De Rossi, G. Wallace, D. Diamond, Smart Nanotextiles: A Review of Materials and Applications, MRS Bulletin vol. 32 (2007) 434—442.

[4] M. Gerard, A. Chaubey, B.D. Malhotra, Application of conducting polymers to biosensors, Biosens. Bioelectron. vol. 17 (2002) 345—359.

[5] K. Rebrin, N.F. Sheppard, G. Steil, Use of subcutaneous interstitial fluid glucose to estimate blood glucose: revisiting delay and sensor offset, J. Diabetes Sci. Technol. vol. 4 (2010) 1087—1098.

[6] M. Chu, T. Shirai, D. Takahashi, T. Arakawa, H. Kudo, K. Sano, et al., Biomedical soft contact-lens sensor for in situ ocular biomonitoring of tear contents, Biomed. Microdevices vol. 13 (2011) 603—611.

[7] E. Papacosta, G.P. Nassis, Saliva as a tool for monitoring steroid, peptide and immune markers in sport and exercise science, J. Sci. Med. Sport vol. 14 (2011) 424—434.

[8] B. Schazmann, D. Morris, C. Slater, S. Beirne, C. Fay, R. Reuveny, et al., A wearable electrochemical sensor for the real-time measurement of sweat sodium concentration, Anal. Methods vol. 2 (2010) 342—348.

[9] S. Coyle, F. Benito-Lopez, T. Radu, K.T. Lau, D. Diamond, Fibers and fabrics for chemical and biological sensing, J. Text. App vol. 14 (2010).

[10] J. Massie, K. Gaskin, P. Van Asperen, B. Wilcken, Sweat testing following newborn screening for cystic fibrosis, Pediatr. Pulmonol. vol. 29 (2000) 452—456.

[11] D.P. Lima, D.G. Diniz, S.A.S. Moimaz, D.r.H. Sumida, A.C.u. Okamoto, Saliva: reflection of the body, Int. J. Infect. Dis. vol. 14 (2010) e184—e188.

[12] K.-H. Kim, S.A. Jahan, E. Kabir, A review of breath analysis for diagnosis of human health, Trends Anal. Chem. vol. 33 (2012) 1—8.

[13] D. Hill, R. Binions, Breath Analysis for Medical Diagnosis, Int. J. Smart Sens. Intell. Syst. vol. 5 (2012) 401—440.

[14] Y. Ohashi, M. Dogru, K. Tsubota, Laboratory findings in tear fluid analysis, Clin. Chim. Acta. vol. 369 (2006) 17—28.

[15] G. Voirin, J. Luprano, S.p. Pasche, S. Angeloni, R.a. Ischer, M. Liley, Wearable biosensors for monitoring wound healing, Adv. Sci. Technol. vol. 57 (2009) 80—87.

[16] G. Whitesides, Solving problems, Lab Chip vol. 10 (2010) 2317–2318.

[17] S. Xing, J. Jiang, T. Pan, Interfacial microfluidic transport on micropatterned superhydrophobic textile, Lab Chip vol. 13 (2013) 1937–1947.

[18] A. Nilghaz, D.H.B. Wicaksono, D. Gustiono, F.A. Abdul Majid, E. Supriyanto, M.R. Abdul Kadir, Flexible microfluidic cloth-based analytical devices using a low-cost wax patterning technique, Lab Chip vol. 12 (2012) 209–218.

[19] P. Bhandari, T. Narahari, D. Dendukuri, Fab-Chips': a versatile, fabric-based platform for low-cost, rapid and multiplexed diagnostics, Lab Chip vol. 11 (2011) 2493–2499.

[20] D. Morris, S. Coyle, Y. Wu, K.T. Lau, G. Wallace, D. Diamond, Bio-sensing textile based patch with integrated optical detection system for sweat monitoring, Sens. Actuators B vol. 139 (2009) 231–236.

[21] S. Coyle, K. Lau, N. Moyna, D. Diamond, F. Di Francesco, D. Constanzo, et al., BIOTEX-Biosensing textiles for personalised healthcare management, IEEE Trans. Inf. Technol. BioMed. vol. 14 (2010) 364–370.

[22] V.F. Curto, S. Coyle, R. Byrne, N. Angelov, D. Diamond, F. Benito-Lopez, Concept and development of an autonomous wearable micro-fluidic platform for real time pH sweat analysis, Sens. Actuators B Chem. vol. 175 (2012) 263–270.

[23] V.F. Curto, C. Fay, S. Coyle, R. Byrne, C. O'Toole, C. Barry, et al., Real-time sweat pH monitoring based on a wearable chemical barcode micro-fluidic platform incorporating ionic liquids, Sens. Actuators B Chem. vol. 171-172 (2012) 1327–1334.

[24] S. Chandrasekhar, L.K. Iyer, J.P. Panchal, E.M. Topp, J.B. Cannon, V.V. Ranade, Microarrays and microneedle arrays for delivery of peptides, proteins, vaccines and other applications, Expert Opin. Drug Deliv. vol. 10 (2013) 1155–1170.

[25] H. Suzuki, T. Tokuda, K. Kobayashi, A disposable "intelligent mosquito" with a reversible sampling mechanism using the volume-phase transition of a gel, Sens. Actuators B vol. 83 (2002) 53–59.

[26] K. Sakaguchi, Y. Hirota, N. Hashimoto, W. Ogawa, T. Hamaguchi, T. Matsuo, et al., Evaluation of a minimally invasive system for measuring glucose area under the curve during oral glucose tolerance tests: usefulness of sweat monitoring for precise measurement, J. Diabetes Sci. Technol. vol. 7 (2013) 678–688.

[27] P.R. Miller, S.D. Gittard, T.L. Edwards, D.M. Lopez, X. Xiao, D.R. Wheeler, et al., Integrated carbon fiber electrodes within hollow polymer microneedles for transdermal electrochemical sensing, Biomicrofluidics vol. 5 (2011).

[28] J.R. Windmiller, N. Zhou, M.-C. Chuang, G. Valdes-Ramirez, P. Santhosh, P.R. Miller, et al., Microneedle array-based carbon paste amperometric sensors and biosensors, Analyst vol. 136 (2011) 1846–1851.

[29] P.R. Miller, S.A. Skoog, T.L. Edwards, D.M. Lopez, D.R. Wheeler, D.C. Arango, et al., Multiplexed microneedle-based biosensor array for characterization of metabolic acidosis, Talanta vol. 88 (2012) 739–742.

[30] C. O'Mahony, F. Pini, A. Blake, C. Webster, J. O'Brien, K.G. McCarthy, Microneedle-based electrodes with integrated through-silicon via for biopotential recording, Sens. Actuators A Phys. vol. 186 (2012) 130–136.

[31] L.M. Strambini, A. Longo, A. Diligenti, G. Barillaro, A minimally invasive microchip for transdermal injection/sampling applications, Lab Chip vol. 12 (2012) 3370–3379.

[32] < https://mylapka.com/bam >, (Last Accessed: 23.06.14).

[33] < http://www.breathresearch.com/ >, (Last Accessed: 23.06.14).

[34] A. Nilghaz, D.R. Ballerini, W. Shen, Exploration of microfluidic devices based on multi-filament threads and textiles: A review, Biomicrofluidics vol. 7 (2013) 051501.

[35] X. Li, J. Tian, W. Shen, Thread as a Versatile Material for Low-Cost Microfluidic Diagnostics, ACS Appl. Mater. Interfaces vol. 2 (2009) 1–6, 2010/01/27.

[36] M. Reches, K.A. Mirica, R. Dasgupta, M.D. Dickey, M.J. Butte, G.M. Whitesides, Thread as a Matrix for Biomedical Assays, ACS Appl. Mater. Interfaces vol. 2 (2010) 1722–1728, 2010/06/23.

[37] R. Badugu, J.R. Lakowicz, C.D. Geddes, A Glucose Sensing Contact Lens: A Non-Invasive Technique for Continuous Physiological Glucose Monitoring, J. Fluores vol. 13 (2003) 371–374.

[38] T. Guinovart, M. Parrilla, G.A. Crespo, F.X. Rius, F.J. Andrade, Potentiometric sensors using cotton yarns, carbon nanotubes and polymeric membranes, Analyst vol. 138 (2013) 5208–5215.

[39] A.J. Bandodkar, V.W.S. Hung, W. Jia, G. Valdes-Ramirez, J.R. Windmiller, A.G. Martinez, et al., Tattoo-based potentiometric ion-selective sensors for epidermal pH monitoring, Analyst vol. 138 (2013) 123–128.

[40] W. Jia, A.J. Bandodkar, G. Valdos-Ramirez, J.R. Windmiller, Z. Yang, J. Ramirez, et al., Electrochemical Tattoo Biosensors for Real-Time Noninvasive Lactate Monitoring in Human Perspiration, Anal. Chem. vol. 85 (2013) 6553–6560, 2013/07/16.

[41] D. Khodagholy, V.F. Curto, K.J. Fraser, M. Gurfinkel, R. Byrne, D. Diamond, et al., Organic electrochemical transistor incorporating an ionogel as a solid state electrolyte for lactate sensing, J. Mater. Chem. vol. 22 (2012) 4440–4443.

[42] Y.-L. Yang, M.-C. Chuang, S.-L. Lou, J. Wang, Thick-film textile-based amperometric sensors and biosensors, Analyst vol. 135 (2010) 1230–1234.

[43] S. Claramunt, O. Monereo, M. Boix, R. Leghrib, J.D. Prades, A. Cornet, et al., Flexible gas sensor array with an embedded heater based on metal decorated carbon nanofibres, Sens. Actuators B Chem. vol. 187 (2013) 401–406.

[44] I.V. Anoshkin, A.G. Nasibulin, P.R. Mudimela, M. He, V. Ermolov, E.I. Kauppinen, Single-walled carbon nanotube networks for ethanol vapor sensing applications, Nano Res. vol. 6 (2013) 77–86.

[45] J. Li, H. Ng, H. Chen, Carbon nanotubes and nanowires for biological sensing, first ed., Protein Nanotechnology, vol. 300, Humana Press, 2005, pp. 191–223.

[46] J.H.M. Bergmann, A.H. McGregor, Body-Worn Sensor Design: What Do Patients and Clinicians Want? Ann. Biomed. Eng. vol. 39 (2011) 2299–2312.

[47] M. Patterson, S. Galloway, M.A. Nimmo, Variations in regional sweat composition in normal human males, Exp. Physiol. vol. 85 (2000) 869–876.

[48] V. Oncescu, D. O'Dell, D. Erickson, Smartphone based health accessory for colorimetric detection of biomarkers in sweat and saliva, Lab Chip vol. 13 (2013) 3232–3238.

[49] < http://www.bgstar.com/web/ibgstar >, (Last Accessed: 23.06.14).

[50] < http://www.ginerinc.com/ >, (Last Accessed: 23.06.14).

[51] < http://www.mc10inc.com >, (Last Accessed: 23.06.14).

[52] C. Ataman, T. Kinkeldei, G. Mattana, A. Vasquez Quintero, F. Molina-Lopez, J. Courbat, et al., A robust platform for textile integrated gas sensors, Sens. Actuators B Chem. vol. 177 (2013) 1053–1061.

[53] T. Kinkeldei, C. Zysset, N. Muenzenrieder, G. Troester, An electronic nose on flexible substrates integrated into a smart textile, Sens. Actuators B Chem. vol. 174 (2012) 81–86.

[54] R. A. Potyrailo, C. Surman, W. G. Morris, and S. Go," Selective detection of chemical species in liquids and gases using radio-frequency identification (RFID) sensors," in Solid-State Sensors, Actuators and Microsystems Conference, 2009. TRANSDUCERS 2009. International, 2009, pp. 1650-1653.

[55] R.A. Potyrailo, C. Surman, A passive radio-frequency identification (RFID) gas sensor with self-correction against fluctuations of ambient temperature, Sens. Actuators B Chem. vol. 185 (2013) 587–593.

[56] M.S. Mannoor, H. Tao, J.D. Clayton, A. Sengupta, D.L. Kaplan, R.R. Naik, et al., Graphene-Based Wireless Bacteria Detection on Tooth Enamel, Nat. Commun. vol. 3 (2012).

[57] < http://www.menarinidiag.co.uk/Products/continuous_glucose_monitoring >, (Last Accessed 23.06.14).

[58] Markets and Markets. (2013), World Sensors Market in Healthcare Applications.

[59] P.J. Derbyshire, H. Barr, F. Davis, S.P.J. Higson, Lactate in human sweat: a critical review of research to the present day, J. Physiol. Sci. vol. 62 (2012) 429–440.

[60] C. Chen, K.D. Campbell, I. Negi, R.A. Iglesias, P. Owens, N. Tao, et al., A new sensor for the assessment of personal exposure to volatile organic compounds, Atmos. Environ. vol. 54 (2012).

Wearable Inertial Sensors and Their Applications

Toshiyo Tamura

Osaka Electro-Communication University, Neyagawa, Japan

1. INTRODUCTION

Wearable inertial sensors are the most common wearable devices for the measurement of motion and physical activities associated with daily living. The combination of an accelerometer and a gyroscopic sensor is particularly effective for evaluating motion, and their small size makes them easy to wear on various parts of the body.

2. WEARABLE INERTIAL SENSORS

In this section, measurement parameters for accelerometers, gyroscopic sensors, and magnetic sensors, as well as monitoring principles for these sensors, are reviewed.

2.1 Principles of Inertial Sensors

A coordinate system must be defined prior to deployment of an inertial sensor for body motion measurement. If the origin of a moving coordinate system has acceleration (A_0) and rotates with a gyroscopic angular velocity (ω), and if a mass (m) has the position vector (r') and the velocity (u') with respect to the moving coordinate system, then the inertial force observed is

$$mA' = -mA_0 + 2mu' \times \omega + m\omega \times (r' \times \omega) + mr' \times \frac{d\omega}{dt} \tag{1}$$

where′ indicates a variable corresponding to the moving coordinate system. The right side of the equation has four terms arising from the angular accelerations that correspond to the linear inertial, Coriolis, centrifugal, and apparent forces. Thus, the moving coordinate system is complex, and signal handling must be taken into account.

2.2 Accelerometers

Acceleration can be derived from the first derivative of the velocity or the second derivative of the displacement. However, differentiation of the signal usually increases noise. Thus, direct measurement of acceleration is often easier and more convenient. The acceleration of linear motion (α) is, according to Newton's second law, the force (F) acting on a mass (m):

$$F = m\alpha \tag{2}$$

However, the apparent forces, such as the centrifugal and Coriolis forces, may also appear as described in section 2.1 above.

Many types of accelerometers with different specifications are commercially available. For example, accelerometers are relatively cheap and reliable for use as shock sensors in automobiles. Novel MEMS technology is used to produce small and sensitive accelerometers. Overall, the correct type of accelerometer must be selected for each specific application.

Beam-type accelerometers are the most sensitive in the body motion acceleration range. In a beam-type accelerometer, an elastic beam is fixed to a base at one end, and a mass, called the seismic mass, is attached to the other end, as shown in Figure 1(a). When the seismic mass is accelerated, a force proportional to the mass times the acceleration occurs, and the beam bends elastically in response to the force. To avoid resonant oscillation after transient input, an adequate damping coefficient must be built into the mechanical system. Instead of a beam, a diaphragm, spring, or any other elastic material can be used in the accelerometer.

Determining the amplitude and direction of acceleration in three-dimensional (3D) space requires a triaxial accelerometer, as shown in Figure 1(b). Triaxial accelerometers based on MEMS technology are commercially available.

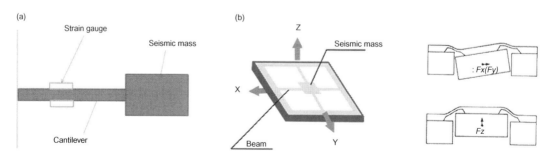

FIGURE 1   (a) Uniaxial accelerometer and (b) tri-axial accelerometer.

The displacement of the seismic mass can be detected using several methods, including those based on piezoresistive, piezoelectric effects, or capacitance. Often, a semiconductor strain gauge or capacitance-displacement accelerometer is used to measure motion in humans and animals, because it is small and relatively inexpensive.

Typically, a piezoresistive element is a strain gauge bonded to or incorporated into a mass-loaded cantilever beam. As the beam bends in response to acceleration, the resistance changes. Four individual sensing elements are arranged in a Wheatstone bridge configuration that provides output while canceling the cross-axis, temperature, and other spurious inputs. Using micromachining (MEMs) technology, piezoresistors can be easily implanted in the support beams connecting the seismic mass to the frame or support structure to enable high sensitivity.

Piezoelectric accelerometers are commonly used when only the time-varying components of acceleration require measurement. Extremely low power consumption, simple detection circuits, high sensitivity, and inherent temperature stability characterize piezoelectric accelerometers. A polarization voltage occurs in the piezoelectric material that is proportional to its deformation. The polarity of the polarization voltage depends on the molecular structure of the material. Figure 2 presents an example of a bimorph configuration beam containing two piezoelectric elements with different polarities designed to produce a double or differential output. A triaxial bimorph also has been developed [1].

The terminal voltage of the piezoelectric sensor is due to the charge that arises due to the flexion of the piezoelectric element and its capacitance. The terminal voltage is proportional to the stored charge; more accurately, a charge amplifier can be used to measure the generated charge, as shown in Figure 3.

When the input capacitance, which includes the capacitance of the piezoelectric element and the stray capacitance, is represented by C_d, then the following relationship is valid:

$$Q = C_d V_i + C_f(V_i - V) \qquad (3)$$

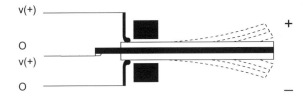

v(+)

O
v(+)

O

+

−

FIGURE 2   Configuration of a bimorph.

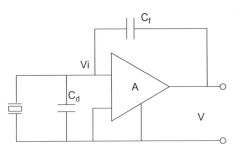

C_f

V_i

C_d

A

V

FIGURE 3   A charged amplifier.

$$V = -AV_i \tag{4}$$

where Q is the generated charge and V_i, V, and A are the input voltage, output voltage, and the gain of the amplifier, respectively. If $A \gg 1$ and $AC \gg C_i$, then

$$V = -Q/C \tag{5}$$

In this case, the output voltage is proportional to the generated charge, regardless of the input capacitance.

Acceleration signals are used for a wide range of measurements, including assessment of balance, gait and sit-to-stand transfers, classification of movements, easing of physical activities, and estimation of metabolic energy expenditure.

2.3 Gyroscopic Sensors

Angular velocity can be measured with a gyroscope, which consists of a spinning wheel mounted on a movable frame. When the wheel is spinning, it tends to retain its initial orientation in space, regardless of the central forces applied to it. When the direction of the axis is externally altered, a torque proportional to the rotation rate of the axis of inclination arises, which can be used to detect angular velocity. An example of this type of transducer is the dynamically tuned gyroscope.

Typically, micromachined gyroscopes are specialized vibrating accelerometers that measure Coriolis forces (Figure 4). A basic vibratory gyroscope consists of a proof mass mounted on a suspension that allows the proof mass to move in two orthogonal directions. To generate a Coriolis force, the proof mass must be in motion. To this end, the proof mass is electronically forced to oscillate in a direction parallel to the chip surface. If the gyroscope chip is rotated about the axis perpendicular to the chip surface, then a Coriolis force causes the proof mass to be deflected in the second direction. The amplitude

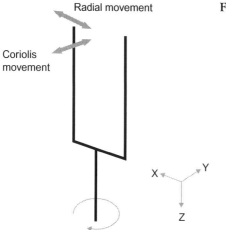

FIGURE 4   The Coriolis force.

of this oscillatory deflection is proportional to the rate of rotation, so that capacitive sensing, as in the case of the accelerometer discussed above, can be used to produce a voltage proportional to the angular rotation rate.

A tuning fork directly detects the angular velocity in a vibratory gyroscopic sensor. The tines of the tuning fork are piezoelectrically excited perpendicular to the wafer surface. When the tuning fork is rotated about the axis parallel to the tines at an angular velocity Ω, the **Coriolis effect** produces a torque proportional to Ω:

$$a_{cor} = 2V_{pm} \times \Omega \tag{6}$$

where a_{cor} is the Coriolis acceleration and V_{pm} is the velocity of the proof mass. Figure 5 presents an example of a tuning fork gyroscopic sensor. The angular rate input axis is parallel to the wafer surface, and the tines are excited with piezoelectric actuators. Due to the Coriolis effect, an angular velocity parallel to the axis of the stem generates a periodic torque, which results in a torsional oscillation of the stem. The torsional oscillation is detected with an implanted piezoresistor located in the middle of the stem. A slot in the center of the stem enhances the shear stress at the read-out piezoresistor position, thereby resulting in a higher sensitivity.

An angular velocity sensor with two or three axes is also possible, as shown in Figure 6 [2]. The triaxial micro-angular velocity sensor is mainly fabricated by the silicon-on-insulator (SOI) technique, and it operates to detect the three-axes angular velocities. The outer ring is driven by the rotational comb electrodes to rotate, counterclockwise and clockwise alternatively, around the z-axis. Once the gyroscope is perturbed by Coriolis acceleration resulting from external rotation excitation around the y-axis, the outer ring responds to tilt in the direction of the x-axis. On the other hand, the inner-disc is forced to oscillate about the y-axis if the external rotation excitation is about the x-axis. All the tilts along x-axis or y-axis will result in the change of voltage output across the corresponding capacitors.

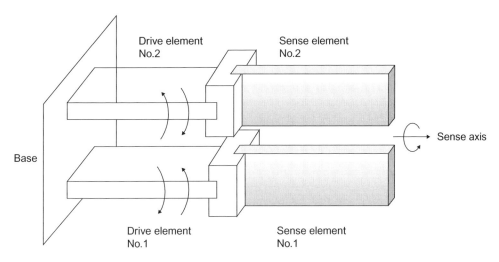

FIGURE 5 Configuration of a tuning fork.

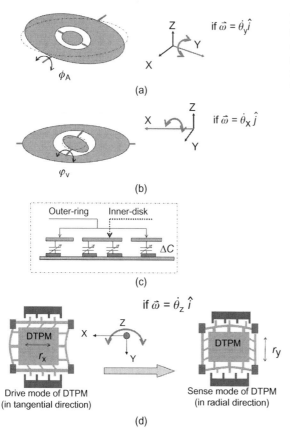

FIGURE 6 Sensing mode motions under Coriolis effects: (a) sensing mode of outer-ring, (b) sensing mode of inner-disk, (c) side-view of sensing electrodes, (d) driving mode and sensing mode of distributed translational proof mass (DTPM). *(With permission [2])*

Similarly, if the external angular excitation is about the z-axis, then the distributed translational proof mass will move in the radial direction and be detected by the comb electrodes.

2.4 Magnetic Sensors

2.4.1 *The Hall Effect*

Magnetic sensors, which are available with different sensitivities, are used for precise measurements of body movement. Common magnetic sensors are based on the Hall effect (Figure 7), which generates magnetic impedance and magnetic resistance as a result of the interaction between moving electronic carriers and an external magnetic field. In metals, these carriers are electrons. When an electron moves through a magnetic field, it is subjected to a sideways force:

$$F = qvB \tag{7}$$

where q is electronic charge, v is the speed of an electron, and B is the magnetic field.

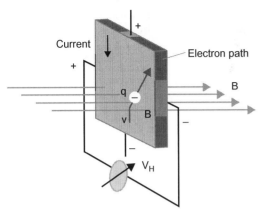

FIGURE 7   The Hall effect.

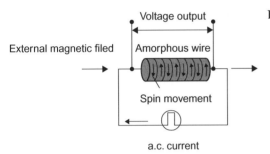

FIGURE 8   A magnetoimpedance sensor.

2.4.2 *Magnetoimpedance Sensors*

Magnetoimpedance sensors are based on the magnetoimpedance (MI) effect in amorphous wires and are typically fabricated with complementary metal-oxide-semiconductor (CMOS) multi-vibrator integrated circuitry [3,4]. Amorphous materials are non-crystalline; they have a uniform internal structure and typically exhibit ideal soft magnetic properties. As shown in Figure 8, the MI effect occurs when a pulse current is applied to an amorphous wire, generating a dramatic change in the impedance (Z) according to a minute external magnetic field. Because the magnetic permeability changes with the external magnetic field (H_{ex}) in the axial direction, it is possible to detect the strength of the external magnetic field from the change in Z, as follows:

$$Z = \frac{a}{2\sqrt{2\rho}} R_{dc}(1+j)\sqrt{\omega\mu(H_{ex})}\cdots \tag{8}$$

where a is the diameter of the amorphous wire, ρ is the specific electrical resistance, R_{dc} is the DC resistance, and ω is the angular velocity.

Amorphous wires exhibit the MI effect because of their unique magnetic domain structure in which the arrangement of the wire surface spin is lined up in the circumferential direction. This arrangement results in a larger change in circumferential magnetic permeability μ, and the MI effect is maximized.

One shortcoming of MI sensors is that, because the wire impedance itself changes symmetrically with the polarity of the external magnetic field, the direction of the magnetic field cannot be determined, and it is difficult to obtain output linearity. However, by winding a pickup coil around the amorphous wire and detecting the induction voltage, only the imaginary number components of the formula are detected, the output properties become linear, and it becomes possible to determine the polarity of the magnetic field direction.

2.4.3 Magnetoresistance Sensors

When a perpendicular magnetic field is applied to a classical semiconductor, such as an InSb plate surface, the resistance increases. This effect is magnetoresistance (MR), and it depends on the electron mobility of the material. The resistance under a magnetic field R_B is

$$R_B = R_0 \frac{\rho_B}{\rho_0}(1 + m(\mu B)^2) \tag{9}$$

where R_0 is the resistance under a non-magnetic field, μ is the magnetic permeability, ρ_B/ρ_0 is the specific relative resistance, B is the magnetic field, and m is a geometric factor consisting of the length and width of the material. If a magnetic field is not applied, the current flows straight through the InSb plate. However, if magnetic flux is applied, a Lorentz force proportional to the magnetic flux density will deflect the current path. The practical use of this principle in magnetic sensing is limited by low MR levels.

Anisotropic magnetoresistance (AMR) sensors are also commercially available. AMR, which is the magnetic field-induced change in scattering due to the atomic orbitals, is a common characteristic of ferromagnetic materials. AMR is maximized when both directions are parallel and minimized when both directions are perpendicular. Typical MR levels are close to 1%, and the response is sufficiently linear to enable the use of AMR devices in practical applications.

2.4.4 Giant Magnetoresistance Sensors

The electric current in a magnetic multilayer consisting of a sequence of thin magnetic layers separated by equally thin non-magnetic metallic layers is strongly influenced by the relative orientation of the magnetizations of the magnetic layers (Figure 9). The cause of this large resistance variation, also called giant magnetoresistance (GMR), is attributed to the scattering of the electrons at the layer interfaces. Thus, any structure with metal-magnetic interfaces is a candidate for displaying GMR, and significant efforts are underway to identify structures to enhance the GMR effect. The use of GMR in conjunction with a three-axis accelerometer, a three-axis gyroscope, and a three-axis magnetometer results in a nine-degrees-of-freedom inertial measurement system. However, few studies based

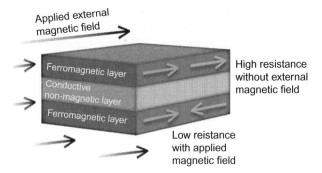

FIGURE 9 GMR sensors consist of a ferromagnetic material that changes resistance in a magnetic field. This property lets it trigger an electric circuit under sufficient magnetic field strength.

on nine-degrees-of-freedom inertial measurement exist.

ANGLE MEASUREMENT

Gyroscope signals are integrated to produce angle estimates. In general, the orientation angle derived from the angular velocity ω_{gy} measured by a gyroscope is given as

$$\varphi_k = \int_t (\omega_{gy} \times A + \omega_{off})dt + \varphi_{off} \cdots \tag{10}$$

where ω_{gy} is the gyroscope output signal, A is the proportional factor, ω_{off} is the angular velocity offset, and φ_{off} is the initial integration offset (at t = 0) [5]. Accelerometers are used to compensate for slowly occurring errors in the gyroscope integral. A combination of error correcting (auto-resetting) and offset correction (auto-nulling) techniques has been proposed [6,7].

Using combined signals from accelerometers, gyroscopes, and magnetometers, the Kalman-based fusion algorithm has been applied to obtain dynamic orientations and positions of human body segments [8–10]. Furthermore, use of segment orientations have been proposed to visualize 3D gait from accelerometer and gyroscopic measurements in a global coordinate system [11]. In addition, a Gaussian particle filter has been used with wearable inertial sensors to evaluate the maximum angle of a walking cycle [12].

3. OBTAINED PARAMETERS FROM INERTIA SENSORS

3.1 Mathematical Analyses

Signals from inertial sensors can be used to determine several measures, including walking speed, various root mean square (RMS) values, cadence, of steps, and stride lengths.

Instead of using a stopwatch, inertial sensors, such as accelerometers and gyroscopes, have been used to estimate walking speed, which is an important measure of walking performance in daily life [13]. The experimental design included both treadmill walking and over-ground walking at both predetermined and preferred speeds. The algorithms for estimation of walking speed can be classified as either the direct integration of a human gait model or the abstraction model. The abstraction model is a "black box" model that employs a neural network to describe the complex relationship between the sensor measurements and the walking speed.

RMS values provide information on the average magnitude of accelerations, angular velocities, and magnetic flux in each direction. The acceleration signal during the walking phase shows balance during gait [14]. The RMS values can be determined as shown in Eq. (11), in which N corresponds to the number of in the signal X_i, and X_i is the measured amplitude of acceleration, angular velocity, and magnetic field.

$$X_{rms} = \sqrt{\frac{1}{N}\sum_{t=0}^{N-1}(X_i^2)} \tag{11}$$

One stride, or one complete gait cycle, is defined as the time between one heel strike to the next heel strike of the same foot. Thus, each stride comprises two steps, with each step covering the period from one heel strike to the heel strike of the contralateral limb. The number of steps and the number of strides is determined using the acceleration signal, as shown in Figure 10. The stride frequency, or cadence, is the number of full cycles per minute. Cadence is correlated positively with gait and can be used to measure the walking performance [15]. The average step time, average stride time, and cadence are calculated from the following equations:

$$\text{Average step} = \frac{Walk\ time[s]}{Number\ of\ gait\ step} \tag{12}$$

$$\text{Average stride} = \frac{Walk\ time[s]}{Number\ of\ gait\ stride} \tag{13}$$

$$\text{Cadence} = \frac{Number\ of\ steps}{Walk\ time[min]} \tag{14}$$

An autocorrelation function for vertical acceleration has been calculated and evaluated using leg symmetry [16]:

$$\text{Ad}(m) = \frac{1}{N-|m|}\sum_{j=1}^{N-|m|} x(i)\cdot x(i+m) \tag{15}$$

where N is the total number of samples and m is the time lag expressed as the number of samples. When the autocorrelation of the acceleration signal is computed during the gait, the first peak of Ad(m), Ad1, reflects the regularity of the acceleration between consecutive steps, which can be interpreted as a measure of the symmetry between left and right leg control. The second peak of Ad(m), Ad2, reflects the regularity of consecutive strides.

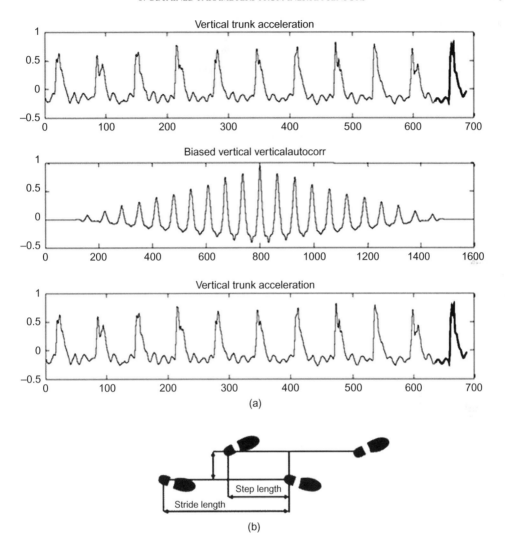

FIGURE 10   Gait cycle characteristics estimated by autocorrelation function. (a) Acceleration signal, biased correlation, and unbiased-unbiased correlation shown in Eq. 15, and first peak corresponded to step regularity and second peak represented stride regularity. (b) Definition of step and stride.

Higher Ad1 and Ad2 values reflect higher step and stride regularity [16–18]. The walking cadence can be obtained from the time lag between Ad1 and Ad2.

The gait parameters that can be extracted from nine-degrees-of-freedom inertial sensors are summarized in Table 1.

Static and dynamic activities can be estimated using accelerometry. The amplitude and RMS values are measures to differentiate normal and abnormal walking, as well as age-related functions [19].

TABLE 1 Possible Parameters for Wearable Inertial Sensors

Parameters	Signals
Velocity	Triaxial acceleration
Number of steps	Triaxial acceleration
Number of strides	Triaxial acceleration
Time	Triaxial acceleration
Average stride time	Triaxial acceleration
Average stride time	Triaxial acceleration
Cadence	Triaxial acceleration
Root mean square (RMS) values	Triaxial acceleration, triaxial gyroscopic velocity, triaxial magnetic field
Regularity of steps	Triaxial acceleration
Regularity of strides	Triaxial acceleration
Angle	Combined with triaxial acceleration, triaxial angular velocity, and triaxial magnetic field

3.2 Comparison between Rehabilitation Score and Acceleration

The signals obtained from wearable inertial sensors are often used for qualitative evaluation in clinical practice. However, there are also several semi-quantitative scores that can be obtained from wearable inertial sensors. Specifically, the activity of daily life (ADL) score is most often used in clinical practice. While basic ADL categories have been suggested, that which specifically constitutes a particular ADL in a particular environment for a particular person may vary. In basic ADL determinations, normal human functions, such as feeding ourselves, bathing, dressing, grooming, work, homemaking, and leisure have been counted, accumulated, and compared to obtain a total score. Other instrumental ADL scores may also be included, including those that are not fundamental functions, such as managing money, shopping, and transportation, because these functions often determine whether individuals may live independently in a community [20].

Several evaluation tools, such as the Katz ADL score and the Lawton Instrumental ADL (IADL) scale, are available to evaluate patients. The Katz ADL score comprises six individual activities, and the maximum total score is six. Few studies of the relationship between ADL and inertial parameters have been conducted [21−23]. Improvements in the interpretation of inertial parameters are critical to future rehabilitation studies.

4. APPLICATIONS FOR WEARABLE MOTION SENSORS

The use of acceleration signals was discussed as early as the late 1980s. Acceleration signals at the lower back, which is close to the center of gravity in a typical human, were monitored, and different signals were obtained for patients with various diseases. With the development of integrated circuit (IC) technology, the cost of accelerometers, gyroscopes, and magnetic sensors has greatly reduced, and these instruments have been used in several rehabilitation studies.

4.1 Fall Risk Assessment with Rehabilitation Battery

Various tools for assessment of subjective and objective falling risk have been developed. Subjective methods, such as diaries, questionnaires, and surveys, are inexpensive; however, these tools depend on the experience of therapists and caregivers, individual observation, and subjective interpretation, all of which lead to inconsistency in assessment results [23]. Some standard tests for risk assessment also require subjective judgments. For example, the "timed up and go" (TUG) test is a simple test of the ability to perform a sequence of basic activities. The American and British Geriatrics Societies recommend using the TUG test as part of a screening battery for fall risk (American Guidelines). Traditionally, the test is scored by manually recording the time taken to rise from a standardized chair, walk three meters, turn around, walk back, turn, and sit back down in the chair [24]. This test has been extensively used to assess balance and mobility in the elderly [25–27]. The test is considered simple because it uses total test time as a threshold for prediction of fall risk in the elderly [28]. TUG test research suggests a 13.5-s duration for completion of the whole test as the threshold to classify fallers and non-fallers [29]. Distinguishing postural transitions in the TUG test, however, depends on subjective judgment that counts the time taken for each posture transition.

Several studies have suggested using wearable inertial sensors to classify the fall risk among the elderly and improve the consistency of results. In recent studies, researchers used three wearable inertial sensors to determine the phase transition, which were attached at the waist dorsally and also at both the right and left thighs [30–32]. A single accelerometer at the track was also used to assess the fall risk [33]. Figure 11 shows a typical example of phase classification at the waist by accelerometer and gyro sensor.

Another previous study investigated the test-retest reliability of an instrumented TUG test and found temporal gait parameters to be the most reliable measures, with mixed results for spatial gait parameters and sit-stand variables [34,35]. These objective measures enabled more complete, sensitive TUG-based fall risk assessment.

The Berg Balance Scale (BBS) is used to evaluate the balance control of elderly individuals. However, the BBS also requires subjective observation and determination for scoring some test items [36].

Acceleration signals can be used to measure gait accurately and can be applied to studies of walking, sit-to-stand performance, and walking performance. An accelerometer mounted on the torso reflects the periodic movements while walking, including speeding up and slowing down, as well as vertical, lateral, and anteroposterior moves [37]. The evaluation of walking performance typically requires either relatively long straight walking surfaces of 20–40 m [37,38] or treadmill walking [39]. However, for rehabilitation evaluation, distances of 6–10 m may be sufficient for evaluating walking performance. At distances greater than 10 m, patients in rehabilitation may tire. Instead, in rehabilitation

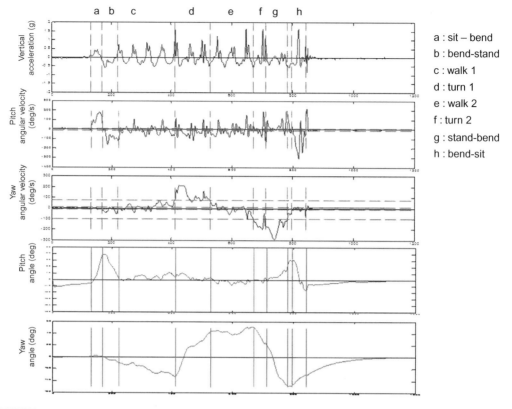

FIGURE 11 An example of TUG test at waist with accelerometer and gyro sensor.

settings, the time required for an individual to walk 10 m without assistance is measured; the time required to walk 6 m is also measured to evaluate acceleration and deceleration.

Acceleration measurements have also been compared with a variety of clinical balance tests, including variants of Romberg's tests, heel–toe straight-line walking, and a functional reach test, as well as quantitative motor coordination tests, including rapid stepping tests and heel–toe transitions [40]. Other common batteries for fall risk evaluation are sit-to-stand transitions [41], standing posture sway, and the left-right alternating level step test. The inertial parameters for these could be obtained using sensors to assess the fall risk [42–45]. In general, inertial sensors are promising sensors for fall risk assessment. Future studies should focus on evaluation of fall risk using different techniques to determine the most promising sensor sites and optimal predictive variables.

4.2 Fall Detection

Falls and fall-induced injuries are major public-health problems among elderly individuals. Many methods and programs to prevent fall injuries already exist, including regular exercise, vitamin D and calcium supplementation, the withdrawal of psychotropic medication, environmental hazards assessment and modification, hip protectors, and multifactorial preventive programs for the simultaneous assessment and reduction of many of the predisposing and situational risk factors. One such risk of falling assessment is the TUG test described above.

Ambulatory measures have many potential uses, including the objective assessment of mobility in a clinical or home environment over a prolonged period (e.g., the 24-h ambulatory mobility monitor). Ambulatory monitors may be used for initial assessment or as an adjunct to monitoring progress during and after rehabilitation. Research is currently underway to evaluate accelerometers for fall-detection medical devices that may be used to prevent falls and consequent injuries in the elderly population.

Automated fall detectors have been developed to support independent living and safety. These detectors are based mostly on accelerometers attached to the body. Currently, most fall-detection devices are prototypes that are still in the research stage.

Fall-detection algorithms detect different phases of a fall event: (1) motion prior to impact based on high velocity [46,47] and fast postural changes or free falls [48,49]; and (2) the impact itself based on high acceleration [50,51], a rapid change in acceleration and end posture [52,53], or reduced general activity after the impact [51]. In earlier studies, Kangas et al. found that a waist-worn triaxial accelerometer with a simple algorithm was sufficient for fall detection [54,55].

4.3 Quantitative Evaluation of Hemiplegic Patients

A semi-quantitative score based on a physical therapist's observation is used as a standard for evaluation of hemiplegic patients. Hemiplegic legs and arms are observed and scored from 1 to 6 on the Bronstrom scale [56]. The physical therapist scores the patient's performance while walking or executing upper-arm movements. However, judgments among experienced therapists may vary, because no standard criteria for assessing hemiplegia exist.

Instead, acceleration and gyroscopic sensors can be used to quantify hemiplegic patient performance. The walking performance of hemiplegic patients has been typically evaluated based on the symmetry and regularity of the hemiplegic site. The gyroscopic sensor has been used to detect the gait phase of hemiplegic walking [31,33,57]. The extensibility of hemiplegic patients' lower limbs was significantly lower than that of normal subjects, and the hemiplegic gait exhibited poor balance compared with normal gait [58,59]. The influence of rehabilitation training of hemiplegic gaits could also be analyzed quantitatively.

4.4 Clinical Assessment for Parkinson's Disease

Inertial sensors have also been applied to patients with various gait abnormalities, including the dyskinesia associated with Parkinson's disease (PD). The use of a sensor-based system to monitor PD is promising for improvement of the clinical management of PD patients. A portable triaxial accelerometer attached at the shoulder was used to monitor the severity of the dyskinesia, which occurs as an uncomfortable side effect of PD medication. The acceleration signal and severity of the dyskinesia are highly correlated [60].

Several scores exist for evaluation of PD, including the Unified Parkinson's Disease Rating Scale (UPDRS) III. Comparison studies of UPDRS scores and parameters from inertial sensors have been carried out by several researchers [61–64]. A separate study investigated the correlation between gait parameters and a motor score based on modified abnormal involuntary movement (AIM) and Goetz scales [65]. Two accelerometers placed over the middle of the lower back using a semi-elastic belt recorded craniocaudal and side-to-side accelerations. The PD gait was characterized by a reduction in walking velocity, which was explained by reductions in stride frequency and step length. In addition, a reduction in walking regularity and craniocaudal activity also was noted. Walking regularity and craniocaudal activity were strongly correlated with the motor score, thereby suggesting that gait regularity and craniocaudal activity are particularly suitable for characterizing a stabilized PD gait.

4.5 Energy Expenditure

Current accelerometers can estimate the energy expenditure associated with physical activity. Over the past several decades, the integral of the acceleration signal per unit time has been assumed to be proportional to the oxygen consumption [66,67]. A small portable accelerometer was developed to estimate the energy expenditure of daily activities. The oxygen consumption of 14 different activities was measured in 21 subjects, each wearing the portable accelerometer at the waist. The reproducibility of accelerometer readings was high (four subjects, 14 activities; r = 0.94). The standard error of the oxygen consumption (V_{O2}) estimate from the accelerometer, based on 21 subjects and 14 activities, was 6.6 mL min^{-1} kg^{-1}. This method provides an epidemiological way to evaluate and control environments and to create individual activities to obtain good correlation.

Physical activities are characterized by intensity, type, duration, and frequency. The absolute intensity of physical activities is the energy expenditure of the activity (EEact). Many researchers use the minute-to-minute EEact values predicted by accelometers to classify daily activities with predetermined threshold values and to enable the evaluation of the duration and frequency of different activities. Various output measurements from accelerometers must be calibrated with well-measured EEact values produced by different tasks. Additionally, the intensity of the accelerometer signal can be used to estimate metabolic equivalents (METs) in adults and children. Recent technological developments include advancements in both linear and nonlinear analytical modeling approaches. Reference energy expenditure data may be obtained via either indirect calorimetry or deuterium-labeled water techniques in a well-controlled environment.

Most linear validation studies have evaluated the correlation coefficient between the activity counts from the monitor and energy expenditures measured using indirect calorimeters. For example, level walking exhibited high correlation, but the correlation coefficient associated with housekeeping activity was very low. In individual monitoring, a physical activity monitor may underestimate certain physical activities while overestimating others; thus, the total sum of the predicted EEact is often comparable to the measured overall EEact.

Nonlinear approaches have also been attempted. A two-component (vertical and horizontal) power model was developed to translate individual activity counts:

$$\text{EEact} = a(\sqrt{A_a{}^2 + A_l{}^2})^{p1} + bA_v{}^{p2} \tag{16}$$

where A_v is the vertical acceleration count, and counts in the A_a and A_l terms are combined in the anterior-posterior and lateral directions, i.e., in the horizontal plane. The coefficients a, b, p1, and p2 are determined by a traditional unconstrained nonlinear optimization algorithm for each individual [68]. Compared with the linear model, the correlation was improved, and the underestimation was reduced from 50% to 3% [69]. Thus, the nonlinear model improves the accuracy of the EEact prediction; however, the nonlinear model is sometimes unstable.

5. PRACTICAL CONSIDERATIONS FOR WEARABLE INERTIAL SENSOR APPLICATIONS IN CLINICAL PRACTICE AND FUTURE RESEARCH DIRECTIONS

Wearable inertial sensors with appropriate specifications have been used in clinical applications. The critical issue for successful clinical use of wearable inertial sensors is sensor selection, which must be considered on a case-by-case basis. Acceleration, angular velocity, and magnetic sensors with a maximum nine-degrees-of-freedom are commercially available. Higher accuracy is not necessarily required for rehabilitation purposes. In clinical settings, simple signal interpretation is required for both physicians and patients. For more widespread popularization of wearable inertial sensors in clinical practice, development of improved evidenced-based interpretation is proposed.

References

[1] Q. Zou, W. Tan, E.S. Kim, G.E. Loeb, Single- and triaxial piezoelectric-bimorph accelerometers, J. Microelectromechnical Syst. 17 (1) (2008) 45–57.
[2] N.-C. Tsai, C.-Y. Sue, Design and analysis of a tri-axis gyroscope micromachined by surface fabrication, IEEE Sensors J. 8 (12) (2008) 1933–1940.
[3] K. Mohri, L.V. Panina, T. Uchiyama, K. Bushida, M. Noda, Sensitive and quick response micro magnetic sensor utilizing magneto-impedance in Co-rich amorphous wires, IEEE Trans. Magnetic. 31 (2) (1995) 1266–1275.
[4] K. Mohri, T. Uchiyama, L.P. Shen, C.M. Cai, L.V. Panina, Sensitive micro magnetic sensor family utilizing magneto-impedance (M) and stress-impedance (SI) effects for intelligent measurements and controls, Sen. Actuators A Phys. 91 (1–2) (2001) 85–90.
[5] J.E. Bortz, A new mathematical formulation for strapdowm inertial navigation, IEEE Trans. Aerosp. Electron. Syst. 7 (1971) 61–66.
[6] R.E. Mayagoitia, A.V. Nene, P.H. Veltink, Accelerometer and rate gyroscope measurement of kinematics: an inexpensive alternative to optical analysis systems, J. Biomech. 35 (4) (2002) 537–542.
[7] R. Williamson, B.J. Andrews, Detecting absolute human knee angle and angular velocity using accelerometers and rate gyroscopes, Med. Biol. Emg. Comput. 39 (2001) 1–9.

[8] R. Zhu, Z. Zhou, A real-time articulated human motion tracking using tri-axial internal/magnetic sensors package, IEEE Trans. Neural Syst. Rehabil. Eng. 12 (2) (2004) 295–302.

[9] A.M. Sabatini, Quaternion-based extended Kalman filter for determining orientation by inertial and magnetic sensing, IEEE Trans. Biomed. Eng. 53 (2006) 1346–1356.

[10] G. Cooper, I. Sheret, L. McMillian, K. Siliverdis, N. Sha, D. Hodgins, et al., Inertial sensor-based knee flexion/extension angle estimation, J. Biomech. 42 (16) (2009) 2678–2685.

[11] R. Takeda, S. Tadano, A. Natorigawa, M. Todoh, S. Yoshinari, Gait posture estimation using wearable acceleration and gyro sensors, J. Biomech. 42 (2009) 2486–2494.

[12] Z. Zhang, Z. Huang, J. Wu, Ambulatory hip angle estimation using Gaussian particle filter, J. Signal Process. Syst. 58 (2010) 341–357.

[13] S. Yang, Q. Li, Inertial sensor-based methods in walking speed estimation: a systematic review, Sensors 12 (2012) 6012–6116.

[14] M. Henriksen, H. Lund, R. Moe-Nilssen, H. Bliddal, B. Danneskiod-Samsoe, Test-retest reliability of trunk accelerometric gait analysis, Gait. Posture. 19 (3) (2004) 288–297.

[15] Y.-R. Yang, Y.-Y. Lee, S.-J. Cheng, P.-Y. Lin, R.-Y. Wang, Relationships between gait and dynamic balance in early Parkinson's disease, Gait. Posture. 27 (4) (2008) 611–615.

[16] R. Moe-Nilssen, J.L. Helbostad, Estimation of gait cycle characteristics by trunk accelerometry, J. Biomech. 37 (2004) 121–126.

[17] A. Tura, M. Raggi, L. Rocchi, A.G. Cutti, L. Chiari, Gait symmetry and regularity in transfemoral amputees assessed by trunk accelerations, J. Neuroeng. Rehabil. 7 (2010) 4.

[18] A. Tura, L. Rocchi, L. Chiari, Recommended number of strides for automatic assessment of gait symmetry and regularity in above-knee amputees by means of accelerometry and autocorrelation analysis, J. Neuroeng. Rehabil. 9 (2012) 11.

[19] H.J. Yack, R.C. Berger, Dynamic stability in the elderly: identifying a possible measure, J. Gerontol. 48 (1993) M225–M230.

[20] S.S. Roley, J.V. DeLany, C.J. Barrows, American occupational therapy association committee of practice, "Occupational therapy practice framework: domain and practice," second ed. Am. J. Occup. Ther. 62 (6) (2008) 625–683.

[21] M.J. Mathie, A.C. Closter, N.H. Lovel, B.G. Veller, S.R. Lord, A. Tiedemann, Accelerometry: providing an integrated, practical method for long-term, ambulatory monitoring of human movement, J. Telemed. Telecare 10 (2004) 144–151.

[22] M.N. Nyan, F.E.H. Tay, M. Manimaran, K.H.W. Seah, Garment-based detection of falls and activities of daily living using 3-axis MEMS accelerometer, J. Phys. Conf. Ser. 34 (2006) 1059.

[23] T. Tamura, M. Sekine, H. Miyoshi, Y. Kuwae, T. Fujimoto, Wearable inertia sensor application in the rehabilitation field, Adv. Sci. Tech. 85 (2013) 28–32.

[24] G.A.L. Meijer, K.R. Westerterp, F.M.H. Verhoeven, H.B.M. Koper, F. Hoor, Methods to assess physical activity with special reference to motion sensors and accelerometers, IEEE. Trans. Biomed. Eng. 38 (1991) 221–229.

[25] American Geriatrics Society, British Geriatrics Society, and American Academy of Orthopaedic Surgeons Panel on Falls Prevention, Guidelines for the prevention of falls in older persons, J. Am. Geriatr. Soc. 49 (2001) 664–672.

[26] D. Podsiadlo, S. Richardson, The timed-up-&-go: a test of basic functional mobility for frail elderly persons, J. Am. Geriatr. Soc. 39 (1991) 142–148.

[27] K.O. Berg, B.E. Maki, J.I. Williams, P.J. Holliday, S.L. Wood-Dauphinee, Clinical and laboratory measures of postural balance in an elderly population, Arch. Phys. Med. Rehabil. 73 (11) (1992) 1073–1080.

[28] C. Zampieri, A. Salarian, P. Carlson-Kuhta, K. Aminian, J.G. Nutt, F.B. Horak, The instrumented timed up and go test: potential outcome measure for disease modifying therapies in Parkinson's disease, J. Neurol. Neurosurg. Psych. 81 (2) (2010) 171–176.

[29] G. Thrane, R.M. Joakimsen, E. Thornquist, The association between timed up and go test and history of falls: The tromso study, BMC. Geriatr. 7 (1) (2007) 1.

[30] A. Shumway-Cook, S. Brauer, M. Woollacott, Predicting the probability for falls in community-dwelling older adults using the timed up & go test, Phys. Ther. 80 (9) (2000) 896–903.

[31] Y. Higashi, K. Yamakoshi, T. Fujimoto, M. Sekine, T. Tamura, 2008. Quantitative evaluation of movement using the timed up-and-go test, IEEE Eng Med Biol Mag. 27(4):38–46.

[32] B.R. Greene, A. O'Donovan, R. Romero-Ortuno, L. Cogan, C.N. Scanaill, R.A. Kenny, Quantitative falls risk assessment using the timed up and go test, IEEE Trans. Biomed. Eng. 57 (12) (2010) 2918–2926.

[33] N.A. Zakaria, Y. Kuwae, T. Tamura, K. Mnato, S. Kanaya, Quantitative analysis of fall risk using TUG test computer methods, Biomech. Biomed. Eng. (2013). Available from: http://dx.doi.org/doi:10.1080/10255842.2013.805211.

[34] A. Salarian, F.B. Horak, C. Zampieri, P. Carlson-Kuhta, J.G. Nutt, K. Aminian, iTUG, a sensitive and reliable measure of mobility, IEEE Trans. Neural Syst. Rehabil. Eng. 18 (3) (2010) 303–310.

[35] A. Weiss, T. Herman, M. Plotnik, M. Brozgol, N. Giladi, J.M. Hausdorff, An instrumented timed up and go: the added value of an accelerometer for identifying fall risk in idiopathic fallers, Physiol. Meas. 32 (2011) 2003–2018.

[36] K.O. Berg, S.L. Wood-Dauphinee, J.I. Williams, D. Gayton, Measuring balance in elderly: Preliminary development of an instrument, Physiother. Can. 41 (1989) 304–311.

[37] A. Auvinet, G. Berrut, C. Touzard, L. Moutel, N. Collet, D. Chaleil, et al., Reference data for normal subjects obtained with an accelerometric device, Gait Posture 16 (2002) 124–134.

[38] H.B. Menz, S.R. Lord, R.C. Fitzpatrick, Age-related differences in walking stability, Age Aging 32 (2003) 137–142.

[39] W. Zijlstra, A.L. Hof, Assessment of spatio-temporal gait parameters from trunk accelerations during human walking, Gait Posture 18 (2003) 1–10.

[40] C.Y. Cho, G. Kamen, Detecting balance deficits in frequent fallers using clinical and quantitative evaluation tools, J. Amer. Geriatr. Soc. 46 (1998) 426–430.

[41] R. Ganea, A. Paraschiv-Ionescu, C. Büla, S. Rochat, K. Aminian, Multi-parametric evaluation of sit-to-stand and stand-to-sit transitions in elderly people, Med. Eng. Phys. 33 (2011) 1086–1093.

[42] T. Shany, S.J. Redmond, M.R. Narayanan, N.H. Lovell, Sensors-based wearable systems for monitoring of human movement and falls, IEEE Sens. J. 12 (2012) 658–670.

[43] T. Shany, S.J. Redmond, M. Marschollek, N.H. Lovell, Assessing fall risk using wearable sensors: a practical discussion, J. Gerontol. Geriatr. 45 (2012) 694–706.

[44] W. Tao, T. Liu, R. Xheng, H. Feng, Gait analysis using wearable sensors, Sensors 12 (2012) 2255–2283.

[45] J. Howcroft, J. Kofman, E.D. Lemaire, Review of fall risk assessment in geriatric populations using inertial sensors, J. Neuro. Eng. Rehabil. 10 (2013) 91.

[46] U. Lindemann, A. Hock, M. Stuber, W. Keck, C. Becker, Evaluation of a fall detector based on accelerometers: a pilot study, Med. Biol. Eng. Comput. 43 (2005) 548–551.

[47] A.K. Bourke, K.J. O'Donovan, G. Ólaighin, The identification of vertical velocity profiles using an inertial sensor to investigate pre-impact detection of falls, Med. Eng. Phys. 30 (7) (2008) 937–946.

[48] N. Noury, P. Rumeau, A.K. Bourke, G. Ólaighin, J.E. Lundy, A proposal for the classification and evaluation of fall detectors, IRBM 29 (6) (2008) 340–349.

[49] A.K. Bourke, J.V. O'Brien, G.M. Lyons, Evaluation of a threshold-based tri-axial accelerometer fall detection algorithm, Gait Posture 26 (2) (2007) 194–199.

[50] D.M. Karantonis, M.R. Narayanan, M. Mathie, N.H. Lovell, B.G. Celler, Implementation of a real-time human movement classifier using a triaxial accelerometer for ambulatory monitoring, IEEE Trans. Inf. Technol. Biomed. 10 (2006) 156–167.

[51] A.K. Bourke, G.M. Lyons, A threshold-based fall-detection algorithm using a bi-axial gyroscope sensor, Med. Biol. Eng. Comput. 30 (1) (2008) 84–90.

[52] A. Dinh, Y. Shi, D. Teng, A. Ralhan, L. Chen, V.D. Bello-Haas, et al., A Fall and near-fall assessment and evaluation system, Open Biomed. Eng. J. 3 (2009) 1–7.

[53] M. Kangas, A. Konttila, P. Lindgren, I. Winblad, T. Jämsä, Comparison of low-complexity fall detection algorithms for body attached accelerometer, Gait Posture 28 (2008) 285–291.

[54] M. Kangas, I. Vikman, J. Wiklander, P. Lindgren, L. Nyberg, T. Jämsä, Sensitivity and specificity of fall detection in people aged 40 years and over, Gait Posture 29 (4) (2009) 571–574.

[55] A. Sixsmith, N. Johnson, A smart sensor to detect the falls of the elderly, IEEE Pervasive Comput. 3 (2) (2004) 42–47.

[56] S. Brunnstrom, Movement Therapy in Hemiplegia: A Neurophysiological Approach, Harper & Row, New York, New York, 1970.

[57] N. Abaid, P. Cappa, E. Palermo, M. Petrarca, M. Porfiri, Gait detection in children with and without hemiplegia using single-axis wearable gyroscopes, PLOS One 8 (9) (2013) 73152.

[58] Y. Guo, D. Wu, G. Liu, G. Zhoo, B. Huang, L. Wang, A low-cost body inertia-sensing network for practical gait discrimination of hemiplegia patients, Telemed. E-Health 18 (10) (2012) 748−754.

[59] Y. Guo, G. Zhao, Q. Liu, Z. Mei, K. Ivanov, L. Wang, Balance and knee extensibility evaluation of hemiplegic gait using an inertia body sensor network, Biomed. Eng. Online 12 (2013) 83.

[60] S.Y. Chang, C.F. Lai, H.C. Josh Chao, J.H. Park, Y.M. Huang, An environmental-adaptive fall detection system on mobile device, J. Med. Syst. 35 (5) (2011) 1299−1312.

[61] A. Salarian, H. Russmann, F.J.G. Vingerhoets, P.R. Burkhard, K. Aminian, Ambulatory monitoring of physical activities in patients with Parkinson's disease, IEEE Trans. Biomed. Eng. 54 (12) (2007) 2296−2299.

[62] C. Zampieri, A. Salarian, P. Carlson-Kuhta, K. Aminian, J.G. Nutt, F.B. Horak, The instrumented timed up and go test: Potential outcome measure for disease modifying therapies in Parkinson's disease, J. Neurol. Neurosurg. Psychiatry 81 (2) (2010) 171−176.

[63] A. Sant'Anna, A. Salarian, N. Wickström, A new measure of movement symmetry in early Parkinson's disease patients using symbolic processing of inertial sensor data, IEEE Trans. Biomed. Eng. 58 (7) (2011) 2127−2135.

[64] M. Manchi, L. King, A. Salarian, L. Holmstrim, J. Mcnames, F.B. Horak, Mobility lab to assess balance and gait with synchronized body-worn sensors, J. Bioeng. Biomed. Sci. 2012 (2012) S1.

[65] A.J. Manson, P. Brown, J.D. O'Sullivan, P. Asselman, D. Buckwell, A.J. Lees, An ambulatory dyskinesia mmonitor, J. Neurol. Neurosurg. Psychiatry 68 (1) (2000) 96−201.

[66] T.C. Wong, J.G. Webster, H.J. Montoye, R. Washburn, Portable accelerometer device for measuring human energy expenditure, IEEE Trans. Biomed. Eng. 28 (6) (1981) 467−471.

[67] H.J. Montoye, R. Washburn, S. Servais, A. Ertl, J.G. Webster, F.J. Nagle, Estimation of energy expenditure by a portable accelerometer, Med. Sci. Sports Exerc. 15 (5) (1983) 403−407.

[68] K.Y. Chen, M. Sun, Improving energy expenditure estimation by using triaxial accelerometer, J. Appl. Physiol. 83 (1997) 2112−2122.

[69] K.Y. Chen, D.R. Bassett Jr., The technology of accelerometry-based activity monitors: Current and future, Med. Sci. Sports. Exerc. 37 (Suppl. 11) (2005) S490−S500.

Application of Optical Heart Rate Monitoring

Mathieu Lemay, Mattia Bertschi, Josep Sola, Philippe Renevey,
Jakub Parak, and Ilkka Korhonen

Tampere University of Technology, Tampere, Finland

1. INTRODUCTION

Since the introduction of portable devices in 1957 by Dr. Norman Holter, heart rate (HR) monitors have been extensively used in clinical practices as a diagnostic and prognostic tool, mainly for cardiovascular diseases. In parallel, the use of HR monitoring devices by sport physiologists to analyze the response of the body to exercise or training stress and to evaluate the training level of athletes has gained popularity. With the increasing interest of consumers to health and wellness monitoring, an explosion of commercialized HR monitors based on various technologies has been observed during the last decade. This chapter will give an overview of the different techniques actually known to monitor HR, focusing in particular on the so-called photoplethysmographic (PPG) technique. The history of the PPG technique, its basics, including measurement principles, measurement sites, quality factors and its applications to sport, fitness, daily life, and healthcare, will be addressed in respective sections.

The heart is the muscle in charge of pumping the blood throughout the blood vessels, namely the arteries and the veins carrying oxygenated and deoxygenated blood, respectively. The regulation of the rate at which the heart is pumping is mainly controlled by the autonomic nervous system. This system comprises sympathetic and parasympathetic (or vagal) functions, which are partially complementary. The sympathetic nervous system is responsible for increasing HR, constricting blood vessels, and other "stress" responses, while the parasympathetic system promotes maintenance of the body at rest (e.g., slowing down HR) [1]. The autonomic modulation of HR can be altered by health status, training and over-training conditions [2], mental stress and other anxiety forms, cardiac disorders, as well as other critical illnesses or injuries [3]. Therefore, the analysis of HR and its variability (HRV) provide quantitative information on the modulation of cardiac parasympathetic and sympathetic nerve inputs, and consequently constitutes one of the major tools to evaluate the health conditions of a subject, to improve the training and recovery of athletes, or on the medical side, to diagnose important diseases such as autonomous neuropathy, cardiac arrhythmia, or infarction, etc.

There are different techniques to non-invasively monitor HR. The following section summarizes the most popular ones; namely, the techniques based on bio-potential, electric-acoustic, ultrasound, and bio-electrical measurements. The novel, but nonetheless promising, approach based on PPG is finally introduced.

Electrocardiography (ECG) is a bio-potential technique aiming at monitoring the electrical activity of the heart, and constitutes the gold-standard technique to monitor HR [4]. In short, the contraction/relaxation cycle of heart cells is associated with periods of electrical depolarization (increasing the potential) and repolarization (decreasing the potential), respectively. These periods induce local electrical dipoles, which generate surface potentials. ECG monitors the resulting surface potentials observed on specific thorax locations. The delay between the depolarization/repolarization of different regions of the heart produces the typical PQRST waves that characterize the ECG signal of a healthy subject (Figure 1).

The HR is habitually expressed in beats per minute (bpm). However, its value is derived from consecutive heartbeat intervals expressed in milliseconds and measured on ECG signals. On ECG signals, heartbeat interval is defined as the time delay between two consecutive heartbeats. It is usually measured as the time delay between two R-wave

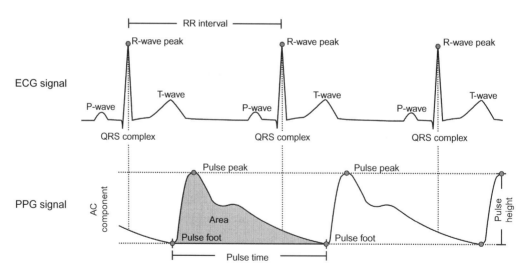

FIGURE 1   Typical synchronized electrocardiogram (ECG) and photoplethysmographic (PPG) waveforms and their respective components.

peaks and commonly referred as RR intervals (see Figure 1). These R-wave peaks, which denote the early depolarization of the ventricles (lower heart chambers), are correlated with the heart contraction [5]. Conventionally, these time series of heartbeat intervals are converted into bpm as follows: $(1/RR\ interval\ values \times 60)$. If desired, these heartbeat interval series are analyzed in the time- and frequency-domains to extract HRV features correlated to the influence of the parasympathetic and sympathetic nervous systems [6]. Physiological complex indicators such as stress level, physical recovery and sleep quality can also be derived from these HRV features [7].

ECG signals are classically acquired by placing silver/silver chloride electrodes on defined anatomical locations, and by connecting them to monitoring platforms. When the ECG monitoring systems are wearable, they are specifically referred as Holter systems. A large variety of alternative devices that monitor averaged HR values (one HR value over a specific time window) or heartbeat intervals based on bio-potential measurements exists. Most of them are employed during sport or daily activities (see [8–10] for examples of commercial devices). Their bio-potential sensors are based on gel, dry, or textile electrode principles [11]. There is also a family of strapless/wireless devices that temporally estimates fingertip bio-potentials using sensors embedded into watches (including Timex's Health Touch Plus [12] and Salutron's SmartHealth [13]).

Phonocardiography (PCG) is an electric-acoustic technology that allows the measurement and analysis of heartbeat sounds. The closing of the atrioventricular valves produces a characteristic sound labeled S1, which is easily identified from PCG signals. These S1 events can be used to estimate heartbeat intervals and consequently to estimate HR. *Echocardiography* is a sonogram (ultrasound measurements) of the heart that monitors the tissue and blood motions related to the pumping action of the heart. It is usually used to visually diagnose specific heart diseases such as heart defect or valve dysfunctions.

Impedance cardiography (ICG) is a bio-electrical technique aimed at monitoring cardio-related displacements of blood within the thorax by injecting electrical currents within the thoracic cavity and measuring resulting changes in the voltages [14]. This technique is usually used to estimate the amount of blood ejection. To our knowledge, no commercialized PCG-, echocardiography-, or ICG-based devices are dedicated to HR monitoring.

Photoplethysmography (PPG) is an optical technology aimed at measuring tissue light propagation changes during cardiac cycle. Its most popular application is monitoring of subject's oxygen saturation (pulse oximetry) [15]. For this purpose, two wavelength lights are used to estimate the arterial blood absorbance, which is linked to the blood oxygenation level. Various commercialized devices are available, including Nonin's [16], Masimo's [17], and Covidien's [18] devices. An extension of this application, namely near-infrared spectroscopy, estimates both oxygenation and deoxygenation of blood on a peripheral scale such as tissue and bones. In the HR monitoring context, the measurement of volumetric changes of a microvascular bed of tissue due to blood flow is the target [19]. This measure brings information on arterial pulsatility content (see Figure 1). Wearable HR monitoring devices using this technology are already available on the market (including Nonin's Onyx 2 [16], MIO's Alpha [20], Basis [21], and Impact Sports Technology's ePulse 2 [22] products).

2. PHOTOPLETHYSMOGRAPHY BASICS

2.1 History

Early in 1936, two independent research groups in New Jersey and Stanford explored the use of a non-invasive optical instrument to assess blood volume changes in rabbit ears [19]. One year later, a first study on the use of PPG to measure blood volume changes in human fingers was published by the team of Alrick Hertzman in St. Louis (US), paving the way toward the introduction of PPG in human monitoring. At that point, the term photoplethysmography was adopted in order to etymologically depict a new technique that could measure changes of volume (*plethysmography*) by optical means (*photo*).

For several decades PPG technology was restrained to physiological studies, kept apart from actual clinical practices. The fact that bulky light sources and sophisticated processing/visualization means were required to acquire and interpret the PPG signals limited wider regular use [19].

It was not until the later appearance of light-emitting diode (LED) technology in 1962 that PPG techniques raised the enthusiasm of a new generation of researchers. The fact that a PPG optical setting could be simplified to a simple LED and a photodetector opened the door to dozens of out-of-lab applications. In particular, the introduction of PPG in clinical routine was undeniably triggered by the development of a so-called pulse oximeter in 1972 by a team of engineers at Nihon Kohden labs [23]. Pulse oximetry was created as a non-invasive spectrometric technique that could provide first-ever real-time estimates on arterial blood gas content by simply placing an optical probe around the fingertip. Since then, pulse oximeters have penetrated many single operating theaters, intensive care units, and practitioners' offices, creating a worldwide market of over a billion dollars. While the original dual-wavelength pulse oximeters of Nihon Kohden provided estimates of arterial oxygen saturation (SpO_2), Masimo Corporation recursively improved the pulse oximetry technology by introducing the concept of perfusion index (a PPG-derived estimate of arterial pulsatility) in 1995, and the concept of pleth variability index (a PPG-derived estimate pulse pressure variation) in 2007. Finally, in 2011, Masimo commercialized the first multi-wavelength pulse oximeter providing simultaneous estimates of arterial saturation on oxygen, carboxyhemoglobin, and methemoglobin.

The introduction of pulse oximetry in clinical routine is undoubtedly associated with the generalized acceptance of PPG as a non-invasive monitoring technique being low cost, unobtrusive, and easy to use. In particular, during the past decades, PPG-derived approaches have been investigated for the assessment of parameters such as HR [24], blood pressure via volume unloading techniques [25], blood pressure via pulse transit time techniques [26], and endothelial dysfunction, among others. However, until very recently, its application to ambulatory HR monitoring was limited due to PPG's sensitivity to movement artifacts.

2.2 Measurement Principles

PPG technique relies on illuminating a living tissue with a light beam, capturing a portion of the light that has propagated through the living tissue, and analyzing said captured light, depicting functional or structural information on the tissue. The attenuation of light, from the light beam (source) to the photodetector (signal), is typically modeled by the Beer-Lambert law. This law states that in a homogeneous medium, light intensity decays exponentially as a function of path length (l) and light absorption coefficient (α) corresponding to medium properties at a specific wavelength. Accordingly, and assuming the intensity of an injected monochromatic light beam being I_0, one expects the intensity of the transmitted light through the medium to be

$$I = I_0 e^{-\alpha l} \tag{1}$$

The properties of the Beer-Lambert law are valid if more than one substance absorbs light in the medium or if a succession of several media is foreseen. In both cases, each absorber contributes to the total absorbance, as a sum of the individual absorbance. The Beer-Lambert law suggests that the sum of the transmitted and absorbed light is equal to the incident light (Figure 2). Reflection at medium surface as well as other physical processes (e.g., light scattering) are not contemplated by this model. The Beer-Lambert law helps in understanding the absorbance of light traveling through homogeneous layers. However, the blood and other biological tissues are not homogeneous, quite contrary, and therefore absorption of light passing through is not simply proportional to the concentration of hemoglobin and to the optical path length. Blood is an inhomogeneous liquid exhibiting a nonlinear absorbance of light. Absorbance and scattering varies during the cardiac cycle with respect to the orientation of red blood cells during the contraction and relaxation periods of the heart. Absorbance is increased because of light reflection at the skin surface and multiple scattering effects, causing the deviation of the light beam from its initial direction. Furthermore, skin and other tissues are in homogeneous, and the variation of their structures and shapes (mainly due to movement) cause complex changes in the light reflection and absorption.

A living tissue can be modeled as a concatenation of several media, each one being characterized by a different path length and light absorption coefficient. Assuming now

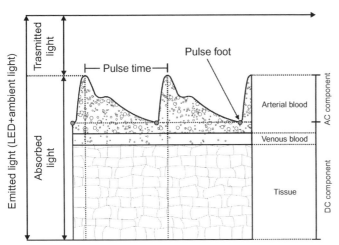

FIGURE 2 Simplified representation of the components of the PPG signal. The AC component due to pulsating arterial blood absorption and the DC component due to a sum of arterial blood, venous blood, and other tissues are displayed *(adapted from [15])*.

that at least one of the illuminated media represents an artery or a vein, each time the heart is beating, a blood pressure pulse is generated and propagates in this blood vessel. When a local increase of the blood pressure occurs, it modifies both the geometry (due to volume change) and the properties (due to changes in blood composition and concentration) of the medium representing the blood vessel. This results in an increase of the light absorption and an attenuation of the transmitted light intensity. The volumetric changes of venous and arterial blood highly contribute to the observed PPG signal variations. These variations are commonly divided into two components universally referred to as AC and DC components (see Figure 2). This nomenclature derives from the electrical engineering domain, where AC indicates a periodically varying level of voltage and DC a static level of voltage. Similarly, in PPG signals, AC refers to the pulsatile arterial blood, while DC refers to the "constant" light absorption due to tissue, venous blood, and diastolic volume of the arterial blood. In reality, the DC component is not constant but varies slowly, typically over several heartbeats. The main factors affecting the DC fluctuations are respiratory and vasomotor activities, and thermoregulation (as described in a later section).

The transmitted light captured by the photodetector might come from two different modes or pathways, as shown in Figure 3. In "transmission" mode, the tissue is illuminated at one side and the light transmitted through it is gathered at the other side. Unfortunately, not all body locations are prone to be monitored via transmission PPG measurements. When aiming at performing PPG analysis at body locations such as the forehead, the sternum, or the ankle, the emitted light is completely absorbed before reaching the opposite side of the body. In these conditions, an alternative operational configuration is available: the "reflectance" mode. In reflectance, the light source is placed next to the detector onto the skin surface, and the predominant light interaction is that of scattering.

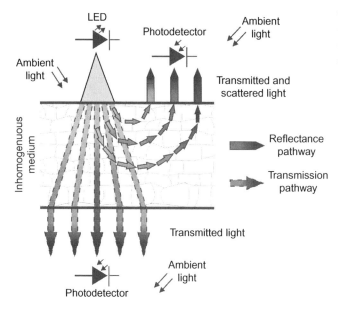

FIGURE 3 Transmission versus reflectance light modes: the roles of absorption and scattering mechanisms *(adapted from [15])*.

The wavelength of the injected light beam is also of paramount importance in the interaction of light and tissues. Each tissue constituent exhibits a specific optical behavior when traversed by a precise wavelength. The absorption spectrum shown in Figure 4 represents the optical behavior in terms of coefficient of absorption/extinction of a particular molecule with respect to light wavelength. The main constituent of tissue, namely water (H_2O), depicts in its absorption spectrum a window that allows wavelengths shorter than 950 nm to be transmitted more efficiently. Melanin is another constituent of tissue that strongly absorbs light wavelengths shorter than 500 nm, and its skin concentration depends on skin pigmentation. Hemoglobin (Hb) is the principal constituent of blood, whose absorbing characteristics change with its chemical binding. Hb molecules that are not able to bind reversibly with molecular oxygen (O_2) are called dysfunctional hemoglobin (e.g., methemoglobin, carboxyhemoglobin, and sulfhemoglobin). Functional hemoglobin is called oxyhemoglobin (HbO_2) if it is fully saturated with oxygen (i.e., carrying four O_2 molecules) and reduced Hb if it is not fully saturated. In healthy persons, most of the Hb molecules are of the functional type.

The choice of the wavelength at which absorbance is monitored is a trade-off and depends on the targeted application, but is usually in the 510 to 920 nm range corresponding to green and infrared lights, respectively. Measurements done on light skins and at normal ambient temperature (around 20°C) have shown that reflected green light has an advantage in terms of AC/DC component ratio over reflected infrared light [27], and might therefore be more suitable for ambulatory monitoring applications. The longer the wavelength is, the deeper the light penetrates, and the scattering effects associated with infrared light in deeper tissues produce a more complex reflected signal. However, in cold ambient conditions, blood microcirculation dramatically decreases and it becomes an

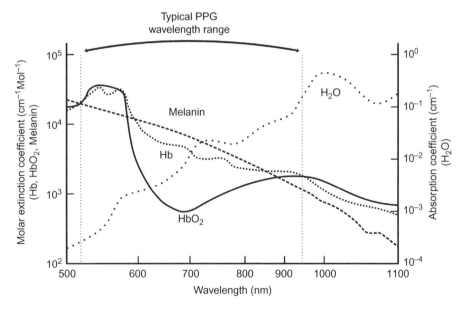

FIGURE 4   Absorption and molar extinction coefficients of main biological tissue constituents (H_2O, Hb, HbO_2, and Melanin) at 500 to 1100 nm window wavelengths.

advantage to reach deeper tissues. The dark skin pigmentation (high melanin concentration) strongly absorbed wavelengths shorter than 650 nm. In these two conditions, infrared light is desired. Therefore, selection of an optimal wavelength for ambulatory HR monitoring depends on targeted applications and usage conditions, and is a compromise between competing factors (vulnerability to artifacts vs. sensitivity during poor skin perfusion).

The measurement of PPG signals requires the use of at least one light-emitting and one light-receiving means. LED is the preferred light source for HR monitoring mainly because of its compactness, low cost, simplicity in use, and limited power consumption within a narrow bandwidth. Common photodetectors (light sensors) used in PPG measurements are photodiodes, but photocells and phototransistors can also be used. The signal coming from the photodetector is pre-amplified, filtered, and digitalized at a fixed sampling frequency, usually of about 25 Hz. For more details regarding the electronic schematic, see [15].

As described earlier, the PPG signal is a combination of DC and AC components as displayed by Figure 1 and Figure 2. The DC component, related to static component, has slow baseline variations and defines the pulse foot. The AC component is characterized by a sequence of PPG waveforms. The shape of the PPG waveform is not unique, but always depicts an inflow phase characterized by a steep rising wave and a runoff phase characterized by a slow decrease of the amplitude. Several parameters can be extracted from the PPG waveform to describe the AC component, and the most commonly used, displayed in Figure 1, are the pulse foot, defined as the lowest amplitude at the beginning of each pulse; the pulse peak, defined as the maximum amplitude between two pulse feet; the pulse height, defined as the amplitude difference between the peak and foot of the pulse; and the pulse time, defined as the time lasted between two following pulse feet (or peak). Other parameters can be found in the literature to evaluate additional physiological parameters, namely pulse area, propagation time, crest time, and inclination time [28].

2.3 Measurement Sites

The most common measurement sites for transmission mode PPG are the fingertip and earlobe, but other measurement sites are also possible, e.g., toes. For the reflectance mode, more variations in the measurement site are possible, including the forearm, the wrist, the ankle, the forehead, and the torso [24,29–31]. One of the practical differences between these modes is that the transmission mode sensors usually use a cuff or a clip to attach the sensor. This causes a probe-dependent increase in transmural pressure, potentially sufficient to collapse the low-pressure venous system, and hence slow changes in the local peripheral blood volume, leading finally to suppression of venous oscillations. The reflectance mode sensors are usually attached to the patient and hence may not apply enough pressure to collapse the veins. Therefore, this mode may be more beneficial if the venous components of PPG variations are of interest [29], and may also be applied to areas that are less affected by vasoconstriction [31]. However, if attachment of the reflectance mode PPG sensor is done by applying pressure to the tissue, it may suffer from the same phenomenon. Optimal measurement site depends, again, on the targeted application. Different anatomical sites differ in terms of density of microvasculature close to skin, skin thickness and structure, tissue characteristics below the skin (e.g., amount and structure of fat tissue, muscles, large vessels, and tendons), and amount of movements during typical physical activities. Furthermore, usability and user acceptance issues play an important role in this selection. For example, arm and forearm are less prone to movement artifacts and variations in environmental temperature as compared to wrist [32], while wrist may be more accepted as a location by the consumers for long-term use.

2.4 Factors Affecting the Quality of Signal

PPG measurements derive from a complex interaction between light propagation and tissue characteristics. In the following section, an overview model of the PPG phenomena is illustrated. As depicted by Figure 5, the PPG signal derived from the PPG phenomenon is determined by three families of factors: sensing, cardiovascular, and biological factors.

PPG measurements are highly determined by the implemented sensing setup: the amount and nature of emitted light, the coupling between the skin, tissue, and the optical probe, and the response of the photodetector will influence the measured PPG signal.

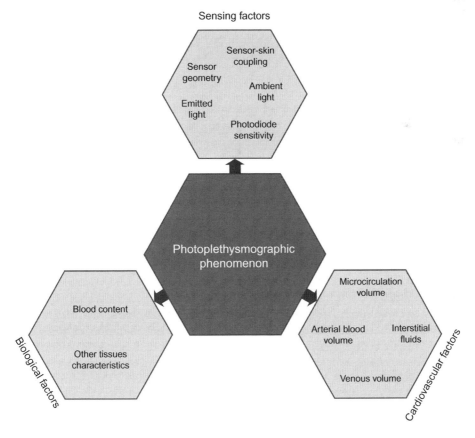

FIGURE 5   An overview model of the PPG phenomena and its three families of factors that influence PPG signal.

Sensor geometry and ergonomics constitute an important factor that will also highly determine the so-called optical shunting effect, which is the amount of direct light traveling from the light emitter toward the photodetector without penetrating the biological tissue. Perturbations due to ambient light on PPG measurements are to be minimized as well by an adequate sensor design. Optimal distance between light emitters and receivers is to be chosen as a trade-off between the desired depths of tissue penetration and the achievable light intensity to be injected into the skin. Larger distances will depict deeper penetration of photons through the scattering process while smaller distances will require smaller light intensities to achieve a reasonable amount of photons reaching the photodiode. Empirical studies have shown that optimal separation distances are in the range of 6 to 10 mm for infrared light [33] and ~2 mm for green light [34]. Reflectance PPG measurements should be avoided in regions where large arteries induce pulsatility in all surrounding tissues [35]. In some body locations, reflectance PPG relying on light absorption (instead of scattering) can be achieved as well. In such operational configurations a bone surface is used to reflect part of the injected photons back.

Concerning the biological factors, the inherent tissue characteristics such as blood content and skin pigmentation have important impacts on the amplitude of the measured PPG signal by modifying the absorption and scattering properties of the tissue. Time-variant cardiovascular factors influenced by body position, age, and cardiovascular stresses also influence the nature and morphology of the measured PPG waveforms, and more precisely, the AC component behavior.

The next section addresses the problem related to PPG motion artifacts in the context of HR monitoring. The different sources of motion artifacts related or not to the sensing factors are detailed and appropriate optomechanical and signal-processing designs are proposed.

2.5 Motion Artifact Minimization and Removal

Due to its measurement principle, the PPG signals are quite sensitive to motion artifacts. The origins of these artifacts are multiples and often occur concomitantly. Three different causes can generate the motion artifacts.

2.5.1 Tissue Modifications Due to Movements

Voluntary or involuntary movements produce modifications of the inner tissues (e.g., motion of the muscles and tendons, and compression or dilatation of the tissues) that change the content of the tissues spanned by the light and thus modify the received signals. The motion-related acceleration or the gravity also affects especially the shape of the soft tissues (e.g., fat) and changes the distribution of the fluids in the tissue due to inertial forces. These factors result in changes along the optical path and modify the received optical signals. These modifications depend on the location of the optical probes on the body (e.g., an ear-located sensor is less prone to be affected by such artifacts than a wrist-located sensor) as well as the mechanism for how the sensor is attached to the skin (sensor pressure, mechanical support provided by the sensor housing, and possible strap, etc.).

2.5.2 *Relative Motion of the Sensor-Skin Interface*

The optical probe is attached to the skin via some binding (e.g., clutch, strap, adhesive media, and clothes). This link is not perfectly rigid and operates therefore as a mass-spring system. A local or global movement of the body generates accelerations that may produce a displacement of the sensor relative to the skin surface. This displacement changes the optical path of light and, as the tissues are generally not homogeneous, modifies the optical signals.

2.5.3 *Changes in the Pressure between the Optical Probe and the Skin*

The pressure applied by the probe on the skin surface modulates the amplitude of the received signals. As the skin surface is not perfectly flat, an initial increase of the pressure results in an augmentation of the pulsating component of the PPG due to an improvement of the optical interface between the probe and the skin. When the applied pressure exceeds some threshold value the amplitude of the pulsating component decreases due to squashing of blood vessels. The accelerations produced by motion induce variations in the skin-probe pressure, and may cause re-distribution of the fluids within the tissue and result in modulation of the optical-signal amplitude and waveform. The importance of this type of perturbation is directly related to the mass of the probe and to the mechanical properties of the body-attachment solution.

In order to cope with the motion-related artifacts, it is mandatory to model the effects of motion on the optical signals. Most of the actual methods postulate that motion artifacts are additives. This assumption is seductive because it allows formulating simple signal-processing techniques to reduce or to remove the artifact components from PPG signals. However, its validity is limited. The combination of multiplicative and additive models, possibly with nonlinear relations, is certainly more representative of the real relations that exist between the motion and the motion artifacts observed in the PPG signals (see Figure 6). In this context, the observed signal $y(t)$ is represented by the sum of the pulse component $s(t)$ and a weighted multiplication of both pulse and motion components $s(t)$ and $m(t)$ as described by the following equation: $y(t) = (1 + \alpha \cdots m(t)) \cdots s(t)$, and the observed frequency components Y are described by the sum of pulse components S and the weighted convolution of pulse and motion components S and M as described by the following equation: $Y = S + \alpha \cdots S * M$.

Typical use cases of an HR monitoring system allow classifying the motion artifacts into three different categories. The first category is related to rhythmical motions generally produced during endurance activities such as walking, running, and biking. Under such circumstances, motion artifacts behave as a stationary process and the PPG signals can be enhanced by signal-processing techniques. The two other categories are related to non-rhythmical motions and are categorized as intermittent or continuous. These two categories of motion artifacts are generally more difficult to cope with, especially the continuous one. The successful estimation of the HR with optical systems during activity is dependent on two main factors: the optomechanical design of the probe and the signal-processing algorithms.

2.6 Optomechanical Design

The design of the optical probe has to take into account different constraints in order to minimize the effects of the motion on the PPG signals. To reduce motion artifacts

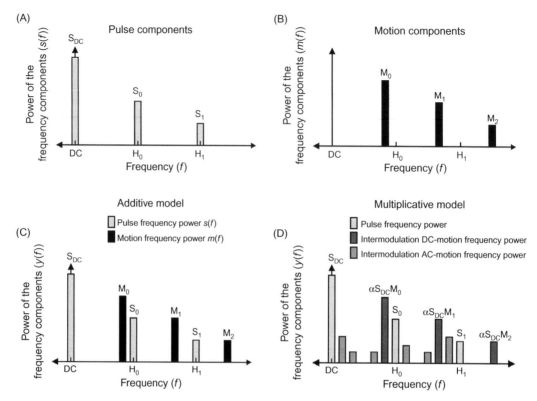

FIGURE 6   (a) Frequency components S of the optical pulsating signal $s(t)$. (b) Frequency components M of the optical motion-artifact signal $m(t)$. (c) Frequency components Y of the observed signal $y(t)$ following an additive model. (d) Frequency components Y of the observed signal $y(t)$ following a multiplicative model.

produced by the inertial forces induced by the motion (relative skin-sensor displacements and pressure variations), the mass of the measurement system has to be minimized. The probe can also be designed such that friction force takes place between the sensor and the skin surface to reduce the relative displacement. The attachment of the sensor to the body has to avoid insufficient stiffness that increases its sensitivity to motion artifacts. The attachment also has to provide an adequate pressure of the sensor on the skin surface resulting in an optimal optical interface. The probe fixation should also ensure that the applied pressure is not excessive, which could possibly result in blood-vessel clutching and discomfort. Finally, the distribution of the blood vessels in the tissues is generally non-uniform, and small displacements of the optical emitter-receiver pair can result in drastic changes in the amplitude of the pulsating component (AC component) and therefore increase sensitivity of the optical signal to motion-artifact corruption. The optomechanical design of the probe is therefore of tremendous importance concerning the sensitivity to motion artifacts.

2.7 Dedicated Signal Processing

No optomechanical design alone is able to reduce the sensitivity to motion artifacts to a suitable level: a robust estimation of the HR requires as well the implementation of a signal processing algorithm. Typically, the processing scheme of the optical signals involves three main steps, namely the enhancement, the spectral estimation, and the robust estimation of HR.

The PPG signal enhancement consists of suppressing, or at least reducing, the motion artifacts in the observed signals while preserving the pulsating component. When only one PPG signal is available without other signals of other sources, the possibility of enhancement is limited. Under such circumstances, the enhancement is restricted to the acceptation of uncorrupted segments and to the removal of motion-corrupted segments. This selection is generally obtained by the analysis of the morphological properties of the signal such as the amplitude and its stability.

In order to facilitate the removal of the motion artifacts in the PPG signal, it is possible to use extra signals that contain information on the skin and/or sensor motions. A possible solution is to obtain a motion reference signal by using an extra light emitter at a different wavelength. The wavelength has to be selected such that its sensibility to the optical attenuation of the blood is minimal (outside of the typical PPG wavelength range). This will ensure that the extra signal mostly contains motion artifacts. For optimal efficiency, the probe has to be designed such that the pulse-measurement and the motion-measurement wavelengths share the same optical path. The integration of pressure sensors in the probe is also possible. Another possibility consists of the addition of 3D accelerometers to combine optical and direct sensor-motion signals.

Different algorithmic approaches combine the optical and the motion signals to reduce the artifacts. The simplest approach consists of using the motion reference signal to discard motion-corrupted segments. More sophisticated approaches are based on the assumption that motion artifacts follow an additive model, as described before. Different approaches have been developed to identify the relation existing between the motion reference signal and the motion component present in the pulse signal. The first approach consists of estimating the spectrum of both optical and motion signals, then identifying the spectral peaks that are related to the motion present in both signals, and keeping the non-motion peaks present only in the optical signal by filtering processes. The main limitation of these approaches is that it can only be used when the motion artifacts are rhythmical.

Another approach uses an adaptive filter [36] to find the model parameters that map the motion signal on the motion components present in the optical signal. At the end, these motion components are subtracted from the optical signals [37]. This approach is more robust because it does not require rhythmical motion. Practically, it works on rhythmical and also on limited non-rhythmical motion conditions.

Finally, an efficient method that is suitable for rhythmical motion consists of the estimation of the fundamental frequency of the motion from a reference signal and the use of notch filters centered on the harmonics of this frequency to remove the artifacts in the pulse signals [38].

Different HR estimation approaches might be applied to PPG signals once the optical signals have been enhanced. These methods are divided into two categories: the ones operating in the frequency or the ones in the time domains. Frequency domain approaches estimate the spectral density of the signal using either a non-parametric (fast Fourier, discrete cosine, or wavelet transforms) or parametric methods (autoregressive model). In order to ensure a rejection of the erroneous estimations, every measurement is associated with an index of reliability. For the frequency domain approaches, the reliability index is generally the entropy value of the spectrum (high reliability when only one dominant peak is present in the spectrum and low reliability when the spectrum contains several possible dominant frequencies).

The approaches operating in the time domain are divided into two categories that are based on the detection of events related to the heartbeat or on the tracking of the instantaneous dominant frequency. Event detection-based approaches consist of the detection of characteristic events related to heartbeats in the PPG signals, typically maxima, minima, or zero-crossings. The temporal intervals between these events expressing heartbeat intervals are used to estimate the HR. The reliability of the measurement is estimated from the dispersion of the values of these intervals. Finally, adaptive frequency tracking approaches are based on a model whose parameters are adapted to track the dominant frequency observed in the PPG signals. With the appropriate filter setting, the tracked frequency expressed averaged HR values. The reliability is estimated from the ratio of the energy present at the dominant frequency over the energy of the whole signal.

Finally, the cardiac frequency values resulting from the spectral analysis or the event temporal intervals are processed to obtain the current HR values usually expressed in bpm. Different formulations such as Bayesian estimation, reliability-dependent autoregressive estimation, outlier rejection, and a model of the dynamics of the cardiovascular system can be used to obtain the final HR estimate.

3. APPLICATIONS

3.1 Sport and Fitness

Monitoring HR during exercise is especially useful in endurance training, professional training planning, or fitness workout. Maximal oxygen uptake and energy expenditure can be accurately estimated from HR measurements [39]. HR monitoring in real time during training allows a user to control his training intensity accurately to optimize training and to avoid too low or too high training loads. Also, training effect, i.e., excess post-exercise oxygen consumption, can be accurately estimated from HR recorded during the training session. Furthermore, use of HR monitoring devices adds motivation for users to exercise [40].

The original idea for wearable HR monitoring dedicated to sports came from cross-country skiing training in the late 1970s. Professor Seppo Säynäjäkangas from the University of Oulu made a prototype of a wearable HR monitor that used a wired connection. In 1983 the first chest-strap wearable HR monitor was produced by Polar Electro. This device consisted of two parts: a watch receiver and a transmitter on an ECG-based chest strap [41]. Today, ECG-based chest straps are widely used for monitoring HR during sports, and their annual sales exceed 10 million pieces worldwide. Chest-strap-based HR monitors provide relatively accurate monitoring of HR, but they suffer from reduced comfort from the chest strap, especially for female users. In addition, their reliability may be compromised with dry skin, dirty electrodes, or poor strap placement.

Optical monitoring of HR has been recently introduced as an alternative to overcome especially the usability and user acceptance-related challenges in chest-strap HR monitors. Optical monitoring of HR during sports may be done from different body positions, including ear [42,43], forearm [44], and wrist [20]. Most of the commercialized devices use a green light source (one or several LEDs) combined with a single photodetector. It has been shown that green light has a better signal-to-noise ratio for AC components during movement than longer wavelengths [17]. Table 1 gives the characteristics of two typical devices designed for sports based on optical HR monitoring, namely the Scosche's RHYTHM and the Mio's Alpha (Figure 7).

The reliability of the currently available devices has not been studied widely. Figure 8 shows an example of two optical HR monitors (forearm-located Scosche's RHYTHM and wrist-located Mio Alpha) as compared with reference HR (ECG-based chest strap) during walking and running on a treadmill with increasing load. The examples show that these devices may provide high-quality HR monitoring during exercise, but significant errors are also possible. Possible reasons for poor performance in some cases may include poor device attachment (sensor-skin contact), poor skin perfusion, or algorithm failures.

Figure 9 shows an example of optical HR monitoring during cross-country skiing with the same ECG-based chest strap reference and PPG-based devices (forearm-located Scosche's RHYTHM and wrist-located Mio Alpha). The device performance during this 48 minutes of cross-country skiing results in RMSE (normalized correlation) values of 13.12 bpm (0.76) for the forearm-located device and 18.07 bpm (0.47) for the wrist-located device. Both optical HR monitors fail at the beginning of the exercise due to low skin temperature, but work reliably once skin temperature and thereafter blood perfusion close to skin increased.

Current solutions are mostly based on green light with the advantage of providing robustness against motion artifacts and the disadvantage of being sensitive to poor skin perfusion. As displayed by Figure 8 and Figure 9, these solutions do not yet reach the reliability of chest-strap HR monitoring in a wide range of conditions. Cold conditions especially appear challenging for the PPG HR technique (see Figure 9). Furthermore, it is likely

TABLE 1 Technological Description and Features of Representative Devices for *ad hoc* Performance Tests

Features	Scosche's RHYTHM	Mio's Alpha
Source light color and #LEDs	2 infrared LEDs	2 green LEDs
Number of photo detectors	1	1
Location	Forearm	Wrist
Band type	Textile	Plastic rubber
Wireless connectivity	Bluetooth	Bluetooth/ANT +
Data storing	Without memory	Without memory
Display	No display	Dot − Matrix LCD

Available devices differ in terms of sensor location, exact design of the sensor element (wavelength(s) used, number of LEDs and photodetectors, LED − PD spacing), and algorithms to extract HR from motion-disturbed signal.

FIGURE 7   (a) Scosche's RHYTHM device located at the forearm with a view of the light sources (extremities) and photodetector (middle) design (dashed line). (b) Mio's Alpha device located at the wrist with a view of the light sources (extremities) and photodetector (middle) design (dashed line). Scosche uses two infrared LEDs and Mio uses two green LEDs.

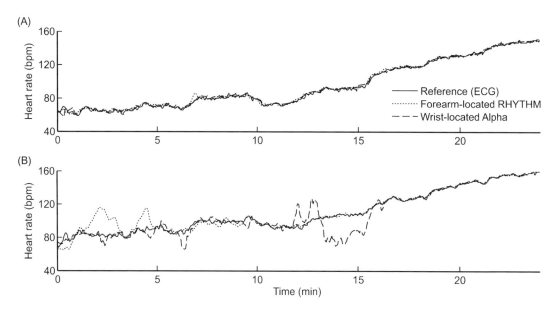

FIGURE 8   Optical HR monitoring from forearm and wrist locations during treadmill walking and running with increasing load, as compared to chest-strap ECG system on healthy volunteers. (a) Example of good quality experiment (RMSE values of 1.79 and 1.73; correlation values of 0.99 and 0.99, respectively). (b) Example of poor quality experiment (RMSE values of 7.27 and 9.60 and correlation values of 0.96 and 0.94, respectively).

that any sport that includes vigorous movements of the body part where the sensor is attached will be challenging for optical HR monitoring (e.g., racket games for wrist- or forearm-located sensors). However, if these challenges could be overcome, optical HR monitoring during sports could become an attractive alternative to a broad range of fitness consumers.

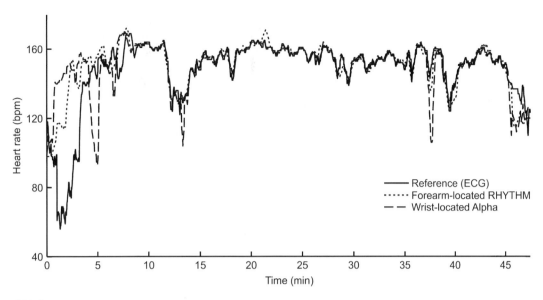

FIGURE 9 Optical HR monitoring from forearm and wrist locations during cross-country skiing as compared to chest-strap ECG system (RMSE values of 13.12 and 18.08 and correlation values of 0.76 and 0.47, respectively).

3.2 Daily Life

Consumer interest in wearable sensors beyond sports is increasing rapidly. Monitoring of movement (acceleration) during daily life allows quantifying patterns and amount of physical activity, step count, and rough estimates of energy expenditure, while monitoring during sleep allows estimation of sleep duration and to some extent sleep quality. Typical sensor solutions are accelerometry-based devices that are worn on the wrist or trunk. HR monitoring during daily life would allow more accurate estimation of physical activity and energy expenditure [45], but also physiological stress and recovery [46]. However, chest-strap or electrode-based solutions are not widely acceptable for long-term use.

Optical HR monitoring is potentially more acceptable to users due to its potentially better wearability and unobtrusiveness. However, reaching continuous reliable optical HR monitoring during daily life is challenging. While relatively reliable HR monitoring during sports may be achieved with current solutions (see previous section) these solutions may not be directly applicable to daily life. Comfortable, snug, and fit-to-the-skin sensors are essential for reliable monitoring as poor sensor contact drastically increases motion artifacts. However, long-term continuous sensor and strap contact require a solution that allows skin ventilation and does not compress the vascular bed. For example, optical HR monitoring designs targeting sport applications may not be acceptable for daily-life users. Finally, optical sensing requires significant power due to inherent power consumption of LEDs and required circuitry, and significant attention needs to be paid to solutions to extend battery life beyond current solutions to reach full 24/7 monitoring.

Today, some solutions for extending optical heart rate monitoring to daily life exist [21]. However, continuous reliable monitoring of heart rate during daily life has not yet been reached.

3.3 HRV Applications

The analysis of HR and its variability (HRV) has been the subject of numerous clinical studies concerning cardiological diseases, sleep analysis and apnea, physiologic phenomena, pharmacological responses, and risk stratification. However, a clinical consensus has been reached only in two scenarios: (1) the prediction of risk after acute myocardial infraction and (2) early detection of diabetic neuropathy. In patients following acute myocardial infraction, depressed HRV has proven to be a good predictor of mortality and arrhythmic complications, and independent of other established factors [47]. As for the assessment of diabetic autonomic neuropathy, the short- and long-term HRV analyses have proven to be accurate in its early detection [48]. Other promising studies have also investigated the potential of HRV in other cardiological diseases such as hypertension [49], congestive heart failure (insufficient pump action) [50–52], arrhythmias [53–55], and sudden death or cardiac arrest [56,57]. All of these studies are based on gold-standard HRV features. In order to extract these features, the techniques used to monitor HR have to be able to detect and record accurate heartbeat time locations (e.g., the timing of R wave peaks from ECG signals).

Although PPG monitoring devices are medically accepted as a means to assess average heart rate values, little is known about the reliability of PPG signals to extract these relevant HRV features [58,59]. However, one thing is certain, the methodology associated with the PPG post-processing is important. Successive heartbeat detection, motion artifact correction, normal beat detection, and uniformly distributed heartbeat interval processes have to be applied before extracting the desired HRV features.

Figure 10 shows an example of the correlation between heartbeats estimated from ECG and PPG signals. In this example, the heartbeat detection algorithm was obtained from a multi-channel first derivative signal in which the superior envelope was estimated. An adaptive threshold approach was applied to this superior envelope to detect heartbeat locations. In order to analyze the behavior of the autonomous nervous system, any ectopic, premature beats, or outliers were also rejected from the heartbeat time series. The entire post-processing approach is described in detail by Arberet et al. [60]. This example shows that with a high-quality PPG signal it is possible to extract HRV time series that closely resemble that extracted from ECG.

An interesting study [24] evaluated the heartbeat detection performance of a reflectance PPG sensor integrated into a wrist device in the context of sleep monitoring of subjects affected by chronic mountain sickness and sleep-disordered breathing. Figure 11 provides a Bland-Altman plot summary of the comparison between ECG- and PPG-estimated heartbeat time series from the analyzed dataset (N = 26 subject, $\approx 930'000$ heartbeats). The overall mean absolute error and its standard deviation ($\mu \pm \sigma$) when estimating RR intervals are 0.05 ± 17.96 ms.

Once the proper algorithm is applied to detect the normal heartbeat locations, a uniformly resampled process must be applied to the heartbeat time series before any

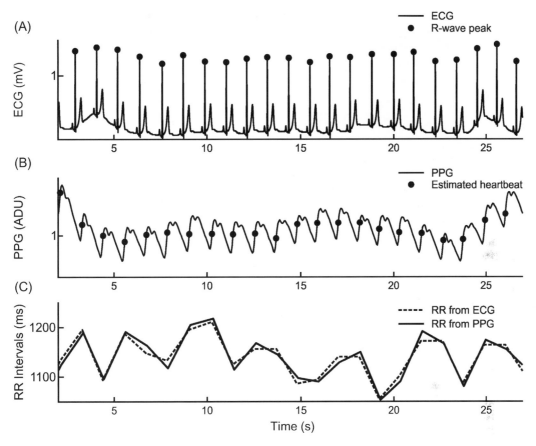

FIGURE 10 Illustration of typical R-wave peak detection (cardiac muscle contraction) observed from ECG signals (a), the corresponding heartbeats detected on PPG signals (b) and the resulting heartbeat intervals from both origins (b) [60].

time- and frequency-domain HRV feature extraction; state-of-the-art guidelines suggest resampling at 4 Hz [4]. Then, the gold-standard HRV features can be computed from the resulting uniformly resampled RR signals. These HRV features might be estimated over 5-minute or 24-hour segments, depending on the physiological behavior in observation. In the time domain, the most common variable to calculate is the standard deviation of the heartbeat intervals, labeled SDNN, which stands for standard deviation (SD) of the normal beat (NN) intervals. The standard deviation of differences between adjacent heartbeat intervals, labeled SDSD, which stands for standard deviation (SD) of the heartbeat standard deviation (SD) values, is another gold-standard feature generally used in HRV analysis. Figure 12 displays an example of the evolution of these two features estimated from ECG and PPG signals.

In the frequency domain, various non-parametric and parametric methods exist to estimate the frequency components [4], the most popular one being the fitting of an

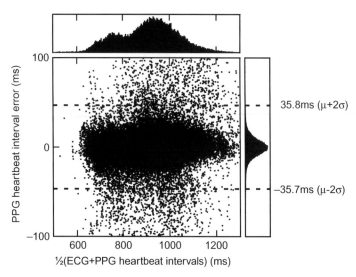

FIGURE 11 Bland-Altman plot comparing reference ECG-derived heartbeat intervals (RR intervals as measured by ECG) to associated PPG-derived heartbeat intervals (as estimated by the PPG-wrist device). The entire dataset contains a total of $N \approx 933k$ heartbeats from 26 subjects affected by chronic mountain sickness and sleep-disorder breathing. The overall error μ is 0.05 ± 17.96 ms and its standard deviation σ is 18 ± 2 ms [24].

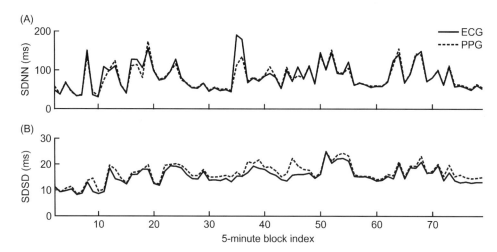

FIGURE 12 Time-domain HRV features (SDNN and SDSD) estimated from ECG and PPG signals. The correlation (normalized absolute error values) are 0.93 (0.15) and 0.95 (0.11), respectively [60].

auto-regressive model of a given order (usually 12) to the heartbeat interval time series. A spectral analysis is then applied to estimate the powers located in the very low, low and high frequencies (0.003-0.04, 0.04−0.15, 0.15−0.4 Hz, respectively). The ratio between low- and high-frequency powers is usually added to provide information on the relationship between sympathetic (low) and parasympathetic (high) nerve activities. Figure 13 displays an example of the evolution of the described four frequency-domain HRV features estimated from ECG and PPG signals. One can observe that the general behavior of the HRV features estimated from PPG signals is well correlated to the ones estimated from ECG signals.

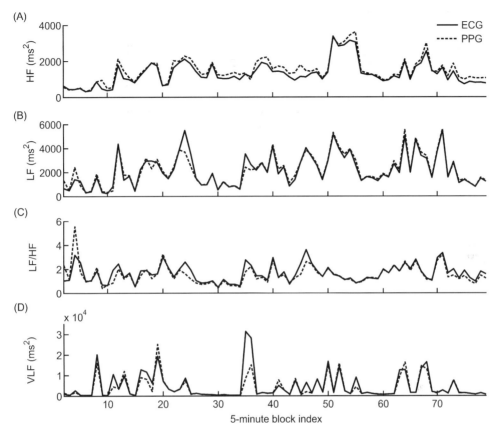

FIGURE 13 Frequency-domain HRV features (HF, LF, LF/HF, and VLF) estimated from ECG and PPG signals. The correlation (normalized absolute error) values are 0.98 (0.16), 0.96 (0.14), 0.84 (0.25), and 0.86 (0.43), respectively [60].

Clinical applications of PPG-based HR monitoring devices are at the embryonic stage. Most of the HR-based clinical studies based on Holter and other wearable HR monitoring devices could be reproduced using a less cumbersome, inconspicuous, and more daily-life long-term appropriate PPG-wrist device. The advantage of such devices is also related to the accuracy of the diagnostic. Cumbersome systems affect the behavior of the subjects, especially during night monitoring. Therefore, a light and well-integrated PPG device might in the future provide novel insights into long-term cardiovascular regulation mechanisms.

4. CONCLUSION AND OUTLOOK

This chapter aimed at providing an overview of the current developments in the field of optical HR monitoring. Starting from the basics of the PPG technology, this chapter highlighted the importance of fully understanding the PPG opto-electrical phenomenon. Because of the complex interactions between mechanical sensor parts, optical properties of the sensors, and particularities of living tissues, the design of optical HR monitors should be driven by a thorough analysis of the specific PPG configuration to be implemented. Different trade-offs are to be expected, for instance, when targeting a long-term wellness monitor to be used during sleep, than when targeting an HR monitor for short-term use during a marathon or sport. The optical-electrical-mechanical design of any PPG-based sensor is thus a critical issue that might compromise the usability of the HR monitor.

A second important aspect covered by this chapter is the design of a signal-processing strategy that allows deriving the desired health indexes from the raw PPG signals. The take-home-message of this chapter is that there is no universal algorithm to be implemented within an optical HR monitor, and deep understanding of the underlying physiological problem will guide the developer toward the optimal algorithmic configuration.

Finally, this chapter covered the most recent advances of optical HR monitoring in two different fields: from the monitoring of athletes toward hospitalized patients. The goal of the provided material was to demonstrate that, supported by an optimized design, the PPG technology can cover a very large spectrum of applications going from very precise HRV analysis of sleep to HR monitoring during cross-country skiing.

What is preparing the future? Imagination. The technological floor is now set in the optical heart rate monitoring domain: the principal elements of the PPG puzzle are gradually becoming reality and how to combine them in order to meet particular demands is currently the main challenge. The explosion of smartphones and smartwatches offers an excellent perspective to these techniques: optical monitoring is non-obtrusive and comfortable, and can be easily integrated in such devices. Smart textiles, integrating functional garments and patches, are other promising technologies where optical HR could play an important role. And last but not least, contactless sensors, using a modified PPG-technology, could monitor computer users via webcams.

In conclusion, optical HR monitors have the potential to become one of the central technologies to support the development of the twenty-first century's health and well-being assessment revolution. The technology is there; the implementation is in your field.

NOMENCLATURE

AC Alternating current
BPM Beats per minute
DC Direct current
ECG Electrocardiogram
H_2O Water
Hb Hemoglobin
HbO_2 Oxyhemoglobin
HR Heart rate
HRV Heart rate variability
ICG Impedance cardiography
LED light-emitting diode
ms Millisecond
O_2 Oxygen

PCG Phonocardiography
PPG Photoplethysmography
RMSE Root-mean-square error

Acknowledgments

This chapter was made possible by grants from the Wilsdorf Foundation and the Swiss National Science Foundation. The authors would like to thanks the collaborators of CHUV (Claudio Sartori *et al.*) responsible for data acquisition. The authors would also like to thank all CSEM's collaborators who were involved in the design and development of the wrist-located PPG device.

References

[1] P. Brodal, The Central Nervous System: Structure and Function, third ed, Oxford University Press, Inc, New York, 2004.

[2] A.E. Aubert, B. Seps, F. Beckers, Heart rate variability in athletes, Sports Med. 33 (2003) 889–919.

[3] M. Malik, J. Bigger, A. Camm, R. Kleiger, A. Malliani, A. Moss, et al., Heart rate variability: Standards of measurement, physiological interpretation, and clinical use, Eur. Heart J. 17 (1996) 354–381.

[4] M.H. Crawford, S.J. Bernstein, P.C. Deedwania, J.P. DiMarco, K.J. Ferrick, A. Garson, et al., ACC/AHA guidelines for ambulatory electrocardiography: executive summary and recommendations: a report of the American College of Cardiology/American Heart Association task force on practice guidelines, Circulation 100 (1999) 886–893.

[5] G.-X. Yan, R.S. Lankipalli, J.F. Burke, S. Musco, P.R. Kowey, Ventricular repolarization components on the electrocardiogram, J. Am. Coll. Cardiol. 42 (2003) 401–409.

[6] J. Sztajzel, Heart rate variability: a noninvasive electrocardiographic method to measure the autonomic nervous system, Swiss. Med. wkly. 134 (2004) 514–522.

[7] M.H. Bonnet, D.L. Arand, Heart rate variability: sleep stage, time of night, and arousal influences, Electroencephalogr. Clin. Neurophysiol. 102 (1997) 390–396.

[8] <http://www.polar.com>, (Accessed Last: 23.06.14).

[9] <http://www.suunto.com>, (Accessed Last: 23.06.14).

[10] <http://www.garmin.com>, (Accessed Last: 23.06.14).

[11] J.G. Webster, The measurement, instrumentation and sensors6 handbook, first ed., CRC Press, Inc., Boca Raton, 1999.

[12] Timex Group USA Inc. (2013). Heart Rate Monitor | Global Timex. [Online]. Available: <http://global.timex.com/collections/heart-rate-monitor>. (Last Accessed: 07.07.14).

[13] Salutron Inc. (2013). Smart Health-Salutron. [Online]. Available: <http://salutron.com/smart-health>. (Last Accessed: 07.07.14).

[14] M.J.E. Parry, J. McFetridge-Durdle, Ambulatory Impedance Cardiography, Nurs. Res. 55 (2006) 283–291.

[15] J.G. Webster, Design of Pulse Oximeters, first ed., IOP Publishing Ltd, London, 1997.

[16] Nonin Medical Inc. (2013). Nonin - Pulse Oximeter (Sp02) Monitoring Solutions. [Online]. Available: <http://www.nonin.com/pulseoximetry>. (Last Accessed: 07.07.14).

[17] Masimo Corporation. (2013). Masimo - close to the heart. [Online]. Available: <http://www.masimo.com>. (Last Accessed: 07.07.14).

[18] Covidien. (2013). Nellcor Pulse Oximetry Monitoring from Covidien. [Online]. Available: <http://www.covidien.com>. (Last Accessed: 07.07.14).

[19] J. Allen, Photoplethysmography and its application in clinical physiological measurement, Physiol. Meas. 28 (2007) R1.

[20] Physical Enterprises Inc. (2013). MIO Global. [Online]. Available: <http://www.mioglobal.com>. (Last Accessed: 07.07.14).

[21] BASIS Science Inc. (2013). Basis — health and heart rate monitor for wellness and fitness. [Online]. Available: <http://www.mybasis.com>. (Last Accessed: 07.07.14).

[22] Impact Sports Technologies. (2013). ePulse2. [Online]. Available:<http://www.impactsports.com/epulseii.html>. (Last Accessed: 07.07.14).

[23] J.W. Severinghaus, Takuo Aoyagi: discovery of pulse oximetry, Anesth. Analg. 105 (2007) S1–S4.

[24] P. Renevey, J. Sola, P. Theurillat, M. Bertschi, J. Krauss, A. Daniela, et al., Validation of a wrist monitor for accurate estimation of RR intervals during sleep, Conf. Proc. IEEE Eng. Med. Biolol. Soc. (2013) 5493–5496.

[25] Penaz, J. (1973). Photoelectric measurement of blood pressure, volume and flow in the finger. In proceedings Digest of the 10th international conference on medical and biological engineering, 104.

[26] J. Sola, M. Proenca, D. Ferrario, J.-A. Porchet, A. Falhi, O. Grossenbacher, et al., Non-invasive and non-occlusive blood pressure estimation via a chest sensor, IEEE Trans. Biomed. Eng. (2013).

[27] Y. Maeda, M. Sekine, T. Tamura, A. Moriya, T. Suzuki, K. Kameyama, Comparison of reflected green light and infrared photoplethysmography, Conf. Proc. IEEE Eng. Med. Biol. Soc. (2008) 2270–2272.

[28] G. Natalini, A. Rosano, M.E. Franceschetti, P. Facchetti, A. Bernardini, Variations in arterial blood pressure and photoplethysmography during mechanical ventilation, Anesth. Analg. 103 (2006) 1182–1188.

[29] L. Nilsson, A. Johansson, S. Kalman, Respiration can be monitored by photoplethysmography with high sensitivity and specificity regardless of anaesthesia and ventilatory mode, Acta. Anaesthesiol. Scand. 49 (2005) 1157–1162.

[30] B. Jönsson, C. Laurent, T. Skau, L.-G. Lindberg, A new probe for ankle systolic pressure measurement using photoplethysmography (PPG), Ann. Biomed. Eng. 33 (2005) 232–239.

[31] K.H. Shelley, D. Tamai, D. Jablonka, M. Gesquiere, R.G. Stout, D.G. Silverman, The effect of venous pulsation on the forehead pulse oximeter wave form as a possible source of error in Spo2 calculation, Anesth. Analg. 100 (2005) 743–747.

[32] Y. Maeda, M. Sekine, T. Tamura, Relationship between measurement site and motion artifacts in wearable reflected photoplethysmography, J. Med. Syst. 35 (2011) 969–976.

[33] Y. Mendelson, B.D. Ochs, Noninvasive pulse oximetry utilizing skin reflectance: noninvasive pulse oximetry utilizing skin reflectance, IEEE Trans. Biomed. Eng. 35 (1988) 798–805.

[34] Huang, F., Yuan, P., Lin, K., Chang, H. and Tsai, C. (2011). Analysis of reflectance photoplethysmograph sensors. in proceedings World Academy of Science, Engineering and Technology (59) 1266–1269.

[35] J.L. Reuss, Arterial pulsatility and the modeling of reflectance pulse oximetry, Conf. Proc. IEEE Eng. Med. Biol. Soc. 3 (2003) 1901–1904.

[36] S.S. Haykin, Adaptive Filter Theory, forth ed., Prentice Hall, New Jersey, 2001.

[37] Renevey, P., Vetter, R., Krauss, J., Celka, P., and Depeursinge, Y. (2001). Wrist-located pulse detection using IR signals, activity and nonlinear artifact cancellation. In proceedings 23rd Annual International Conference of the IEEE Engineering in Medicine and Biology Society (3), 3030–3033.

[38] B. Lee, Y. Kee, J. Han, W.J. Yi, Adaptive comb filtering for motion artifact reduction from PPG with a structure of adaptive lattice IIR notch filter, Conf. Proc. IEEE Eng. Med. Biol. Soc. (2011) 7937–7940.

[39] J. Achten, A.E. Jeukendrup, Heart rate monitoring: applications and limitations, Sports Med. 33 (2003) 517–538.

[40] A. Ahtinen, J. Mantyjarvi, J. Hakkila, Using heart rate monitors for personal wellness--the user experience perspective, Conf. Proc. IEEE Eng. Med. Biol. Soc. (2008) 1591–1597.

[41] R.M.T. Laukkanen, P.K. Virtanen, Heart rate monitors: state of the art, J. Sports Sci. 16 (1998) 3–7.

[42] El-Khoury, M., Sola, J., Neuman, V. and Krauss, J. (2007). Portable SpO2 monitor: a fast response approach. in proceedings IEEE International Conference on Portable Information Devices, 1–5.

[43] Valencell Inc. (2013). PerformTek®. [Online]. Available: <http://www.valencell.com>. (Last Accessed: 07.07.14).

[44] Scosche Industries. (2013). RHYTHM - The Best Heart Rate Monitor for iPhone. [Online]. Available: <http://www.scosche.com/rhythm>. (Last Accessed: 07.07.14).

[45] P.G. Montgomery, D.J. Green, N. Etxebarria, D.B. Pyne, P.U. Saunders, C.L. Minahan, Validation of heart rate monitor-based predictions of oxygen uptake and energy expenditure, J. Strength Cond. Res. 23 (2009) 1489–1495.

[46] A. Uusitalo, T. Mets, K. Martinmäki, S. Mauno, U. Kinnunen, H. Rusko, Heart rate variability related to effort at work, Appl. Ergon. 42 (2011) 830–838.

[47] R.E. Kleiger, J.P. Miller, J.T. Bigger, A.J. Moss, Decreased heart rate variability and its association with increased mortality after acute myocardial infarction, Am. J. Cardiol. 59 (1987) 256–262.

[48] F. Bellavere, I. Balzani, G. De Masi, M. Carraro, P. Carenza, C. Cobelli, et al., Power spectral analysis of heart-rate variations improves assessment of diabetic cardiac autonomic neuropathy, Diabetes 41 (1992) 633—640.

[49] S. Guzzetti, S. Dassi, M. Pecis, R. Casati, A.M. Masu, P. Longoni, et al., Altered pattern of circadian neural control of heart period in mild hypertension, J. Hypertens. 9 (1991) 831—838.

[50] G. Casolo, E. Balli, T. Taddei, J. Amuhasi, C. Gori, Decreased spontaneous heart rate variability in congestive heart failure, Am. J. Cardiol. 64 (1989) 1162—1167.

[51] P.F. Binkley, G.J. Haas, R.C. Starling, E. Nunziata, P.A. Hatton, C.V. Leier, et al., Sustained augmentation of parasympathetic tone with angiotensin-converting enzyme inhibition in patients with congestive heart failure, Am. J. Cardiol. 21 (1993) 655—661.

[52] J.N. Townend, J.N. West, M.K. Davies, W.A. Littler, Effect of quinapril on blood pressure and heart rate in congestive heart failure, Am. J. Cardiol. 69 (1992) 1587—1590.

[53] H.V. Huikuri, J.O. Valkama, K.E. Airaksinen, T. Seppänen, K.M. Kessler, J.T. Takkunen, et al., Frequency domain measures of heart rate variability before the onset of nonsustained and sustained ventricular tachycardia in patients with coronary artery disease, Circulation 87 (1993) 1220—1228.

[54] M.G. Tsipouras, D.I. Fotiadis, D. Sideris, An arrhythmia classification system based on the RR-interval signal, Artif. Intell. Med. 33 (2005) 237—250.

[55] N. Al-Rawahi, M.S. Green, Diagnosis of supraventricular tachycardia, J. Assoc. Physicians India 55 (2007) 21—24.

[56] C.M. Dougherty, R.L. Burr, Comparison of heart rate variability in survivors and nonsurvivors of sudden cardiac arrest, Am. J. Cardiol. 70 (1992) 441—448.

[57] A. Algra, J.G. Tijssen, J.R. Roelandt, J. Pool, J. Lubsen, "Heart rate variability from 24-hour electrocardiography and the 2-year risk for sudden death,", Circulation vol. 88 (1993) 180—185.

[58] K.H. Shelley, Photoplethysmography: beyond the calculation of arterial oxygen saturation and heart rate, Anesth. Analg. 105 (2007) S31—S36.

[59] P. Dehkordi, A. Garde, W. Karlen, D. Wensley, J.M. Ansermino, G.A. Dumont, Pulse rate variability compared with heart rate variability in children with and without sleep disordered breathing, Conf. Proc. IEEE Eng. Med. Biol. Soc. (2013) 6563—6566.

[60] Arberet, S., Lemay, M., Renevey, P., Sola, J., Grossenbacher, O., Andries, D., Sartori, C., Bertschi M, (2013). Photoplethysmography-based ambulatory heartbeat monitoring embedded into a dedicated bracelet. In proceedings 40th Conference on Computers in Cardiology, (in press).

Measurement of Energy Expenditure by Body-worn Heat-flow Sensors

Neil Szuminsky[1], John Dykstra[1], and Ed Melanson[2]

[1]MetaLogics Corporation, Minneapolis, Minnesota, USA, [2]University of Colorado Anschutz
Medical Campus, Denver, Colorado, USA

1. INTRODUCTION

Measurement of energy expenditure (EE) in free-living environments is often a goal for consumers as well as health and research professionals. However, accurately measuring the number of calories burned in free-living individuals (i.e., total energy expenditure) is difficult. While there are many body-worn devices that report calories, almost all of these devices determine energy expenditure by measuring one or more surrogate parameters, most commonly some form of motion. Systems that simply measure motion can misreport the energy expenditure in situations where there either is no motion, or the detected movement can be caused by different activities with very different energy expenditures. Motion-based sensors are known to be insensitive to non-ambulatory activities, especially those like upper body activities or stationary bike use. The accuracy of the energy expenditure estimate can be improved in these systems by improving estimation of activity level or type. This is often accomplished by use of three axis accelerometers, but has also been done by using additional measurement parameters. These additional parameters allow more complex algorithmic estimation schemes to be utilized, but are still relying on the correlation between motion and energy expenditure.

Clinically validated techniques for the measurement of energy expenditures require measurement of oxygen consumption (indirect calorimetry) or heat production (direct calorimetry). Indirect calorimeters are the most common type of calorimeters in use. They are referred to as indirect calorimeters because they in fact do not directly measure calories, but rather measure the oxygen consumption and carbon dioxide production that occurs during metabolism. A direct calorimeter is designed to measure the heat produced during

FIGURE 1   Prototype of MetaLogics personal calorie monitor.

metabolism. Since a calorie is in fact a measure of heat, these systems do directly measure kilocalories. Most direct calorimeter systems were designed as rooms with sophisticated systems for measuring all forms of heat exchange from the body. The use of the past tense is intentional as direct calorimeters, despite being the original measurement systems for energy expenditure, have largely been supplanted by indirect calorimeters in the laboratory setting due to a number of practical advantages in the research environment.

There are, or have been, body-worn calorimeters based on both measurement techniques. Body-worn, portable indirect calorimeters are available, but with the need for a mask, tubing, and fairly large electronics, these devices are more easily viewed as portable "backpack" versions of metabolic carts rather than true body-worn devices. "Suit direct calorimeters," which measure heat flux from the skin surface with a web of water tubes connected to an external cart, similarly don't live up to the image of a body-worn sensor. Direct calorimetry by way of heat-flux sensors mounted directly to the skin would appear to be more easily classified as body-worn sensors. Historically, a few such systems were developed, but found limited use. The inability of these early systems to measure evaporative heat losses (i.e., perspiration), which restricted their use to low activity levels, is the most likely reason for that limited use. It should be noted that neither of these generally accepted methods (indirect and direct calorimetry) are currently feasible for use in a free-living environment and therefore their use has been limited mostly to clinical research environments.

A novel, body-worn direct calorimeter (Figure 1) was designed to overcome the limitations associated with the reliable measurement of conductive, radiant and convective, and evaporative heat losses. A commercial version of this device is now available.[1] This wearable sensor was achieved by designing a heat-flow gauge for this specific application. The gauge was designed to have minimal impact on the skin surface (factors such as occluded area and thermal resistance being considered) and the addition of a membrane covering the heat-flow gauge, which allows perspiration to be transported onto the gauge surface and evaporate.

2. ENERGY EXPENDITURE BACKGROUND

Energy expenditure at first glance should be simply related to the amount of work performed during a particular activity, and it would seem that simple physics would allow for a

[1] The MetaLogics Personal Calorie Monitor, MetaLogics Corporation, 1550 Utica Avenue South, Suite 770, Minneapolis, MN, 55416.

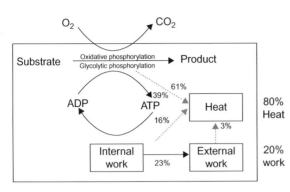

FIGURE 2 Simplified view of energy efficiency of ATP production and utilization. When substrate is used to convert ADP to ATP via oxidative and glycolytic phosphorylation, only 39% of energy is captured in ATP. As ATP is utilized, only 30% of its captured energy produces work. In both cases, remaining fraction is lost as heat. Net is a 20% metabolic efficiency.

precise calculation. For example, lift a 5 kg weight 30 cm and you would expend a known amount of energy that can be expressed through the simple formula we learned in Physics 101:

$$E = mgh$$

(m is mass, g is gravity, and h is distance lifted).

Solving this for the above example results in 14.7 J, or 3.51 calories required. However, in the human body there are internal processes required to allow the performance of that external work. Everything from the thought processes used in planning the movement, to the movement of blood to feed nourishment to the muscles, to actually performing the movement requires energy expenditure.

The primary currency for energy expenditure within the body is adenosine triphosphate (ATP). When an action potential is generated in the neuron to initiate the movement, ATP is consumed (i.e., dephosphorylated to ADP) to restore the resting membrane potential. The contraction of the muscle cells in response to the firing of the neuron consumes ATP, as does the pumping of blood through the muscle to provide glucose and oxygen to the tissue and removing lactic acid and other metabolic wastes. Even maintenance of the hemoglobin in the red cells to allow it to transport oxygen requires energy. All of this "internal" work is required in order to perform that external work of lifting the weight. Thus, knowing ATP consumption would allow a much more accurate determination of energy expenditure for the activity than the physics calculation. Of course, the direct measure of intra-cellular ATP consumption *in vivo* is neither practical nor feasible for measuring total energy expenditure.

Fortunately, there are two aspects of ATP consumption in metabolic processes that allow it to be measured and used to determine energy expenditure (Figure 2). The first is that the formation of the ATP from ADP is proportional to the oxygen consumed by an organism. The second is that significant heat energy is released as ATP is dephosphorylated. Indirect calorimetry takes advantage of the former and measures the oxygen consumed, while direct calorimetry, by measuring the heat generated, takes advantage of the latter aspect.

The production of ATP occurs by the catabolic metabolism of a number of substrates. Carbohydrates, especially glucose, are usually considered as the primary energy substrate, and often are. These substrates drive the resynthesis of ATP via the glycolysis pathway, Kreb's cycle, and oxidative phosphorylation [1]. Conversion of one molecule of glucose through the pathways consumes six oxygen molecules and produces six molecules of carbon dioxide, while generating as many as 38 molecules of ATP. If the substrate is a fat, the picture is a bit more complicated since fats are not as uniform in composition as carbohydrates. The actual size of the fatty acid molecules found in fat dictates the actual oxygen consumption and carbon dioxide production, but in general, utilization of fats consumes approximately three oxygen molecules for every two carbon dioxides produced. The ratio of carbon dioxide produced to oxygen consumed is known as the respiratory quotient (RQ) and is an indication of the substrates being utilized during metabolism. For carbohydrates, RQ is 1 but for lipids it can be as low as 0.7.

The number of ATP molecules produced also varies by the substrates being utilized. Utilization of carbohydrates for ATP production is the most efficient, providing 5.047 kcal for each liter of oxygen consumed. Fat utilization, on the other hand, only produces 4.686 kcal for each liter of oxygen. Thus, to accurately determine ATP production it is important to measure both oxygen consumption and carbon dioxide production. When measured in expired air, the ratio of the volume of CO_2 produced (VCO_2) to volume of oxygen consumed (VO_2) is termed the respiratory exchange ratio (RER). Indirect calorimeters measure the volume of respired air and the concentration of oxygen in the expired air to determine the volume of oxygen consumed and, in most cases, they also measure the concentration of carbon dioxide expired to determine the substrate utilized. These values are then used to calculate the energy expenditure needed for the regeneration of ATP utilized during the metabolic processes.

While indirect calorimetry relies on determining what was "burned" to replace the ATPs consumed during metabolic activity, direct calorimetry relies on measuring the heat released during the cellular metabolism. When ATP is used as an energy source to drive a metabolic process, only a fraction of the energy used to create the high-energy phosphate bond is converted to actual work. Most of the energy, approximately 80%, is dissipated as heat (Figure 2). As homeotherms, our body's goal of maintaining a constant core temperature dictates that it balance this metabolic heat gained from the use of ATP with heat losses from the body (i.e., the body seeks to achieve thermal balance). The heat side of the thermal balance ledger consists mostly of metabolic processes, but also includes heat gained from the environment. The body can lose heat by several means, including conduction, radiation, convection, and evaporation. At rest, radiative heat loss (which allows infrared photography to work) and convective losses predominate. With increased activities or in an environment warmer than the body, evaporation provides the mechanism for heat loss. The body has a number of control mechanisms for actively altering heat loss from the body, including controlling capillary beds of the extremities and the activation of sweat glands.

While the body attempts to achieve thermal balance (Figure 3), it is not always in that state. This is in part due to the time constants involved in generation and dissipation of heat. Also, the thermal balance is dictated by the body's goal of maintaining a constant core temperature, not necessarily a constant mean body temperature. We experience this in cool or cold environments as the body shunts circulation away from the periphery allowing the temperature of those tissues to fall (cold hands) in order to maintain a constant core temperature (Figure 4). Thus, the body's average temperature may be cooler than the core temperature, which is typically the case at rest in normal room temperature environments. The cooler average body temperature represents a thermal heat sink that the body can dump

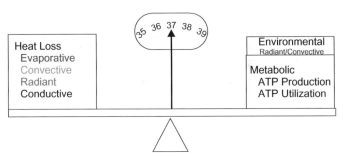

FIGURE 3 Thermal balance: In order to maintain a set core temperature, the thermal inputs — metabolic and environmental — must be balanced with thermal losses by conductive, radiant, convective, and evaporative heat loss.

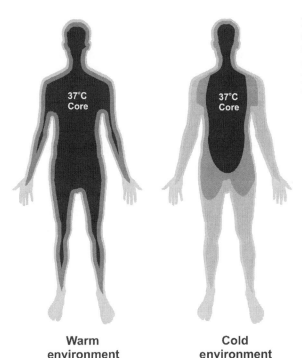

FIGURE 4 Mean body temperature: As homeotherms, humans maintain a fairly constant core temperature, but not all tissues are maintained at that temperature, which results in the average temperature of the entire body being lower than the core temperature.

heat into, causing the average body temperature to rise. This means that generated heat may not need to be dissipated to the environment, at least not right away. As a result, heat loss usually lags behind production. In addition, the core temperature set point is typically 37°C, but this temperature can and does change. The net result is that thermal balance should be considered as occurring over time as opposed to being instantaneous.

Direct calorimetry relies on the body's desire for thermal balance and accurate measurement of all four forms of heat flux from the body. Historically, direct calorimeters have generally been constructed as specialized rooms with the walls, ceiling, and floor designed to measure heat flow by what is known as gradient layer calorimetry. The room measures the conductive, radiant, and convective heat fluxes directly, and by measuring the increase in water vapor that occurs with evaporation, one can determine the evaporative heat losses. The expense of building these special rooms and inherent limitations, such as the need to account for heat flux from other sources in the room (e.g., a personal computer), has allowed comparatively less expensive and more practical indirect calorimetry systems to supplant direct calorimetry in the research and clinical environments.

3. EXAMPLES OF BODY-WORN DEVICES

3.1 Motion-Based Estimation of Energy Expenditure

There are a large number of body-worn physical activity monitors that measure body movements and estimate EE from the movement data. The movement measures can range from the pendulum-based step counting found in simple pedometers to multi-axis accelerometers. Irrespective of the way in which movement is measured, these devices utilize correlations between activities and energy expenditure. These devices are typically worn on the waist. For a number of activities, such as walking, this correlation is quite strong. These correlations are so strong that a simple pendulum mechanism worn at the hip can detect and count strides and provide reliable energy expenditure walking on a flat surface. It, however, cannot provide a good estimate when worn while walking up a hill nor for activities that generate energy expenditure without steps, such as bicycle riding. Accelerometers, especially those that measure acceleration in multiple axes, overcome these limitations since they are able to detect motions other than stride, which allows a more sophisticated correlation of movement to energy expenditure that can be utilized for calculating energy expenditure. The performance of these devices also benefits from the fact that acceleration has a stronger correlation with energy expenditure than motion (e.g., stride) [2]. Positioning of motion-based devices can, however, influence their performance. For instance, it has been shown that triaxial accelerometers worn on the hip, a common and convenient mounting position, cannot distinguish between sedentary and light intensity physical activities [3].

The underlying algorithms for estimating energy expenditure with multi-axis accelerometers impact their performance. The earliest algorithms, and perhaps the simplest, rely on just counting movement events over a period of time. A more sophisticated approach that has been used utilizes underlying characteristics of the motion data, such as coefficient of variation or other features, to classify the activity. The energy expenditure can then be calculated using correlations for specific activities, which can improve the estimate. Even more sophisticated approaches, such as the use of hidden Markov models and neural networks, are also being used to identify activities from the raw acceleration data. By improving the classification of the activity, these approaches again improve on the estimation by using more appropriate correlations.

There are a number of commercially available accelerometer-based devices. These include uniaxial Biotrainer and Actigraph, and triaxial devices such as Actical and Tracmor. More recent market entries include devices from Fitbit, Nike, Jawbone, and several others. It has been shown that these devices often underestimate energy expenditure for some activities and overestimate it for others, but generally underpredict daily EE [4]. This is most likely due to the differences in energy expenditures that occur for tasks that produce similar movements and accelerations. In fact, performance of accelerometer-based devices has been shown to improve when the activity is set as a variable in the correlation. Despite the shortcomings, accelerometer-based devices are used extensively by consumers and researchers. A comprehensive review of many activity monitors [4] is available.

3.2 Indirect Calorimeters

Measurement of energy expenditure with an indirect calorimeter relies on the known proportionality of oxygen consumption and carbon dioxide production from the substrates used for energy production within the cells of the body. As a result, these systems require capturing at least the volume and oxygen content of the exhaled air. Thus these systems require a face mask or mouthpiece be used to reliably determine the entire volume of exhaled air as well as the necessary sensor(s) to measure the gas concentration(s) in order to derive VO_2 (and VCO_2). As a result, these devices do not easily fit into the typical image of a body-worn sensor, but instead may more accurately be described as handheld or portable wearable indirect calorimeter systems. Two examples of these types of devices include the Cosmed K4b2 and the MicroLife Bodygem.

3.2.1 Cosmed K4b2

The K4b2 (Figure 5) is a miniaturized high-performance respiratory gas analysis system, in essence a miniature version of a standard laboratory indirect calorimeter that is small and light enough to be carried by the user. It utilizes a face mask with a turbine flowmeter and an air-sampling system with a galvanic fuel cell oxygen sensor and non-dispersive infrared carbon dioxide sensor in an electronics enclosure that is strapped to the body. Specifications for these components are equivalent in performance to laboratory indirect calorimeters, but in a significantly smaller and lighter configuration. Since the K4b2 requires a facemask (or mouthpiece/noseclip), it restricts some of the activities it can be used for, and the length of time it can be used, making it impractical for free-living energy expenditure measurements.

3.2.2 MicroLife Bodygem

A handheld device, the Bodygem (also marketed as Medgem) utilizes an oxygen sensor based on the fluorescent quenching of light by ruthenium. The device is designed for assessment of resting metabolic rate (RMR). Since during rest it is can be assumed that the metabolic substrate will be constant, this allows measurement of just oxygen as there is no need to determine REQ. The device assumes a mixed fuel use and utilizes a fixed REQ of 0.85. This device is limited to measurement of RMR and therefore cannot be considered a device for measuring total EE.

3.3 Direct Calorimeters

A body-worn direct calorimeter by definition must measure the heat flux from the body in order to determine energy expenditure. Historically, two approaches have been utilized for measuring heat flows: water-cooled suits and heat-flow gauges.

FIGURE 5 Cosmed K4b2. While allowing for indirect calorimetry outside of laboratory setting, does not fit image of body-worn sensor and is not practical for free-living assessments.

3.3.1 Historical Water-Cooled Suits

These systems utilize a garment constructed with numerous tubes (Figure 6) and require significant water-chilling equipment. The water circulated throughout the tubing of the suit is warmed by the heat generated within the body. By measuring the temperature of the entering and exiting water and the volume of flow, it is possible to calculate the energy expenditure necessary to warm the coolant. When the heat removed is equal to the metabolic heat generated, the wearer experiences thermal comfort. The suits can be operated outside this paradigm, and studies have used coolant temperatures as low as 5°C. These suits typically often only cover the trunk as full-body coverage can be impractical. A major limitation of this method is that the suit itself disrupts the body's normal heat flux by occluding the skin. This issue, coupled with the need for the user to be tethered to a device to cool, circulate, and record the temperature of the water, limits their utility for monitoring energy expenditure in a free-living environment. Water-cooled suits do, however, still find use in thermal maintenance systems, such as astronaut spacesuits.

3.3.2 Historical Heat-Flow Gauges

Direct calorimetry systems based on heat-flow gauges fit better into the image of a body-worn device. The heat-flow gauges (HFG), which are easily worn on the skin, work

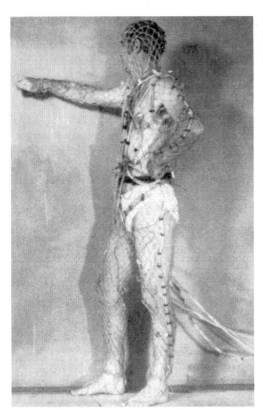

FIGURE 6 Suit calorimeter from Web (used with permission): Circulating water in tubing woven into suit is used to maintain thermal comfort; supporting cart with water heat exchanger is missing from image.

by measuring the small thermal difference that develops when heat flows across an insulator. Measuring the temperature difference and knowing the heat-flow resistance of the insulator allows calculation of the heat flow. Most HFGs utilize a pair of thermocouples (TCs) to measure the temperature difference, wired in series rather than discretely so that the output of the pair of TCs is directly proportional to the temperature difference (Figure 7) and thus the heat flow. This approach is more accurate than measuring the skin and air-side temperatures and calculating the difference. The voltage output of the HFG is then directly proportional to the heat flux, which has units of energy per time per area. There are numerous heat flux units, including joule/sec/m^2 (equivalent to watt/m^2) and kcal/hr/m^2, that are suitable for human calorimetry (1 joule/sec/m^2 = 1 watt/ m^2 = 0.8598 kcal/hr/m^2 = 0.317 BTU/hr/ft^2).

Since the HFG adds thermal resistance to the surface it's measuring, thus altering the local heat flow, the thermal resistance of the HFG will generally be kept small to minimize the effect. Minimizing the thermal resistance will, however, decrease the temperature difference across the insulator and thus the voltage difference between the pair of thermocouples. This can render the signal generated too small to measure. To increase the sensitivity, a ladder of TC pairs is typically used, with each TC pair stacked end-to-end like a string of batteries to increase the voltage produced for a given heat flux (Figure 8). This stacked

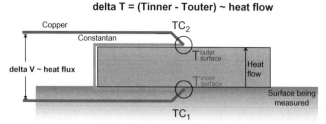

FIGURE 7 Basis for heat-flow gauge operation: Measurement of temperature difference on either side of an insulator is proportional to heat flow. Most heat-flow gauges utilize thermocouples (TC) in a serial arrangement where voltage difference (delta V) is proportional to heat flow.

FIGURE 8 Thermopile structure: Most heat-flow gauges utilize a ladder of thermocouple pairs arranged so that outputs of each pair (top:bottom) are additive, thereby increasing output voltage.

arrangement of TCs is commonly referred to as a thermopile. By using the ladder of TC pairs, signals in the uV per kcal/hr/m^2 range can easily be generated from the HFG while still maintaining low thermal resistance.

A number of research studies using HFGs on the body to measure heat flux or as a calorimeter were published in the 1980s and 1990s. In 1983 Layton et al. [5] reported use of an array of commercially available HFGs attached to the body underneath a cooling tube network "suit calorimeter." A range of heat flows was apparently obtained by altering the temperature of the suit coolant rather than by altering the physical activity of the individuals. The researchers measured heat flux from multiple body segments by both the HFG and the suit calorimeter and calculated an overall heat flux from a sum that weighted heat fluxes based upon body surface area. The output of the HFGs compared to the suit calorimeter showed good agreement. Danielsson and English both published studies in 1990 on regional heat flux measurements using HFGs. Danielsson produced his gauges in-house [6], while English utilized commercially available gauges [7]. These studies focused on the measurement of the heat flux and did not utilize the measurement to obtain energy expenditure. None of these studies mentioned measurement of evaporative heat flux.

The apparent inability of the HFGs used in these studies to measure evaporative losses is mostly due to the geometry of the HFGs that were used. The HFGs that were used were fairly large both in terms of surface area and also in terms of thickness, both of which can alter the heat flow from the site. A large surface area sensor might not be of concern in the typical *in vitro* applications of HFGs, but when mounted on the body, the occlusion of the skin surface can alter perspiration at the site and potentially alter local skin temperature. By altering skin temperature, the heat flux would also be altered. A thick sensor, by projecting above the surface being measured, will also alter the air currents over the surface and thus the convective heat loss at that site. The errors caused by the thermal resistance of the HFG and changes to convective losses measured from the human body were aptly described in literature of the period by Ducharme [8] (Figure 9). Another potential source

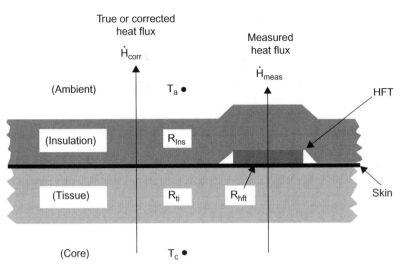

True or corrected heat flux

$\dot{H}_{corr}$

Measured heat flux

$\dot{H}_{meas}$

(Ambient) T_a •

HFT

(Insulation) R_{Ins}

(Tissue) R_{ti} R_{hft}

Skin

(Core) T_c •

FIGURE 9 From [8] (used with permission): Heat-flow gauge alters surface on which it is placed.

of error using a body-worn direct calorimeter is that it can only measure heat flux, not total energy. As a result it must rely on an estimation of the total body surface area to determine total body heat loss.

3.4 Body Media

This commercially available body-worn system has a heat-flow sensor among other sensing modalities. The description of the heat-flow measurement portion of the device states that it uses thermistors on the skin and air side of a "heat pipe" to measure what is described as a convective heat flux. The device also measures galvanic skin resistance, which is influenced by perspiration production that should correlate with evaporative losses, thus permitting an indirect estimation of evaporative loss. In addition, the device utilizes a three-axis accelerometer to measure arm motion. Energy expenditure estimation is performed by proprietary algorithms that combine information from these sensors.

3.5 MetaLogics Personal Calorie Monitor

The MetaLogics Personal Calorie Monitor (PCM) (Figure 1) has been designed as a body-worn direct calorimeter by directly measuring all four forms of heat flux from the body surface. To do so, a number of careful design details were considered. The heat flux gauge in the PCM is designed to minimize its impact on the skin surface and to allow it to effectively measure evaporative losses, something commercially available HFGs are not designed to do and generally are not capable of doing. Several design features allow transport of perspiration onto the impermeable HFG so that evaporative heat loss can be measured. In addition, careful attention was given to the supporting structures of the device to minimize impact on measured heat flux at the site the PCM is worn.

4. DESIGN CONSIDERATIONS

Utilizing a thin and flexible HFG has impacts on several areas of the performance. The flexibility of the HFG allows it to conform to the body surface to help insure reliable contact with the skin. By choosing a 7 mm thick HFG, the alteration to convective losses are minimized. This choice also has the benefit of reducing the thermal resistance of the device. But the original intent for choosing a thin HFG was to minimize the height to facilitate transport of perspiration onto the HFG. The HFG itself is impervious to perspiration, so in order to measure evaporative heat flow, perspiration must find its way onto the surface of the gauge. The millimeters thick HFGs used by Layton, Danielsson, and English present a wall that would block the movement of perspiration onto the HFG surface, thus preventing the HFG from experiencing the same evaporation as the surrounding skin surface. A thinner device would hopefully provide only a speed bump by comparison, allowing an easier movement of perspiration onto the HFG surface.

The thickness of the HFG isn't the only dimension that affects the ability to reliably measure evaporative losses. The other is the width. If perspiration has to migrate across a large dimension, as is found in typical commercially available HFGs (Figure 10), a nonuniform evaporation profile can result — higher at the outer edges compared to the center (Figure 11). To minimize this, the width of the HFG was reduced (Figure 12). The reduced width also minimized the occluded area.

An additional design element proved necessary to ensure perspiration transport onto the surface of the heat-flow gauge, especially at insensible levels of perspiration. This

FIGURE 10   A commercially available heat-flow gauge. Note large size.

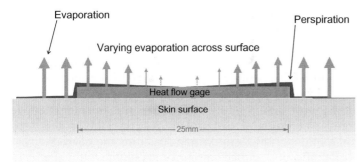

FIGURE 11 Evaporation gradient. Varying evaporation across surface of large heat-flow gauge as perspiration evaporates as it moves across surface.

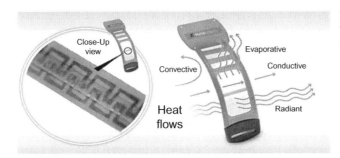

FIGURE 12 Custom heat-flow gauge: Thin and narrow to minimize changes to evaporation across the surface.

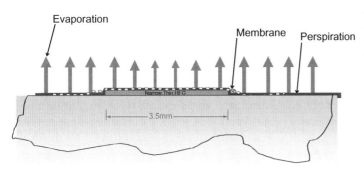

FIGURE 13 Evaporative membrane. Evaporation gradient reduced by narrow width and use of a membrane to facilitate transportation across the surface.

design element took the form of a membrane over the HFG to facilitate the transport of the perspiration horizontally without significantly altering the surface area for evaporation (Figure 13), which combined with the narrow width, would produce a more uniform evaporation profile. This membrane was carefully chosen so that it would neither act like a sponge and hold perspiration nor artificially increase the wicking of perspiration such as fibers with increased surface area (e.g., Coolmax®). This is important since any change that would alter when evaporation took place (delaying and/or prolonging it) would decouple the heat flow sensed from that of the adjacent skin surface changes. Likewise, an increased rate of evaporation would significantly alter the evaporative heat flux from the HFG compared to the skin surface. A number of properties of proposed membrane materials were considered during selection. These include wetting angle, wicking properties, thickness, and robustness. Candidate membranes were then tested by constructing sensors and comparing their energy expenditure output to indirect calorimetry during periods of insensible and frank perspiration. Eventually, a membrane was chosen that is matched to the perfusion rate of the skin itself, and this characteristic was validated through testing.

These design elements produced an HFG assembly that is sensitive to conductive, radiant, convective, and evaporative heat losses. The design of the supporting structures for the HFG also was considered. One of the most significant design features was the height above the skin surface in order to minimize alteration of air currents over the HFG. If the heat-flow gauge was less than a several millimeter thick supporting structure, the microenvironment within the walls would have significantly reduced air movement that would alter convective heat loss. Also, a tall supporting structure would potentially act as a dam, either blocking perspiration outside the sensor area from moving across the HFG or trapping perspiration within the sensing area instead of allowing it to move away. In either case, the heat flux felt by the HFG would not match the surrounding skin surface. The HFG assembly therefore utilizes an extremely thin supporting structure to minimize these effects.

The location of the HFG in relationship to the electronics enclosure was also carefully considered. It was assumed that the enclosure's occlusion of the skin surface would alter the local skin temperature and also produce an area of increased perspiration around its edges, as perspiration produced under the enclosure would migrate outwards. To minimize these possible effects, the HFG was placed at least 3 cm from the electronics, as can be seen in Figure 1. This also has the benefit of isolating the HFG from any heat generated by the electronics, although these are small considering the micro-ampere current operation of the device. The housing for the integrated circuit components was also kept as small as possible to further minimize any effect on the skin temperature or perspiration near the area of heat flux measurement.

Finally, the calculations of energy expenditure needed to reflect that the heat flux measured by the PCM (kcal/hr/m^2) needs to be converted to rate (kcal/hr) by multiplying by the body surface area (BSA). Also, since the heat that is measured is only the portion of the ATP derived energy that is dissipated as heat, the measured heat output must therefore be corrected for the amount of energy that produces work rather than heat, i.e., the metabolic efficiency (ME). Therefore, the PCM calculates energy expenditure by multiplying the measured flux times the BSA times ME:

$$\frac{kcal}{hr} = Total\ Heat\ Flux \left(\frac{\frac{kcal}{hr}}{m^2} \right) * BSA * \frac{1}{100\% - ME\%}$$

Assuming metabolic efficiency of 25%, the *ME* correction factor would be 1.33.

5. PERFORMANCE

In a properly designed device, the active area of the HFG should have the same heat losses, including evaporative losses, as the surrounding skin surface, which results in the skin and sensor having the same temperature. When infrared images of the sensor mounted on the skin surface were examined, the active area was not discernable but the electronics housing was easily seen (Figure 14). This was validated through all levels of activity, including during heavy exercise and frank perspiration.

Infrared images also validated that the PCM was measuring evaporative losses. During exercise, it was observed that the skin surface for many participants would initially warm from resting temperatures, then fall due to the onset of perspiration (Figure 15). If the sensor only measured convective and radiant heat flux from the exposed skin surface (since exposed to air, conductive losses would not be occurring) the sensor output would fall due to the cooling. However, the output of the HFG increased during the cooling phase, indicating that it was detecting the evaporative heat flux that was cooling the skin surface. It was also noted that on the IR images the active area of the sensor was essentially invisible, which indicates that its surface was matching that of the surrounding skin.

The sensor output was also compared to the rate of perspiration production using a custom transepidermal water loss instrument that measured perspiration production by measuring the increase in humidity in an air stream passing over the skin surface. The comparison had to be limited to the onset of frank perspiration since presence of visible unevaporated perspiration indicates evaporation has reached maximum levels at the ambient condition. The results of this testing are shown in Figure 16. The heat flux (in kcal/hr/m^2) and perspiration (in arbitrary units) both independently scaled to align the plots. As can be seen, the heat flux tracks the rate of perspiration.

During development, comparisons were conducted against indirect calorimetry for relatively short trials of approximately 45 to 60 minutes. The typical trial consisted of a period of rest, followed by 2 or 3 levels of activity and a final period of rest. This type of testing was used to evaluate and select the evaporative membrane. Two activities, walking/running on a treadmill and pedaling a stationary bike under various resistance loads, were used most commonly, although more treadmill tests were conducted since it was easier to get consistent activity levels across trials and subjects compared to the stationary bike.

An additional goal of the early validations was to determine if a single sensor located at a single body site could be utilized since the intent of the PCM development was a

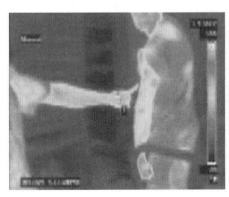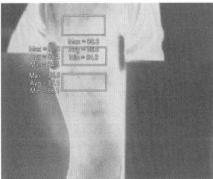

FIGURE 14 Heat-flow gauge mimics skin surface: Sensor on two different subjects, at rest (right) and just after exercising (left). The active area of heat-flow gauge (being pointed to on left image, within center box on right) is essentially invisible, suggesting that it mimics skin surface. Note that electronics housing and elements of Velcro strap are visible as cooler objects to the right or left of the sensor area.

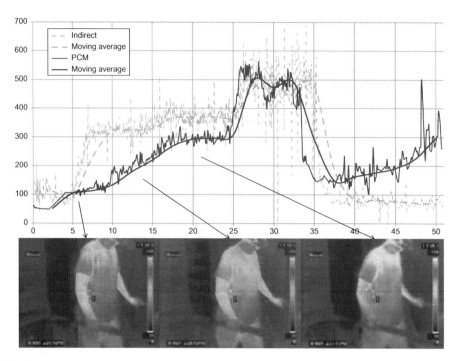

FIGURE 15   Infrared images show surface cooling due to evaporation: Infrared images were captured at minutes 6, 13, and 21; notice extensive cooling that has occurred at the 21 minute mark due to evaporation of perspiration. PCM (solid lines) shows continued increase in heat flux, which would only occur with detection of evaporative heat loss as cooling skin surface would reduce convective and radiant. Dotted lines are the indirect calorimeter measurement.

consumer product. For these studies, the subjects were fitted with multiple sensors typically located on the chest and back, the upper arm and thigh, but also in some cases additional sites including forearm, calf, and forehead. These studies showed that a sensor mounted on or near the trunk, specifically on the chest or back, upper arm or thigh, showed better correlation with indirect calorimetry. The upper arm was chosen over the other sites for a number of reasons, including difficulties with maintaining position on the thigh, and actual or perceived issues with wearing a chest strap.

6. VALIDATIONS

6.1 Comparison to Metabolic Cart

An early study [9] compared a prototype PCM to indirect calorimetry for several levels of energy expenditure using treadmill, stepping, and stationary bike. In this study, seven healthy male subjects (age = 21.57 ± 5.06 years, BMI = 22.37 ± 1.91 kg/m^2) wore sensors on the upper arm, thigh, chest, and back, and the activity periods were long enough for the

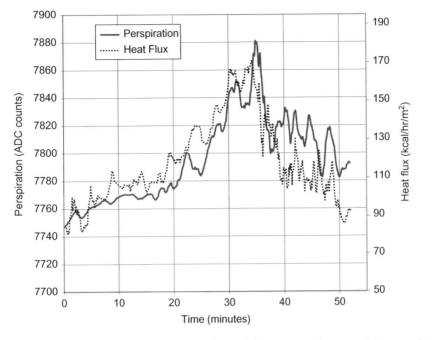

FIGURE 16   Transepidermal water loss: Heat flow and TEWL (from custom instrument) show tracking.

heat flux to reach near steady state. The area-weighted sum of the sensors at steady state was then compared to the indirect calorimeter output at the same time point. The PCM showed good correlation to the metabolic cart for walking (44.42 ± 6.12 (IC) vs. 42.46 ± 16.89 kcal (PCM)), stepping (47.26 ± 5.61 (IC) vs. 43.23 ± 18.48 kcal (PCM)), and cycling (43.06 ± 4.65 (IC) vs. 43.08 ± 25.85 kcal (PCM)).

Another early study [10] with 20 subjects (age $= 21.5 \pm 3.38$ years; BMI $= 23.3 \pm 3.55$ kg/m2) included walking, stepping, cycling, and sliding. Two activities, stepping and sliding, were chosen because they are problematic to accurately measure energy expenditure with the available devices using accelerometers. This study utilized a single PCM mounted on the upper arm for measurement. The sensor was calibrated utilizing treadmill data from several of the subjects. When this calibration was applied to all data from the subjects, no significant differences between the indirect calorimeter and the PCMs were found for all four activities (Figure 17).

6.2 Comparison to Room Calorimeter

Since the PCM relies on thermal balance, and the PCM and metabolic cart measure distinct aspects of energy expenditure with very different time frames, comparison of the PCM to the metabolic cart may not always be favorable. The thermal lag means that the PCM is more likely to reflect earlier activities (i.e., activities that occurred prior to the session) than the indirect measurement, which could cause the PCM to over-report. Similarly, thermal storage can delay release of heat generated during an activity, causing the PCM to under-report energy expenditure. Ideally, longer time periods for the comparisons would

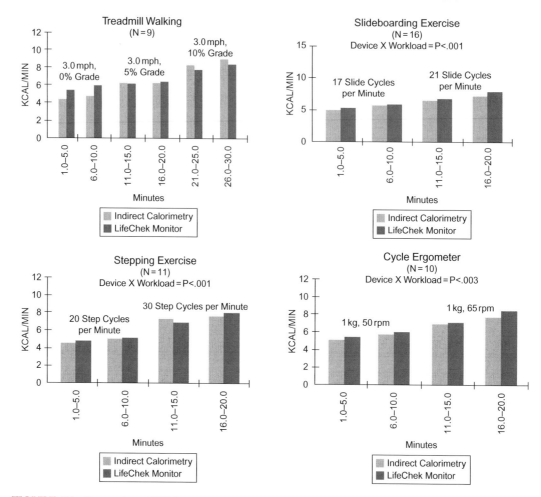

FIGURE 17   Comparison of PCM to indirect calorimeter measurements during various activities data from [10].

allow thermal equilibrium to be achieved. However, most subjects don't tolerate the mask or mouthpiece required for the metabolic cart for more than one hour. A room calorimeter, on the other hand, allows for much longer sessions since the subjects are not required to wear a mask. Access to a room calorimeter was provided by the University of Colorado at Denver's Anschutz Medical Center, allowing validation to be performed across longer timeframes [11].

A small number of subjects from an ongoing study at the UCD metabolic chamber were fitted with the PCM (identified as LifeChek calorie sensor) and were housed in the room calorimeter for a 24-hr period. During this time, there were several short exercise bouts, but otherwise they were mostly sedentary. Due to durability issues with these early prototype devices, not all subjects were able to wear the PCM for the full 24-hr period (average time was 17 hrs). There was good correlation (r = 0.98) in total EE between room and PCM

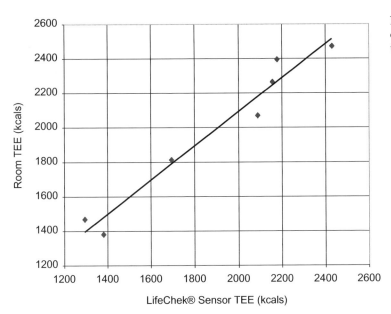

FIGURE 18 Total energy expenditure of PCM compared to room calorimeter (from [11]).

(Figure 18). In addition, minute-by-minute comparisons were made, and the measured EE for the PCM and room for one subject are shown in Figure 19.

Another study at UCD [12] examined the performance in thermal neutral conditions. For these trials, 34 subjects were in the room calorimeter for five hours wearing clothing appropriate to the thermal neutral condition. There were two 20-minute bouts of walking on the treadmill at one hour and three hours after entry into the room. The performance in the thermal neutral condition showed an overall good correlation with the room (Room mean EE 547.0 kcal and PCM mean EE 546.8 kcal).

There are additional studies ongoing and planned to address several aspects of device performance. This includes testing at different ambient temperatures and longer intervals. There will also be studies on the effect of clothing.

7. CONCLUSION

The accurate measurement of energy expenditure in a free-living environment is something that has been highly sought after by both researchers and clinicians who understand the need for accuracy, especially when using this data in clinical interventions such as weight-management programs. However, it is well documented in the literature that current body-worn devices are lacking in precision for a variety of reasons, as described above. This is especially true for those devices that estimate energy expenditure by relying on measurement of activity. Of the two methods that measure energy expenditure, it appears that only direct calorimetry may be the most practical for consumer use.

The MetaLogics Corp Personal Calorie Monitor was designed to measure all four forms of heat flow from the body. As such, it has the potential to be the first body-worn direct

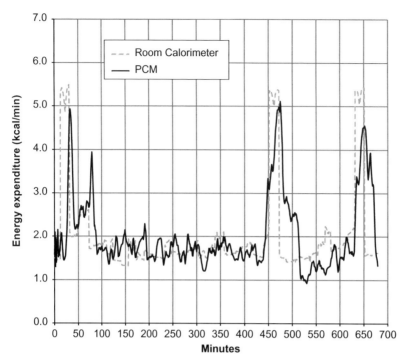

FIGURE 19 Minute-by-minute energy expenditure from PCM compared to room calorimeter [11].

calorimeter commercially available that is suitable for accurately measuring total energy expenditure in free-living environments. By measuring heat loss from the body, even from a single site, it is able to measure the energy expenditures from activities that accelerometer-based systems fail to detect. However, its reliance on thermal balance places its temporal resolution somewhere between indirect calorimeters and doubly labeled water. Activity bouts are detectable, but may not be discretely measurable due to time delays between heat production and heat losses. Other potential sources of error using a single HFG can be temporary disturbances of local environment, which do not represent the entire body, and which can be caused by events such as leaning against a cold surface or occluding heat flux in the are of the HFG with a heavy insulator (e.g., during sleep the gauge is positioned between the users skin and the mattress for an extended period and heat flux is reduced in that particular are of the body). Similarly, adding or removing clothing may temporarily disturb an HFG. However, the clothing, if appropriate for the thermal conditions, will generally simply allow the body to maintain a comfortable skin temperature at a given heat loss. Future studies are planned to validate this.

Glossary

Heat flux the amount of heat energy transferred across a surface expressed as a heat rate per unit area. In SI units this is $Watt/m^2$ but can be expressed in other units including $kcal/hr/m^2$

Heat-flow gauge any of a number of transducers that produce an electrical signal proportional to heat flux through the device.

References

[1] S.K. Powers, E.T. Howley, Exercise Physiology: Theory and Application to Fitness and Performance, seventh ed., McGraw-Hill, New York, 2009 Chapter 3: Bioenergetics, pp. 22—49.

[2] K.Y. Chen, D.R. Bassett Jr., The technology of accelerometry-based activity monitors: current and future, Med. Sci. Sports Exerc. 37 (2005) S490—S500.

[3] S.L. Kozey Keadle, A. Libertine, K. Lyden, J.W. Staudenmayer, P.S. Freedson, Validation of wearable monitors for assessing sedentary behavior, Med. Sci. Sports Exerc. 43 (8) (2011) 1561—1567.

[4] G. Plasqui, K.R. Westerterp, Physical activity assessment with accelerometers: an evaluation against doubly labeled water, Obesity (Silver Spring) 15 (2007) 2371—2379.

[5] R.P. Layton, W.H. Mints Jr, J.F. Annis, M.J. Rack, P. Webb, Calorimetry with heat flux transducers: comparison with a suit calorimeter, J. Appl. Physiol. 54 (1983) 1361—1367.

[6] U. Danielsson, Convective heat transfer measured directly with a heat flux sensor, J. Appl. Physiol. 68 (1990) 1275—1281.

[7] M.J. English, C. Farmer, W.A. Scott, Heat loss in exposed volunteers, J. Trauma. 30 (1990) 422—425.

[8] J. Frim, M.B. Ducharme, Heat flux transducer measurement error: a simplified view, J. Appl. Physiol. 74 (1993) 2040—2044.

[9] J.M. Jakicic, J. Kang, R.J. Robertson, R.R. Wing, F.L. Goss, Validity of a portable heat sensing system to measure energy expenditure, Med. Sci. Sports Exerc. 25 (Suppl. 6) (1993).

[10] C. Winters, K. Lagally, J.M. Jakicic, R.R. Wing, R.J. Robertson, Estimated energy expenditure using KAL-X heat flux monitor during several modes of physical activity. (Abstract), Med. Sci. Sports Exerc. 30 (Suppl. 5) (1998) S134.

[11] E. Melanson, J. Dykstra, N. Szuminsky, A novel approach for measuring energy expenditure in free-living humans, Conf. Proc. IEEE Eng. Med. Biol. Soc. 2009 (2008) 6873—6877.

[12] K. Lyden, T. Swibas, V. Catenacci, R. Guo, N. Szuminsky, E. Melanson, et al., Device to measure free-living energy expenditure, Med. Sci. Sports Exerc. 45 (5S) (2013) 99—101.

3.1

Knitted Electronic Textiles

Rita Paradiso, Laura Caldani, and Maria Pacelli

Smartex s.r.l., Pisa, Italy

1. FROM FIBERS TO TEXTILE SENSORS

From the time of birth, fabric is the first and the most natural interface for the body, a soft, warm, and reassuring material that protects our skin from the environment. Clothing usually covers more than 80% of the skin, which is why textile material can be seen as the most appropriate interface to implement new sensorial and interactive functions. Functions like sensing, transmission, and energy generation are implementable through textile technology. Functional yarns and fibers can be used to manufacture garments where electrical and computing properties are combined with the traditional mechanical characteristics of fabric, giving rise to electronic textile (e-textile) platforms that are mechanically comparable with the textiles that are normally used to produce our garments.

Electrical conductivity is the main physical property that is capable of transforming a textile material into a sensing material and that plays an important role in the development of e-textile apparels. Conductive fabrics can be used as bioelectrodes or (when combined with elastomers) as piezoresistive sensors that are capable of sensing biomechanical variables. Several different methods can be used to construct an electrically conductive textile, starting from the integration of metal monofilaments into the yarn, the enrichment of the fiber with conductive components, and the coating of man-made fibers with a conductive layer, to the printing of conductive pigments onto the fabric surface.

A yarn can be defined as a linear assembly of fibers or filaments arranged into a continuous strand, with textile characteristics such as tenacity and flexibility. Conductive yarns are generally made with conductive inorganic components combined with traditional textile fibers. Metals can be used in the form of fibers blended in the pre-spinning stage or in the form of filaments that can be mixed with other yarns at doubling, knitting, and weaving stages (Figure 1(a), (b)). Conductive bicomponent fibers composed by a matrix polymer and a conductive layer can be manufactured using the conjugate fiber spinning

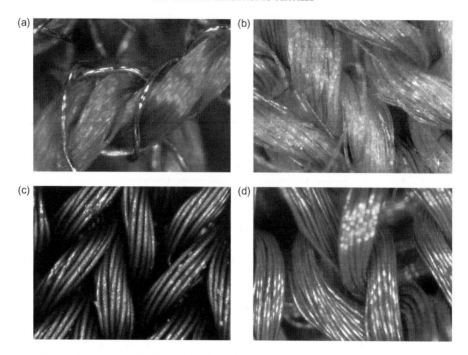

FIGURE 1 Conductive yarns and fabrics: (a) enlarged view of a fabric made with a conductive yarn obtained with stainless steel wires twisted around cotton and elite fibers; (b) enlarged view of a fabric made with conductive stainless steel fibers blended with polyester fibers; (c) enlarged view of a fabric realized with conductive bicomponent fibers; and (d) enlarged view of a fabric realized with silver-coated fibers.

technology. In this case, the conductive layer contains a densely embedded carbon black or white metal compound as conductive particles. Such fibers can be blended with other fibers to make the resultant fabric sufficiently conductive. Conductive nylon and polyester are available in the form of filaments and staple fibers (short fibers that need to be spun into yarn, Figure 1(c)). Another possibility is the use of coated conductive fibers, as shown in Figure 1(d). The coating can be applied through various techniques. Highly conductive fibers can be produced by metallic or galvanic coating, but these methods have some limitations in terms of adhesion and corrosion resistance and suitability of the substrate, while metallic salt coatings have some limitations in terms of conductivity.

In conventional textile production, metal components in the form of fibers, filaments, or particles are typically used for technical applications such as shielding and antistatic protection (work clothes, dust-free garments, school uniforms, dress suits, sweaters, carpets, upholsteries, car seats, blankets, curtains, and static-free brushes for cleaning office equipment), in bacteriostatic applications (for apparel and furnishings), as well as for fashion since the presence of metal changes the mechanical properties of fabric, creating a wrinkle or shaping effect. Pure stainless steel slivers can be blended with fibers such as polyamide, polyester, and cotton at the spinning mill to obtain electrically conductive yarns in a wide range of yarn counts. Such yarns and fibers are corrosion-resistant, inert, and stable, and can guarantee long life.

From the perspective of conductivity, manufacturability, and textile handling, silver-plated fibers may be considered the best option for the production of conductive fabric sensors to be worn close to the body. However, poor washability and poor resistance to strain due to the development of stress cracks during cladding, as well as sweat oxidization problems, make their life span shorter than fibers and yarns made of stainless steel. Several more stable products based on silver have appeared on the market in the last decade (XStatic [1], Shieldex [2]) and have been used for sensing applications, such as the women's SuperNova Seamless Glide bra by Adidas and the H2 heart rate sensor by POLAR.

Another important property required for sensing applications is elastic recovery of the fabric, which is the result of combined use of elastic and functional fibers. The elastomeric fibers are those fibers that possess extremely high elongations at break and that recover fully and rapidly from high elongations. These fibers are normally used in applications where high elasticity is necessary within the textile structure. An elastic yarn may consist solely of a number of elastomeric fibers combined to make a "bare" elastomeric yarn, such as spandex, where each fiber is made up of many smaller individual fibers that adhere to one another due to the natural stickiness of their surface [3]. Alternatively, the yarn may use the elastic strand as a core in a composite yarn having inelastic staple fibers as an outer covering. A yarn such as this is said to be "core spun." The use of a core spun elastomeric yarn in the fabric improves appearance, handling characteristics, shrinkage control, color fastness, control over elongation, and the power of recovery. In addition, the outer covering, which may be composed of natural fibers such as cotton, man-made fibers such as polyester or polyamide or a combination of both, provides additional breathability to the fabric in which it is used.

Conductive elastic yarns can be manufactured through different processes. Usually, the final structure comprises at least one elastic core thread, at least one electrically conductive thread that is wound around the core thread or is core spun around the elastomer, and, if it is needed to insulate the conductive filament, one non-electrically conductive yarn that is wound around the whole core structure. Elastic components are used for the production of stretchable fabric that allows creating apparels or garments capable of fitting the body shape as a second skin. This is an important property as a textile sensing surface collects the information from the human body by fitting it in a comfortable and unobtrusive way. Stretch fabrics are manufactured by using elastic yarns or by mixing elastomeric filaments with other natural or man-made fibers at doubling, knitting, and weaving stages. The result is a fabric that is not only stretchable, but may also have the desirable characteristics of wrinkle-resistance, transpirability, and washability.

2. THE INTERLACED NETWORK

Fabrics can be manufactured with different technologies from knitting to weaving. The path followed by a single yarn in the fabric is different according to the production process. Knitting methodology consists of loops called stitches that are pulled through each other. The active stitches are held on a needle until another loop can be passed through them and a single yarn is building the whole structure. Weaving is a process where two distinct sets of yarns or threads, called the warp and the filling or weft, are interlaced with each other to form a fabric or cloth. The warp threads run lengthways on the piece of cloth, and the weft runs across from side to side.

When yarns are combined into the fabric, the resulting structure is a network where the contacts among the single filaments are random, and the fibers are untidily assembled. Compared to weaving that requires running the same yarn along the weft and the warp direction, knitting technologies are more suitable for smart fabrics. First, knitting provides the possibility to select a desired functional yarn and confine it in a specific region of the fabric according to a precise architecture. Second, knitted fabric is usually highly elastic (it stretches easily) and readily drapeable (it hangs and folds nicely), and it is porous. This means that it is breathable and comfortable when worn next to the skin. Finally, knitted fabric can be made in the required garment shape.

Machines and looms for knitting can be defined according to needle arrangement. Needles are fixed on metal structures that can have a circular or linear shape. In knitting machines the needles are free to move individually, while in knitting looms the movement of needles is done jointly (Figure 2). The set of needles and metal structures is known as the "bed." Both machines and looms can be equipped with single or double beds.

Computerized flat knitting and circular knitting machines allow operating each needle individually. Factors such as knitting speed, yarn tension, cone size, batch difference, and humidity can contribute to variations in loop formation. A flat knitting machine is very flexible, allowing complex stitch designs, such as jacquard, plated, and intarsia patterns, double jersey, shaped knitting, and precise width adjustment. This type of machine is,

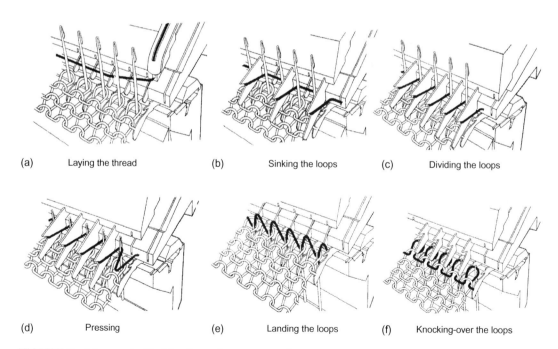

(a)   Laying the thread   (b)   Sinking the loops   (c)   Dividing the loops

(d)   Pressing   (e)   Landing the loops   (f)   Knocking-over the loops

FIGURE 2   Movement of knitting loom elements to produce a course of loop [4].

however, relatively slow when compared to circular machines. Machines are usually equipped with several independently motorized yarn carriers that allow the electronic selection of each single needle. For an e-textile application, intarsia is one of the most important knitting techniques used to create patterns with multiple diversified yarns. Double-bed knitting machines can also be used to realize the double-jersey technique where two layers of fabric are knit simultaneously. The fabrics may be inseparable, as in interlock knitted fabrics, or they can be knit as two unconnected fabrics, as in tubular knitted fabric.

Another interesting feature of knitting is the possibility of handling two different yarns concurrently with the same needle and overlapping, known as the *vanisè* technique or plated knitting. With this technique a metal yarn can be covered (plated) with another one (Figure 3(a)), and with the second series of needles, the external side of the fabric can also be protected. Using a combination of different knitting techniques offers the possibility of designing and implementing a sort of logic circuit. With conductive patterns that can be on a specific side of the fabric, these patterns can be connected through fabric tracks where the conductive elements are running inside the fabric, invisible from both the sides of the fabric, in a multi-layered structure (Figure 3(b)).

Circular knitting, and in particular the seamless technique, provides comfortable, stretchable, well-fitting, and adherent garments, which makes this technology suitable for sensing applications where adherence, elasticity, and comfort are the main requirements. The seamless technique comes from the fusion of two fields, hosiery, and knitting, and allows the production of tubular fabrics without seams, laid-in elastic yarns inserted in the welt bands and equipped with areas having gradual compression. Therefore, the garments knitted on seamless machines merge comfort with functional performance as they allow the creation of different stitches such as rib, net, jacquard, piquet, stripes, and laces, as well as pre-shaped structures, hidden supports, pockets, collars. and hoods. It is possible to implement seamless systems, where both electrodes and sensors are knitted in the same production step, through intarsia technology. The main difference between circular knitting and flat knitting is that the latter can combine intarsia and double knitting, while the

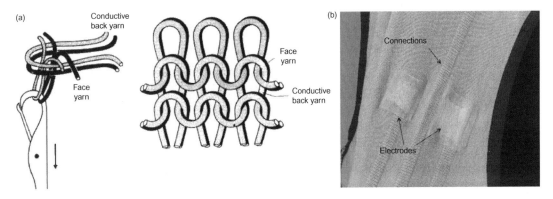

FIGURE 3 Plated knitting: (a) plating in weft knitting [4] and (b) fabric connections. The conductive yarn is knitted inside the multi-layer structure and fabric electrodes with membrane.

former can only handle these two processes separately. However, seamless technology is unique in combining elasticity and comfort of the fabric with low production costs.

Different knitting techniques can be used to provide different sensing functionalities. In particular, piezoresistive fabric sensors can be implemented using knitting techniques by combining conductive and elastic yarns, and by an industrial serigraphy screenprinting process. Circular knitting machines (such as Santoni machines) can be used for the production of piezoresistive fabric sensors due to the high elastic recovery of fabric made with this technology. Conductive bicomponent fibers yarn, such as polyamide loaded with carbon particles, is one of the yarns typically used in combination with one or more elastomers to implement these sensors. Piezoresistive fabric sensors change the electrical resistance according to strain, and the variation of the electrical properties occurs due to the different path of the electrical current inside the fabric structure. Usually this property can be observed in stretchable fabric, where a mechanical solicitation affects the flow of carriers inside the structure. When the conductivity of the yarn is due to the presence of conductive particles (like in a bicomponent fiber), the elongation of the yarn produces different distribution of the conductive particles in the structure of the yarn and in the fabric, leading to a modification of charge transport mechanism. The interconnection among fibers and stitches is affected by the mechanical deformation, as can be seen in Figure 4. The elongation of the fabric modifies the distance among the stitches as well as the arrangement of the fibers in the yarn.

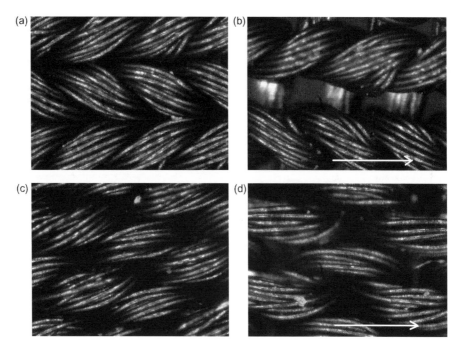

FIGURE 4   Interconnections in fabric under mechanical deformation: (a) front view, rest condition; (b) front view, under mechanical strain along the arrow direction; (c) back view, rest condition; and (d) back view, under mechanical solicitation, fibers are elongated along the arrow direction.

The path of charges inside the textile structure is correlated to the distance among the conductive components of the network. When the structure of the yarn and the resulting fabric is modified by a mechanical solicitation, the fabric sensor behaves as a strain-gauge transducer within the limits of sensor elasticity such that it does not break or permanently deform. Modeling of the charge transport mechanisms inside the textile structure is not easy, but when the structure of the yarn and the resulting fabric is kept planar to the extent possible, the fabric sensor can be considered a strain-gauge transducer. The different architecture of the yarn and resulting fabric confers a different response to the whole sensor. A recent study [5] demonstrated that a small change in the structure of the fabric, due to a different geometry of the conductive yarn, results in a dramatic change of the functionality and sensitivity of the final fabric sensor.

The lithographic technique is normally used to implement piezoresistive-coated sensors. During the process, a rubber or silicone solution containing conductive particles is applied to the fabric, then, after the removal of excessive rubber material, the conductive elements are immobilized in the structure through treatment at high temperature. This technology provides both sensors and wiring by using the same elastic material and avoids the use of obtrusive metallic wires, which may hamper the movements of the kinematic chain. The mechanical properties of the final product are affected by the speed of the coating process, the viscosity of the solution, and the capability of the material used as substrate to adsorb it [6]. The viscoelastic properties of the fabric substrate affects the mechanical response of the textile sensor. The hysteresis effects can be reduced by acting at the level of the textile structure, as well as by increasing the elastic properties of fabric. These fabrics behave as strain-gauge sensors [7] and show piezoresistive properties similar to knitted fabric sensors. In both cases, the increase of elastic properties in the fabric is directly proportional to the increase in piezoresistive properties.

3. TEXTILE SENSORS FOR PHYSIOLOGICAL STATE MONITORING

A wearable sensor system based on textile technology, where sensors are implemented with fibers and yarns, requires construction of a fabric containing shaped regions where sensors are located on a specific part on the body. Thus, the textile-sensing interface has to be a garment tightly fitting the body, to avoid any possible mismatch between the body and the sensors. The fabric will act as a second skin, and it has to be elastic and comfortable. By using a flat knitting and seamless knitting technique it is possible to confine specific yarns in defined regions of the fabric and at the same time to process different yarns together according to a desired topology.

Sensors, electrodes, and connections can be fully integrated in the fabric and produced in one single step by combining conductive and non-conductive yarns [8]. A combination of intarsia and double-knitting techniques allows the production of double layers, using the external non-conductive part to isolate the electrode from the environment. Use of another yarn in vanisè configuration allows multi-layered structures, where the conductive surface is sandwiched between two insulated textile surfaces. The same conductive yarn can be used for the electrodes as well as for the implementation of the connections, as seen in Figure 3.

Conductive textile is the basic material for the detection of electrical signals. In standard clinical practice, electrodes located on specific parts of the body are used to measure electrical potentials of biological origin. Biopotentials occur due to electrochemical activity of the cells, where the electrical activity is caused by differences in ion concentrations within the body. There are several diagnostic applications of biopotentials: electrocardiography (ECG), electroencephalography (EEG), electromyography (EMG), and electrooculography (EOG). Biopotential electrodes convert ionic conduction to electronic conduction and are used for measuring electric potential of biological origin, or to transmit electrical energy to and from a human subject.

Textile electrodes can be used to detect a variety of biological signals [9] as well as to measure body impedance and skin conductance. Piezoresistive sensors can be used to monitor respiratory activity at the thorax and abdominal level as well as movement of joints. Other important parameters like temperature or SpO_2 can simultaneously be measured by means of transducers embedded into the fabric. The list of signals detectable with non- invasive textile or wearable sensors varies according to a specific application, but some of the most important information about the physiological state of the user is provided by cardio-pulmonary activity.

Electrodes can be classified according to conduction mechanism as "perfectly" polarizable, characterized by a capacitive effect, with no charges flowing between the electrode and tissue (i.e., stainless steel electrodes), and "perfectly" non-polarizable, characterized by a resistive effect, with free charges flowing between the electrode and tissue (i.e., Ag/AgCl). Electrodes are applied directly on the body surface; in order to get a good skin contact, a thin layer of electrolyte is usually applied between the skin and the electrode. Electrodes can be manufactured using different materials. In the standard practice, pregeled disposable electrodes are typically made using Ag/AgCl and are employed in long-term ambulatory monitoring and when high stability is needed. An alternative solution, compatible with textile technology, is represented by stainless steel polarizable electrodes; in this case, an electrolyte containing salts of low corrosion potential may be required.

If a pair of electrodes is in an electrolyte and one moves with respect to the other, a potential difference appears across the electrodes. This potential difference is known as a motion artifact and it acts as a source of noise and interference in biopotential measurements. Motion artifacts are greatly attenuated if the electrode is separated from the skin surface by a layer of conductive gel or paste. Any mechanical disturbances caused by relative motion between the electrode and the skin are damped by the gel layer, and their effect on the signal is limited. Moreover, artifacts can be reduced by reducing the skin impedance, such as removing the upper layers of the skin via abrasion. The standard practice foresees cleansing of the skin with solvents and use of a conductive paste for reducing skin impedance.

Conductive fabric electrodes can be used together with hydrogel membranes that act as electrolyte by reducing the contact resistance between the skin and the electrode. In order to improve the stability of the contact, the hydrogel membrane is used in the form of a patch that is adhesive on both sides as shown in Figure 5; this feature improves the quality of signal by reducing motion artifacts and the effects of the contact impedance [10,11].

Since the use of textile materials for wearable sensing culminates with the implementation of imperceptible monitoring systems, it should not be any different from traditional

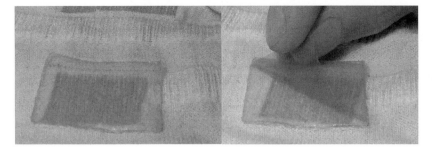

FIGURE 5   Textile electrode with a hydrogel membrane.

garments or fabric, and the use of a material like gel is not desirable. Several factors have to be considered to implement electrodes able to sense without hydrogel. First, such an electrode cannot acquire information of clinical quality, instead it can give daily indication of health status and the physical behavior of the user in a context that cannot be monitored in another way, without interfering with the user's daily life. Second, the sensor locations should be selected according to ergonomic criteria at the locations that minimize motion artifacts. Third, contact with the skin should be improved by utilizing the elastic properties of the garment. Fourth, local environmental conditions in terms of temperature and humidity in the electrode region should be manipulated to increase sweat production, e.g., by decreasing the breathable properties of the fabric. Sweat can act as an electrolyte to decrease skin-electrode contact impedance. Finally, the fabric should be designed to fit tightly to the skin.

4. BIOMECHANICAL SENSING

Knitted piezoresistive fabric (KPF) sensors are based on the piezoresistive effect that is characterized by the change of the electrical resistance when the sensor experiences a mechanical strain. This effect occurs due to the change of current path in the fabric, due to the variation of the conductive contacts between the filaments inside the yarn, and due to the deformation of the fabric loops during the applied strain. Typically, these sensors are characterized by electro-dynamic tests where a pre-defined strain is applied with controlled amplitude. The electrical-resistance variation of the KPF sensors is monotonously correlated to the strain amplitude.

In order to define the electrical characteristic curve of KPF sensors, a set of data was acquired in static condition and analyzed in terms of electrical resistance, sensitivity, hysteresis effect, and repeatability [5]. Two kinds of piezoresistive sensors, called T_KPF and P_KPF sensors, were compared with the goal of demonstrating how the structure of conductive yarn influences the electrical response of the fabric sensor. The characteristic curve of the P_KPF sensor is shown in Figure 6.

The T_KPF sensor is made by using the conductive yarn after the texturization process with the aim of improving the mechanical properties. The P_KPF sensor is made with the same conductive yarn with an ordered structure in which each filament appears in a

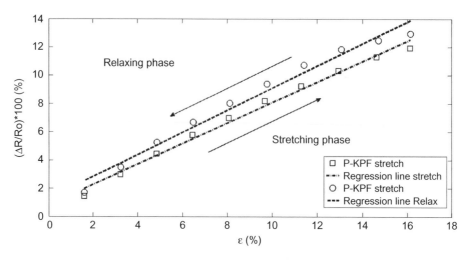

FIGURE 6 The calibration curves of the P_KPF strain sensor; the experimental data are represented by the markers and the linear fit result is shown by the dotted lines.

parallel configuration. Figure 7 illustrates the hysteresis effects evaluated on both kinds of KPF sensors. Rectangular sensors with dimensions of 10 mm x 62 mm have been strained along the longer side to a maximum elongation of 1 mm at 0.25 Hz. The results show that the hysteresis effect is less pronounced in the P_KPF sensor. In fact, the maximum percentage of the measured hysteresis error of the P_KPF sensor is less than 10% compared with that of the T_KPF sensor, which is less than 40%.

Piezoresistive textile sensors have been used to detect the movements of the human body in a wide range of configurations, for monitoring of hand, wrist, elbow, and knee articulation [7]. To test the capability of fabric sensors to evaluate joints movements, they were compared with standard electrogoniometers during simultaneous motor acquisitions. KPF sensors have also been applied to detect the movement of the fingers in combination with other kinds of sensors with the aim of developing the first integrated prototype for the simultaneous acquisition of gesture and physiological signals as a new gestural interface [12].

A new generation of wearable goniometers is under study in the framework of the European project Interaction. The electrical response of the wearable goniometer is based on the different response in terms of piezoresistive effect of the two KPF layers during the flexion and extension phase. For example, in the flexion phase, the length of the KPF layer close to the joint is constant, while the external KPF layer elongates in response to the flexion of the joint. In such sensors the variation of electrical resistance doesn't depend on the elongation effect, but only on the angle of the joint [13].

KPF sensors can also be used to detect breathing activity, in particular the plethysmography signal, by measuring the mechanical variation of the rib cage in the inspiration and expiration phases of respiration. To verify the correlation between the change in electrical resistance of KPF sensors as a function of respiration, the electrical behavior of a KPF sensor has been compared with the signal obtained from a standard ambulatory

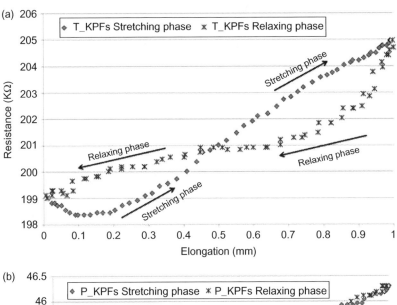

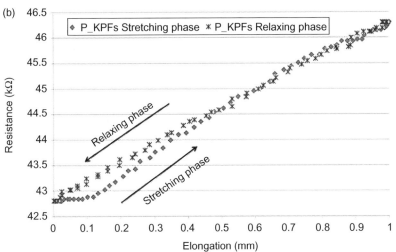

FIGURE 7 The hysteresis behavior of the knitted piezoresistive fabric sensors: (a) T_KPF sensor and (b) P_KPF sensor.

plethysmography sensor (Biopac Lab System, Inc.) by placing both kinds of sensors in the thoracic position during normal respiratory activity. The breathing rate calculated from the breathing sensor signals produced comparable results [14].

5. NON-INVASIVE SWEAT MONITORING BY TEXTILE SENSORS

A sweat analysis system that can be easily integrated into fabric for real-time analysis of sweat during exercise has been developed in the framework of the Biotex project [15]. This project aimed at the development of sensing patches adapted to different targeted body fluids and biological specimens. Most of the sensors were developed for implementation in a textile material and easy integration in a patch or a garment. The remaining sensors were developed taking into account textile compatibility. Sensors based on optical, electrochemical, and electrical principles (impedance monitoring) were explored for sensing:

1. Sweat monitoring: relative quantity (i.e., perspiration rate), over salinity (i.e., conductivity), specific ions (K + , Na + , Cl-, Mg + ; Ca +), pH, and organics
2. Infection detection through blood and body liquid monitoring for burnt persons
3. Oxygen saturation of blood for medical, sport and security applications

A number of textile sensors were distributed around the body. The sensor garment combined a multi-parametric patch for sweat analysis measures, such as pH, sodium, and conductivity with other sensing modules, including perspiration rate, ECG, respiration, and blood oximetry sensors.

A special textile-based platform with fluid-handling properties was designed and used to collect and analyze sweat samples. Sweat collection and flow were controlled by a passive textile pump based on capillary action, which has been implemented by combining hydrophilic and hydrophobic fabrics with highly absorbent material [16]. A textile channel was created using hydrophilic material, while the absorbent was placed at the end of the channel. The absorbent controls the fluid flow, drawing the sweat toward it along the length of the channel (Figure 8). In this way, a continuous flow of sweat enters the channel where it is analyzed by the sensors and then travels toward the absorbent where it is stored. A location on the lower back was chosen to collect sweat as this is an unobtrusive location for sensor placement during exercise.

The sensor garment also contained a humidity sensor and ionic concentration sensors, allowing the monitoring of the loss of liquid from subjects and to alarm in the case of dehydration. The skin is a complex structure, but for the modeling of perspiration, it can be approximated by a homogenous flat surface that continuously emits water vapor. With this assumption, Fick's first law of diffusion can be used to calculate sweat rate from the gradient of humidity measured by a pair of wearable humidity sensors located at two

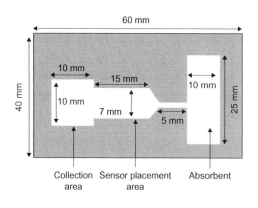
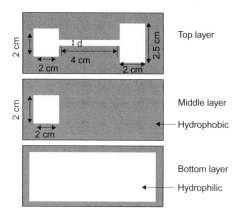

FIGURE 8   Textile passive fluid pump.

different distances from the skin. A textile humidity sensor can be implemented by sandwiching a hydrophilic insulating film between two conductive fabrics (capacitor plates). A conductive yarn (70% polyester/30% stainless steel, Bekintex) was used to manufacture the woven conductive fabric. An example of the sensor is shown in Figure 9. The capacitance change was sensed and used to derive the sweat-rate gradient over time.

The calibration curve of the textile humidity sensor is shown in Figure 9 with the corresponding curve of a commercial sensor (Philips H1). The commercial sensor is characterized by an almost linear calibration curve and a shorter response time, though the variation in capacitance is much more limited, with a dynamic range of about 30 pF. The textile sensor shows a much larger variation in capacitance, with a dynamic range that is more than one order of magnitude higher than the commercial sensor (3.5 nF), but most of this variation only occurs when the relative humidity is above 50%. The textile sensor performs reasonably well when the sensor is placed between a membrane and the skin (where high humidity was expected), but needs improvement in the low humidity region.

The sodium sensor was fabricated on a flexible kapton surface. This electrochemical sensor measures the open-circuit potential between a reference gold electrode and an ion-selective electrode (solid contact ion-selective electrode), which is a function of the sodium concentration. The sodium-selective electrode is made of a gold contact covered with a polymeric membrane that contains polypyrrole, plasticizer, ionophore, and ion exchanger. The polymeric ion-selective membrane requires a conditioning period before use (12 h in 1 mM NaCl) and must be calibrated before use using solutions with known concentrations of sodium (20, 40, 60, and 80 mM). The sensor is reusable and can be washed before reuse.

The conductivity sensor measures sweat conductivity, which is a function of the type and concentration of the ions, typically ranging from 2 to 15 mS/cm, with an average conductivity of 5 mS/cm for the sweat of healthy individuals [17]. For a single electrolyte solution, conductivity is an empirical function of concentration, often linear over a limited

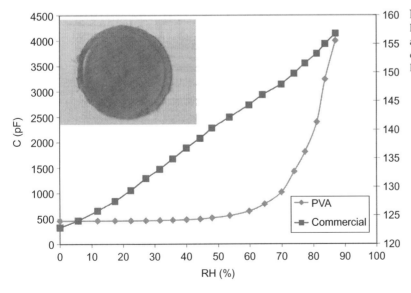

FIGURE 9 Textile humidity sensor (upper left) and its calibration curve compared to a commercial humidity sensor [13].

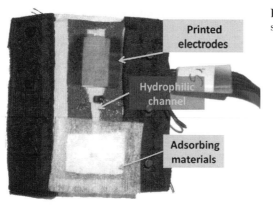

FIGURE 10   A prototype of the passive pump and sweat sensor.

range if temperature and geometry of the electrodes are constant. In the case of a multi-component electrolyte solution, the conductivity is equal to the sum of the conductivities of its individual ions. The conductivity sensor consisted of electrodes fabricated on a flexible plastic patch that also contained the sodium sensor. As conductivity and sodium measurements are temperature dependent, a temperature sensor (Analog Device ADT7301) was included within the system to compensate for temperature changes. The Kapton patch was then placed across fluidic channel. An improved version of this sensor was implemented in the scope of Proetex project [18], where the electrodes for a sodium sensor were printed and manufactured directly on the fabric, as shown in Figure 10.

A special band worn around the abdomen was designed for the integration of the sweat sensors. Both the humidity and the ionic sensors were kept inside special pockets designed to allow easy recharge of the sensors.

6. SMART FABRICS AND INTERACTIVE TEXTILE PLATFORMS FOR REMOTE MONITORING

Textile sensors, such as those described earlier in this chapter, can be used for remote monitoring of physiological parameters. The first ever e-textile platform addressing clinical rehabilitation for cardiac patients was implemented in the scope of the Wealthy project [8]. The developed wearable-integrated system was able to acquire, simultaneously and in a natural environment, a set of physiological parameters such as electrocardiogram, respiratory activity, posture, temperature, and movement index. Figure 11 shows the core of the system, a knit fabric where sensors and connections are fully integrated and the conductive fibers are woven with stretchable yarns. The Wealthy sensing platform comprises: 1) six textile ECG electrodes, implemented as a double-fabric layer with the conductive part in contact with the skin; 2) three larger textile electrodes for the impedance measurement; 3) textile connections implemented as a multi-layered structure with the conductive fabric sandwiched between two insulating layers of textile; 4) two temperature transducers embedded in the garment; and 5) an integrated connector for the SpO2 sensor [19]. The fabric was made with a yarn containing stainless steel monofilament (Figure 1(a)).

(A) (B)

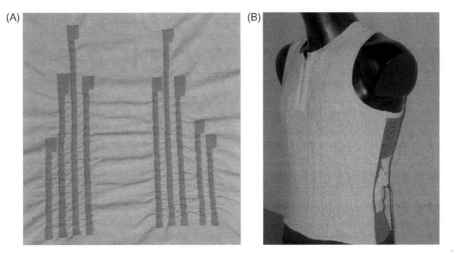

FIGURE 11   Wealthy textile system: (a) knit fabric, the fabric connections, and the fabric electrodes; and (b) the final system.

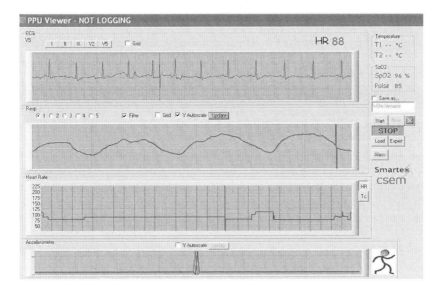

FIGURE 12   Typical set of signals acquired by the Wealthy platform.

Figure 3(b) illustrates particularities of the connections and the electrodes. Typical signals acquired by this system are shown in Figure 12.

Building on experience acquired within the Wealthy project and aiming at the implementation of a garment as unobtrusive as possible, a simplified platform, called the Wearable Wellness System (WWS), has been developed for long-term physiological monitoring and preventative healthcare. The WWS takes into account the limitations of wet electrodes used in the Wealthy platform that required the use of hydrogel for each measurement session. The WWS sensing shirt comprises two textile dry electrodes for ECG detection, and one textile piezoresistive sensor for detection of respiratory activity [20]. All textile sensors are fully integrated and manufactured as a one-step process. A fabric bus connects the shirt to a dedicated electronic device (developed in the frame of several projects, from MyHeart [21] to Psyche [22]) that is able to acquire and wirelessly transmit physiological signals. Posture (lying or standing) and level of activity can be monitored through a 3-D accelerometer integrated in this electronic device.

Other wearable textile platforms have been developed for applications in the field of security and protection. The latest generation of equipment and uniforms is characterized by functions like sensing, communication, and alerting. This is the paradigm of E-Sponder [23], a project that aims at the implantation of a service delivery platform based on a suite of real-time data-centric technologies and applications. The fusion of field-derived data within a central system provides information analysis, communication support, and decision support to first responders that act during crises occurring in critical infrastructure or elsewhere. Special equipment was developed to be worn by the first responders that is capable of monitoring the position and physiological parameters (ECG, breathing rate, and skin temperature) of rescuers in the field. The equipment comprises a shirt and a jacket, with sensing regions and support for embedded electronics. The garments have been designed according to the outcomes of the ergonomic trials in which rescue personnel tested the comfort and functionality of the wearable system. For the underwear shirt, the fabric and all the accessories exhibit fire-retardant characteristics. Four electrodes (two for ECG monitoring and two for respiratory monitoring by impedance measurement) are embedded in an elastic band to guarantee their adherence to the skin during action.

An instrumented jacket was produced by modifying a standard garment to hold the portable electronic devices developed during the project. A smartphone is used to receive the signals coming from the shirt and transmit them along with positioning information in real time. Special internal pockets, closed by fire-retardant zips and connected by a track of fabric integrating a flexible cable for device charging, are used to hold the smartphone and the positioning sensor. External pockets were added to hold external temperature and gas sensors.

7. SYSTEM FOR REMOTE REHABILITATION

Within the scope of the European project MyHeart [24], a neurological rehabilitation system (NRS) supporting stroke patients in the performance of speech and motor rehabilitation therapy was designed and manufactured. The NRS was designed for hospital and home use. When hospitalized, patients are trained to use the system. After hospital discharge, patients can use the rehabilitation device at home (or in a long-stay ward) and continue exercises with the help of a caregiver, relative, or nurse.

NRS is based on a textile-sensing platform capable of detecting body movement and a software package able to recognize if a set of movements performed during the rehabilitation exercise is correct. The system was designed to assist patients remotely during home rehabilitation or to support simultaneous sessions with several patients in a clinic. During rehabilitation, patients perform a set of exercises designed to recover functionality. Patients review a tutorial video about the exercise they are asked to perform. Next, they don the sensing garment with the help of a caregiver. After a calibration phase, the motion-recognition software starts to provide real-time feedback on the progress and accuracy of exercises by displaying symbols such as colored bars and a smiling or frowning face. The movements have to be repeated until the assigned exercise suit is performed correctly or time expires.

Motion recognition in NRS is based on printed piezoresistive fabric sensors [25]. Movement of the joints of the upper limb is monitored through 29 textile sensors placed on a shirt. The NRS garment monitors shoulder, elbow, and wrist joints, and can monitor the lateral abduction and adduction of the arm, the 90° flexion of the arm in the sagittal plane, external rotation of the arm with flexed elbow, forearm flex-extension, and prono-supination. Functions such as eating and combing were also recognized. To insulate the sensing material from the external environment and from the body, sensors were sandwiched between two layers of fabric, and shaped printed patterns were embedded in the elastic garment with the printed face in contact with the garment fabric, as shown in Figure 13.

Use of textile electrodes for EMG detection and functional electrical stimulation (FES) therapy has been explored in the Tremor project [26]. A sleeve integrating multi-electrode patches has been designed to help patients affected by movement disorders. The design of the sensing part was based on the results of ergonomic tests as well as on the functionality required to detect the presence of tremor and to control it through FES therapy. The platform combined two separate electrode patches with different distribution of the electrodes, corresponding to the different muscles of the arm [20]. The electrodes were round, 1 cm in

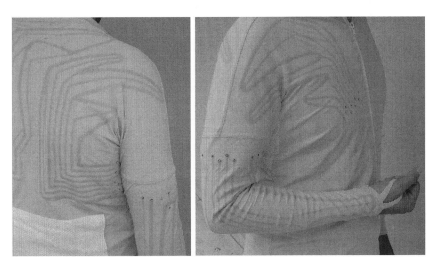

FIGURE 13   The neuro-rehabilitation platform.

diameter and 1 cm of inter-electrode spacing. Size and shape were chosen to reduce the risks of edge effects that can increase pain during the FES.

The number of stainless steel electrodes was redundant to guarantee the maximum coverage of the region to stimulate, and it depended on the arm's size. For a size S of the upper arm, 28 sensing units were deployed in two matrices composed by four rows alternating with three and four electrodes. The larger muscle area of the lower arm was covered by 42 electrodes distributed in two matrices with six rows alternating with three and four electrodes. Three sleeve sizes were developed in order to permit the stimulation of larger arms, adding one electrode to each row to increase the sensing area. Figure 14 shows the largest sleeve size; it covers two-thirds of the proximal forearm. The chosen electrode structure allowed for easier placement of the hydrogel membranes on the electrodes. To improve coupling with the skin, use of an electrolyte in the form of hydrogel was necessary to reduce pain during FES therapy. For EMG measurements, the electrodes were moistened with a small amount of water before each measurement.

This sensing platform combines fabric electrodes and biomechanical textile sensors to perform EMG detection and recognition and to provide FES when needed [27]. The system demonstrated promising preliminary results as a possible alternative for remote rehabilitation therapy [28].

The latest generation of a wearable system for stroke rehabilitation (under development in the scope of the Interaction project) combines electronics devices (inertial sensors and portable electronics) with textile sensors (wearable goniometers and EMG sensors). The goal is development of a system able to sense the motor functions in stroke patients in a remote environment without any supervision. The wearable system is comprised of a shirt and trousers. The male and female versions of the garment have been tested on patients to evaluate their level of acceptability. The fabric that has been selected for the implementation of the garment is thermally comfortable as well as very soft and able to slip on the skin. Moreover, the use of zips and elastic strings linked to the slider facilitates the wearing of the shirts. The garments have been designed taking into account future implementation of the sensors: a more elastic and heavier fabric has been inserted in both shirt and trouser to hold the sensors and the electronics.

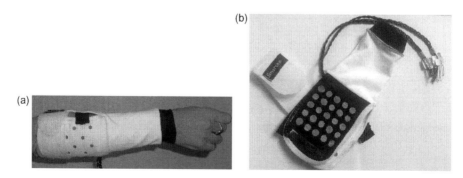

FIGURE 14   System for the forearm designed to be tested for both EMG measurements and FES therapy: (a) outer side and (b) inner side.

8. SYSTEMS FOR EMOTIONAL STATE ASSESSMENT

The monitoring solutions presented in the previous sections can be used the whole day. Monitoring can be performed during the day, during the night, or during special selected time-frames. For instance, more detailed monitoring can be done in the morning and before sleep. Such monitoring can acquire data from the cardiopulmonary system as well as weight, blood pressure, and activity and information about therapy, pain, mood, and any other relevant data.

One of the more interesting applications concerns the use of smart fabric and interactive textile (SFIT) systems to push people toward a healthier lifestyle. It is important to help people acquire awareness of their overall health, to motivate them to become active in staying healthy and feeling well. SFIT technology may potentially provide innovative solutions to manage lack of sleep, stress, inactivity, metabolic syndrome, and mood disorders. The European project Psyche focused on the development of a personal, cost-effective, multi-parametric monitoring system based on textile platforms and portable sensing devices for the long-term and short-term acquisition of data from patients affected by mood disorders. The sensing platform developed during the project (Figure 15) allows identification of the triggers that precede every episode the patient has suffered.

The project relies on the previously described Wearable Wellness System (WWS) to collect physiological data that includes heart rate variability, respiratory rate, activity, and

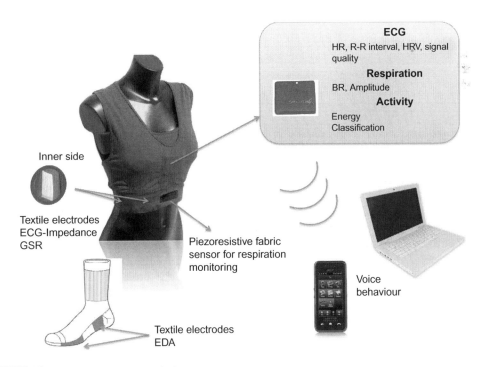

ECG
HR, R-R interval, HRV, signal quality
Respiration
BR, Amplitude
Activity
Energy Classification

Inner side

Textile electrodes
ECG-Impedance
GSR

Piezoresistive fabric sensor for respiration monitoring

Voice behaviour

Textile electrodes
EDA

FIGURE 15   Psyche multi-sensing platform.

movement. The WWSs have been distributed among healthy volunteers and individuals with mood disorders for the acquisition of data that will serve as reference for the database [22]. Biochemical measurements, voice analysis for emotional assessment [29], and the detection of attitudinal indicators (social interaction, daily activity, productivity, emotional perception) are also being considered as predictive metrics. The reference database also foresees inclusion of sleep-pattern data, measures of cardiovascular and respiratory functions, electro-dermal response, and stress-related hormones, including change in the diurnal variations of all these measurements [30]. The system is also able to gather subjective data through the smartphone where clinical questionnaires can be completed by patients.

The Psyche system also provides the chance for patients to establish communication with other patients in similar situations. The communication tool could be used to implement communities of users and to implement online group therapies mediated by a health professional involved in the long-term healthcare of bipolar patients. The use of the WWS sensing platform is voluntary: the smartphone allows the patient to activate and deactivate the sensors. The feedback from the patient is mediated by the physician, who can remotely assess symptoms and trends.

The final validation study of the Psyche system is ongoing in three different research centers. Fifteen complete platforms have been distributed, and for each platform a redundant number of WWSs have been manufactured to cover all sizes and models. So far more than 30 patients have been enrolled in the study. Each week the WWS is used for two nights to collect the full set of physiological parameters, while the smartphone is used during the whole week for the behavioral information. Data are automatically sent to a remote server through the smartphone. Preliminary results on data collected in the first phase of the project have already been published and are extremely encouraging [31,32].

9. CONCLUSIONS

Electronic textiles are a key enabling factor in the creation of SFIT-based systems, which are conceived as the integration of textile and non-textile sensors, computing capabilities, and an interactive communication network. SFIT systems rely on advances in such fields as material processing (fibers and polymers), microelectronics, signal processing, nanotechnologies, and telecommunication. Textile is the common platform where smart fibers are integrated, where the properties of the material are augmented through a combination of chemical processes, and where the structure of the fabric allows the use of redundant sensor configurations. The examples presented in this chapter illustrate applications based on textile platforms in which the fabric is enriched with new functionalities while maintaining the mechanical properties that make the material comfortable, conformable, and pleasant to the touch. The full potential of e-textile technology has not yet fully been explored and, in the future, novel flexible structures will be developed that will be sensitive to new electric, chemical, and biological variables. Within these structures properties such as toughness, elasticity, breathability, and appearance will merge with sensing capabilities. Meanwhile, the wearable e-textile platforms that have already been implemented are providing more and more data, generating information about people living normal lives. Future research will analyze these data and generate new strategies for the treatment of diseases and implement new tools for remote assistance. Knowledge in fields like medicine and physiology will also increase as well as the consciousness of people in maintaining healthy lifestyles and managing chronic diseases.

Glossary

ECG Electrocardiograph
EMG Electromyograph
EOG Electrooculograph
EEG Electroencephalograph
G Gauge factor (G) or strain factor is the percent of change in resistance per unit strain.
SFIT Smart fabrics and interactive textile
FR First responder
KPF Knitted piezoresistive fabric
T_KPF Textured knitted piezoresistive fabric
P_KPF Parallel knitted piezoresistive fabric

References

[1] <http://www.x-staticperformance.com>, (Last Accessed: 24.06.14).

[2] <http://www.shieldextrading.net>, (Last Accessed: 24.06.14).

[3] How spandex is made - material, manufacture, making, history, used, processing, parts, composition, structure, product, History, Raw Materials, The Manufacturing Process of spandex, Quality Control. [Online]. Available: <http://www.madehow.com/Volume-4/Spandex.html>. [Accessed 20.01.14].

[4] J. David. Spencer,"Knitting Techonology", Published by Woodhead Publishing Limited, Abington Hall, Abington Cambridge CB1 6AH, England, Third edition 2001.

[5] M. Pacelli, L. Caldani, R. Paradiso, Performances evaluation of piezoresistive fabric sensors as function of yarn structure, Conf. Proc. IEEE Eng. Med. Biol. Soc. 2013 (2013) 6502–6505.

[6] M. Pacelli, L. Caldani, R. Paradiso, Textile piezoresistive sensors for biomechanical variables monitoring, Conf. Proc. IEEE Eng. Med. Biol. Soc. 1 (2006) 5358–5361.

[7] F. Lorussi, E.P. Scilingo, M. Tesconi, A. Tognetti, D.D.e. Rossi, Strain sensing fabric for hand posture and gesture monitoring, IEEE Trans. Inf. Technol. Biomed. 9 (3) (2005) 372–381.

[8] R. Paradiso, G. Loriga, N. Taccini, A wearable health care system based on knitted integrated sensors, IEEE Trans. Inf. Technol. Biomed. 9 (3) (2005) 337–344.

[9] E.P. Scilingo, A. Gemignani, R. Paradiso, N. Taccini, B. Ghelarducci, D.D.e. Rossi, Performance evaluation of sensing fabrics for monitoring physiological and biomechanical variables, IEEE Trans. Inf. Technol. Biomed. 9 (3) (2005) 345–352.

[10] R. Paradiso, A. Gemignani, E.P. Scilingo, D. De-Rossi, Knitted bioclothes for cardiopulmonary monitoring, Proc. 25th Annu. Int. Conf. IEEE Eng. Med. Biol. Soc., 2003 4 (2003) 3720–3723.

[11] A. Helal, M. Mokhtari, B. Abdulrazak, distributor Smart Technology for Aging, Disability, and Independence: Computer and Engineering Design and Applications, Wiley; John Wiley, Hoboken, N.J.; Chichester, 2006

[12] N. Carbonaro, A. Greco, G. Anania, G. Dalle Mura, A. Tognetti, E. P. Scilingo, D. De Rossi, and A. Lanatà, Unobtrusive Physiological and Gesture Wearable Acquisition System: A Preliminary Study on Behavioral and Emotional Correlations, presented at the GLOBAL HEALTH 2012, The First International Conference on Global Health Challenges, 2012, pp. 88–92.

[13] G.D. Mura, F. Lorussi, A. Tognetti, G. Anania, N. Carbonaro, M. Pacelli, et al., Piezoresistive goniometer network for sensing gloves, in: L.M.R. Romero (Ed.), XIII Mediterranean Conference on Medical and Biological Engineering and Computing 2013, Springer International Publishing, 2014, pp. 1547–1550.

[14] G. Loriga, N. Taccini, D. De Rossi, R. Paradiso, Textile sensing interfaces for cardiopulmonary signs monitoring, Conf. Proc. IEEE Eng. Med. Biol. Soc. 7 (2005) 7349–7352.

[15] S. Coyle, K.-T. Lau, N. Moyna, D. O'Gorman, D. Diamond, F. Di Francesco, et al., BIOTEX--biosensing textiles for personalised healthcare management, IEEE Trans. Inf. Technol. Biomed. 14 (2) (2010) 364–370.

[16] G. Wallace, D. Diamond, K. T. Lau, S. Coyle, Y. Wu, and D. Morris, Flow analysis apparatus and method US20080213133 A104-Sep-2008.

[17] T.S. Licht, M. Stern, H. Shwachman, Measurement of the electrical conductivity of sweat its application to the study of cystic fibrosis of the pancreas, Clin. Chem. 3 (1) (1957) 37–48.

[18] D. Curone, E.L. Secco, L. Caldani, A. Lanatà, R. Paradiso, A. Tognetti, et al., Assessment of sensing fire fighters uniforms for physiological parameter measurement in harsh environment, IEEE Trans. Inf. Technol. Biomed. 16 (3) (2012) 501–511.

[19] <http://www.nonin.com>, (Last Accessed: 24.06.14).

[20] R. Paradiso, L. Caldani, Electronic textile platforms for monitoring in a natural environment, Res. J. Text. Apparel 14 (4) (2010).

[21] R. Paradiso, D. De Rossi, Advances in textile technologies for unobtrusive monitoring of vital parameters and movements, 28th Ann. Int. Conf. IEEE Eng. Med. Biol. Soc., 2006. EMBS '06 (2006) 392–395.

[22] R. Paradiso, A.M. Bianchi, K. Lau, E.P. Scilingo, PSYCHE: personalised monitoring systems for care in mental health, Conf. Proc. IEEE Eng. Med. Biol. Soc. 2010 (2010) 3602–3605.

[23] <http://www.e-sponder.eu>, (Last Accessed: 24.06.14).

[24] T. Giorgino, P. Tormene, G. Maggioni, C. Pistarini, S. Quaglini, Wireless support to poststroke rehabilitation: MyHeart's neurological rehabilitation concept, IEEE Trans. Inf. Technol. Biomed. 13 (6) (2009) 1012–1018.

[25] A. Tognetti, F. Lorussi, R. Bartalesi, S. Quaglini, M. Tesconi, G. Zupone, et al., Wearable kinesthetic system for capturing and classifying upper limb gesture in post-stroke rehabilitation, J. Neuroeng. Rehabil. 2 (2005) 8.

[26] J.A. Gallego, E. Rocon, J. Ibañez, J.L. Dideriksen, A.D. Koutsou, R. Paradiso, et al., A soft wearable robot for tremor assessment and suppression, 2011 IEEE Int. Conf. Rob. Autom. (ICRA) (2011) 2249–2254.

[27] L. Caldani, M. Pacelli, D. Farina, R. Paradiso, E-textile platforms for rehabilitation, Conf. Proc. IEEE Eng. Med. Biol. Soc. 2010 (2010) 5181–5184.

[28] L. Caldani, C. Mancuso, R. Paradiso, E-textile platform for movement disorder treatment, in: J.L. Pons, D. Torricelli, M. Pajaro (Eds.), Converging Clinical and Engineering Research on Neurorehabilitation, Springer Berlin Heidelberg, 2013, pp. 1049–1053.

[29] G. Valenza, C. Gentili, A. Lanatà, E.P. Scilingo, Mood recognition in bipolar patients through the PSYCHE platform: preliminary evaluations and perspectives, Artif. Intell. Med. 57 (1) (2013) 49–58.

[30] A. Greco, A. Lanatà, G. Valenza, G. Rota, N. Vanello, E.P. Scilingo, On the deconvolution analysis of electrodermal activity in bipolar patients, Conf. Proc. IEEE Eng. Med. Biol. Soc. 2012 (2012) 6691–6694.

[31] M. Migliorini, S. Mariani, A.M. Bianchi, Decision tree for smart feature extraction from sleep HR in bipolar patients, Conf. Proc. IEEE Eng. Med. Biol. Soc. 2013 (2013) 5033–5036.

[32] S. Mariani, M. Migliorini, G. Tacchino, C. Gentili, G. Bertschy, S. Werner, et al., Clinical state assessment in bipolar patients by means of HRV features obtained with a sensorized T-shirt, Conf. Proc. IEEE Eng. Med. Biol. Soc. 2012 (2012) 2240–2243.

3.2

Woven Electronic Textiles

Tomohiro Kuroda[1], Hideya Takahashi[2] and Atsuji Masuda[3]

[1]Kyoto University Hospital, Kyoto, Japan, [2]Osaka City University, Osaka, Japan,
[3]Industrial Technology Center of Fukui Prefecture, Fukui, Japan

1. INTRODUCTION

Wearable sensors are designed to be unobtrusively and seamlessly worn in everyday life. Therefore, most of the wearable sensors should be embedded into clothing. Although some of the wearable sensor designs rely on mounting electronic components on ready-made clothes using existing or added pockets and compartments, a better approach is to develop clothing based on fabrics with embedded electronics. The fabrics with embedded electronics are called e-textiles, electronic textiles or smart textiles [1]. Establishing an industrial production process to manufacture e-textiles is needed to make wearable sensors a commodity. This production process must be available for complex textiles since e-textiles are much more complex than conventional fabrics. Additionally, the production process should be available for both prototyping and mass production, just as with modern microchip production processes.

A common academic approach to develop e-textiles is embroidery. Embroidery is a handy method for prototyping simple e-textiles, as it requires just a single sewing machine. On the other hand, embroidery has clear drawbacks for mass production of complex e-textiles, because embroidery cannot be used for implementation of multi-layer circuits and the production speed is quite limited. Another often utilized approach is knitting [2]. However, knitted products are bulkier than woven products, which have more flow and can be made much thinner.

Weaving has been the most common approach to producing fabrics throughout history. Mass production of textile fabrics with graphical patterns of certain complexity was

FIGURE 1 A portrait of Joseph Jacquard woven by a Jacquard loom.

established as an industrial process in the eighteenth century when Joseph Jacquard (Figure 1) invented an automated loom controlled by punched cards. The Industrial Revolution, derived from Jacquard's invention, introduced the concept of "programming" by punched card, which also counted as an information revolution [3]. The long history of industrial development after Jacquard's invention has enabled the textile industry to produce complex-structured fabrics, such as multi-layered fabrics.

The key to implementation of machine-woven electronic textiles is using an appropriate conductive, or functional yarn available for weaving machine, and in establishing a weaving process for such non-standard yarn. This chapter provides several snapshots of frontier research on yarn production and development of the production process as well as basics of the textile industry. This chapter also provides several example prototypes of e-textiles created by weaving.

2. TEXTILES

Fabric, a flat fibrous structure, can be classified into three groups, non-woven cloth, knitted fabric, and textiles, according to its structure and production process.

A textile is composed by orthogonally crossing yarns, called warp and weft. Despite its rather simple structure, the variety, density, and crossing patterns of yarns derives textures with various natures. Textile made by hygroscopic fibers with open weave is good for summer clothing, whereas a textile with stiff structure of thin and strong fibers is preferable for tents. Thus, to produce a textile with certain nature, we need to find the best combination of structure and materials.

The following sections provide an overview of basic technologies of yarns and textile weave meant for e-textiles.

2.1 Yarn

Yarn is a long continuous length of interlocked fibers. Conventional yarns used for textiles are made of natural fibers such as silk, cotton, wool, and synthetic fibers such as polyester and polyamide. Special purpose textiles designed for industrial applications utilize specially designed fibers such as carbon fiber or aramid fiber. E-textile requires both conventional yarn and special yarn with conductive fibers or with fiber-mounted electronic devices. For example, carbon nanotube-plated yarn [4] and metalized aramid fiber [5] may be used in a textile heater.

Among various ways to produce conductive yarn, conductive yarn suitable for e-textile can be produced from two types of fibers, filamentous conductive metal (metal fiber) and conventional fiber plated with conductive metal. As the nature of metal fiber, such as its elasticity, is different from conventional fiber, various yarn-production techniques are applied to make the yarn applicable for machine weaving. For example, Figure 2 shows a conductive yarn produced by entwining metal fiber around conventional yarn, and Figure 3 shows a yarn composed of metal fiber braids. On the other hand, metal-plated yarn is easily applicable for machine weaving, as its nature is similar to conventional yarn. However, sometimes the metal plating comes off during the textile production process, and the metal-plated yarns lose conductivity. Several metal-plated yarns are available commercially [6]. Metal-based conductive yarn with low resistance is preferable to carry

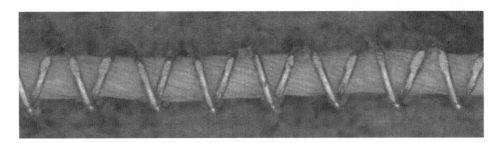

FIGURE 2   Entwining metal fiber around conventional (polyester) yarn.

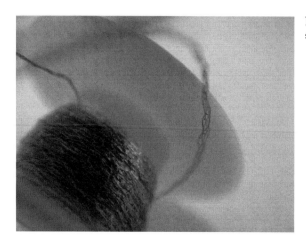

FIGURE 3 Metal fiber braids of three 40 μm stainless-steel fibers.

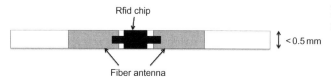

FIGURE 4 Conceptual sketch of RFID fiber.

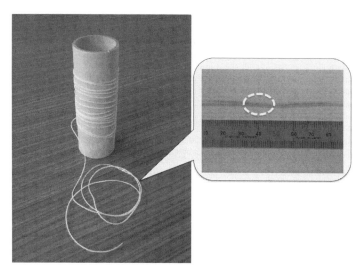

FIGURE 5 RFID fiber: IC chip is mounted within the yellow circle.

electric signals. On the other hand, carbon-based fibers, whose resistance is relatively high, can be applicable for heating.

Recently, electronic devices such as IC (integrated circuit) chips or LEDs (light emitting diode) were mounted onto or embedded into fibers. Figures 4 and 5 show monofilament fiber of 1.5 mm embedding RFID (radio frequency identification) chips. This fiber is

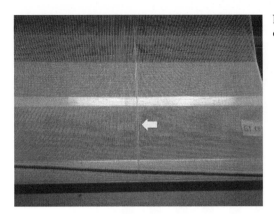

FIGURE 6 Textile embedding RFID: RFID fiber is embedded where marked with the arrow.

slightly thicker than ordinal yarn because of a resin layer covering the fiber. However, the nature of the fiber is similar to ordinal polyester, which is waterproof and available for machine weaving. With a high-strength fiber such as aramid as the core, the resulting RFID fiber becomes high strength as well [7]. The authors developed RFID fibers operating in different frequency bands, such as 2.45 GHz SHF RFID and 860 MHz to 960 MHz UHF-band RFID. The electronic textiles with embedded RFID fibers can be used for uniform management, counterfeit prevention, and even cotton management during the surgical operation (Figure 6).

2.2 Textile Weaves

A textile is composed of orthogonally crossing yarns, called warp and weft. The crossing pattern is called the textile weave and the minimal repeating pattern is called the weave repeat. The most basic of the three fundamental types of textile weave, plain weave, is composed so that the warp and weft are aligned to form a simple criss-cross pattern. Each weft goes over a warp, then under the next one. The pattern alternates as shown in Figure 7. As the weave repeat of plain weave is simply a crossing pattern of warp and weft, innumerable patterns are available.

Production of multi-layered structures in woven textiles is also an easy task and can be created by piling up warp or weft or both of them. Simple weave repeat enables us to pile up many threads. Figure 8 shows an example of weave repeat of two-layered textile, and Figure 9 shows the cross-section of five-layered textile. By introducing conductive yarn into multi-layered textile, we can embed a topologically complicated circuit with unconnected trace crossings within a textile. Figure 10 shows a simple LED switching circuit, with equivalent woven pattern suitable for multi-layered textile shown in Figure 11. As marked by the circle in the middle of the image, this circuit has intersecting but unconnected traces that cannot be implemented in a one-layered textile. Figure 12 shows the resulting prototype. In the prototype, intersecting but isolated traces are implemented by placing two conductive yarns in skew position within a two-layered textile. The layered structure makes the circuit durable to the deformations caused by body motion or washing [8].

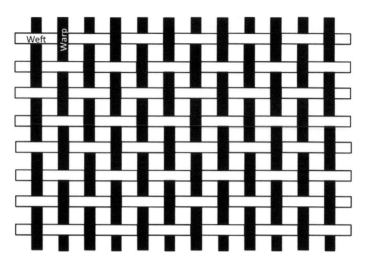

FIGURE 7   Illustration of plain weave.

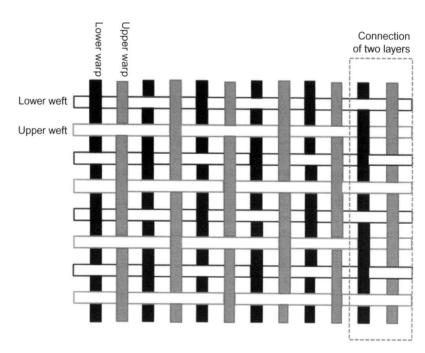

FIGURE 8   Illustration of multi-layered textile.

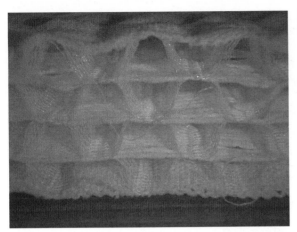

FIGURE 9 Cross-section of multi-layered textile.

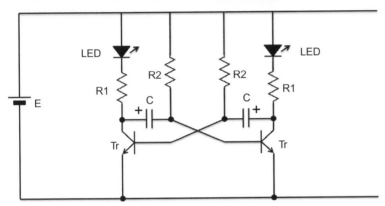

FIGURE 10 A stable multi-vibrator circuit.

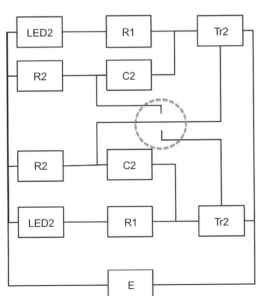

FIGURE 11 Equivalent pattern for e-textile.

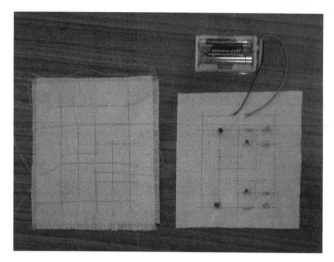

FIGURE 12   The resulting prototype.

FIGURE 13   Rapier loom.

2.3 Looms

A loom (Figure 13) is a textile-production machine. The loom controls two required motions: shedding motion to open warp (Figure 14), and insertion motion to insert weft between the raised and lowered warps (Figure 15). Looms can be classified by the variety of the two motions.

Common shedding motions enable most complicated textile weaves such as dobby and Jacquard. Common textiles 1 m wide use thousands of warps. The dobby loom handles warps in groups, whereas the Jacquard loom controls each warp.

The common dobby loom has 12 to 30 heald frames (Figure 16). Each heald frame handles one group of warps. For example, a dobby loom with ten heald frames handling 1,000

FIGURE 14 Shedding motion of rapier loom.

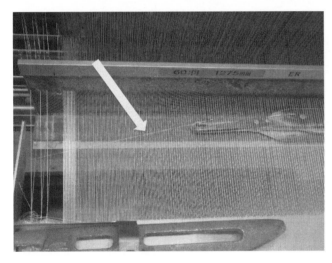

FIGURE 15 Insertion motion of rapier loom: White weft (indicated by the arrow) is inserted between warps.

warps produces textile of 10,000 warp ends wide with ten repeats of the weave going across. As the dobby loom can manage heald frame so flexibly, the dobby loom can produce complicated textile weaves.

The Jacquard loom (Figure 17) can handle each warp individually. This mechanism enables the Jacquard loom to produce large designs or complicated patterns the dobby loom cannot. Figure 18 shows an example of a large pattern produced by a Jacquard loom. Similar to the dobby loom, the number of independently controlled heddles (heddle set) defines the pattern complexity of the Jacquard loom. The number of heddles of commercially available Jacquard looms varies from hundreds to thousands. To weave a 5,000 warp ends-wide textile, a Jacquard loom with 5,000 heddle sets can handle all warps independently, and a Jacquard loom with 1,000 heddle sets repeats the same weave five times.

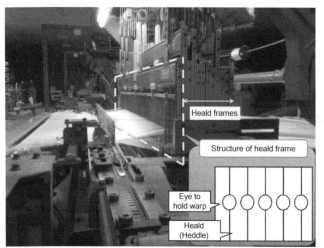

FIGURE 16 Heald frame of a dobby loom.

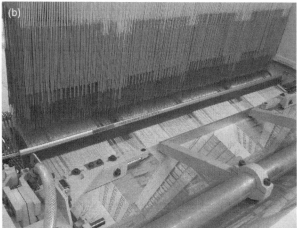

FIGURE 17 Jacquard loom. Left: overview, right: zoomed image of shedding warps.

Figure 18 shows an example of an 8,000 warp ends-wide textile using a Jacquard loom with 4,000 heddle sets. So the same pattern repeats twice along weft.

Common insertion motions available for various wefts are shuttle and rapier.

A shuttle loom inserts the weft into a boat-shaped case called the shuttle (Figure 19) and throws the shuttle between raised and lowered warp threads as shown in Figure 14. As the shuttle loom inserts the weft thread with a spool, the loom inserts the weft without

FIGURE 18   An example of a textile produced using a Jacquard loom.

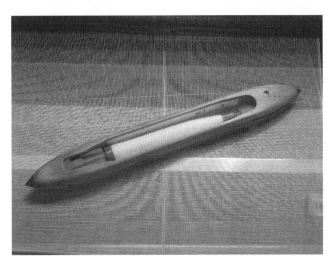

FIGURE 19   The shuttle.

cutting it. Therefore, the shuttle loom enables embedding of complicated circuits created from a conductive yarn weft. On the other hand, as the weft must be reeled up into the shuttle beforehand, productivity of the shuttle loom is low.

A rapier loom inserts weft using a long tape or a pole with a thread-holding mechanism, called a rapier, on its end (Figure 20). The rapier hooks and pulls the weft from one end to the other. Thus, the look and motion of the rapier is similar to a double-edged sword, just as its name suggests. Unlike a shuttle loom, the weft in the rapier loom must be cut on each insertion. On the other hand, the direct-hooking mechanism of the rapier loom allows for handling of thick yarns, fancy yarns (yarns with uneven thickness or loops), or elastic yarns. Therefore, due to its flexibility, a rapier loom is used for expensive textiles or complicated industrial materials.

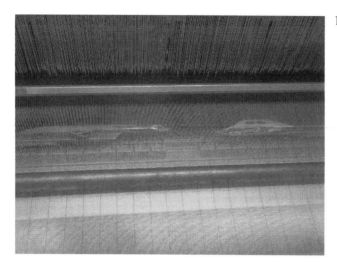

FIGURE 20   Rapier loom.

---

3. APPLICATIONS

This chapter covers four typical e-textile applications.

Section 3.1 introduces an application of a simple textile with conductive yarn. Section 3.2 introduces an application of a simple structured multi-layer textile with both conductive and non-conductive yarn, and section 3.3 presents an example of a complicated circuit woven by Jacquard and shuttle. Finally, section 3.4 introduces an example of a textile with device-embedded yarn.

3.1 Touchpad

This section introduces a fabric touchpad [9] that can be mounted on clothing (Figure 21) as an input interface for wearable computers. The touchpad is implemented by a simple weave with a conductive yarn.

Figure 22 shows the composition of the touchpad. The touchpad is composed of a square-shaped textile made of conductive yarns (conductive textile) and electrodes attached to the four edges of the textile. Same voltage e is applied to each electrode through signal pickup resistors. The conductive textile can be regarded as an equivalent to a resistor array as shown in Figure 23. Once the user touches the conductive textile with his/her fingertip, a capacitor is created between the fingertip and the textile and weak electric current is sent through the human body and the capacitor. The electric current through each of the pickup resistors varies due to the position of the fingertip. Therefore, the position of the fingertip can be pinpointed from the voltages measured at the pickup resistors R_1 and R_2.

Figure 24 shows the measurement mechanism of the finger position along the x-axis. When the size of the touchpad is $L \times L$, the position of the fingertip is x, the resistances are

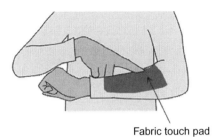

Fabric touch pad

FIGURE 21   Conceptual sketch of a fabric touchpad.

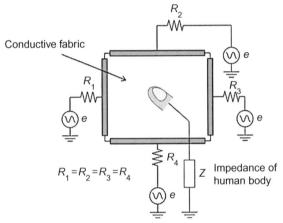

Conductive fabric

$R_1 = R_2 = R_3 = R_4$

Impedance of human body

FIGURE   22   The   composition   of   a   fabric touchpad.

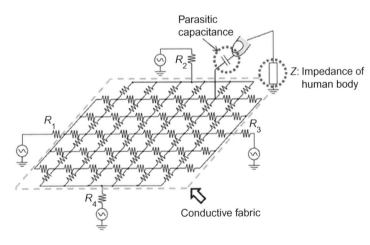

Parasitic capacitance

R_2

Z: Impedance of human body

R_1

R_3

R_4

Conductive fabric

FIGURE 23   The equivalent circuit of a fabric touchpad.

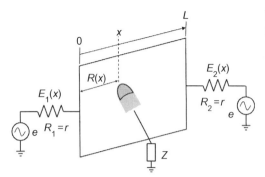

FIGURE 24 The mechanism to measure the fingertip position along the x-axis.

$R_1 = R_2 = R_3 = R_4 = r$, the impedance of the human body is Z, and the resistance between the fingertip and the resistor on the left end of the touchpad is $R(x)$. The voltages $E_1(x)$ on the resistance R_1 and $E_3(x)$ on the resistance R_3 are defined as follows:

$$E_1(x) = \frac{r(r + R(L) + R(x))}{- \{R(x)\}^2 + R(L) \cdot R(x) + R(L) \cdot (Z + r) + 2Z \cdot r + r^2} \cdot e \tag{1}$$

$$E_3(x) = E_1(L - x) \tag{2}$$

Thus, the position of fingertip along the x-axis, x, is given as

$$x = \frac{R(x)}{R(L)} \cdot L = \frac{E_1(L - x)}{E_1(x) + E_1(L - x)} \cdot \left(1 + \frac{2r}{R(L)}\right) - \frac{r}{R(L)} \tag{3}$$

The position of the fingertip along the y-axis, y, is given in the same manner as

$$E_2(y) = \frac{r(r + R(L) + R(y))}{- \{R(y)\}^2 + R(L) \cdot R(y) + R(L) \cdot (Z + r) + 2Z \cdot r + r^2} \cdot e \tag{4}$$

$$E_4(y) = E_2(L - y) \tag{5}$$

$$y = \frac{R(y)}{R(L)} \cdot L = \frac{E_2(L - y)}{E_2(y) + E_2(L - y)} \cdot \left(1 + \frac{2r}{R(L)}\right) - \frac{r}{R(L)} \tag{6}$$

Figure 25 shows the prototype. The prototype is a patch of $12\,\text{cm} \times 10\,\text{cm}$ conductive textile with four electrodes.

This prototype provides functionality typically available in a standard single-touch touchpad, such as pointing, selecting, clicking, and dragging. Figure 26(b) and (c) shows the measured voltage while the fingertip moves along the x-axis and y-axis defined as Figure 26 (a). The results in Figure 26 indicate that the measured voltage accurately pinpoints the position of the fingertip on the pad.

Because the prototype touchpad is composed of a plain-weave textile and resistors, the deformation of the textile does not harm its measurement performance, and the sensor is easy to wash and to mount on any clothes.

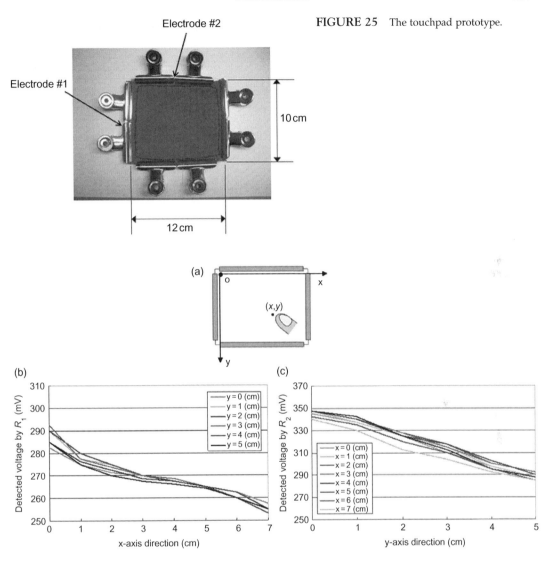

FIGURE 25   The touchpad prototype.

FIGURE 26   The property of the prototype: (a) given coordinate, (b) property along the x-axis and (c) property along the y-axis.

3.2 Textile Switch

This section introduces a textile switch [10] implemented through use of a multi-layered fabric with a simple structure. Figure 27 shows a sketch of a cross-section of the textile switch. Mounting multiple conductive yarns into multi-layered textile allows alignment of the yarns into a skewed position. The conductive yarns contact each other when the textile is pressed to crush the gap between the layers.

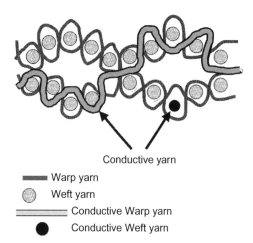

FIGURE 27   Conceptual sketch of cross-section of contact sensor.

Conductive yarn

▬ Warp yarn
◯ Weft yarn
▭ Conductive Warp yarn
● Conductive Weft yarn

The pressure required for the switching operation can be controlled by the materials of the yarns or the structure of the textile. The characteristic response time of the prototype textile switch is only 100 μs, which allows for switching action up to 30 Hz [11]. A cyclic pressure test, applying a pressure of 20 kPa at a frequency of 20 Hz reveals that the durability of the prototype is more than 1 million switching cycles.

The prototype is thin and flexible, and can be mounted anywhere; for example as a room entrance counter that can detect direction of passing people by utilizing a multi-section textile sensor [12].

3.3 Textile Electrodes

This section provides an example of an e-textile with a complicated weave pattern. A Jacquard loom with shuttle enables the creation of complicated circuit patterns from a single continuous conductive yarn as discussed in section 2. Japanese traditional weaving, NISHIJIN [13], is a good example of this type of weaving.

Figure 28 shows a prototype of an electrode for electrocardiogram (ECG) signal acquisition [14]. The conductive yarn is composed of three 40 μm stainless braids as shown in Figure 3. The electrode textile is woven by TSUZURE (brocade), which is a manual weaving operation with shuttles. The three circles and one dot are woven from independent conductive yarns, and each figure is woven from a continuous yarn. Thus, the electric properties of the textile are favorable for measuring weak bio-signals. If the electrode was composed of several independent conductive yarns contacting each other, then the weak bio-signals may be significantly attenuated at the connections between yarns.

Figure 29 shows another example: a twelve-lead ECG vest. The textile is produced by NUIWAKE; the machine weaving using a Jacquard loom and shuttle producing a TSUZURE-like weave. Here again, each electrode is composed of continuous conductive yarn (twisted silver-plated polyester and rayon). Figure 30 shows the obtained ECG signal.

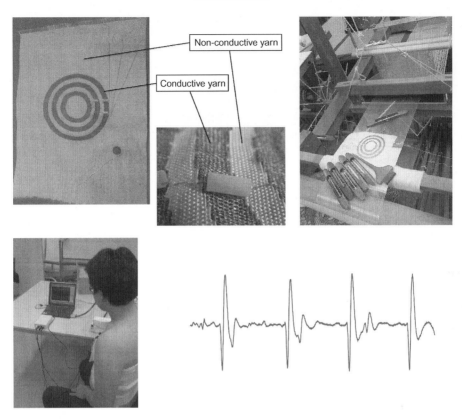

FIGURE 28 Simple electrode textile for electrocardiogram.

3.4 RFID Textiles

This section introduces textiles with electronic devices embedded into the yarn. RFID is widely used as a non-contact communication device. Passive RFID tags are typically thin and tiny, powered by the electromagnetic waves emitted by the transceiver, and are widely applied in the apparel industry, such as in uniform management. The conventional method to introduce an RFID tag into clothing is to attach it as a price tag or to sew the tag into the clothing. Embedding RFID tags during textile manufacturing may be preferable. This section presents two different methods to embed RFID tags into textiles.

The first approach is to weave a specially developed RFID fiber as shown in Figure 5, or a tape-shaped RFID yarn as shown in Figure 31 into the textile. The authors developed RFID yarns embedding 2.45 GHz SHF band, 13.56 MHz HF band, and 860 MHz to 960 MHz UHF band RFID tags. To insert this yarn as weft, the authors also designed and developed a new rapier loom as shown in Figure 32 [15]. The developed loom has both standard rapier and special-purpose rapier to insert tape-shaped RFID yarn. The loom switches between these two rapiers repeatedly and produces RFID textiles, as shown in Figure 33, automatically and continually. The produced RFID textile can be laminated processed into tile carpet (Figure 34) through a standard cloth-manufacturing process [16].

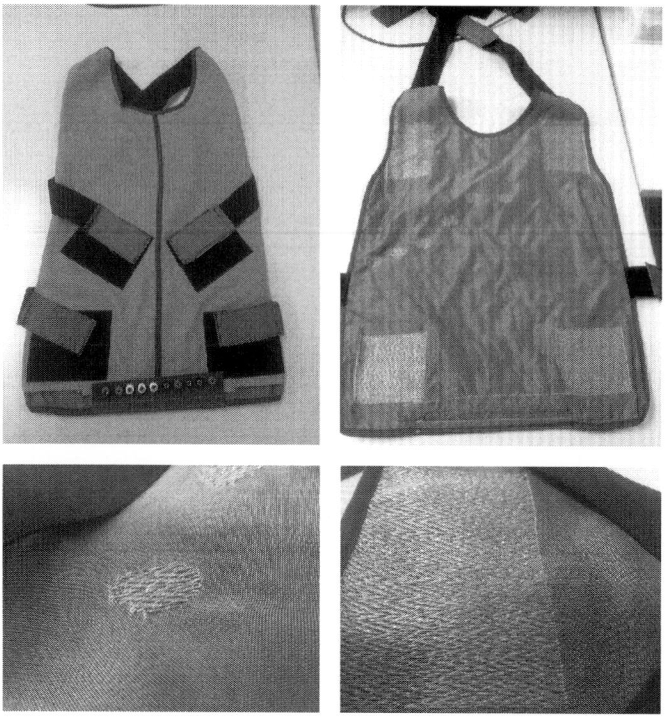

FIGURE 29   Twelve-lead ECG vest.

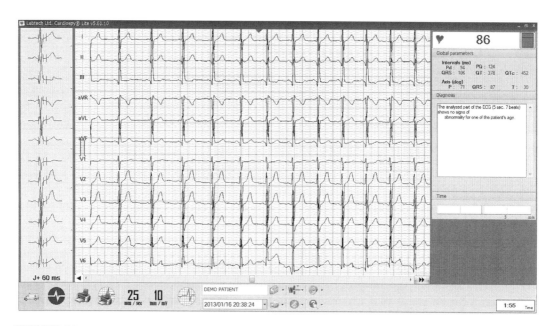

FIGURE 30   Obtained ECG by the twelve-lead ECG vest.

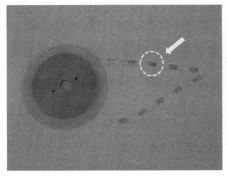 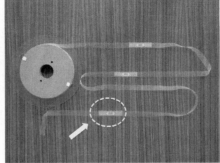

FIGURE 31   Tape-shaped RFID yarn.

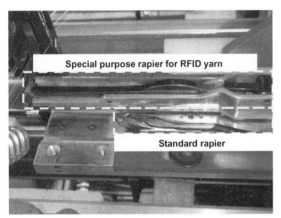

FIGURE 32   The special rapier loom for RFID yarn insertion.

Another approach is to directly embed RFID chips into textiles [17]. Figure 35 shows an example of an embedded washable UHF-band linen tag [18] integrated into the textile during the production process. The RFID tag is inserted into a pocket within two-layered textile using a slightly modified standard loom. Since the tag is fully integrated with the textile, it is impossible to remove the tag without cutting the textile.

RFID was originally designed for logistics and item tracking. The apparel industry primarily utilizes RFID for logistics [19], such as laundry service or uniform management. However, several other applications have been proposed.

One such application is positioning. By using a database of positions of multiple RFID tags embedded into the floor, a moving object such as a robot with an RFID reader can position itself [20].

Another possible application is in nursing, such as in urination detection for disposable diapers [21]. Figure 36 shows an RFID fiber embedded into a disposable diaper. As the RFID fiber shown in Figure 5 is just as soft as conventional yarn, unlike conventional RFID tags, the fiber won't decrease the touch and feel of the diaper [22,23]. Figure 37 shows the mechanism to detect urination. As illustrated in the upper portion of Figure 37,

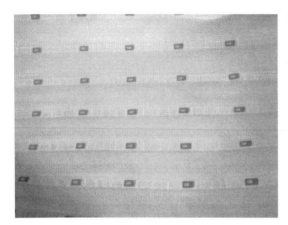

FIGURE 33   RFID textile (HF band).

FIGURE 34   RFD tile carpet.

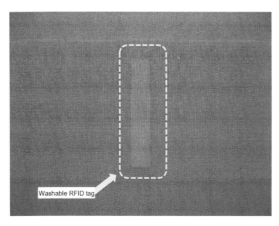

FIGURE 35   RFID textile with washable UHF-band linen tag.

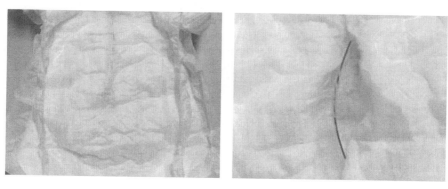

FIGURE 36   Paper diaper with embedded RFID fiber.

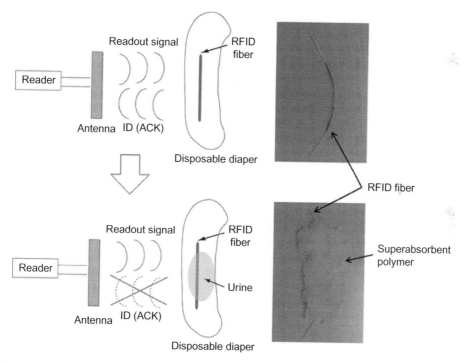

FIGURE 37   Principle of urination detection.

the RFID fiber responds with its own ID whenever it is activated by the reader, which also works as the power source. When the diaper is wet, as illustrated in the lower sketch of Figure 37, a high-molecular absorber that is full of moisture changes the antenna property of the RFID fiber and thus prevents it from responding to the reader's signal, causing the system to detect urination. The reader antenna can be placed under the bed pad or seat of a wheelchair.

4. SUMMARY

This chapter presented electronic textiles mainly produced by weaving. The weaving enables textiles to embed complicated circuits as shown by the NISHIJIN electrode example or specially designed electronic yarns as shown by the RFID textile example. Thus, the weaving technique can be a silver bullet to enrich a variety of e-textiles: it may be used to implement textiles with embedded LEDs, embedded physical or chemical sensors, and even textiles with integrated electronic circuits.

Wearable sensors made of such enriched e-textiles coupled with a ubiquitous computing environment may provide innovative ubiquitous information services. Figure 38 shows an example of such a health service. A twelve-lead wearable vest coupled with a sensor network with indoor localization capabilities [24,25] enables patients after cardiac events to perform rehabilitation under the eyes of a computational health condition assessment system, which continuously monitors vital signs and calls medical personnel in the case of an emergency.

The enriched e-textile may even change the paradigm of information services. Our daily life is full of textiles: carpets, sleeping mats, wallpaper, etc. Thus, e-textiles can be used not only for wearable sensors but also for housing materials. Once e-textiles acquire sensing, computational, or communication capabilities, they may enable our living environment to provide information services. An intelligent sleeping mat or sofa could check your health

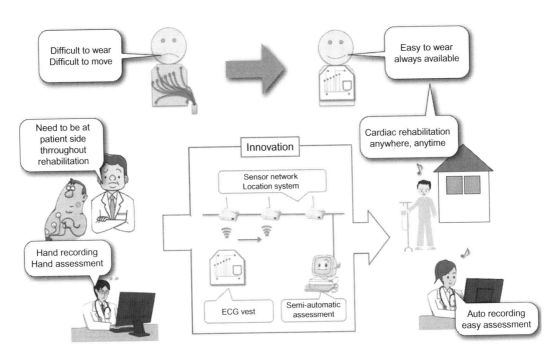

FIGURE 38   A rehabilitation system for cardiac patients with wearable sensor vest and ubiquitous sensor network.

condition throughout the day to report to your doctor, and intelligent wallpaper with air pollution, temperature, and moisture sensors could initiate ventilation and air-conditioning to keep air fresh and confortable.

Overall, e-textiles will likely penetrate most of the aspects of our daily lives in the very near future, and mass production of e-textiles by weaving techniques is one of the enabling factors for their broad dissemination.

Glossary

Brocade a class of richly decorative shuttle-woven fabric. The decorative pattern is created by non-structural supplementary weft to give the appearance of an embroidered pattern.
Embroidery decorating fabric with a needle and a thread of yarn.
Fabric a material made of fibers.
Heddle A component of a loom, which holds a warp through its eye to raise and lower it.
Knitted fabric fabric consisting of a number of consecutive rows of loops.
Non-woven cloth fabric made from long fibers bonded together by chemical, mechanical, heat, or solvent treatment.
Skew position (Mathematics) Neither parallel nor intersecting placement of two straight lines within three dimensional space.
(Woven) Textile fabric consisting of two distinct sets of yarns interlaced orthogonally.
Textile heater textile able to generate heat from electric power supply.
Warp threads that run lengthwise in a woven fabric.
Weft threads interlaced through warp in a woven fabric.

References

[1] M. Suh. E-Textiles for wearability: review on electrical and mechanical properties. Textile World 2010. <http://www.textileworld.com/Articles/2010/June/Textile_News/E-Textiles_For_Wearability-Review_On_Electrical_And_Mechanical_Properties>.
[2] G. Loriga, N. Taccini, M. Pacelli, R. Paradiso, Flat knitted sensors for respiration monitoring, IEEE Indust. Electro. Mag. 1 (2) (2008) 4–7.
[3] J. Al-Khalili, Order and disorder, BBC (2012).
[4] Matsubun Textile Co. Ltd. CNTEC. <http://www.matsubun.co.jp/product/cntec.html> (Japanese). (Last Accessed: 29.06.14).
[5] N. Martinez, K. Hisada, I. Tabata, et al., The effectiveness of thermal treatment for development of conductive metalized aramid fiber using supercritical fluid dioxide − fiber-metal adhesive strength improvement −, J. Supercrit. Fluids 56 (3) (2011) 322–329.
[6] Nihon Sanmo Dyeing Co. Ltd. Thunderon. <http://www.sanmo.co.jp/technology/Function/thunderon1.html> (Japanese). (Last Accessed: 29.06.14).
[7] H. Harii, M. Sekiguchi, Y. Iwasaki, et al., The development of RFID fiber, Proc. Textile Machinery Soc. Jap. Ann. Conf. (2012) 210–211 (Japanese).
[8] Y. Hayashi, M. Yoshida, H. Iijima et al. Development of the textile circuit board for the smart textile (part 2). Industrial technology center of Fukui pref. Research Report 2008: 25; 19–22 (Japanese).
[9] S. Fujii, H. Takahashi, Textile touch pad for wearable computer input device, Proc. Textile Machinery Soc. Jap. Ann. Conf. (2010) 72–73 (Japanese).
[10] A Masuda, T. Murakami, K. Kondo. Pressure sensor sheet. Jap. Pat. 4273233. 2009.
[11] A. Masuda, T. Murakami, K. Kondo, The respondent properties of the textile pressure sensors −the evaluation of compression properties of the multi-layer textile, J. Text. Eng. 56 (6) (2010) 3–5 (Japanese).
[12] A. Masuda, T. Murakami. Moving object detection system. Jap. Pat. 5233055. 2013.
[13] NISHIJIN textile industrial association. The history of NISHIJIN. <http://www.nishijin.or.jp/eng/eng.htm>. (Last Accessed: 29.06.14).

[14] T. Kuroda, K. Hiran, K. Sugimura, et al., Applying NISHIJIN historical textile technique for e-tetile, Proc. IEEE Ann. Conf. Eng. Med. Biol. Soc. (2013) 1226—1229.

[15] A. Masuda, T. Murakami, H. Ijima, The development of weaving machine for IC-tag, SENNI (M. Textile Machinery Soc. Jap.) 64 (7) (2011) 426—430 (Japanese).

[16] K. Hashimoto, The development of RFID textile, SENNI (M. Textile Machinery Soc. Jap.) 64 (7) (2011) 421—425.

[17] H. Iijima, A. Masuda, T. Murakami, et al., The development of IC tag insertion machine, Mach. Soc. Jap. Ann. Conf. (2012) 222—223 (Japanese).

[18] Fujitsu Frontech North America. UHF RFID Tag WT-A521/A522. <http://www.fujitsu.com/downloads/COMP/ffna/rfid/rfid_wt-a521_datasheet.pdf>. (Last Accessed: 29.06.14).

[19] M.C. O'Cornnor, Positek RFID offering UHF system for tracking linens, textiles, RFID J. (2010)<http://www.rfidjournal.com/articles/view?7535>. (Last Accessed: 29.06.14).

[20] T. Murakami Masuda, H. Takeuchi, et al., Development of radio frequency identification textile and map-making system, J. Textile Eng. 57 (1) (2011) 9—13 (Japanese).

[21] T. Kan. A study on RFID sensor system for detection of volume of urinary incontinence. Master thesis, Grad. Schol. Eng. Osaka City Univ. 2013 (Japanese).

[22] A. Tanaka, Y. Nakagawa, K. Kitamura, et al., A wireless self-powered urinary incontinence sensor system, Proc. IEEE Sens. (2009) 1674—1677.

[23] R. Takahashi, K. Yamada, Y. Ohno, et al., Development of new type incontinence sensor using RFID tag, Proc. Sound Music Comput. Conf. (2010) 2695—2700.

[24] T. Kuroda, H. Noma, C. Naito, et al., Prototyping sensor network system for automatic vital signs collection, Method Inform. Med. (2013).

[25] T. Kuroda, T. Takemura, H. Noma, et al., Impact of position tracking on the outpatient navigation system, Proc. IEEE Eng. Ann. Conf. Med. Biol. Soc. (2012) 6104—6106.

Flexible Electronics from Foils to Textiles: Materials, Devices, and Assembly

Giovanni Salvatore and Gerhard Tröster

ETH Zürich, Zürich, Switzerland

1. INTRODUCTION

While silicon is bulky and rigid, plastic electronics are soft, deformable, and lightweight. Devices based on this new technology are stretchable, twistable, and deformable into curvilinear shapes, thereby enabling applications that would be impossible to achieve by using the hard electronics of today. Rollable displays [1], conformable sensors [2], plastic solar cells [3] and flexible batteries [4] promise to change our daily lives like Complementary Metal-Oxide-Semiconductor (CMOS) technology did in the past. So far, the effective commercialization of such technologies has been mainly prevented by cost and performance constraints. However, for some applications, like medical or implantable devices, the specific functionalities provided by flexible, biocompatible, conformable, and light plastic electronics are much more important than the aforementioned obstacles, and future scenarios can be realistically foreseen.

Flexible electronic circuits are essential for the realization of such systems. Historically, progress in electronics has been dominated by a development path first noted by Gordon Moore in 1965: the number of transistors in silicon microprocessors tends to double every 18–24 months, mainly through reductions in the sizes of the transistors. This observation, known as Moore's law, led to miniaturization, power and cost reduction, and has guided the efforts of the microelectronics industry over the last decades. Continued downscaling represents an important future for electronics, but not the only one. In the early 1990s, a completely different class of electronics emerged to meet a need that was impossible to address with silicon wafer technology: active matrix circuits for switching pixels in liquid crystal displays. The development of such technology is definitely not dictated by Moore's law and, besides costs and performance, other requirements count. Here, the primary scaling metric is overall size: bigger is better because large area coverage means large video screens. From flexible displays, the scope has expanded to include more compelling and more technically challenging opportunities in biomedical devices which are minimally invasive and which can be worn or implanted. To meet these requirements new design strategies, materials, and fabrication schemes have to be considered and investigated.

Here, mechanical design is as important as circuit design. Devices and products often exploit curvilinear, ergonomic, or biologically inspired layouts. In the last few years, remarkable advancements have been achieved in the study and processing of materials that combine both good mechanical and electrical properties. This includes organic and graphene-like materials, carbon nanotubes (CNTs), inorganic nano-membranes or nanowires, and amorphous oxides. Finally, standard micro-electronics techniques have been complemented by new process schemes like printing, bottom-up self-assembling, and materials transferring from one substrate to another.

Mechanically flexible devices can be achieved by direct fabrication on plastic foil [5], by peeling off a polymer layer spin coated on a rigid substrate [6], by dissolving a sacrificial layer that liberates the thin membrane [7], or by spalling the thin top layer from a crystalline silicon wafer after device fabrication [8]. Each approach presents advantages and suffers limitations that have consequences in the cost and performance of the final components. The limitations are mainly due to the low-thermal budget and to the feature ultimate resolution. A common feature of any of these approaches is the use of thin films, which is a prerequisite to achieving highly bendable electronics. The minimum-bending radius is usually on the order of millimeters, and induced strain during bending is the main cause of failure in the active layers of devices. Smaller bending radius (<1 mm) can be achieved either by using materials with intrinsic mechanical capabilities, which are able to stand high-strain levels [9], by encapsulating the electronics on the zero-strain plane [7,10] or by using very thin substrates [11]. In these cases bending radii smaller than 100 μm can be reached [10−12].

This chapter begins with an overview of semiconductor materials used for the fabrication of thin-film transistors. The focus is on thin-film transistors and circuits that are fabricated directly on polyimide foils and based on amorphous indium gallium zinc oxide (a-IGZO) and high-k dielectric. Particular attention is devoted to the characterization of the device performance when mechanical strain is applied, and to the design rules that minimize the impact of deformation. The main goal of the study remains the identification of the critical technological parameters to improve the speed and achieve > 100 MHz operation. The last part of the chapter shows an original approach to integrating electronics into textiles that can be used in medical applications or in environmental monitoring. Finally, we draw conclusions about future trends in flexible electronics.

2. THIN-FILM TRANSISTORS: MATERIALS AND TECHNOLOGIES

The transistor is the basic building block of any digital and analog circuit. However, depending on the application, the evaluation of its performance changes and often some parameters are more important than others.

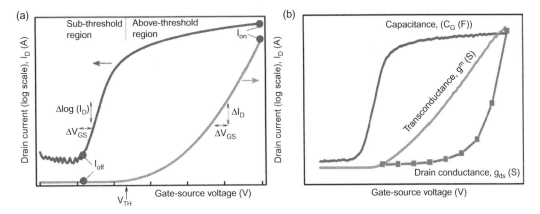

FIGURE 1   Drain current (a) and transconductance, drain conductance, and capacitance (b) as function of gate-source voltage of a thin-film transistor. Subthreshold swing, on/off current ratio, and threshold voltage stability are the most important parameters for digital logic. Transconductance and drain conductance are important factors in analog design. Transconductance and gate capacitance set the maximum operating frequency of the device.

CMOS digital logic is based on silicon complementary metal oxide semiconductors and consists of both n- and p-channel transistors that can switch (Figure 1(a)) between the on-state (with a large on-current, Ion, at $V_{GS} = V_{DD}$, where V_{DD} is the maximum voltage supplied to the device) and the off-state (with a small off-current, Ioff, at $V_{GS} = 0$). Thus, any device that is to be used in a similar design scheme must have excellent switching capabilities, on/off current ratio (Ion/Ioff) greater than 10^5, and symmetrical threshold voltages, i.e., $V_{TH,n} = -V_{TH,p}$. An obstacle that is usually encountered in thin-film technology is the difficulty in realizing complementary logic because of the lack of p- and n-type semiconductors with comparable values of mobility. Organic materials suffer from poor n-type semiconductors, while for amorphous oxide the opposite is true.

In a radio frequency (RF) circuit, switching capabilities are not required *per se*. For high-speed applications, thin-film transistors (TFTs) should respond quickly to variations in V_{GS}; this practically means short gates and fast carriers in the channel. The series resistances between the channel and the source and drain terminals are also important, and their adverse impact on the device becomes more pronounced as the gate length decreases. The transient frequency, f_T, is the most widely used figure of merit for RF devices and is defined as the frequency at which the current gain reaches the unity.

The transient frequency can be maximized by making the intrinsic transconductance, g_m, as large as possible and making the drain conductance, g_{ds}, the gate capacitance (which consists of the channel, C_{ch}, and gate to drain/source overlapping capacitances, C_{GD}/C_{GS}), and source/drain contact resistances (R_S/R_D) as small as possible (Table 1). However, the values of all these quantities vary with the voltage operating point (Figure 1 (b)). An optimum design consists of finding a compromise between the g_m and g_{ds}, which usually increase with the increase of V_{GS}. Drain-current saturation is also necessary to maximize the intrinsic voltage gain, Gint = g_m/g_{ds}, which has become a popular figure of merit for mixed-signal circuits.

TABLE 1   Definition of the Most Significant Parameters in a Thin-film Transistor

Quantity	Definition
Transconductance (S)	$g_m = \frac{\partial I_D}{\partial V_{GS}}\big\vert_{V_{DS}=const} = \frac{W}{L}C_{ox}\mu(V_{GS}-V_{TH})^2$
Drain conductance (S)	$g_{ds} = \frac{\partial I_D}{\partial V_{DS}}\big\vert_{V_{GS}=const}$
Subthreshold swing (mV/dec)	$SS = \frac{\partial \log(I_D)}{\partial V_{GS}}$
Gate capacitance (F)	$c_g = \frac{\partial Q_{ch}}{\partial V_{GS}}\big\vert_{V_S=V_D=0} = C_{ch} + C_{GD} + C_{GS}$
Transient frequency (Hz)	$f_T = \frac{g_m}{2\pi}\frac{1}{(C_{GS}+C_{GD})[1+g_{ds}(R_S+R_D)]+C_{DS}g_m(R_S+R_D)}$

Channel-length scaling is the most straightforward approach to achieving high-frequency operation, but it is very challenging, especially when printing techniques are used and in the case of direct fabrication on foils. This explains the continuous search for high mobility materials that can be processed at low temperature and deposited in thin films.

3. REVIEW OF SEMICONDUCTORS EMPLOYED IN FLEXIBLE ELECTRONICS

In recent years, remarkable advancements have been achieved in the study and processing of materials that exhibit good electrical properties and that can be processed at low temperature and be deposited in thin films. This includes organic [13] and graphene-like materials [14], carbon nanotubes (CNTs) [15,16], and inorganic materials like nanowires/ribbons/membranes [7,17] and amorphous oxides [18–20]. Here, the attention is mostly focused on one distinguishing feature, the mobility. As previously mentioned, field-effective mobility is very important both in digital and analog circuits; in fact, it has an impact on the on-current level, on the transconductance, and hence on the maximum operating frequency of the device (Table 1). Figure 2 shows the mobility range of the three classes of semiconductors that are usually used in the fabrications of flexible thin-film transistors.

The study of organic materials has been going on for decades, mainly driven by the quest for inexpensive, large-area flexible devices. Organic semiconductors comprise polymeric and molecular units and exhibit the common feature of π-conjugated bonds, which are responsible for the optical and electrical properties. The solid-state structure of such materials is based on weak interactions (Van der Waals) between neighboring molecules/polymer chains, and sets the properties from conductors to insulators. From a physical perspective, the crystal structure of the material should be as pure as possible (impurities act as traps) and provide sufficient overlap of frontier orbitals to allow efficient charge migration. Moreover, the highest occupied/lowest unoccupied orbital (HOMO/LUMO)

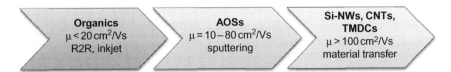

FIGURE 2 Semiconductors employed in flexible electronics classified according to field-effective mobility μ. Organic materials exhibit low mobility values but offer the possibility of large-area fabrication being compatible with solution processes and printing techniques such as inkjet and roll-to-roll (R2R). Inorganic materials, such as silicon nanowires (Si-NWs), carbon nanotubes (CNTs), and transition metal dichalcogenides (TMDCs), have high mobility but cannot be directly grown on plastic substrate because of the high temperature required. Amorphous oxide semiconductors (AOSs) offer a good trade-off between mobility, which is about 20 cm^2/Vs (at room temperature), and large-area deposition (sputtering or solution process).

should be at levels accessible at reasonable electric fields. Finally, since the transport occurs along the direction of the intermolecular π-π stacking, the molecules should be oriented parallel to the substrate normal. Among organic materials, single crystals are the most interesting from a performance point of view. Sundar et al. demonstrated a mobility of 15 cm^2/Vs in a transistor fabricated by laminating the elastomeric device stamp against a rubrene crystal [21]. Single crystals can also be grown by physical vapor deposition directly on the destination substrate and then processed by standard microelectronic techniques (lithography, etching, lift-off). However, organics are particularly interesting because they can be processed in solution form and are compatible with printing techniques such as inkjet [22] and R2R [23]. In this case, the mobility is in the order of 1 cm^2/Vs or less. Nevertheless, this makes them suitable for large-area electronics, hence, for solar cells [3] or for the realization of large-scale circuits that operate in the KHz range. Process reliability, stability to mixture and environment, and improvement of transport capabilities remain the main challenges for the technology.

Carbon nanotubes (CNTs) [15,16], graphene, and transition metal dichalcogenides (TMDCs) are good candidates for high-performance flexible electronics [14,24]. In particular, 2-D materials have attracted great attention thanks to the possibility of being grown in large areas, their small thickness which ensures high bendability, and their good transport properties. Pristine graphene lacks a bandgap, hence, field-effect transistors cannot be effectively switched off and have low on/off switching ratios. Bandgaps can be engineered using nanostructuring [25], chemical functionalization [26], and applying a high electric field to bilayer grapheme [27], but these methods add complexity and diminish mobility. In contrast, several 2-D TMDCs, such as MoS$_2$, MoSe$_2$, WS$_2$, and WSe$_2$, possess sizable bandgaps around 1 to 2 eV and field-effective mobility larger than 100 cm^2/Vs, allowing applications such as transistors, photodetectors, and electroluminescent devices. One obstacle in the adoption of such materials for flexible electronics is that they require high growth temperature. Thus, in order to be integrated onto plastic substrate they must be transferred from the support used for the growth to the substrate where electronics will be realized. In recent years enormous advancements have been registered in this field. Spin coating, liquid [28] or mechanical exfoliation [24] imprinting techniques have been complemented by other approaches that enable the transfer of a large portion of materials from one substrate to another [29,30]. Cao et al. [16] demonstrated transistors based on dense and perfectly aligned single-walled CNTs with a mobility of 480 cm^2/Vs, while in the case of grapheme-based TFTs a mobility of 203 cm^2/Vs has been reported. The work on transition-metal dichalcogenides is more recent, and flexible transistors made up of MoS$_2$, which is the most widely studied material, show a mobility that ranges between 15 cm^2/Vs and 50 cm^2/Vs [30]. The identification, the film uniformity, the placement and alignment, and the synthesis of good quality materials on a large scale, of both CNTs and graphene-like materials, are the main challenges for the realization of high-performance devices.

Among inorganic materials, amorphous silicon (a-Si) has been the most widely employed material for flexible displays. However, its mobility of about 1 cm^2/Vs has been outperformed by the above-listed materials and by amorphous oxides. However, inorganic materials also include Si and III-V nano-wires [6,17], nano-membranes, and nano-ribbons [7], which are usually derived from high-quality wafers and then transferred onto plastic substrate by printing or spin coating. A more recent approach consists of spalling technology, enabling the large-area transfer of ultra-thin body silicon devices to a plastic substrate at room temperature [8]. In all these cases performance is similar to bulk silicon, and mobility can reach about 300 cm^2/Vs, which enables circuits operating in the GHz regime. Besides performance, such schemes enable circuitry complexity, which, at the moment, cannot be achieved in any other way. However, the transfer of nanomaterials, often in small amounts and on small areas, poses obstacles to the realization of devices on a large scale and at competitive costs.

Amorphous oxides will be discussed in the following section. The importance of such materials is proved by its use in the switching matrix of large-area displays.

4. THIN-FILM TRANSISTORS BASED ON A-IGZO

Amorphous oxides have been studied for decades, and initial attempts occurred as early as the 1960s. However, only forty years later, with the work of Hosono and Nomura [18], a significant worldwide interest appeared, especially driven by active matrix for organic light emitting diodes (AMOLED) technology, both in academia and industry.

The impressive progress, in relatively short time, occurring in the study, processing, and use of such materials, is due to some key features:

- They can be produced at low temperature (even at room temperature) [18] and hence they are compatible with the low thermal budget of plastic substrates.
- They have highly smooth surfaces, characteristic of amorphous structures, and no grain boundaries, thereby obviating the primary limitation of mobility in polycrystalline semiconductors. This offers advantages for integration and for high-quality interfaces [20].
- They exhibit good transport capabilities (mobility $\approx 10-80c$ m^2/Vs) that do not depend on the degree of film disorder [18].
- They are compatible with large-area fabrication since they can be deposited by RF sputtering [31] or more recently by solution process [20].
- They are transparent in the visible spectrum [32].

A week point of this class of materials is that p-type semiconductors do not exhibit transport properties comparable to the n-type. In fact, until now, despite the increasing interest, there has been no report of p-type oxide TFTs that show values of mobility similar to n-TFTs.

Concerns are also often raised about the stability of such materials. The environmental and long-term stability of oxide semiconductor-based transistors is influenced by the absorption and desorption of oxygen (O_2) and water (H_2O) molecules from the ambient atmosphere. These O_2 and H_2O molecules can act as acceptors in a-IGZO. Consequently, shifts in the threshold voltage can occur. A possible way to increase the long-term stability of IGZO TFTs is the encapsulation of the channel by a passivation layer or an additional metallic top gate [33,34].

4.1 Thin-Film Transistor Fabrication and Characterization

The low-temperature processing of these materials enable the fabrication of devices directly on plastic foils [31]. Different device architecture can be adopted [20] (staggered or coplanar bottom or top gate). Here, the focus is on staggered bottom-gate devices because of the easy processing and enhanced electrical properties. The cross-section including materials and layer thicknesses of an amorphous indium gallium zinc oxide, a-IGZO, TFT is shown in Figure 3(a). The fabrication involves five photolithography mask steps, and is carried out at a maximum process temperature of 150 °C [5,19]. Here, IGZO is deposited by room temperature RF magnetron sputtering, the Al_2O_3 gate insulator and device passivation layers are deposited by atomic layer deposition, and metal contacts are e-beam evaporated. A micrograph of a fully processed device is shown in Figure 3(b).

An IGZO transfer characteristic measured under ambient conditions is shown in Figure 4(a). Typical performance parameters of flexible IGZO TFTs are: linear field effect mobility = 15 cm^2/Vs, saturation field effect mobility = 14 cm^2/Vs, threshold voltage = 0.33 V, on/off current ratio = 2.7 × 10^6, and subthreshold swing (inverse of sub-threshold slope) = 150 mV/dec, extrapolated using the Shichman and Hodges equations to model the MOSFET current. The gate-source leakage current is always smaller than

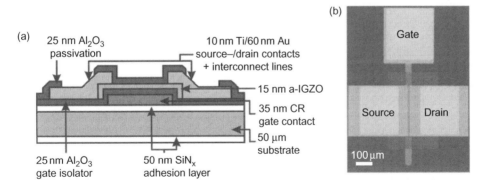

FIGURE 3   (a) Cross-section of a flexible IGZO TFT and (b) micrograph of a fully processed IGZO TFT (W/L ratio is 280 μm/10 μm).

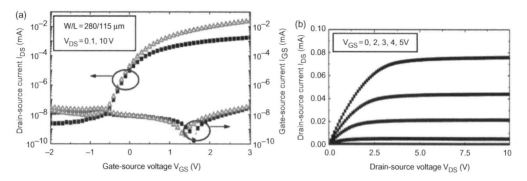

FIGURE 4 (a) Typical IGZO TFT transfer characteristic measured at source-drain voltages of 0.1 V (linear regime) and 10 V (saturation regime). (b) Typical a-IGZO TFT output characteristic with a maximum saturation current of about 75 μA.

0.1 nA. The family curve plot of the same IGZO TFT is shown in Figure 4(b). Here, a clear saturation behavior for high drain-source voltages is visible. Corresponding to the extracted threshold voltage, the drain-source current is saturated above a drain source voltage of ≈ 4.5 V (for an applied gate-source voltage of 5 V).

The influence of electrical stress on IGZO TFTs, and, therefore, their electrical stability, can be determined by gate-bias stress measurements: the transfer characteristic of a device is measured and, afterwards, a gate-bias stress field (E_{Str}) of 1×10^8 V/m is applied to the gate contact (with respect to the 25 nm thick gate insulator, this corresponds to a gate-source voltage of 2.5 V). Finally, after the stress field is turned off again, the TFT transfer characteristic is measured for a second time. This procedure (performed in the darkness) has to be repeated for different stress times (t_{Str}), whereas a new TFT has to be used for each measurement. The measured transfer characteristics can then be used to extract the threshold voltage V_{TH}, before and after the gate-bias stress was applied, and therefore the threshold voltage shift (ΔV_{TH}) induced by electrical stress. For a gate-bias stress time of 2,000 s, the threshold voltage of the flexible IGZO TFT shown in Figure 3 is increased by ≈ 86 mV. Due to the Al_2O_3 passivation, IGZO TFTs are by up to two orders of magnitude more stable than standard hydrogenated amorphous silicon (a-Si:H) TFTs. This is shown in Figure S1 (supplementary figures), which draws the ΔV_{TH} for a-IGZO TFTs and standard a-Si:H TFTs. Over the whole measurement range the influence of electrical stress on a-Si:H TFT threshold voltage is more than one order of magnitude larger than the influence on a-IGZO TFTs.

The thin-film transistors can be characterized also for their AC performance. The use of a ground-signal-ground layout for the transistor (Figure 5(a)) enables the AC characterization by using a standard two-port network analyzer. Port one and port two of the network analyzer can be connected to the gate and the drain contact of the TFT (Figure 5(b)), whereas the source is grounded. All AC measurements are performed in the saturation regime of the flexible a-IGZO TFTs, therefore bias voltages of $V_{GS} = 2$ V and $V_{DS} = 2$ V are applied using an external DC voltage source. The AC peak-to-peak voltage amplitude during the measurements is 100 mV. Prior to all measurements the system must be calibrated using a standard calibration procedure (short, open, 50 Ω load).

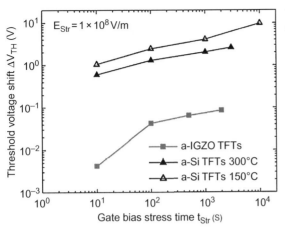

FIGURE S1 Electrical stability of IGZO TFTs compared to the stability of standard a-Si:H TFT's, measured at a constant stress field E_{Str} of 1×10^8 V/m.

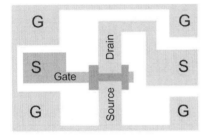
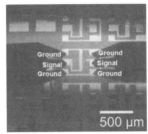

FIGURE 5 (a) Ground-signal-ground layout of a transistor and (b) optical microscope image of a contacted transistor with the network analyzer probes.

Figure 6 shows the measured S-parameters for a flexible a-IGZO TFT with a channel length of 1 μm.

To determine the transit frequency f_T, S-parameters are used to calculate the frequency-dependent current-gain h_{21}. The result of a typical device is plotted in Figure 7. The unity-gain frequency of the absolute value $|h_{21}|$ of the current gain defining f_T is extracted to be 47 MHz. An alternative way to extract f_T is the Gummel method and based on the inverse slope of the imaginary part of h_{21} at small frequencies.

4.2 Influence of Mechanical Strain

To determine the influence of mechanical strain on the performance of flexible IGZO TFTs, the devices can be attached on a reusable carrier substrate and loaded between two parallel plates in a custom-built bending tester [35]. This bending tester allows the electrical characterization of TFTs while bent to arbitrary radii.

The effects of mechanical strain induced by bending on IGZO TFT transfer characteristic is shown in Figure 8. Here, measurements taken while flat are compared to measurements done while the TFTs are bent to a tensile radius of 3.5 mm (tensile mechanical strain $\varepsilon = 0.7\%$ according to a simple calculation). Bending is performed parallel and perpendicular to the TFT channel, and therefore also parallel and perpendicular to the current flow.

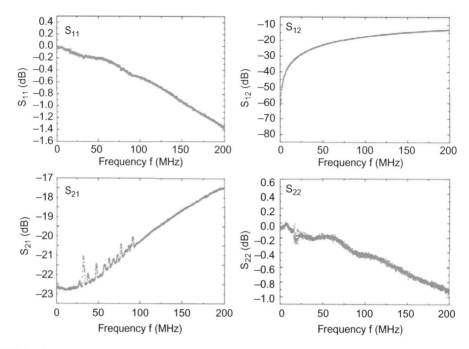

FIGURE 6   S-parameter characterization of a 1 μm long channel thin-film transistor.

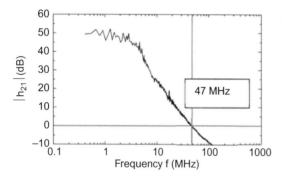

FIGURE 7   Cut-off frequency extraction for a 1 μm long channel thin-film transistor based on IGZO.

The graphs show the different effects of strain depending on the direction relative to the TFT channel. Parallel bending increases the drain current by $\approx 3\%$ (Figure 8(b)). This is mainly due to an increase of the field-effect mobility under tensile strain. At the same time, strain perpendicular to the TFT channel causes significantly larger variations of the drain current. Figure 8(a) shows in particular a reduction of the on-current by $\approx 22\%$, and an increase of the off-current by four orders of magnitude. This performance degradation is caused by the formation of capillary cracks, perpendicular to the applied mechanical strain. An SEM image of a capillary crack is shown in Figure 9(a). The average width of the cracks while the substrate is bent to a radius of 3.5 mm is ≈ 90 nm. The cracks propagate through the chromium, a-IGZO and Al_2O_3 layers. An average distance between two

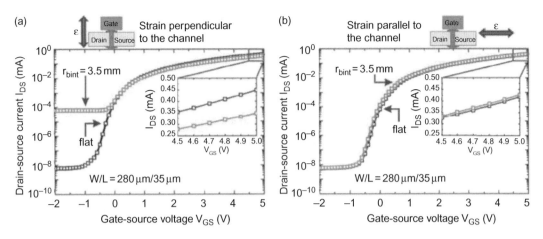

FIGURE 8 IGZO TFT (W/L = 280/35 μm) characterized while flat and with applied strain of 0.72% (bending radius: 3.5 mm) (a) perpendicular and (b) parallel and to the channel.

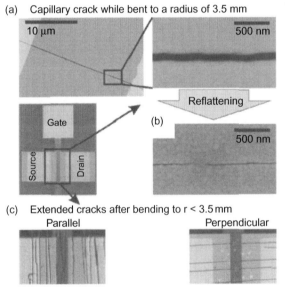

FIGURE 9 (a) SEM images of a capillary crack parallel to the TFT channel induced by tensile bending (ε = 0.72%) perpendicular to the channel. (b) Capillary crack after reflattening the substrate. (c) Extended cracks induced by tensile bending (ε >0.72%) parallel and perpendicular to the channel.

neighboring capillary cracks of ≈100 μm was measured; the minimum measured distance between two cracks is 31 μm. In approximately 50% of all cases only one crack within one TFT channel was visible. Therefore, and because the width of the tested TFT channels is much larger than their length (W/L = 280 μm/35 μm), capillary cracks were only observed during perpendicular bending. Consequently, perpendicular bending disconnects a part of the gate that remains floating. The floating and hence uncontrollable charges in the disconnected part of the gate result in an increased off-current. At the same time the on-current is reduced due to the reduced effective W/L ratio. Since the capillary cracks do not occur

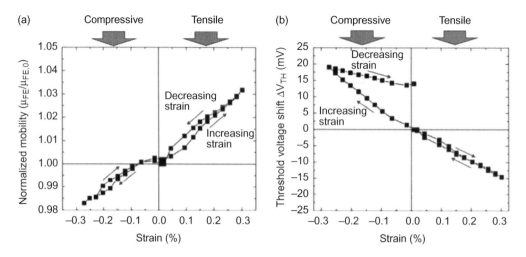

FIGURE 10 (a) Normalized effective mobility and (b) threshold voltage shift for increasing and subsequently decreasing strain. All values are extracted from the same experiment. The time interval between two consecutive measurement points is always 5 minutes.

in specific locations, but on random positions, the curves in Figure 8(a) are exemplary and could also suffer from more or less distinct drain current variations.

In addition to the described current variations, strain-induced cracks have two further effects:

- Reflattening of the substrate reestablishes the initial TFT characteristic. Capillary cracks are still visible on the flat substrate, but their width is drastically reduced. Figure 9(b) shows that the average width of capillary cracks after reflattening is 16 nm. At the same time both edges of the cracks have contact at several locations. Hence, reflattening reestablishes the electrical contact between the separated parts of the gate contact, and the original W/L ratio is recovered.
- Independent from parallel or perpendicular bending, strain >0.7% (bending radius <3.5 mm) permanently destroys the TFTs. This is because of the formation of extended cracks, as shown in Figure 9(c), causes short circuits between different material layers. In this case reflattening does not recover the original TFT characteristic.

The already mentioned influence of mechanical strain on the effective mobility of IGZO TFTs prior to the formation of cracks is shown in Figure 10(a). The graph shows the impact of tensile and compressive bending. Additionally, the strain-induced shift of the threshold voltage is plotted (Figure 10(b)). The effective mobility changes by $+3{:}1\%$ $(-1{:}8\%)$ and V_{TH} is modified by -15 mV $(+19$ mV) under tensile (compressive) bending down to a radius of 8 mm ($\varepsilon \approx 0{:}3\%$). Since the TFTs are not damaged at a bending radius of 8 mm, the TFTs can be reflattened after bending. As seen in Figure 10, reflattening leads to a nearly complete recovery of the performance parameters to their initial values.

This observed shift can be explained by the fact that tensile (compressive) strain modifies the distance between the atoms. Hence, strain decreases (increases) the energy level splitting of the binding and anti-binding orbitals between the atoms in the semiconducting layer [36]. This changes the carrier density at a constant temperature and causes a threshold voltage shift in the observed directions. The increase (decrease) of the effective mobility can be explained by a change of the electron-lattice interaction due to variations of the inter-atomic distance, which results in a change of the effective mass [37] and affects the mobility.

It should be noted that, because the strain is applied parallel to the channel, the channel width is constant during the presented bending experiment, whereas the channel-length variation during bending is small (max. 0.3%). Therefore, the influence of the W/L ratio change on the TFT performance parameters is also small compared to the observed parameter variations. Hence, the described shifts of effective mobility and threshold voltage are mainly due to a change of the IGZO material properties.

4.3 Analog and Digital Circuits Based on a-IGZO

The abovementioned experimental findings, concerning the influence of the direction and strength of mechanical strain in TFTs, can lead to some simple rules for the design of circuits:

- Driver TFTs should be strained only parallel to the channel to avoid capillary cracks and maintain switching performance as long as possible.
- Load TFTs can be oriented in both directions relative to the driver TFTs without compromising the circuit operation, although a perpendicular alignment will increase the load resistance and modify the circuit performance if strained.
- To avoid capillary cracks perpendicular to the channel, the gate length of TFTs strained parallel to the channel should be smaller than the expected minimal distance between two neighboring capillary cracks.

In the next section this guideline will be used to fabricate analog and digital circuits.

4.3.1 Digital Circuits

The digital input and output signals of exemplary a-IGZO inverter and NAND gates under mechanical strain are shown in Figures 11 and 12, respectively. Both gates are measured before bending, while strained to 0.72%, and after bending (reflattened). Additionally, Figure 11(b) shows the analog characteristic of the inverter with load TFT aligned perpendicular to the driver TFT while flat, and strained by 0.72%.

As expected from the TFTs characterization, the digital circuits with all TFTs in parallel are insensitive to bending. On the one hand, tensile bending increases the field effect mobility. At the same time the output levels of NANDs and inverter gates are determined by the ratio between the conductivities of the load and driver TFTs. This ratio remains constant if all TFTs are stained by the same factor and in the same direction. A change of the rise and fall times (t_r and t_f) is not observed either.

Circuits with perpendicular-strained load TFTs show an increase of the load resistance due to the smaller load TFT W/L ratio caused by capillary cracks. This results in a lowering of the output voltage levels and an increase in the rise time.

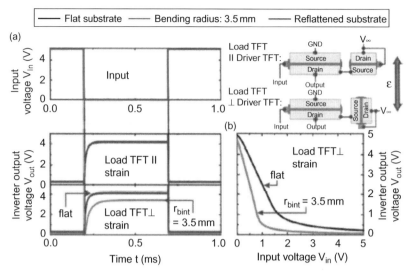

FIGURE 11 (a) Input and output signals of fabricated inverter gates, before bending, with strain applied parallel to the driver TFTs and hence perpendicular and parallel to the load TFTs, and reflattened. (b) DC characteristic measured with a parameter analyzer of the perpendicular-aligned load TFTs while flat and strained (0.72%).

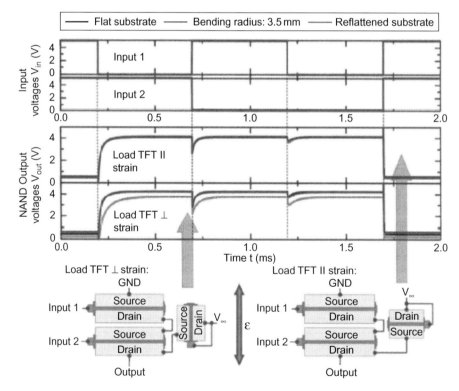

FIGURE 12   Input and output signals of exemplary NAND gates before bending, with strain applied parallel to the driver TFTs and hence perpendicular and parallel to the load TFTs, and reflattened.

(a)

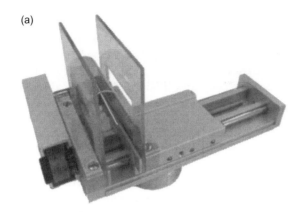

FIGURE 13 (a) Custom-made bending machine and (b) output parameter of NAND gates with strain applied parallel to the driver TFTs and hence perpendicular and parallel to the load TFTs for different numbers of bending cycles.

(b)

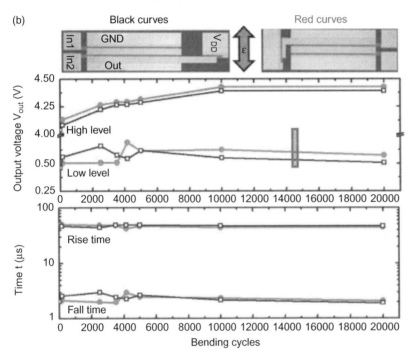

The influence of mechanical strain on a longer time scale can be evaluated by repeatedly bending and reflattening the circuits through the use of a custom-made bending machine (Figure 13(a)). In this case, the minimum bending radius is 4.5 mm. Figure 13(b) shows the NAND parameters during this cycling experiment, measured at the flat substrate after different number of bending cycles. While the rise and fall times as well as the low voltage levels of the output signal are nearly constant, the high voltage level increased by ≈ 0.4 V, independently from the orientation of the load TFTs. Therefore, mainly the driver TFT off-current is modified because of the increase of the threshold voltage by ≈ 100 mV [38].

4.3.2 Analog Circuits

Here, examples of amplifiers based on IGZO thin-film technology are presented and discussed. In particular, the attention is on a transimpedance amplifier (TIA), which is a basic block of a near infrared spectroscopy system (NIRS), which is presented later in the chapter, and on common source and cascode amplifiers that operate above 1 MHz.

Transimpedance amplifiers, which convert a photocurrent into a voltage, are a key component to electrically measure light intensity, e.g., in light barriers or near infrared spectroscopy systems. Figure S2(a) shows a schematic of a mechanically flexible transimpedance amplifier [39]. The amplifier is an enhancement load-gain stage (load resistor Ri) with a feedback resistor Rf [40]. The resistors are made of TFTs with a floating gate to use the channel resistance as a resistor (Figure S2(b)). To determine the performance of the bendable TIA, an input photodiode is connected between the input and ground and applied a supply voltage of 5 V. The input photodiode is stimulated with an LED sending a sinusoidal light signal. This measurement setup resembles future applications of the TIA. Applying a tensile-bending strain of 0.5% reduces the transimpedance gain of the unstrained TIA from 86.5 dBV by 1 dBV to 85.5 dBV, and increases the cut-off frequency from 8.38 to 9.83 kHz (Figure S2(c)).

Simulations based on static and dynamic characteristics of single TFTs can be used to design analog amplifiers [41,42]. Munzenrieder et al. recently demonstrated common source and cascode flexible amplifiers that work above 1 MHz. Schematic circuit diagrams and micrographs of the amplifiers are shown in Figure 14(a),(b). Both amplifiers are designed to work with a V_{DD} of 5 V. The common source amplifier biased at an input voltage $V_{DC} = 1.5$ V exhibits a gain, G, of 6.8 dB, a cut-off frequency, f_C, of 1.2 MHz (Figure 14 (c), black curve) and a power consumption of 690 μW. In order to have a lower power consumption, the cascode amplifier was operated at $V_{DC} = 1.25$ V and at a cascode TFT bias voltage $V_{BIAS} = 2.75$ V. Here, a gain of 7.8 dB, an f_C of 840 kHz (Figure 14(d), black curve),

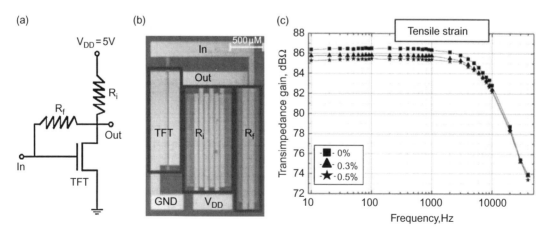

FIGURE S2   (a) Schematic of mechanically flexible TIA made of one TFT, a load resistor Ri, and a feedback resistor Rf; (b) micrograph of fabricated TIA; and (c) bode plots of flexible transimpedance amplifier with 0, 0.3, 0.5% applied tensile strain.

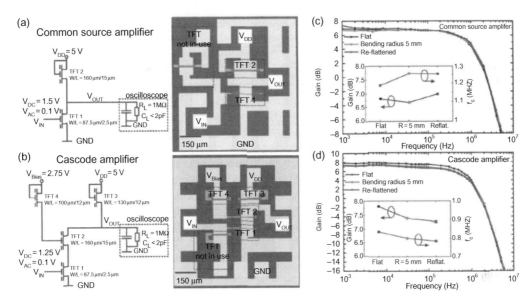

FIGURE 14 Circuit schematics, micrograph, and Bode plots of the fabricated flexible amplifier circuits: (a) Common source amplifier and (b) cascode amplifier. The additional TFTs to bias the inputs were not used. Each circuit occupies an area of $\approx 0.9 \times 0.9$ mm². (c) Frequency response of common source amplifier and of (d) a cascode amplifier. The insets show the evolution of gain and cut-off frequency. The amplifiers are fully operational when strained by 0.52%.

and a power consumption of 395 µW were measured. Higher operating frequencies at the cost of higher power consumptions can be achieved by choosing different bias points.

As in previous cases, the impact of mechanical strain on the flexible amplifiers is evaluated by bending the circuits to a tensile radius of 5 mm parallel to all TFT channels. Figure 14(c),(d) show the resulting Bode plots for the circuits flat, bent, and reflattened. Mainly because of the nearly invariant transconductance ratio and f_T, the measured variations for G and f_C (between flat and bent amplifiers) are less than 4% and 6% for the common source and less than 7% and 6% for the cascode amplifier. Bending to radii <5 mm causes cracks and permanently harms the amplifiers. Experiments also prove that the amplifiers stayed fully operational after 1,000 bending cycles.

5. FURTHER IMPROVEMENTS AND LIMITATIONS

The use of high-mobility materials and channel-length scaling are the most effective strategies to achieve high-speed devices.

As seen in previous sections, the channel and the gate to source/drain overlap length have to be as small as possible to maximize the transient frequency. At the same time, the reduction of the TFT dimensions is limited by the deformation of the free-standing plastic substrate during the fabrication process. This deformation of flexible substrates is caused, for example, by the thermal expansion during the TFT fabrication process, stress induced by the deposition of different material layers, or the absorption of different liquids in the fabrication process. To enable successful alignment of different device layers on a substrate with variable dimensions, tolerances on the photolithography masks are required. These tolerances in general limit the dimensions of flexible TFTs to several micrometers [31]. One possibility to overcome this limitation is to use self-alignment techniques [43]. Since no mask alignment is necessary, self-alignment overcomes the problems related with the deformation of the substrate. In the case of flexible bottom gate TFTs, self-alignment of the source and drain metallization using the gate contact is an approach to reduce the channel length and the gate to source/drain overlaps [44].

Alternative device structures could also offer opportunities to enhance performance. Double-gate architecture is particularly important for digital applications since it positively impacts on gate-channel coupling and, hence, improves the subthreshold swing and transconductance. In a vertical transistor, instead, the channel length is not limited by lithography or printing and could offer the concrete possibility to realize ultra-scaled devices on plastic substrate.

Examples of self-aligned double-gate and vertical transistors are described in the next sections.

5.1 Thin-Film Transistors by Self-Aligned Lithography

The fabrication is similar to the one adopted for conventional lithography. Titanium bottom-gate is structured by lift-off and electrically isolated by a 25 nm thick atomic layer deposited (ALD) Al_2O_3 layer. Amorphous IGZO is deposited by room temperature RF magnetron sputtering. The IGZO and Al_2O_3 layers are structured by two wet etching steps (Figure 15(a). Next, positive photoresist is spin coated and illuminated from the back (through the substrate); here the opaque Ti-gate contacts acted as mask (Figure 15(b)). The subsequent resist development results in a strip of resist,

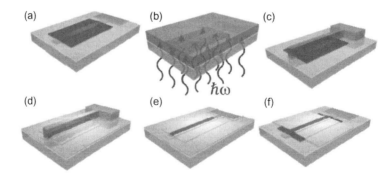

FIGURE 15   Manufacturing process flow of flexible, self-aligned a-IGZO TFTs. (a) Deposition and structuring of the Al_2O_3 gate isolator and the a-IGZO semiconductor and the Ti gate, (b) back-side illumination of photoresist through the substrate using the gate contact as mask, (c) resist development, (d) evaporation of the Cr/Au top metallization, (e) structuring of the TFT channel by lift-off, and (f) top metallization etching to define the channel width and the contacts.

self-aligned on top of the gate contacts (Figure 15(c)). To form source and drain contacts, 10 nm thick Cr and 60 nm thick Au are evaporated (Figure 15(d)). The TFT channel is formed by a lift-off process, which removes a part of the Cr/Au metallization using the self-aligned resist (Figure 15(e)). The width of the TFT is afterward defined by standard photolithography and wet etching (Figure 15(f)). Finally, similar to conventional (non-self-aligned) TFTs, 25 nm of Al_2O_3 identical to the gate insulator are used to passivate the TFT.

An SEM micrograph of a self-aligned TFT is shown in Figure 16(a). Magnifications of the channel region (Figures 16(b), 16(c)) show the dimensions of the TFT channel. TFTs have a channel length of 0.5 μm and a gate to source/drain overlap of 1.55 μm. The overlap is caused by the fact that the resist strip, structured by self-alignment, is narrower than the gate contact itself.

The TFT shown in Figure 16 exhibits the following performance values: An effective field effect mobility of 7.5 cm^2/Vs, a threshold voltage of 0 V, an on/off current ratio of 1.9×10^9, a subthreshold swing of 0.13 V/dec, and a transconductance of 10.42 μS/μm ($V_{GS} = 2$ V). Due to the ALD deposited Al_2O_3 gate isolator, the gate-source leakage current is always smaller than 10^{-13} A/μm. Compared to conventionally fabricated TFTs, reduced effective field-effect mobility is due to the increased contact resistance of self-aligned TFTs (≈ 66 kΩ/μm for conventional TFTs, ≈ 290 kΩ/μm for self-aligned TFTs). However, self-aligned TFTs have two mayor advantages: the reduced gate-source and gate-drain overlaps of 1.55 μm (Figure 17(a)) lead to a reduced gate capacitance of self-aligned TFTs with identical channel length as conventionally fabricated TFTs. Figure 17(b) shows the capacitance voltage characteristic of a conventional and self-aligned flexible IGZO TFT, both with a channel length of 1 μm. Self-alignment decreases the specific gate capacitance from 55 fF/μm to 41 fF/μm.

Furthermore, self-alignment can enable TFTs with channel length smaller than the channel length of conventionally fabricated flexible IGZO TFTs. As a result, the reduced channel length and gate to source/drain overlaps lead to flexible TFTs with superior AC performance. From the plot in Figure 18, the transit frequency of the self-aligned flexible IGZO TFT with a channel length of 0.5 μm is extracted to be 135 MHz.

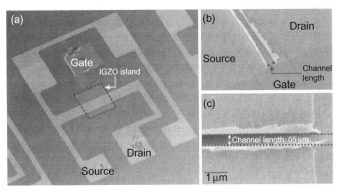

FIGURE 16   (a) SEM micrograph of a self-aligned IGZO TFT. (b) Enlargement of the channel region. (c) Top view of the channel, used to determine the exact channel length and gate to source/drain overlaps.

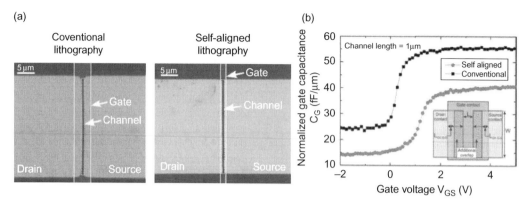

FIGURE 17 (a) Comparison between conventional and self-aligned lithography in the case of fabrication of a 1 μm long channel device. Self-alignment allows reduction of the overlapping length between drain/source and gate and hence reduce the overlapping capacitance. (b) The reduced gate overlap (inset) of self-aligned TFTFs leads to a decrease of the overlap capacities and therefore to an enhancement of the frequency response.

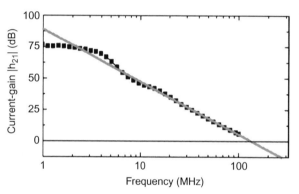

FIGURE 18 Absolute value of h_{21} extracted from S-parameter measurements of a flexible TFT (channel length: 0.5 μm). The extracted f_T is 135 MHz.

5.2 Flexible Double-Gate TFTs

Multi-gate transistors are widely exploited in CMOS technology in order to achieve better switching properties and lower power consumption, especially in ultra-scaled devices [45]. Double-gate (DG) architecture can be beneficial also for flexible devices. Besides performance enhancement, DG architecture also offers better stability to mixture and air exposure [33] and, in the case of two independent gates, the possibility of threshold adjustment. In fact, the top and bottom gate can be either short-circuited or independently biased. Figure S3(a) shows the cross-section of a device in which the two gates are short-circuited as visible from the top view in Figure S3(b) [34]. The fabrication process is very similar to the one of single-gate architecture, but some remarks are needed. Evaporated titanium and chromium provide a sufficient adhesion on polyimide and are used for the bottom and top gate, respectively. Additionally, the work functions of these two materials are comparable ($\Phi_{Ti} = 4.33$ eV, $\Phi_{Cr} = 4.44$ eV), so their influence on the threshold voltage can be neglected. In order to have a symmetric geometry that is then reflected in the

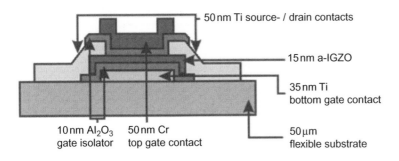

50 nm Ti source- / drain contacts

15 nm a-IGZO

35 nm Ti
bottom gate contact

10 nm Al$_2$O$_3$ 50 nm Cr
gate isolator   top gate contact

50 μm
flexible substrate

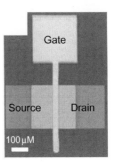

FIGURE S3   (a) Double-gate a-IGZO TFT schematic and (b) micrograph of a fully processed flexible a-IGZO double-gate TFT (W/L = 280 μm/ 10 μm).

operational behavior of the device, 10 nm of ALD deposited Al$_2$O$_3$ is used as a dielectric for both gates.

The device is characterized for its static performance and compared to a single-gate structure that is used as reference. Thanks to the increased gate capacitance (Figure S4(a)), the transconductance (Figure S4(b)) also increases and the subthreshold swing improves by 15 mV/dec, reaching the remarkable value of 69 mV/dec (Figure S4(c)). A shift in the threshold voltage is also observed, probably due to the change of the device geometry and channel formation.

The double-gate TFTs remain fully operational when bent to a radius of 5 mm (Figure S5). The applied tensile strain induces a reduction of the effective field-effect mobility while the subthreshold swing stays constant within the measurement inaccuracies. Compared to previous bending experiments with bottom-gate a-IGZO TFTs, the observed shifts correspond to compressive strain in the TFT channel. This indicates different mechanical properties caused by the changed geometry of the double-gate devices. Bending to even smaller radii is not possible because of the formation of cracks starting in the brittle Cr top-gate contact. The use of more ductile metals like Cu would enable bending radii between 1 mm and 2 mm without the need of modifying the device structure.

5.3 Flexible a-IGZO TFTs with Vertical Channel

To fabricate amorphous semiconductor-based TFTs operating at frequencies > 100 MHz, channel-length scaling in the submicron range is required. This is really challenging, especially in the case of TFTs fabricated directly on free-standing plastic foils. As discussed in previous sections, the self-alignment technique [43,44] has been shown as one option to realize transistors with 500 nm channel length. Nevertheless, self-alignment still relies on photolithographic constrains, which limit further scaling of the length. One attractive alternative is to adopt vertical integration of the TFT structure. In vertical TFTs (VTFTs), the channel length is defined by the thickness of a spacer layer and, therefore, it is no longer limited by lithography resolution and alignment [46]. This allows achieving submicron channel lengths and at the same time, much higher device-packing densities. In the first VTFT structure proposed, the channel length is formed on the vertical sidewall of a

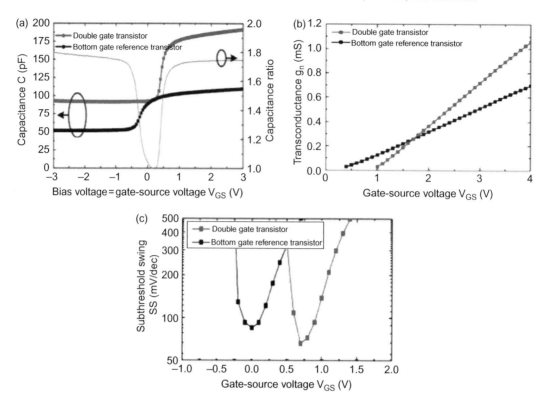

FIGURE S4 Double-gate TFT (W/L = 280 μm/10 μm) performance: (a) C-V measurement of a DG device and of a non-DG device, (b) g_m of both devices, and (c) subthreshold swing, SS.

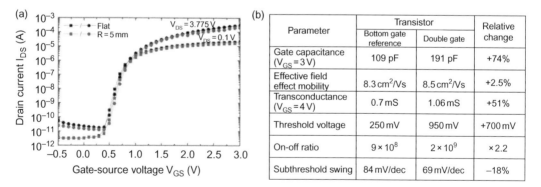

Parameter	Transistor		Relative change
	Bottom gate reference	Double gate	
Gate capacitance (V_{GS} = 3 V)	109 pF	191 pF	+74%
Effective field effect mobility	8.3 cm²/Vs	8.5 cm²/Vs	+2.5%
Transconductance (V_{GS} = 4 V)	0.7 mS	1.06 mS	+51%
Threshold voltage	250 mV	950 mV	+700 mV
On-off ratio	9×10^8	2×10^9	×2.2
Subthreshold swing	84 mV/dec	69 mV/dec	−18%

FIGURE S5 (a) Id-Vg curve of a DG TFT (W/L = 280 μm/35 μm) measured while flat and bent to a tensile radius of 5 mm. (b) The table compares important parameters in single bottom-gate and double-gate transistors.

source/insulator/drain multi-layer structure [46]. In this way, the channel length is precisely defined by the thickness of the insulating film. Using a similar device structure, a-Si: H VTFTs with 100 nm channel lengths [47] and 30 nm-thin SiN_x gate dielectrics [48], as

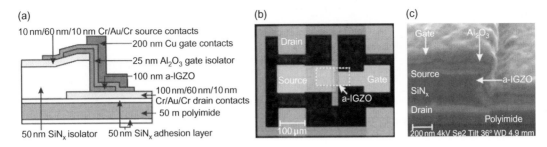

FIGURE 19 Flexible a-IGZO VTFTs (W/L = 60 μm/0.5 μm): (a) schematic cross-section, (b) micrograph, and (c) SEM image of the focused ion beam (FIB) cross-section through the VTFT channel.

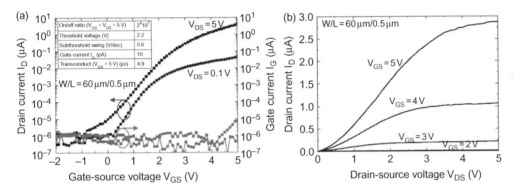

FIGURE S6 Flexible a-IGZO VTFTs: (a) I_{DS}-V_{GS} of VTFTs. Inset: Performance parameters, (b) gate capacitance measurement, and (c) I_{DS}-V_{DS} characteristic.

well as a-IGZO VTFTs with 310 nm channel length [49], have been demonstrated. Recently, another approach has been proposed and applied to ZnO VTFTs [50]. Here, the channel length is defined by the thickness of the gate metal, whereas a combination of conformal and non-conformal deposition processes allows creating the channel and the electrodes. Nevertheless, this approach seems to be not suitable for high-frequency applications, due to the large overlap capacitances between the top electrode and the gate contact.

Using the approach proposed by Uchida et al. [46], the first mechanically flexible amorphous semiconductor-based VTFTs with a channel length of 500 nm have been demonstrated [51]. Figure 19 shows: (a) the device cross-section, (b) a micrograph, and (c) an SEM image of a focused ion beam (FIB) cross-section through the channel of the flexible VTFTs.

Figure S6 shows the I_{DS}-V_{GS} and I_{DS}-V_{DS} characteristics of the VTFTs with the performance parameters (inset Figure S6(a)). From the extracted transconductance $g_m = 4.9 \, \mu S$ ($V_{GS} = V_{DS} = 5 \, V$) and the gate capacitance $C_G = 9.8 \, pF$, a transit frequency $f_T \approx 80 \, kHz$ was estimated. This relatively low value, compared to previously published values for planar flexible a-IGZO TFTs [31,44], is mainly explained by the low g_m and high C_G values extracted. On one hand, the value for g_m is attributed to the relatively high contact resistance $R_C \approx 500 \, k\Omega$ (at the interface between source/drain and IGZO), which is mainly caused by the oxidation of the Cr contacts as well as by the contamination of the metal

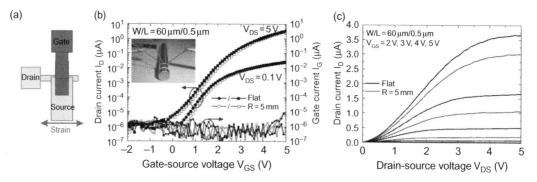

FIGURE S7   Flexible a-IGZO VTFTs (W/L = 60 μm/0.5 μm): (a) schematic image showing the direction of applied tensile strain, (b) transfer, and (c) output characteristic while flat and bent to 5 mm. Inset: Photograph of flexible VTFT contacted and bent around cylindrical rod of 5 mm bending radius.

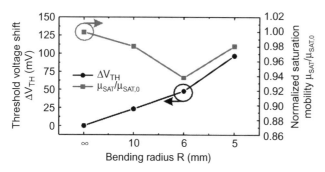

FIGURE S8   Mechanical bendability of flexible a-IGZO VTFTs (W/L = 60 μm/ 0.5 μm): threshold voltage shift ΔV_{TH} and normalized mobility $\mu_{SAT}/\mu_{SAT,0}$ while flat and bent to 10, 6, and 5 mm.

surface during the RIE process. On the other hand, the value for Cg is explained by the presence of a parasitic capacitance between gate contact and IGZO, as visible in Figure 19 (b). Both R_C and C_G can be reduced. On one hand, R_C can be reduced by optimizing the fabrication process and the choice of the source/drain contact materials. Especially, great care has to be taken in structuring the SiN_x sidewall in order to form the vertical channel in a reliable way, and at the same time avoiding any damage to the source/drain contacts. On the other hand, the value for C_G can be reduced by optimizing the layout design with smaller IGZO islands. These modifications can lead to a higher frequency value. Additionally, in the future, methods to fully self-align the gate contact to the source/insulator/drain stack promise to further exploit the advantage of submicron channel lengths.

Flexible VTFTs are characterized while flat and bent to 10, 6, and 5 mm radii, corresponding to tensile strain ε of 0.25%, 0.4%, and 0.5% [52] in the direction shown in Figure S7(a). Figure S7(b),(c) show the I_{DS}-V_{DS} and I_{DS}-V_{GS} characteristics while the device is flat and bent to 5 mm. Bending to 5 mm changes V_{TH} by +97 mV and the saturation field-effect mobility μ_{SAT} by −2%. Figure S8 displays the evolution of ΔV_{TH} and of the normalized μ_{SAT} as a function of the radius. Compared to previous reports on bendable planar a-IGZO devices [35], the observed shifts correspond to compressive strain in the TFT channel. This can be explained by the Poisson effect, which causes compressive strain perpendicular to the applied strain, and thus parallel to the vertical channel. After reflattening, the VTFTs stay fully operational, while bending to smaller radii induces cracks perpendicular to the strain that permanently harm the devices. After 1,000 cycles of repeated bending and reflattening, the VTFTs are fully operational with $\Delta V_{TH} \approx$ -460 mV and almost constant μ_{SAT} ($\pm 0.3\%$).

6. PLASTIC ELECTRONICS FOR SMART TEXTILES

One of the most promising emerging fields in wearable systems is the convergence of electronic components and textiles [53–55]. Such "smart" textiles (also known as electronic or e-textiles) fall into the category of intelligent materials that sense and respond to environmental and physical stimuli. Within the field of wearable computing, smart textile applications range from medical monitoring of physiological signals, including heart rate, and guided training and rehabilitation of athletes, to assistance for emergency first responders and commercial applications integrated into everyday clothing.

Different strategies can be adopted to realize electronic textiles. The simplest one consists of using off-the-shelf components that need to be placed on a substrate or support and then woven or simply attached onto the fabrics. This approach offers rich electronic functionalities but poses concerns regarding comfort and wearability. To maintain essential textile properties, smart textiles are evolving to integrate more electronic functions at the fiber level, but most fibers are limited to a single functionality. A third route consists of combining thin-film flexible electronic devices including sensors and transistors, interconnect lines, and commercial integrated circuits with plastic fibers (e-fibers) that can be woven into textiles using a commercial manufacturing process [55] (Figure S9). Such a scheme tries to combine the comfort of pure textiles and relatively complex electronic

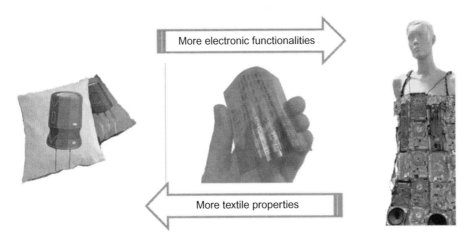

FIGURE S9 Combining electronics functionalities with textile comfort is challenging. The use of bulky and rigid off-the-shelf components offers rich functionalities but poor comfort. The opposite results are achieved by using textile fibers, which have simple and elementary functionalities. A good compromise can be reached if electronics are fabricated on flexible plastic foils that can be subsequently integrated into textiles.

Sensor/actuator
fabrication

Cutting
electronic fibers

Integrating
electronic fibers
into textile

Interconnect line

Temperature sensor

FIGURE 20 Electronics are first fabricated directly on 50 μm thick polyimide foils, then stripes are cut and successively integrated into textiles.

functionalities. Moreover, it requires a multi-disciplinary approach, involving knowledge from the fields of mechanical engineering, materials science, and electrical and textile engineering.

Commercial integrated circuit chips can be attached to plastic foil, and thin-film devices can be fabricated on foils following the same procedure and processes described in the previous pages. After the fabrication, the foils can be cut in strips and then woven into the textile by weaving processes or possibly by industrial manufacturing processes (Figure 20). Interconnection, contact pads, and communication buses can also be realized and integrated into fabrics to ensure the desired functionalities. By using conductive yarns perpendicular to the woven plastic strips, interconnections among several strips within the textile can be realized. At crossing points of conductive yarns and flexible plastic strips, contacts are established using conductive glue or solder. With this approach, it is possible to realize textile integrated bus or array systems [56,57]. During weaving, fibers may be exposed to bending radii much smaller than 1 millimeter and large tensile strains that can be greater than 20% (in this case, TFTs can only survive the weaving process if they are encapsulated or fabricated on a rigid island). The severe mechanical requirements, the difficulties in integrating plastic fibers into large-scale production in the textile industry, and the issues regarding resistance to washing cycles are, at the moment, the main obstacles that preclude large-area production and commercialization of such products.

In the next sections some examples of real e-textiles are discussed with particular emphasis on the limitations imposed by the mechanical constraints due to weaving processes.

6.1 Textile E-nose

Micro- and nano-gas sensors to detect chemical or biological substances are becoming increasingly deployed tools to monitor industrial processes. Measured gases are often complex odors, consisting of a mix of different analytes, and gas sensor arrays, so-called electronic noses (e-noses), are used for their detection. E-noses in combination with classification algorithms enable the determination between odors. Commonly used sensors in e-noses are chemiresistors. They consist of a resistor element that changes its resistance when exposed to a target analyte. Chemiresistors offer the advantage of a simple device architecture and easy signal processing compared to gas-sensitive capacitors or transistors. Existing chemiresistors are made of conductive polymers (CPs) [58] or metal oxides [59].

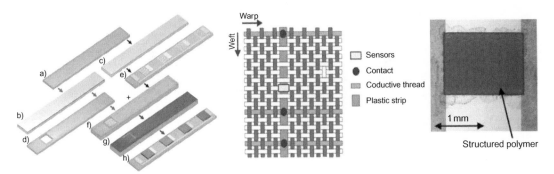

FIGURE 21 Illustration of the fabrication process. PEN or PI substrates were cleaned with acetone/IPA and treated with an oxygen-plasma to promote adhesion. (a) Ti/Au electrodes and contacts are patterned using a standard lift-off technique; (b) The lift-off mask was structured with openings for the sensors by evaporating and patterning a 100 nm thick aluminum layer; (c, e) the pattern for the sensor design consists of an array of four interdigitated finger electrodes representing the gas sensors; (d) openings on the lift-off mask are aligned to the finger electrodes on the sensor substrate using alignment marks; (f) to fix the lift-off mask to the surface of the sensor substrate, IPA drops are used as adhesion promoter; (g) the CP solution is pipetted onto the substrate and spin coated; (h) the lift-off mask is removed, leaving the patterned CP layers on the substrate; (center) schematic graph of the electronic nose system woven into a textile and (right) micrograph of a fabricated gas sensor using the mask lift-off technique.

Inorganic layers can withstand a maximum strain of about 2%, while CPs-based devices continue to function when strained up to 50% [60]; therefore, they are more suitable for integration into textiles. Recent efforts showed the influence of strain on single gas sensors and their performance under bending [61]. Two CP types exist: intrinsically conductive and extrinsically conductive polymers. For the first class, the polymer itself is conductive, while for the second class a conductive filler material is introduced into an insulating polymer to render it conductive. While the latter show lower sensitivity, they outperform intrinsically CP in long-term stability and stability to moisture or oxygen [13].

Kinkeldei et al. demonstrated that it is feasible to integrate low-cost gas sensors into a textile while preserving the functionality [62]. The electronic nose consists of four carbon black/polymer gas sensors fabricated on a flexible polymer substrate. The fabrication scheme is depicted in Figure 21. The strip shown represents one of 40 strips that are processed on a single substrate in one fabrication run. Electrodes are made of Ti/Au and patterned using a standard lift-off technique. The pattern for the sensor design consists of an array of four interdigitated finger electrodes representing the gas sensors. The width of the finger as well as the gap is 20 μm. The sensors are designed to fit on a strip with a width of 2 mm and a length of 50 mm. The e-nose is comprised of four sensors and to address each sensor individually five electrical contacts are necessary: a common ground connected to all sensors and one to access each individual sensor element. The four CP sensing films are deposited by spin coating and patterned with a lift-off technique.

Individual sensor strips were separated by cutting the substrate into 2 mm wide strips. To integrate sensor strips into a smart textile, they were woven into a 45 mm wide and 100 mm long textile band in the weft direction using commercial weaving. Conductive yarns were inserted perpendicular to the strip in the warp direction to access the contact pads on the

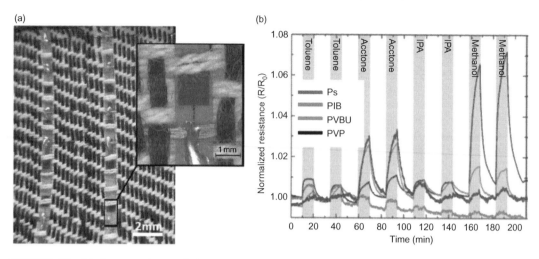

FIGURE 22 (a) Close-up picture of electronic nose strips in a woven textile band. In the inset a micrograph shows one of the four gas sensors on the sensor strip. (b) Influence of solvent exposure to the normalized resistance of four gas sensors (PS, PVBU, PIB, PVP) of an electronic nose. The sensor is repeatedly exposed for 10 min to the solvent vapor (toluene, acetone, IPA, methanol) and 12 min to air. Each of the solvents creates a different sensor signal pattern.

strip. For a mechanically stable contact the yarns were glued to the contact pads. The resistance of the metal threads and contacts can be neglected compared to the resistance of the sensors. A schematic of the textile architecture is shown in Figure 21 (center), and Figure 22(a) shows the plastic strips integrated into the textile. During exposure a single analyte creates a specific signal pattern. As depicted in Figure 22 for each of the inserted solvent vapors the four sensor elements showed different ratios of signal amplitudes, due to the solubility parameter concept. Also, bending created a characteristic ratio of signal amplitudes with the resistance amplitude depending on the sensor curvature. Principle component analysis (PCA) can be used to discriminate between bending and several solvents.

This concept offers the use of cheap sensor and substrate materials together with solution-based deposition methods. The approach shows that it is possible to further integrate sensor technologies into textiles and that the original sensor functionalities are preserved after the textile integration process. The used fabrication methods and materials are further applicable for large area integration of electronics into textiles.

6.2 Textile Integrated Near-Infrared Spectroscopy System

Smart textiles find promising applications in health monitoring and disease diagnosis. An example of a textile integrated near-infrared spectroscopy (NIRS) system is shown in [63]. An NIRS system to monitor blood oxygen saturation sends infrared light into human tissue and measures scattered light intensity. Using two different wavelengths in the near-infrared range allows computing of the oxygen saturation. In previous approaches to realizing a textile-based NIRS system, optical fibers were stitched to textile substrates.

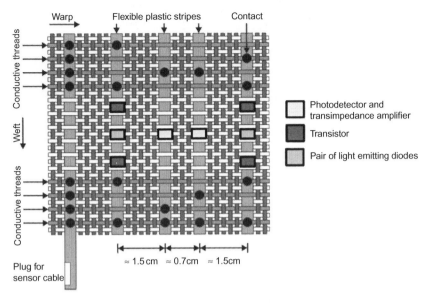

FIGURE 23 Schematic of flexible plastic stripes with photodiodes and LEDs integrated into a woven textile in the weft direction. In the warp direction, insulated copper wires (with a diameter of 71 μm) are integrated to contact the individual stripes. An additional stripe serves as connector rail to connect the textile to control electronics using one plug.

Different from the textile e-nose, the NIRS system consists of thin-film devices fabricated on foils and off-the-shelf components that are attached to the plastic strips. Two light emitting diodes (LEDs) with wavelengths of 760 nm and 870 nm, respectively, are placed on a flexible plastic stripe close to each other. The current of both LEDs is controlled with a transistor that is also located on the flexible plastic stripe. To detect the light that is back-scattered by the hemoglobin in biological tissue, a photodiode in conjunction with a transimpedance amplifier is mounted on a second flexible plastic stripe.

In Figure 23 a schematic of the woven textile with integrated flexible plastic stripes carrying photodiodes or LEDs and transistors is shown. In the warp direction insulated copper wires with a diameter of 71 μm are woven to connect the individual stripes with the connector rail, which provides a plug to establish a connection to the read-out and control electronics.

In Figure 24, a woven NIRS prototype is shown. LEDs, transistor devices, photodiodes, and amplifiers were mounted on the flexible strips using standard micro-fabrication techniques such as wire bonding, soldering, and gluing:

a. LED strip: two LEDs are mounted in conjunction with two transistors as bare dies onto the substrate. All devices were encapsulated using non-transparent epoxy and transparent epoxy. The strip has a length of 6 cm and a width of 2 mm. Figure 24(a) shows an LED strip with the devices and the contact pads for the copper wires.

b. Photodiode strip: a PIN photodiode with an active area of 7.7 mm^2 is mounted onto the strip together with a bare die amplifier in a transimpedance amplifier configuration

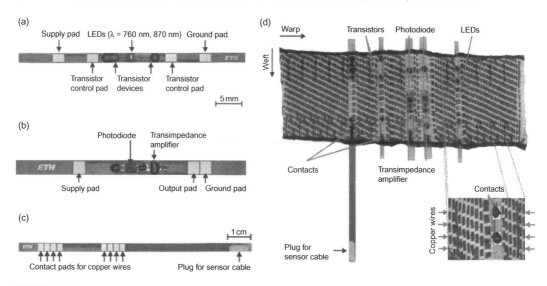

FIGURE 24   (a) LED strip with the LED pair at the center and the two transistors on the left and right side of the LED pair. (b) Photodiode strip with photodiode and transimpedance amplifier. (c) Bus-bar strip with a plug to connect the sensor textile to the control hardware and the contact pads for the copper wires. In (a) and (b) the contact pads are indicated. (d) Woven sensor textile with flexible plastic strips in weft direction carrying LEDs, transistors, photodiodes, and transimpedance amplifiers. In the inset, woven copper wires in the warp direction are visible together with two encapsulated contacts between copper wires and contact pads on a flexible strip.

with a feed-back resistor of 100 kΩ. The transimpedance amplifier and the contacts of the photodiode are covered with non-transparent epoxy. Figure 24(b) depicts a photodiode strip with a length of 6 cm and a width of 4 mm.

c. Bus bar strip: Figure 24(c) shows a bus bar strip to route the eight copper wires within the textile to a single plug.

To fabricate the sensor textile, flexible plastic strips were woven into a 4.5 cm wide and approximately 10 cm long textile band using an industrial narrow fabric-weaving machine (Figure 24(d)).

To test the electrical functionality of the textile-integrated NIRS, the textile is connected to a custom made read-out and control board. This board pulses the LEDs with a frequency of 100 Hz and a duty cycle of 10%. Simultaneously the board samples the voltages of the photodiodes and sends the data to a computer using a USB connection.

Placing the palm of the hand or a finger on the textile and thereby covering the LEDs and photodiodes, the pulse can be detected in correspondence to the heartbeat. The varying baseline is caused by movements of the hand, which changes the light-coupling efficiency between the LEDs/skin and skin/photodiode. Additionally, tissue oxygen saturation (StO_2) and arterial oxygen saturation (SpO_2), change in oxygenated hemoglobin ($\Delta O2Hb$), in deoxygenated hemoglobin (ΔHHb), and in total hemoglobin (ΔtHb) can also be calculated (Figure 25).

The obtained results show that it is possible to detect changes in HHb and O2Hb caused by a venous occlusion. However, textile integration of sensors and actuators influences the NIRS measurements, especially with regards to known limitations of NIRS

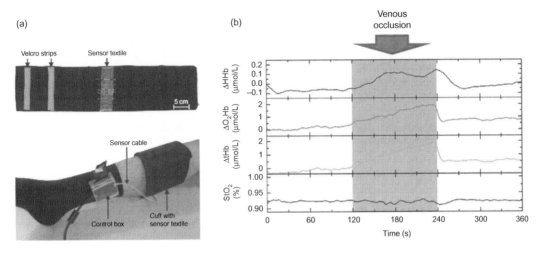

FIGURE 25 (a) The sensor textile is sewn into a textile cuff together with Velcro strips for attaching to the human body. The cuff is strapped to the calf together with the control box. Between the control box and the cuff, the sensor cable is visible. (b) Venous occlusions were performed on the calf for 2 minutes (marked in grey). During the occlusion, the HHb, O2Hb, and tHb concentrations did increase while the tissue oxygen saturation StO2 stayed constant.

systems, such as distance variation between source and detector or motion artifacts [64]. For the textile-integrated NIRS system, the distance between LED and photodiode can vary due to the flexibility of the textile and should be taken into account during data analysis and interpretation. Motion artifacts can change the locations of the LED and detector on the skin. Such effects influencing the signal quality are commonly observed with textile-integrated sensor systems for physiological signals. Examples are demonstrated for ECG measurements [65] and bio-impedance measurements [66]. Nevertheless, the presented NIRS system is a step toward textile integration of NIRS systems enabling long-term monitoring.

7. OUTLOOK AND CONCLUSIONS

In this chapter an overview of materials and technology for flexible electronic circuits was given, mainly focused on semiconductor materials and in particular on a-IGZO. Examples of devices and circuits were provided, and some practical applications such as smart textiles have also been illustrated. It is worth mentioning, however, that some topics, e.g., dielectric materials, were not discussed.

In these last lines, we want to offer an outlook for possible future trends.

As mentioned at the beginning of the chapter, the quest for thin and flexible displays has driven the development of plastic electronics. An evolution in this field is represented by fully transparent displays [32,67]. However, other applications, especially in medicine and biology, have become more and more important over the years. In recent years, extreme flexibility has been achieved by fabricating electronics on very thin substrates.

The use of substrates that are thinner than 10 μm ensures a bending radius in the range of tens of microns [7,11,68]. Such electronics can survive if crumpled, and exhibit a high degree of conformability, which can find application in smart skin [68], brain activity monitoring [2], and implantable devices [69].

The authors believe that future research will be focused on system integration and on trying to make such ultra-flexible wearable/implantable devices energetically autonomous. So far the efforts have been devoted to the development of the technology, which ultimately means trying to boost the performance of a single device. From now on attention will rise to a higher level and be devoted to integrating the required components to build real, functional, and useful systems. At this stage it will be extremely important to have strategies for the powering and communication of such systems. In fact, it is not very realistic to imagine wearable/implantable devices that get power and communicate to the external world through wires. Energy scavengers based on thermoelectric, piezoelectric, photoelectric effects, and near-field power transmission are all possible options to move toward zero-power consumption. A unique and universal solution likely does not exist. The authors do believe that the success of thin-film technology will be inextricably bound to these challenges. However, we are sure the best is yet to come.

References

[1] G. Crawford, Flexible Flat Panel Displays, Wiley, 2005.

[2] D. Khodagholy, T. Doublet, P. Quilichini, M. Gurfinkel, P. Leleux, A. Ghestem, et al., In vivo recordings of brain activity using organic transistors, Nat Commun 4 (2013) 1575, 03/12/online.

[3] F.C. Krebs, S.A. Gevorgyan, J. Alstrup, A roll-to-roll process to flexible polymer solar cells: model studies, manufacture and operational stability studies, J. Mater. Chem. 19 (2009) 5442−5451.

[4] L. Hu, H. Wu, F. La Mantia, Y. Yang, Y. Cui, Thin, flexible secondary Li-ion paper batteries, ACS Nano 4 (2010) 5843−5848.

[5] N. Munzenrieder, C. Zysset, T. Kinkeldei, G. Troster, Design Rules for IGZO Logic Gates on Plastic Foil Enabling Operation at Bending Radii of 3.5 mm, Electron. Devices IEEE Trans. 59 (2012) 2153−2159.

[6] K. Takei, T. Takahashi, J.C. Ho, H. Ko, A.G. Gillies, P.W. Leu, et al., Nanowire active-matrix circuitry for low-voltage macroscale artificial skin, Nat. Mater. 9 (2010) 821−826.

[7] D.-H. Kim, J.-H. Ahn, W.M. Choi, H.-S. Kim, T.-H. Kim, J. Song, et al., Stretchable and Foldable Silicon Integrated Circuits, Science 320 (2008) 507−511, 2008.

[8] D. Shahrjerdi, S.W. Bedell, Extremely Flexible Nanoscale Ultrathin Body Silicon Integrated Circuits on Plastic, Nano Lett. 13 (2013) 315−320, /01/09 2012.

[9] H.T. Yi, M.M. Payne, J.E. Anthony, V. Podzorov, Ultra-flexible solution-processed organic field-effect transistors, Nat. Commun. 3 (2012) 1259, 12/11/online.

[10] T. Sekitani, U. Zschieschang, H. Klauk, T. Someya, Flexible organic transistors and circuits with extreme bending stability, Nat. Mater. 9 (2010) 1015−1022.

[11] M. Kaltenbrunner, M.S. White, E.D. Głowacki, T. Sekitani, T. Someya, N.S. Sariciftci, et al., Ultrathin and lightweight organic solar cells with high flexibility, Nat. Commun. 3 (2012) 770, 04/03/online.

[12] L. Zhang, H. Wang, Y. Zhao, Y. Guo, W. Hu, G. Yu, et al., Substrate-Free Ultra-Flexible Organic Field-Effect Transistors and Five-Stage Ring Oscillators, Adv. Mater. (2013).

[13] J. Lewis, Material challenge for flexible organic devices, Mater. Today 9 (2006) 38−45.

[14] Q.H. Wang, K. Kalantar-Zadeh, A. Kis, J.N. Coleman, M.S. Strano, Electronics and optoelectronics of two-dimensional transition metal dichalcogenides, Nat. Nanotechnol. 7 (2012) 699−712.

[15] S.J. Kang, C. Kocabas, T. Ozel, M. Shim, N. Pimparkar, M.A. Alam, et al., High-performance electronics using dense, perfectly aligned arrays of single-walled carbon nanotubes, Nat. Nanotechnol. 2 (2007) 230−236.

[16] Q. Cao, S.-j. Han, G.S. Tulevski, Y. Zhu, D.D. Lu, W. Haensch, Arrays of single-walled carbon nanotubes with full surface coverage for high-performance electronics, Nat. Nano 8 (2013) 180–186, 03//print.

[17] Z. Fan, J.C. Ho, Z.A. Jacobson, R. Yerushalmi, R.L. Alley, H. Razavi, et al., Wafer-scale assembly of highly ordered semiconductor nanowire arrays by contact printing, Nano Lett. 8 (2008) 20–25.

[18] K. Nomura, H. Ohta, A. Takagi, T. Kamiya, M. Hirano, H. Hosono, Room-temperature fabrication of transparent flexible thin-film transistors using amorphous oxide semiconductors, Nature 432 (2004) 488–492.

[19] N. Munzenrieder, L. Petti, C. Zysset, G.A. Salvatore, T. Kinkeldei, C. Perumal, et al., Flexible a-IGZO TFT amplifier fabricated on a free standing polyimide foil operating at 1.2 MHz while bent to a radius of 5 mm, Electron. Devices Meeting (IEDM) 2012 IEEE Int. (2012) 5.2.1–5.2.4.

[20] E. Fortunato, P. Barquinha, R. Martins, Oxide Semiconductor Thin-Film Transistors: A Review of Recent Advances, Adv. Mater. 24 (2012) 2945–2986.

[21] V.C. Sundar, J. Zaumseil, V. Podzorov, E. Menard, R.L. Willett, T. Someya, et al., Elastomeric transistor stamps: Reversible probing of charge transport in organic crystals, Science 303 (2004) 1644–1646.

[22] H. Sirringhaus, T. Kawase, R. Friend, T. Shimoda, M. Inbasekaran, W. Wu, et al., High-resolution inkjet printing of all-polymer transistor circuits, Science 290 (2000) 2123–2126.

[23] M. Jung, J. Kim, J. Noh, N. Lim, C. Lim, G. Lee, et al., All-printed and roll-to-roll-printable 13.56-MHz-operated 1-bit RF tag on plastic foils, Electron. Devices IEEE Trans. 57 (2010) 571–580.

[24] K. Novoselov, A.K. Geim, S. Morozov, D. Jiang, Y. Zhang, S. Dubonos, et al., Electric field effect in atomically thin carbon films, Science 306 (2004) 666–669.

[25] M.Y. Han, B. Özyilmaz, Y. Zhang, P. Kim, Energy Band-Gap Engineering of Graphene Nanoribbons, Phys. Rev. Lett. 98 (2007) 206805.

[26] R. Balog, B. Jørgensen, L. Nilsson, M. Andersen, E. Rienks, M. Bianchi, et al., "Bandgap opening in graphene induced by patterned hydrogen adsorption," Nat. Mater. 9 (2010) 315–319.

[27] Y. Zhang, T.-T. Tang, C. Girit, Z. Hao, M.C. Martin, A. Zettl, et al., Direct observation of a widely tunable bandgap in bilayer graphene, Nature 459 (2009) 820–823, 06/11/print.

[28] R.J. Smith, P.J. King, M. Lotya, C. Wirtz, U. Khan, S. De, et al., Large-Scale Exfoliation of Inorganic Layered Compounds in Aqueous Surfactant Solutions, Adv. Mater. 23 (2011) 3944–3948.

[29] X. Li, Y. Zhu, W. Cai, M. Borysiak, B. Han, D. Chen, et al., Transfer of large-area graphene films for high-performance transparent conductive electrodes, Nano Lett. 9 (2009) 4359–4363.

[30] G.A. Salvatore, N. Münzenrieder, C. Barraud, L. Petti, C. Zysset, L. Büthe, et al., Fabrication and Transfer of Flexible Few-Layers MoS2 Thin Film Transistors to any arbitrary substrate, ACS Nano (2013).

[31] N. Munzenrieder, L. Petti, C. Zysset, G. Salvatore, T. Kinkeldei, C. Perumal, et al., Flexible a-IGZO TFT amplifier fabricated on a free standing polyimide foil operating at 1.2 MHz while bent to a radius of 5 mm, Electron. Devices Meeting (IEDM) 2012 IEEE Int. (2012) 5.2.1–5.2.4.

[32] K. Nomura, H. Ohta, K. Ueda, T. Kamiya, M. Hirano, H. Hosono, Thin-film transistor fabricated in single-crystalline transparent oxide semiconductor, Science 300 (2003) 1269–1272.

[33] K.-S. Son, J.-S. Jung, K.-H. Lee, T.-S. Kim, J.-S. Park, K. Park, et al., Highly stable double-gate Ga–In–Zn–O thin-film transistor, Electron. Device Lett. IEEE 31 (2010) 812–814.

[34] N. Münzenrieder, C. Zysset, L. Petti, T. Kinkeldei, G.A. Salvatore, G. Tröster, Flexible double gate a-IGZO TFT fabricated on free standing polyimide foil, Solid State Electron. (2013).

[35] N. Munzenrieder, K.H. Cherenack, G. Troster, The effects of mechanical bending and illumination on the performance of flexible IGZO TFTs, Electron. Devices IEEE Trans. 58 (2011) 2041–2048.

[36] I.L. Spain, J. Paauwe, High Pressure Technology: Applications and Processes, vol. 2, CRC Press, 1977.

[37] S. Khan, P.-C. Kuo, A. Jamshidi-Roudbari, M. Hatalis, Effect of uniaxial tensile strain on electrical performance of amorphous IGZO TFTs and circuits on flexible Metal foils, Device Res. Conf. (DRC) (2010) 119–120.

[38] N. Münzenrieder, K. Cherenack, G. Tröster, Testing of flexible InGaZnO-based thin-film transistors under mechanical strain, Eur. Phys. J. Appl. Phys. 55 (2011) 23904.

[39] C. Zysset, N. Münzenrieder, T. Kinkeldei, K. Cherenack, G. Tröster, Indium-gallium-zinc-oxide based mechanically flexible transimpedance amplifier, Electron. Lett. 47 (2011) 691–692.

[40] B. Razavi, Design of high-speed circuits for optical communication systems, Custom Integr. Circuits 2001 IEEE Conf. (2001) 315–322.

WEARABLE SENSORS

[41] H. Kumomi, S. Yaginuma, H. Omura, A. Goyal, A. Sato, M. Watanabe, et al., Materials, devices, and circuits of transparent amorphous-oxide semiconductor, J. Displ. Technol. 5 (2009) 531—540.

[42] C. Perumal, K. Ishida, R. Shabanpour, B.K. Boroujeni, L. Petti, N.S. Munzenrieder, et al., A Compact a-IGZO TFT Model Based on MOSFET SPICE Level = 3 Template for Analog/RF Circuit Designs, Electron. Device Lett. IEEE (2013)1-1 pp.

[43] K. Cherenack, B. Hekmatshoar, J.C. Sturm, S. Wagner, Self-Aligned Amorphous Silicon Thin-Film Transistors Fabricated on Clear Plastic at 300 < formula formulatype = , Electron. Devices IEEE Trans. 57 (2010) 2381—2389.

[44] N. Munzenrieder, L. Petti, C. Zysset, T. Kinkeldei, G.A. Salvatore, G. Troster, Flexible Self-Aligned Amorphous InGaZnO Thin-Film Transistors With Submicrometer Channel Length and a Transit Frequency of 135 MHz, Electron. Devices IEEE Trans. 60 (2013) 2815—2820.

[45] T. Skotnicki, J.A. Hutchby, T.-J. King, H.-S. Wong, F. Boeuf, The end of CMOS scaling: toward the introduction of new materials and structural changes to improve MOSFET performance, Circuits Devices Magazine IEEE 21 (2005) 16—26.

[46] Y. Uchida, Y. Nara, M. Matsumura, Proposed vertical-type amorphous-silicon field-effect transistors, Electron. Device Lett. IEEE 5 (1984) 105—107.

[47] I. Chan, A. Nathan, Amorphous silicon thin-film transistors with 90° vertical nanoscale channel, Appl. Phys. Lett. 86 (2005) 253501—253501-3.

[48] M. Moradi, A. Nathan, H.M. Haverinen, G.E. Jabbour, Vertical Transistor with Ultrathin Silicon Nitride Gate Dielectric, Adv. Mater. 21 (2009) 4505—4510.

[49] S. Ho Rha, J. Jung, Y. Soo Jung, Y. Jang Chung, U. Ki Kim, E. Suk Hwang, et al., Vertically integrated submicron amorphous-$In_2Ga_2ZnO_7$ thin film transistor using a low temperature process, Appl. Phys. Lett. 100 (2012) 203510—203510-5.

[50] S. Nelson, D. Levy, L. Tutt, Defeating the trade-off between process complexity and electrical performance with vertical zinc oxide transistors, Appl. Phys. Lett. 101 (2012) 183503—183503-4.

[51] P.A.L. Petti, N. Munzenrieder, G.A. Salvatore, C. Zysset, A. Frutiger, L. Büthe, et al., Mechanically flexible vertically integrated a-IGOR thin-film transistors with 500 nm channel length fabricated on free standing plastic foil, Accepted Int. Electron. Device Meeting IEDM (2013).

[52] H. Gleskova, S. Wagner, Z. Suo, Failure resistance of amorphous silicon transistors under extreme in-plane strain, Appl. Phys. Lett. 75 (1999) 3011—3013.

[53] D. Marculescu, R. Marculescu, N.H. Zamora, P. Stanley-Marbell, P.K. Khosla, S. Park, et al., Electronic textiles: A platform for pervasive computing, Proc. IEEE 91 (2003) 1995—2018.

[54] R. Paradiso, G. Loriga, N. Taccini, A. Gemignani, B. Ghelarducci, Wealthy, a wearable health-care system: new frontier on etextile, J. Telecommun. Inf. Technol. 4 (2005) 105—113.

[55] K. Cherenack, C. Zysset, T. Kinkeldei, N. Münzenrieder, G. Tröster, Woven electronic fibers with sensing and display functions for smart textiles, Adv. Mater. 22 (2010) 5178—5182.

[56] C. Zysset, N. Munzenrieder, T. Kinkeldei, K. Cherenack, G. Troster, Woven active-matrix display, Electron. Devices IEEE Trans. 59 (2012) 721—728.

[57] C. Zysset, T.W. Kinkeldei, N. Munzenrieder, K. Cherenack, G. Troster, Integration Method for Electronics in Woven Textiles, Compon. Packag. Manuf. Technol. IEEE Trans. 2 (2012) 1107—1117.

[58] J. Janata, M. Josowicz, Conducting polymers in electronic chemical sensors, Nat. Mater. 2 (2003) 19—24.

[59] H.-W. Zan, C.-H. Li, C.-C. Yeh, M.-Z. Dai, H.-F. Meng, C.-C. Tsai, Room-temperature-operated sensitive hybrid gas sensor based on amorphous indium gallium zinc oxide thin-film transistors, Appl. Phys. Lett. 98 (2011) 253503—253503-3.

[60] X.M. Tao, S.M. Shang, W. Zeng, Highly stretchable conductive polymer composited with carbon nanotubes and nanospheres, Adv. Mater. Res. 123 (2010) 109—112.

[61] T. Kinkeldei, C. Zysset, K. Cherenack, G. Troster, A textile integrated sensor system for monitoring humidity and temperature, Solid State Sens. Actuators Microsys. Conf. (TRANSDUCERS) 2011 16th Int. (2011) 1156—1159.

[62] T. Kinkeldei, C. Zysset, N. Münzenrieder, and G. Tröster, "An electronic nose on flexible substrates integrated into a smart textile," Sens. Actuators B Chem., 2012.

[63] C. Zysset, N. Nasseri, L. Büthe, N. Münzenrieder, T. Kinkeldei, L. Petti, et al., Textile integrated sensors and actuators for near-infrared spectroscopy, Opt. Express 21 (2013) 3213—3224.

[64] H.W. Siesler, Y. Ozaki, S. Kawata, and H. M. Heise, Near-infrared spectroscopy: principles, instruments, applications: Wiley. com, 2008.

[65] J. Schumm, S. Axmann, B. Arnrich, G. Tröster, Automatic signal appraisal for unobtrusive ecg measurements, Proc. Biosignal Interpretation Conf. (2009).

[66] G. Medrano, L. Beckmann, N. Zimmermann, T. Grundmann, T. Gries, S. Leonhardt, Bioimpedance spectroscopy with textile electrodes for a continuous monitoring application, 4th International Workshop on Wearable and Implantable Body Sensor Networks (BSN 2007) (2007) 23−28.

[67] M.-S. Lee, K. Lee, S.-Y. Kim, H. Lee, J. Park, K.-H. Choi, et al., High-Performance, Transparent, and Stretchable Electrodes Using Graphene−Metal Nanowire Hybrid Structures, Nano Lett. 13 (2013) 2814−2821, 2013/06/12.

[68] M. Kaltenbrunner, T. Sekitani, J. Reeder, T. Yokota, K. Kuribara, T. Tokuhara, et al., An ultra-lightweight design for imperceptible plastic electronics, Nature 499 (2013) 458−463, 07/25/print.

[69] S.-W. Hwang, H. Tao, D.-H. Kim, H. Cheng, J.-K. Song, E. Rill, et al., A Physically Transient Form of Silicon Electronics, Science 337 (2012) 1640−1644.

4.1

Energy Harvesting at the Human Body

Loreto Mateu, Tobias Dräger, Iker Mayordomo,
and Markus Pollak

Fraunhofer Institute for Integrated Circuits IIS, Nuremberg, Germany

1. INTRODUCTION TO ENERGY HARVESTING SYSTEMS

Energy harvesting power supplies in combination with wearable systems should bring the best scenario to an autonomous wearable system. The minimum desirable achievement would be to extend the time between the replacement or recharge of the batteries. The first issue to address is the selection of the most appropriate energy source for the application under consideration and to take into account the nature of the energy harvesting source (i.e., continuous in time, discontinuous in time, constant in amplitude, etc.). A second issue is to choose carefully the electronic components to be powered in terms of their required power consumption. For this purpose, it is of special interest to take into consideration the supply voltage, the current consumption both in active mode and sleep or standby mode, the start-up time, and the on-time.

An energy harvesting system (Figure 1) is composed of an ambient energy source, energy transducer, power management unit, energy storage element, voltage regulator, and electrical load. The energy transducer converts the ambient energy source (solar, mechanical, thermal, RF energy) into electrical energy. Afterward, a power management unit that rectifies, if necessary, the AC power, and also regulates the voltage value, is required. The energy-storage element is almost always necessary to store the harvested power that will supply the electrical load. Since the energy-storage element and the electrical load can be at different voltage levels, a voltage regulator is necessary. The electrical load is usually composed of one or several sensors, a microcontroller unit, and a RF transceiver that sends wirelessly the sensed data.

In wearable systems, the energy harvesting source can be either the environment (RF, light energy, etc.) or the human body (mechanical, thermal, etc.). There are two ways to

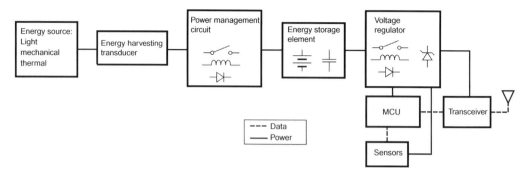

FIGURE 1   Block diagram of an energy harvesting system.

TABLE 1   Summary of Transducers and Power-Management Circuits Required for Several Ambient Energy Sources

Ambient Energy Source	Transducer	Power Management
Light (indoor and outdoor)	Solar cell	DC-DC converter
Mechanical	Piezoelectric, electromagnetic, electrostatic	AC-DC and DC-DC converter
Thermal	Thermoelectric generator	Step-up DC-DC converter

harvest energy from everyday human activity: passively or actively. Passive power is harvested from the user's everyday actions (walking, breathing, body heat, blood pressure, etc.) with unobtrusive techniques, which means that the person does not realize that he is doing any extra work to power an electronic device. Nevertheless, active power is harvested when the user of the electronic device has to do specific work in order to power the device that otherwise would not have been done. This chapter is focused on harvesting energy in a passive way.

The energy transducer depends on the available ambient energy to harvest. Table 1 shows the energy transducer that should be employed with different ambient energy sources. The third column of the table displays the kind of power-management unit required.

A good understanding of the physical principles of the energy harvesting transducer that converts the ambient energy into electrical energy is necessary in order to achieve a good coupling between the ambient energy and the transducer that maximizes the harvested power. The harvested power can also be maximized with the power management unit, which requires an electrical impedance matching between the energy harvesting transducer and the electronics connected. The harvested power makes reference to the power delivered by the transducer, which differs from the power on the electrical load also known as the output power.

As energy-storage elements for energy harvesting systems, rechargeable batteries, supercaps, and capacitors are employed. The parameters to consider when choosing this component are the energy required to be stored, the time between the storage of the electrical energy and its use, the area availables and the maximum current needed by the electrical load.

The following sections give an overview of energy harvesting power supplies feasible for wearable sensors.

2. ENERGY HARVESTING FROM TEMPERATURE GRADIENT AT THE HUMAN BODY

The human body continuously radiates heat. Devices with direct contact with the human body can harvest this wasted energy by means of thermoelectric generators (TEGs). This valuable technology for self-sustaining power supplies consists of a thermocouple module that employs the temperature gradient between the hot (body) and cold (ambient) side of the thermocouple to generate electrical energy. A significant constraint of this solution relies on the relatively low temperature gradient available: in the range of 3 to 5°C between the human body and the environment. Therefore, low voltage differences are provided at the output of the TEG. Specific step-up DC-DC converters have to be used inside a power-management unit for supplying standard integrated circuits since the thermoelectric transducer only provides between 50 to 100 mV/°C in open-circuit conditions. A model of the thermoelectric transducer as well as a figure-of-merit to be able to choose it properly will be provided in this section. Moreover, a topology of a DC-DC converter for converting input voltages on the order of a few hundreds of millivolts into higher output voltages will be presented.

2.1 Thermoelectric Generators

A thermoelectric generator converts the temperature gradient between its hot and cold side into electrical energy. Thermoelectric generators combine a number of p-n junctions, thermally in parallel and electrically in series, as depicted in Figure 2. The Seebeck coefficient of a thermoelectric generator is a measure of the open-circuit voltage generated in response to the temperature gradient across both sides. Most of the thermoelectric generators to be used at room temperature employ bismuth telluride and have a Seebeck coefficient around 40 mV/K to 100 mV/K.

A complete equivalent circuit of a thermoelectric generator is given by Chavez et al. [1]. A simplified version is presented by Lossec et al. [2] where the Peltier effect and the Joule heating effect are no longer included in the model of the thermal behavior [3]. The simplified model of the thermoelectric generator includes a thermal and an electrical part, as shown in Figure 3. The temperature and the heat flow are represented as voltage and current, respectively.

The hot heat source, which is the human body in the present case, is modeled by a voltage source T_b and the cold heat source is modeled by a voltage source T_a. The thermoelectric generator is located between both heat sources and has a temperature gradient fixed by the values T_h and T_c. The thermal resistances are modeled like electrical resistors. Thus, $R_{th,B}$ represents the thermal resistance between the hot heat source and the thermoelectric generator, $R_{th,G}$ is the thermal resistance of the thermoelectric generator, and $R_{th,A}$ models the thermal resistance between the cold side of the thermoelectric generator and the cold heat source.

To improve the heat transfer between the thermoelectric generator and the cold heat source, a heat sink with a thermal resistance $R_{th,A}$ can be employed.

The electrical model is composed of a voltage source with a voltage proportional to the Seebeck coefficient and to the temperature gradient of the thermoelectric generator. The internal electrical resistance of the thermoelectric generator is represented by R_G. Therefore, load R_L will extract the maximum power from the thermoelectric generator if condition

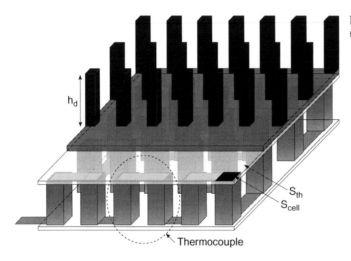

FIGURE 2 Physical structure of a thermoelectric generator [2].

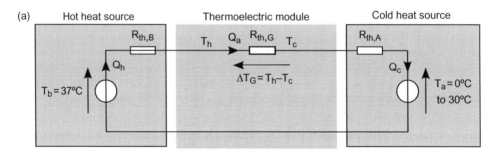

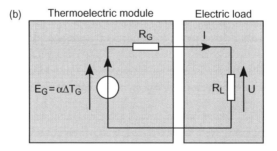

FIGURE 3 Simplified model of the thermal (a) and electrical (b) part of the thermoelectric generator [4].

$R_G = R_L$ is accomplished. It is considered that the output voltage, current, and power delivered by the thermoelectric generator is given at this condition. Thus, the voltage at the maximum power point is half of the open-circuit voltage of the thermoelectric generator.

The parameters required to model thermoelectric modules, which can be found in datasheets, are internal resistance, Seebeck coefficient, thermal conductivity λ_m or thermal resistance R_m, surface of the thermoelectric generator S_{th}, length of the thermocouples l_{th}, thermal conductivity $\lambda_{ceramic}$ of the ceramic surface that surrounds the thermopairs of the thermoelectric module, and height of the ceramic $l_{ceramic}$.

The thermal resistance $R_{th,B}$ is calculated by

$$R_{th,B} = \frac{l_{ceramic}}{\lambda_{ceramic} S_{th}} \tag{1}$$

The thermal resistances $R_{th,H}$ and $R_{th,A}$ are modeled by [4]:

$$R_{th,A} = \frac{1}{h_A S_{th}}; \; R_{th,H} = \frac{1}{k_{H1}(4h_d\sqrt{S_{th}} + k_{H2}S_{th})} \tag{2}$$

where h_d is the fin height of the heat sink and k_{H1} and k_{H2} are constant coefficients associated to the heat sink. If $h_d = 0$, $R_{th,H}$ becomes equal to $R_{th,A}$ since $h_A = k_{H1}k_{H2}$.

The electrical source V_G has a voltage value:

$$V_G = \alpha_m \Delta T_G \tag{3}$$

where the value of ΔT_G is calculated as

$$\Delta T_G = \frac{R_{th,G}}{R_{th,G} + R_{th,A} + R_{th,B}}(T_b - T_a) \tag{4}$$

The output power is maximum if the output electrical load is equal to the internal electrical resistance R_G. Thus, the voltage for this case is

$$V_{MPP} = \frac{V_G}{2} \tag{5}$$

and the maximum output power is given by

$$P_{MPP} = \frac{V_{MPP}^2}{R_G} = \frac{V_G^2}{4R_G} \tag{6}$$

The Interuniversity Microelectronics Center (IMEC) has designed an optimum multi-stage structure design for thermoelectric generators to have at least a minimum output voltage of the TEG, once it is connected to the power-management unit, of 0.7 V. This value could be step-up converted by the state-of the-art converters at that time. Different power-management units to be employed with this multi-stage structure were designed and tested [5–7]. Moreover, the unit is tested to find the location of the wrist that is the optimum for placing a TEG.

Table 2 shows various data of solid-state thermoelements [8]. The figure of merit [9] is calculated as follows from the Seebeck coefficient $\alpha[V/K]$, the electrical conductivity

TABLE 2   Summary Data of Solid-State Thermoelements [8]

Type	l * w * h [mm]	V_{oc}[V]	V_{sc}[A]	R_i[Ω]	α[V/K]	κ[W/K]	P_{100K}[W]	Z[1/C]
TEG-127-150-26	30 × 30 × 3.6	5.83	1.16	3.41	0.054	0.313	1.70	2.7m
TEG-127-150-22	30 × 30 × 4.2	6.05	0.75	5.50	0.056	0.300	1.13	1.9m
TEG-127-200-28	30 × 30 × 4.8	5.10	0.77	6.60	0.056	0.180	1.00	2.6m
TEG-241-150-29	30 × 30 × 3.6	10.36	0.70	10.00	0.096	0.440	1.83	2.1m
TEG-127-175-25	30 × 30 × 3.6	6.43	1.20	3.55	0.059	0.270	1.97	3.6m
TEG-127-175-26	30 × 30 × 2.8	6.37	3.18	1.36	0.059	0.700	5.07	3.7m

V_{oc}: *Open-circuit output voltage.*
V_{sc}: *Short-circuit output current.*
R_i: *Internal resistant.*
α: *Seebeck coefficient.*
κ: *Thermal conductance.*
P_{100K}: *Electrical output power at* $\Delta T = 100K$.
Z: *Figure of merit.*

$\sigma[1/\Omega m]$, the thermal conductivity $\lambda(W/mK)$, and the temperature T as a dimensionless value for the performance of a thermogenerator:

$$ZT = \frac{\sigma \cdot \alpha^2 \cdot T}{\lambda} \tag{7}$$

and with the given parameters of the TEGs of Table 2:

$$ZT = \frac{\alpha^2 \cdot T}{R_i \cdot \kappa}. \tag{8}$$

For energy harvesting applications, where typically small thermal gradients down to 2 K are present, the output voltage of a thermogenerator lies in the range of 100 mV for a 30 × 30 mm device (e.g., TEG 127-200-28). Therefore, the biggest challenge in using the electrical power of a TEG at small temperature differences is to boost these small voltages to a useful level for the electrical load. Step-up converter designs that can manage these tasks are considered in more detail in the next section.

2.2 DC-DC Converter Topologies

The three main DC-DC converter types are depicted in Figure 4. A buck converter is used if a lower voltage from a supply voltage is needed, whereas for greater output voltages a boost converter is employed. If sometimes the output voltage should be lower and sometimes it should be higher, then a buck-boost converter is the best choice. All of the three architectures use an inductor and a capacitor as the energy-storage element. At this point, it has to be mentioned that there exists also linear regulators, which in principle don't need any energy-storage elements. But their output voltage can only be lower than the input voltage and the output current cannot be higher than the input current as is the

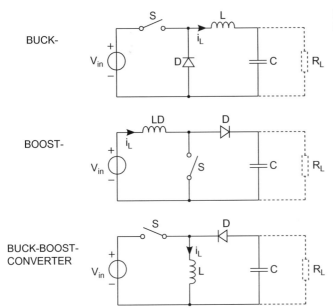

FIGURE 4   Principles of the main DC-DC converter types.

case with buck converters. This means that their efficiency is linearly dependent on the difference between input and output voltage, no matter if the components are considered ideal or not.

Mohan et al. [10] did a comprehensive analysis of these three converter types, which are explained here. Basically three modes of operation are possible with DC-DC converters, depending on the inductor current flow (see Figure 4): Discontinuous mode (DCM), continuous mode (CCM), and boundary between discontinuous and continuous mode. This is shown in Figures 5, 6, and 7 for the case of a buck, boost, and buck-boost converters (see also Figure 4). The buck-boost converter can also be obtained from a series connection of a boost and a buck converter.

The corresponding areas A and B must be equal if ideal components are assumed. It should also be noted that all formulas are only valid for $P_{in} = P_{out}$, i.e., an efficiency of the converters of 100%.

2.3 DC-DC Converter Design for Ultra-low Input Voltages

Today, there are step-up converters for thermoelectric generators and solar cells that have a minimum start-up voltage of 20 mV [11] (see Table 1). There are also bipolar versions of these converters [12] that allow that in the case of thermoelectric generators, the thermal gradient occurs in the opposite direction and that the electrical source delivers a negative voltage.

At low input voltages (less than about 800 mV), for typical boost converter designs with a single coil and active controlled switching transistor it is not possible to start up without an auxiliary voltage. Therefore, these circuits are not appropriate for all energy harvesting

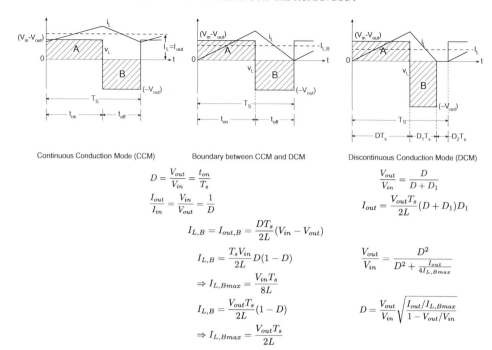

Continuous Conduction Mode (CCM)   Boundary between CCM and DCM   Discontinuous Conduction Mode (DCM)

$$D = \frac{V_{out}}{V_{in}} = \frac{t_{on}}{T_s}$$

$$\frac{I_{out}}{I_{in}} = \frac{V_{in}}{V_{out}} = \frac{1}{D}$$

$$I_{L,B} = I_{out,B} = \frac{DT_s}{2L}(V_{in} - V_{out})$$

$$I_{L,B} = \frac{T_s V_{in}}{2L} D(1 - D)$$

$$\Rightarrow I_{L,Bmax} = \frac{V_{in} T_s}{8L}$$

$$I_{L,B} = \frac{V_{out} T_s}{2L}(1 - D)$$

$$\Rightarrow I_{L,Bmax} = \frac{V_{out} T_s}{2L}$$

$$\frac{V_{out}}{V_{in}} = \frac{D}{D + D_1}$$

$$I_{out} = \frac{V_{out} T_s}{2L}(D + D_1)D_1$$

$$\frac{V_{out}}{V_{in}} = \frac{D^2}{D^2 + \frac{I_{out}}{4I_{L,Bmax}}}$$

$$D = \frac{V_{out}}{V_{in}}\sqrt{\frac{I_{out}/I_{L,Bmax}}{1 - V_{out}/V_{in}}}$$

FIGURE 5   Modes of operation of a DC-DC buck converter and essential formulas.

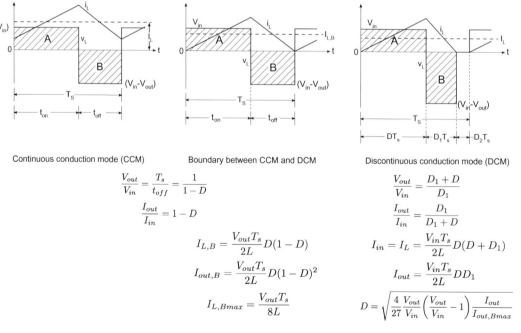

Continuous conduction mode (CCM)   Boundary between CCM and DCM   Discontinuous conduction mode (DCM)

$$\frac{V_{out}}{V_{in}} = \frac{T_s}{t_{off}} = \frac{1}{1 - D}$$

$$\frac{I_{out}}{I_{in}} = 1 - D$$

$$I_{L,B} = \frac{V_{out} T_s}{2L} D(1 - D)$$

$$I_{out,B} = \frac{V_{out} T_s}{2L} D(1 - D)^2$$

$$I_{L,Bmax} = \frac{V_{out} T_s}{8L}$$

$$\frac{V_{out}}{V_{in}} = \frac{D_1 + D}{D_1}$$

$$\frac{I_{out}}{I_{in}} = \frac{D_1}{D_1 + D}$$

$$I_{in} = I_L = \frac{V_{in} T_s}{2L} D(D + D_1)$$

$$I_{out} = \frac{V_{in} T_s}{2L} DD_1$$

$$D = \sqrt{\frac{4}{27}\frac{V_{out}}{V_{in}}\left(\frac{V_{out}}{V_{in}} - 1\right)\frac{I_{out}}{I_{out,Bmax}}}$$

FIGURE 6   Modes of operation of a DC-DC boost converter and essential formulas.

applications. Another approach is based on self-oscillating circuits like the Meissner/ Armstrong oscillator [13]) with the use of a transformer and a normally on active device (e.g., a junction FET or a depletion MOSFETs). An implementation is shown in Figure 8, where basically an oscillator with a rectangular output signal is built up, which is rectified to get a higher output voltage V_{out} with respect to the input V_{in}. This way an operation at input voltages below 100 mV can be established. These circuits can be used either to start up a main boost converter stage [14] or as a complete boost converter design.

The circuit shown in Figure 8 starts up through the saturation of the current flowing in the primary inductance L_1 and the junction FET T_1. As long as this current rises, a voltage

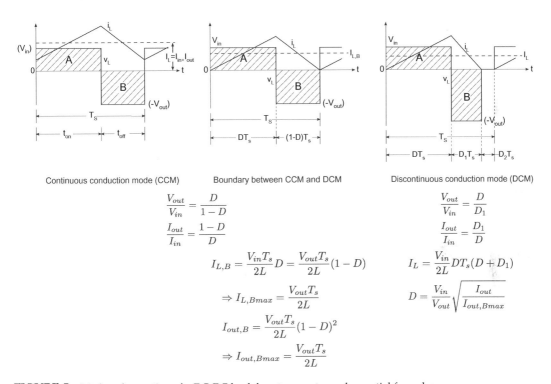

FIGURE 7   Modes of operation of a DC-DC buck-boost converter and essential formulas.

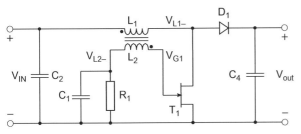

FIGURE 8   Simple self-oscillating boost converter with JFET and coupled inductors.

is induced in the secondary inductance L_2, whereas the gate-source diode is conducting. Therefore, C_1 is charged with a negative voltage V_{L2-}. At some point, the primary current of the transformer saturates due to the ohmic resistance of L_1 and T_1. Consequently, the induced voltage on the secondary side drops to zero and V_{L2} is equal to the gatesource voltage of JFET T_1. This increases its ON-resistance if the absolute value of this negative voltage (referred to as the negative input pin) is close to the pinch-off voltage of JFET T_1. V_{L1} now further decreases due to the falling current through L_1 and finally T_1 shuts OFF completely. From this point the circuit starts oscillating at a frequency determined by the time constant of R_1 and C_1. The disadvantage of the converter is the low efficiency due to the generally high ohmic resistance (typically around 10 Ohms) of junction FETs and of course the need of a transformer. Nevertheless, the step-up converter can be optimized and changed in a simple way to deliver a regulated output voltage and act as a DC-DC converter for ultra-low input voltages without any need for an external start-up voltage from a battery, for example. More information about this topology is available in section S.1 of the supplemental material.

2.3.1 Maximum Power Point Tracking for Impedance Matching

Since some energy transducers like a thermoelectric generator (TEG) based on the Seebeck effect or a solar cell have an internal parasitic ohmic resistance, it is useful to think about matching the load to this resistance in order to get the maximum power out of the transducer. The adaptive matching of a load is considered in the following for the thermogenerator transducer. Nevertheless, this method can be used in an analog way for equivalent energy transducers. If a voltage converter is used and the TEG is modeled as a voltage source V_{OC} with a series resistance R_{TG} (see Figure 9(a)) the resistive load R_L connected to the converter creates an equivalent resistance R_{eq} connected to the TEG. A detailed analysis of this configuration for the output power of the TEG $P_{TG} = \ldots$. If this function is plotted versus the duty cycle D (see in Figure 9(b)) it can be observed again that there exists a global maximum at D_{MPP}. This means that an optimum power point can be found for any resistive load adapting the duty cycle. The same can be done with any other converter type, but in the case of a boost converter R_L always has to be greater than R_{eq}. In the case of employing a buck-boost converter, R_L can be greater or smaller than R_{eq},

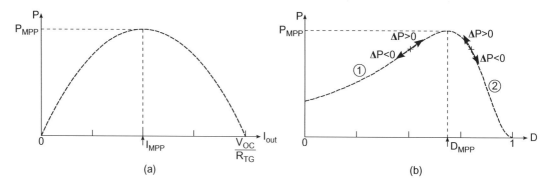

FIGURE 9   Matching a load to an energy transducer: Output power versus (a) output current and (b) duty cycle.

which can be a more flexible solution. Nevertheless, for finding the maximum power point the duty cycle D has to be adjusted, which results in varying the output voltage of the converter in the case of a resistive load R_L. In general, loads like wireless transceivers or sensors are not resistive or need to be supplied by a fixed voltage. In this case, it is not possible to draw the maximum power out of the energy transducer, since the current and thus the power is fixed by the load itself. Nevertheless, if an energy-storage device like a battery or a capacitor is used between the load and the output of the voltage converter, it is still possible to operate the energy transducer in the maximum power point. How that works is shown in detail in section S.1 of the supplemental material.

3. ENERGY HARVESTING FROM FOOT MOTION

Walking is one of the normal human activities that expends energy [15,16]. Different energy harvesting transducers like piezoelectric materials, dielectric elastomers, and rotatory generators have been employed in order to harvest energy from this activity.

The MIT Media Lab has developed several energy harvesting systems that use the low-frequency parasitic power in shoes employing piezoelectric materials. One scheme harvests the energy dissipated in bending the ball of the foot, placing a multilaminar polyvinylidene fluoride (PVDF) bimorph under the insole. The second one consists of harvesting the foot-strike energy using a pre-stressed spring metal strip laminated with a semi-flexible form of lead zirconium titanate (PZT) under the heel. Both devices were excited under a 0.9 Hz walking activity. An average power of 1.3 mW with 250 kΩ load and 8.4 mW with 500 kΩ load was obtained for the PVDF stave and PZT bimorph, respectively [17].

D. Fourie used mode 31 of excitation of piezoelectric elements by placing fifteen strips of PVDF material at the heel of a sneaker. A mean power of 60 μW was obtained with a 470 kΩ resistor [18].

This section is dedicated to piezoelectric materials with a focus on PVDF material. Moreover, piezoelectric structures like bimorph cantilevers resonate usually at some tens to hundreds of Hertz. These structures have a resonant frequency that matches the frequency of the ambient vibration employed as the ambient source. However, body motion has associated mechanical activities with frequencies in the range of units of Hertz and therefore the analysis done for the piezolectric materials is a quasi-static analysis [19,20]. Thus, the piezoelectric transducer does not have to match the frequency of the human motion.

3.1 Physical Principles

Piezoelectric materials are employed as sensors, actuators, and energy harvesting generators. The piezoelectric effect was discovered by Jacques and Pierre Curie in 1880, who found that certain materials, when subjected to mechanical strain, suffered an electrical polarization proportional to the applied strain. The materials that have this property also have the inverse piezoelectric effect that consists of the deformation of the material when an electrical field is applied. Therefore, piezoelectric elements allow conversion between electrical and mechanical energy. This property may be used for harvesting energy from the environment.

In piezoelectric materials, an interaction between mechanical and electrical parameters takes place. The phenomenon of piezoelectricity is described by the piezoelectric constitutive equations:

$$S_i = s_{ij}^E T_j + d_{li} E_l$$
$$D_m = \varepsilon_{mn}^T E_n + d_{mk} T_k \qquad \text{for } i,j,k = 1,\ldots,6 \text{ and } l,m,n = 1,2,3 \qquad (9)$$

where T is the applied mechanical stress [N/m^2], E is the applied electric field [N/C], d corresponds to the piezo strain [(C/m^2)/(N/m^2)], ε^T is the permittivity [F/m] under conditions of constant stress, D is the electric displacement [C/m^2], S is the mechanical strain [m/m], and s^E is the compliance tensor [m^2/N] under conditions of constant electrical field.

In Eq. (9) subscripts correspond to the six directions of the axes, three Cartesian directions plus the shear around the three axes, as shown in Table 3. Repeated subscripts in the products imply a summation over the different components (Einstein notation). Figure 10 shows a piezoelectric PVDF film positioned in its correspondent Cartesian coordinate system. The length of the rod, L, is oriented along axis 1, the width, W, along axis 2, and the thickness, t_c, along axis 3 [21].

TABLE 3   Subscripts of the Reduced Notation for Piezoelectric Constitutive Equations

Reduced Notation	Corresponding Direction of Axes
1	Longitudinal in x direction
2	Longitudinal in y direction
3	Longitudinal in z direction
4	Shear y-z
5	Shear z-x
6	Shear x-y

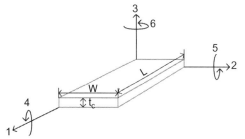

FIGURE 10   Mechanical axis position for piezoelectric materials.

There are other three ways to express the piezoelectric phenomenon, given by the piezoelectric constitutive equations. Each case differs from the others in the independent state variables and the piezoelectric constants employed [20].

Table 4 gives a definition of the piezoelectric constants employed in the four different pairs of piezoelectric equations.

The value for each of the elements of the piezoelectric matrix constants depends on the crystal symmetry of the piezoelectric material. The two most common types of

TABLE 4 Definition of the Piezoelectric Constants

Constant	Definition	S.I. Units
d, piezoelectric charge constant	Dielectric displacement developed when a mechanical stress is developed. E = constant	$\left[\frac{C}{N}\right]$
	Strain developed when an electric field is applied. T = constant	$\left[\frac{m}{V}\right]$
g, piezoelectric voltage constant	Electric field developed when a mechanical stress is applied. D = constant	$\left[\frac{Vm}{N}\right]$
	Strain developed when a dielectric displacement is applied. T = constant	$\left[\frac{m^2}{C}\right]$
e	Dielectric displacement developed when a mechanical strain is applied. E = constant	$\left[\frac{C}{m^2}\right]$
	Mechanical stress developed when an electric field is applied. S = constant	$\left[\frac{N}{Vm}\right]$
h	Electric field developed when a mechanical strain is applied. D = constant	$\left[\frac{V}{m}\right]$
	Mechanical stress developed when a dielectric displacement is applied. S = constant	$\left[\frac{N}{C}\right]$
s^E, compliance	Mechanical strain developed when a mechanical stress is applied. E = constant	$\left[\frac{m^2}{N}\right]$
s^D, compliance	Mechanical strain developed when a mechanical stress is applied. D = constant	$\left[\frac{m^2}{N}\right]$
ε^T, permittivity	Dielectric displacement developed when an electric field is applied. T = constant	$\left[\frac{CV}{m}\right] = \left[\frac{F}{m}\right]$
ε^S, permittivity	Dielectric displacement developed when an electric field is applied. S = constant	$\left[\frac{CV}{m}\right] = \left[\frac{F}{m}\right]$
c^E	Mechanical stress developed when a mechanical strain is applied. E = constant	$\left[\frac{N}{m^2}\right]$
c^D	Mechanical stress developed when a mechanical strain is applied. D = constant	$\left[\frac{N}{m^2}\right]$
β^T	Electrical field developed when a dielectric displacement is applied. T = constant	$\left[\frac{m}{CV}\right] = \left[\frac{m}{F}\right]$
β^S	Electrical field developed when a dielectric displacement is applied. T = constant	$\left[\frac{m}{CV}\right] = \left[\frac{m}{F}\right]$

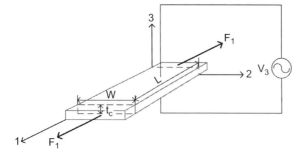

FIGURE 11 Mechanical excitation of the piezo-electric film along axis 1.

piezoelectric materials are PVDF, polyvinylidene fluoride, and PZT, lead zirconate titanate, a ceramic piezoelectric material.

The fact that PVDF has mm2 crystal symmetry and that PVDF films are usually only metallized in the plane perpendicular to direction 3, so that $D_1 = D_2 = 0$, simplifies Eq. (9) into Eq. (10):

$$D_3 = d_{3p}T_p \quad \text{for } p = 1, 2, 3 \tag{10}$$

The previous equation expresses the electric displacement generated in the piezoelectric element in charge mode (i.e., short-circuit between the film connectors and therefore no electric field is applied). In an open circuit, the equivalent expression is given by Eq. (11). This expression shows that only the longitudinal directions and not the shear directions (4, 5, and 6) contribute to the generated voltage. Therefore, there are three possible modes to use the PVDF films: 31, 32, and 33.

$$V_3 = g_{3p}T_p t_c \quad \text{for } p = 1, 2, 3 \tag{11}$$

Figure 10 shows a piezoelectric PVDF film positioned in its correspondent Cartesian coordinate system. The length of the rod, L, is oriented along axis 1, the width, W, along axis 2, and the thickness, t_c, along axis 3 [21].

Piezoelectric materials have different working modes related to the axis where the mechanical excitation and the electrical response are obtained. For example, working mode 31 corresponds to a mechanical excitation along axis 1, whereas the electrical response is obtained in the plane perpendicular to axis 3. Table 5 shows the relation between the piezoelectric constants for PVDF material in the particular case of working mode 31. In Table 5, a new piezoelectric constant k_{31} is introduced that is the piezoelectric coupling constant for PVDF.

Mode 31 corresponds to a mechanical excitation along axis 1, whereas the electrical answer is obtained in the plane perpendicular to axis 3, as shown in Figure 11. In this figure, a mechanical force, F_1, applied along axis 1 causes a strain of the piezoelectric film in the same direction. The mechanical stress generates a voltage in the piezoelectric element in the plane perpendicular to axis 3 since it is in open-circuit. In the case of a piezo-electric element short-circuited, an electrical current instead of a voltage would be obtained. Figure 12 illustrates the case of working mode 33. In this case, the mechanical excitation is applied along axis 3, which causes a strain of the material in the same

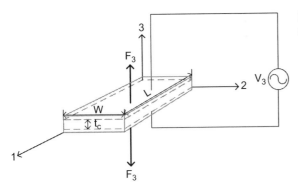

TABLE 5 Relation Between Piezoelectric Constants for PVDF in Mode 31

Relation Between Piezoelectric Constants

$d_{31} = g_{31} \varepsilon_{33}^T$

$s_{11}^E = 1/c_{11}^E$

$d_{31} = e_{31}/c_{11}^E = e_{31} s_{11}^E$

$s_{11}^D = 1/c_{11}^D$

$g_{31} = h_{31}/c_{11}^D = h_{31} s_{11}^D$

$e_{31} = h_{31} \varepsilon_{33}^S$

$\varepsilon_{33}^S = \varepsilon_{33}^T \left(1 - d_{31}^2/\left(S_{11}^E \varepsilon_{33}^T\right)\right) = \varepsilon_{33}^T \left(1 - k_{31}^2\right)$

$S_{11}^D = S_{11}^E \left(1 - d_{31}^2/\left(S_{11}^E \varepsilon_{33}^T\right)\right) = S_{11}^E \left(1 - k_{31}^2\right)$

direction. The electrical answer is obtained, as in the previous case, in the plane perpendicular to axis 3.

Mode 32 is discarded since d_{32} is 10 times less than d_{31} or d_{33}. In mode 31, the stress is applied in direction 1, and in mode 33, the stress is applied in direction 3. The voltage and charge obtained resulting from an applied force in a certain direction, F_1 or F_3, are shown in Table 6.

In a thin PVDF film, the ratio L/t_c is on the order of 1,000, while $d_{31} = 23E - 12 \text{ m/V}$ and $d_{33} = -33E - 12 \text{ m/V}$ [21]. Assuming that $F_1 = F_3$, then V_3 and q_3 for mode 31 will be on the order of 700 times greater than V_3 and q_3 for mode 33. Therefore, for the same mechanical energy input, more electrical energy output is obtained in mode 31 than in mode 33 when the PVDF piezoelectric films are employed. For other piezoelectric materials, mode 33 excitation can be a better solution than mode 31 if the length, width, and thickness of the piezoelectric material are similar [22].

The relations extracted from the piezoelectric constitutive equations between the piezoelectric constants are summarized in Table 5.

An electromechanical model can be derived from the piezoelectric constitutive equations. This model consists of a circuit where mechanical and electrical variables present in the piezoelectric effect are related. Figures 13(a) and 13(b) show an electromechanical

TABLE 6 Voltage V_3 and Charge q_3 Obtained in the Plane Perpendicular to Direction 3 Applying a Mechanical Stress in Direction 1, Mode 31, and in Direction 3, Mode 33

	Mode 31	Mode 33
V_3	$g_{31}\dfrac{F_1}{W}$	$g_{33}\dfrac{F_3}{WL}t_c$
q_3	$d_{31}\dfrac{F_1 L}{t_c}$	$d_{33}F_3$

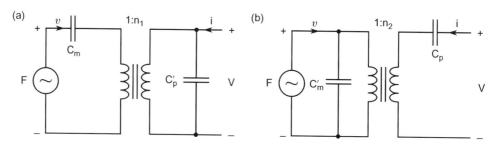

FIGURE 13 Piezoelectric coupling circuits, relating mechanical and electrical magnitudes.

TABLE 7 Relation Between Electrical Components for PVDF in Mode 31.

Relation Between Electrical Components

$$C_{m31} = \frac{LS_{11}^E}{Wt_c}$$

$$C'_{p31} = C_{p31} - \frac{d_{31}^2 L^2}{t_c^2 C_{m31}} = C_p\left(1 - \frac{d_{31}^2}{s_{11}^E \varepsilon_{33}^T}\right) = C_{p31}\left(1 - k_{31}^2\right)$$

$$n_1 = \frac{V}{F_t} = -\frac{C_{m31}t_c}{d_{31}L}$$

$$C'_{m31} = C_{m31}\left(1 - \frac{L2d_{31}^2}{t_c^2 C_p C_{m31}}\right) = C_{m31}\left(1 - k_{31}^2\right)$$

$$C_{p31} = \frac{\varepsilon_{33}^T A}{t_c}$$

$$n_2 = \frac{V_t}{F} = -\frac{d_{31}L}{t_c C_p}$$

piezoelectric model presented in [23,24] where the state variables are F, v, V, and I instead of the state variables (T, $\dot{S}$, E, and $\dot{D}$) employed in the model that appears in [25].

For example, Table 7 shows the relation between the electrical components that appear in Figure 13(a) and 13(b) and the piezoelectric constants for the particular working mode 31. Similar results can be obtained for other working modes and piezoelectric materials.

The analysis done gives as a result an electromechanical piezoelectric model that employs only electrical components (e.g., the force applied over the piezoelectric material is a voltage source). Thus, it is possible to use both the mechanical and the electrical part of the model in an electronic circuit simulator.

3.2 AC-DC Converters

Analysis of the power at the output of a piezoelectric energy harvesting transducer is usually done by connecting a resistor. Then, the value of the resistor is calculated in order to maximize its power. However, this power does not correspond to the harvested power, which is the available electrical power that has been converted from the mechanical excitation. If a resistor is connected as a load just a small fraction of the harvested power will be transferred to the output.

Since the equivalent circuit of a PVDF film is purely capacitive, an inductive load instead of a resistor would be able to extract the maximum electrical power from the piezoelectric transducer. This harvested power would be maximized when the electrical impedance matches the complex conjugate of the mechanical impedance of the piezoelectric element [26]. However, an AC-DC converter must be connected to the PVDF film to rectify its AC power.

There are two possible ways to use an AC-DC converter in this case: a linear or a nonlinear rectifier. The linear rectifier can be implemented with, for example, a diode bridge, a voltage multiplier, or a current multiplier. In this case, no action is done regarding the phase shift between voltage and current and the maximization of the output power is achieved by controlling the voltage magnitude on the piezoelectric electrodes [27]. In contrast to linear techniques, a nonlinear rectifier controls the magnitude and phase of the voltage on the piezoelectric electrodes in order to maximize the harvested power trying to set the voltage and the internal piezoelectric current in phase, simulating a complex conjugate load that matches the impedance of the PVDF film.

A diode bridge can be implemented using Schottky diodes with low forward voltage drop, low capacitance values, and low reverse currents. A forward voltage of 0.25 V at $100\,\mu A$ of current, a reverse current on the order of units or a few tens of nanoamperes, and a capacitance of less than 1 pF are recommended values for the Schotkky diodes to be used with PVDF piezoelectric films. A full-bridge rectifier can also be implemented using four PMOS transistors connected as diodes. However, this configuration has losses associated with the threshold voltages of the transistors [28]. An improvement of this rectifier consists of the connection of the two right PMOS transistors in a cross-coupled configuration (see Figure 14 [28,29]). These two PMOS transistors work in the linear region and compared with the previous full-wave rectifier provide a higher output current at the same output voltage.

Voltage multiplier or current multiplier rectifiers are an alternative to full-wave rectifiers when it is required to increase the voltage or the current and rectify the signal. All of them are called linear rectifiers and afterwards a DC-DC converter is needed. The DC-DC converter can be used to adjust the output voltage and to maximize the power extracted from the piezoelectric transducer.

Ramnadass et al. [30] and Liu [19] compared the output power obtained with a full-bridge rectifier and with a voltage doubler. If no voltage drop is considered at the diodes, both rectifiers provide the same maximum output power. However, when the voltage drop at the diodes is taken into account, the output power obtained with the voltage doubler is higher than with the full-bridge rectifier.

The AC-DC rectifier requires a DC-DC converter to transfer the energy extracted from the PVDF film to the battery (see Figure 15). There is an optimum voltage level on the storage capacitor C_2 for which the power flowing from the piezoelectric transducer is maximum. Thus, the DC-DC converter must work with an appropriate duty cycle to achieve this optimum voltage. Ottman et al. [32,33] used a step-down converter as a DC-DC converter to regulate the voltage on capacitor C_2 to the value that maximizes the power provided by the piezoelectric element. Thus, the optimum input voltage of the converter is

$$V_{C2} = \frac{I_p}{2\omega C_1} = \frac{V_{oc}}{2} \tag{12}$$

where I_p is the peak current, C_1 is the internal capacitance, and V_{oc} is the open circuit voltage of the piezoelectric element.

Nonlinear converters are employed with piezoelectric transducers to increase the harvested energy by implementing a virtual impedance that matches the complex conjugate impedance of the piezoelectric transducer [19,26].

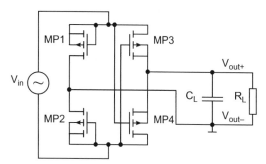

FIGURE 14 Full-wave rectifier with two PMOS diodes and two cross-coupled PMOS transistors [28].

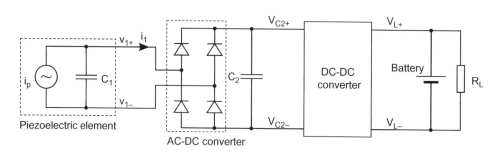

FIGURE 15 Piezoelectric power supply with a diode bridge and a DC-DC converter as power management unit [31].

In nonlinear techniques, when the maximum voltage on the piezoelectric element is achieved, the connection of an inductor with the piezoelectric element creates a resonance circuit with the internal capacitor of the piezoelectric element. This causes the inversion of the piezoelectric voltage in a very short time compared to the mechanical excitation. Well known non-linear techniques are, for example, the parallel synchronized switch harvesting on inductor (parallel SSHI) and the series synchronized switch harvesting on inductor (series SSHI) techniques.

More information about the AC-DC converters for piezoelectric energy harvesters can be found in section S.2 of the supplemental material.

4. WIRELESS ENERGY TRANSMISSION

This section provides an overview of wireless energy transmission, introducing the basic theory and functional principles as well as the state-of-the-art technologies and related standards.

Wireless systems are well known from data transmission applications such as broadcasting, telecommunication, and short-range communication. These technologies basically work by modulating the generated electric, magnetic, or electromagnetic fields. As the main objective of these systems is the transport of information, their performance is mainly focused on the data rate and the system reliability, as well as on the efficiency and power consumption of the respective transmitters and receivers.

Wireless energy transmission systems, also called wireless power transmission (WPT), make use of the same fields and waves to transport power from a transmitter toward a receiver at a distance d, as illustrated in Figure 16. Normally, this can be done using three basic principles: 1) electric field in the near field, 2) magnetic field in the near field, and 3) electromagnetic field in the far field.

When talking about wireless power transmission and specifically about field propagation it is important to differentiate between the near-field and the far-field regions, since the field propagation behavior and the consequent propagation losses strongly differ depending on the region. It is normally considered that for electrically small antennas (i.e., antennas smaller than the frequency wavelength λ), the separation between the far-field and the near-field area is given by $d < \lambda/2\pi$. On the other hand, for electrically large antennas (antenna size D comparable to λ) the far-field region begins with $d > 2D^2/\lambda$ [34].

Within the near field the transmitted magnetic field and the transmitted electrical field are not linked together. The field strength is still relatively strong, but it decreases rapidly with 60 dB per decade as the distance d to the source increases. When reaching the far field, both the electrical and the magnetic fields get linked together, being perpendicular to each other, and form an electromagnetic wave. In this case, the field strength decreases with $1/d^2$. In general, near-field wireless energy transmission systems are used at lower

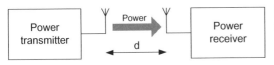

FIGURE 16 Basic wireless power transmission system.

frequencies, achieving high amounts of transmitted power at short distances. On the other hand, far-field systems use higher frequencies for transmission of low amounts of power at long distances.

4.1 Inductive Wireless Energy Transfer in the Near Field

In an inductive energy transmission system both antennas are built with coils. The result is a weak-coupled air gap transformer in which the magnetic field H is used for the energy transfer between them. For optimal results both coils must be placed in parallel (i.e., aligned in one axis), with a short distance between them and no conductive materials in the proximity. The transmitted field strength depends on the current amplitude I of the transmission signal, the radius of the coil r and its number of turns N and can be calculated by means of (13):

$$H(d) = \frac{INr^2}{2\sqrt{(r^2+d^2)^3}} \tag{13}$$

The transmission distance in this kind of system is mainly limited by the antenna coil radius r. Only within this distance an efficient power transmission is possible, as can be seen in Figure 17.

The efficiency of inductive systems is mainly limited by the coupling factor between the two coils. However, it can be increased by tuning the system to resonate at a certain frequency. This is often done with a parallel resonance at the receiver and a serial resonance at the transmitter side. Due to the coupling between receiving and transmitting circuits, load changes at the receiver are also transformed toward the transmitter according to the coupling factor, which influences the whole transmission system performance [35].

4.2 Capacitive Wireless Energy Transfer in the Near Field

In the case of capacitive energy transmission, the transmitting and receiving antennas form an air-filled capacitor. These capacitive systems are meant to be more tolerant toward

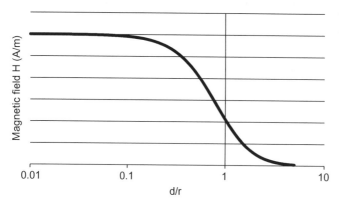

FIGURE 17 Magnetic field strength vs. coil radius and distance.

misalignments between transmitter and receiver when compared to inductive systems. Today, there are some applications using this technology [36]. However, they are not as common as inductive systems. Although both technologies are theoretically adequate for the purpose of energy transmission, the main disadvantage of the capacitive energy transmission is the relatively high field strength that is needed, which results in high voltage potentials and risks for human health. In practical implementations, these risks can be minimized by assuring that the system is only active when a power receiver is placed correctly on the transmitter. This can be done with proper mechanical design as well as with additional communication between the transmitter and the receiver station before initiating the power transmission.

4.3 Electromagnetic Wireless Energy Transmission in the Far Field

The power transmission in the far field normally takes place between a radio frequency (RF) power transmitter and a power receiver, both of them having a transmitting and a receiving antenna, respectively. This scenario is illustrated in Figure 18.

The power that can be received at a distance d from the power transmitter can be estimated as follows:

$$P_{rx} = P_{tx} + G_{tx} - FSL - P + G_{rx}, \tag{14}$$

where P_{tx} is the transmitter output power, G_{tx} and G_{rx} are the transmitter and the receiver antenna gains, respectively, and FSL represents the free space losses for the distance d and is given by

$$FSL = \left(\frac{4\pi d}{\lambda}\right)^2, \tag{15}$$

and finally P is the polarization mismatch between both antennas (e.g., in some applications the transmitters have circular polarized antennas and the receivers use linear antennas, which would involve 3 dB polarization losses).

Figure 19 shows the power received as a function of the distance d for a transmitter operating at 868 MHz with an output power of 30 dBm, 5.15 dBi transmitter antenna gain, 2.15 dBi receiver antenna gain, and 3 dB polarization mismatch between both antennas. As can be seen, for distances longer than 4 m, power levels below −10 dBm (0.1 mW) can be expected.

After the antenna, an RF/DC converter normally combined with a proper matching network is needed. Especially critical is to achieve high efficiencies with the low input power expected. After looking at commercial systems and state-of-the-art RF/DC converters, one can say that typical efficiencies are in the range of 10 to 50%. As a reference, the

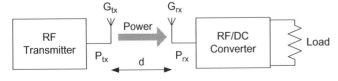

FIGURE 18 Wireless power transmission in the far field.

commercial system Powercast [37] achieves an efficiency of 30% for an input power of −8 dBm at 868 MHz.

4.4 RFID Technology as an Example Application

Typical radio frequency identification (RFID) systems consist of an active reader device and one or more passive transponder devices. For operation, the reader transmits a strong carrier signal that is used on the one hand to transport power toward the transponders and, on the other hand, to transmit data between the reader and the transponder and vice versa. In standard passive RFID systems the transponder has no energy source or storage and cannot operate without the reader.

There are three main types of RFID systems, which are mainly distinguished by the operation frequency: LF (low frequency), HF (high frequency), and UHF (ultra-high frequency). Table 8 provides an overview of the international standards established for each, and their main characteristics. RFID systems are normally used for the identification of objects, animals, and persons. Well-known examples are pet identification (LF), secure

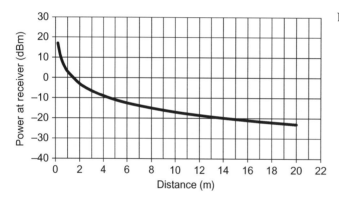

FIGURE 19   Received RF power.

TABLE 8   RFID Frequencies and Characteristics

	Standard	Frequency	Principle	Range	Comments
LF	ISO 18000-1 ISO 18000-2 ISO 11784/85 ISO 14223	119–135 kHz	Inductive	1 m	Very robust, low data rate
HF	ISO 15693 ISO 14443 ISO 18000-3	13.56 MHz	Inductive	1 m	Cryptography, big memory
UHF	ISO 18000-6C EPCGlobal	860–960 MHz	Electromagnetic	15 m	Bulk reading

entrance systems (HF), near field communication (NFC), which is similar to HF-RFID, wireless payments (NFC), and logistic applications (UHF).

Beside these standard applications, RFID technology is also used to power passive wireless sensors and circuits, sometimes with parallel data transmission. Although some commercial applications are available, these applications are not foreseen in the given RFID standards and often make use of proprietary technology and protocols.

4.5 Wireless Power Transmission Regulations

Wireless power transmission is limited by local regulations in terms of frequency use, bandwidth, and power, like any other wireless technology. This is motivated due to human protection on the one hand and to avoid interferences with other electrical systems on the other hand. For that reason, WPT operates mainly at low frequencies (<9 kHz) where no frequency regulation exists, and also within the industrial, scientific, and medical (ISM) radio frequency bands (i.e., 125 kHz, 13.56 MHz, 868 MHz, and 2.4 GHz), which are free to use. Table 9 summarizes the maximum permitted transmitter signal amplitude in Europe, which limits the amount of power that can be transferred with such systems. For the lower frequencies the maximum magnetic field strength is defined at a measurement distance of 10 m from the transmitter; for the higher frequencies the equivalent radiated power (e.r.p.) level is used.

In industrial applications WPT is already used to power moving or rotating receivers where a wired connection is not possible. Beside some standard products within machine automation, these WPT systems are often based on customized and proprietary solutions. Today, efforts are being made to introduce international industrial standards for wireless power transmission, which are mainly focused on consumer electronics. Table 10 shows some of these proposed solutions, standards, and applications.

The first products and components available for some of the standards are presented in Table 10. However, none of them have become ISO/IEEE standards yet, and they also lack a strong market position. Recently, the first commercial integrated solutions for these WPT standards have become available on the market to enable wider use of wireless charging in consumer applications. These solutions focus on short-range transmission with fixed positions and low amounts of power. Typical values are in the range of 1 to 5 watt

TABLE 9 Power Transmission Limits

Frequency Band	Power/Magnetic Field Strength
90–119 kHz	42 dBµA/m
119–135 kHz	66 dBµA/m
135–140 kHz	42 dBµA/m
13.553–13.567 MHz	42 dBµA/m
865.6–867.6 MHz	2 W e.r.p.
2446–2454 MHz MHz	500 mW e.i.r.p.

TABLE 10   WPT Commercial Systems and Standards

Name	Organization	Frequency	Comments
Qi	Wireless Power Con-sortium (WPC)	100–205 kHz	Commercial chip solutions available Commercial
Powermat	Power Matters Alliance (PMA)	277–357 kHz	Commercial charging stations available
A4WP	Alliance for Wireless Power	6.78 MHz	Up to 50 mm transmission distance
NFC Charging	Renasas	13.56 MHz	One-chip solution NFC and charging
Air Voltage	Murata	Unknown	Displacement tolerance
Powercast	Powercast	850–950 MHz	Dedicated power transmitter optional

transmission power at 5 volt voltage within a transmission distance of some millimeters up to one centimeter (5 cm in case of A4WP). The main objective of these systems is to recharge portable devices like mobile phones and cameras easily without additional cables and adapters, just by placing them upon a wireless charging platform. One of the main drawbacks of the technology is the need for optimal conditions (i.e., antenna placement and no influence of surrounding materials) to get the expected results. As a result, current research activities are focused on increasing the efficiency and the tolerance of misplacement [38,39].

4.6 Influence of the Body on the Wireless System

Materials surrounding antennas always have an influence on the performance of the wireless system. Normally, when analyzing the effects of these materials, their permittivity and permeability, which affect the electric field and the magnetic field, respectively, must be considered. High values of these parameters can alter the antenna radiation pattern, the bandwidth, and the frequency of operation (de-tuning of the antenna), as well as affect the impedance matching. The material conductivity is another important parameter that must be considered, since conductive materials can hinder any possibility of electromagnetic transmission, especially as the frequency increases. Apart from these effects, the inherent transmission losses through the material, which are normally given by the dielectric loss tangent, must also be considered.

Several models exist that characterize the dielectric properties of the different parts of the human body. One of the most commonly used is the 4-cole-cole model described in [40]. Software models and phantoms that emulate the human body and allow the design and testing of wireless systems are also available, such as the anatomical data set from the Visible Human Project [41]. Some of the models are approved by the respective regulation agencies so that they can be used to test if the developed systems fulfill the requirements, e.g., in terms of specific absorption rate (SAR), which defines the power absorbed by the human body in watts per kilogram.

If one focuses, for example, on passive UHF RFID, tags that include one or more sensors are already available on the market [42]. With a proper antenna design that deals with the influence of the human body [43], such tags could be used as passive wearable sensors that are low cost, small, and do not need any battery change. However, the extra power needed to power the sensor reduces their read range to 1 or 2 meters. In this sense, energy harvesting could play an important role, providing an extra amount of power for the sensor. Some other ideas regarding wearable RFID tags and proper antennas by means of conductive fabrics that isolate the body from the antenna are suggested in [44]. Battery-less implants that make use of wireless power transmission, for example, by means of inductive coupling technology, are also discussed in [45,46]. In these cases, size restrictions play a key role when designing the wireless system, since they limit the power consumption and the system functionality [45].

5. ENERGY HARVESTING FROM LIGHT

Light is an environmental energy source usually available at most locations. Thus, it is an ambient source to consider for supplying power to a wearable system. A photovoltaic (PV) cell is employed as a transducer to convert light into electricity. Photovoltaic systems are found from the megawatt to the milliwatt range producing electricity for a wide number of applications: from wristwatches to grid-connected PV systems. The power supply of wearable systems with PV cells can be a valid option in the right circumstances. However, long dark periods will require an energy-storage element since extra power will be needed for continuous operation.

The input power density that receives the solar cell from solar light (outdoor) is the incident irradiance, expressed in W/m^2. Another equivalent input parameter for solar cells is the insolation or solar irradiation, which is a measure of the input energy, and is obtained multiplying the irradiance by the time in hours, expressed as $W\,h/m^2$.

Another measure of the input power density for indoor light is illuminance, which is expressed in lux. The conversion of lux into irradiance is not direct and depends on the wavelength of the indoor light employed.

Manufacturers of solar cells use irradiance as the input parameter on their datasheets. However, there are some manufacturers that use illuminance when the application is for indoor light. Outdoors, solar radiation is the available ambient energy source. Solar radiation varies over the earth's surface due to time, month, weather conditions, and location (longitude and latitude). For each location there exists an optimum inclination angle and orientation of the PV solar cells in order to obtain the maximum radiation over the surface of the solar cell [47]. However, this knowledge is of difficult application on a wearable system since the wearable solar cell does not remain static. Indoor light has an irradiance range between 1 and 3 W/m^2 [48]. A solar cell carried by some smart textiles can receive either outdoor solar light, indoor solar light through the windows, or indoor light from a bulb. The power density or irradiance levels in outdoor or indoor conditions are quite different and therefore so is the harvested power. Table 11 provides an estimation of the irradiance levels in different lighting conditions [49].

TABLE 11   Power Density Values for Different Lighting Conditions [49]

Lighting Condition	Power Density (W/m2)
Bright sunlight	500–1000
Overcast sky	50–200
Department store	11–12
Grocery store	10.5–11.5
Meeting rooms	10–11
Office space	9–10.5
Warehouse	2–5.5

5.1 Physical Principles

Light intensity can be measured with photo detectors like photodiodes, photoresistors [50] or phototransistors [51]. Photodiodes and phototransistors provide a drain and a collector emitter current, respectively, proportional to the light intensity, while photoresistors change its resistance proportionally to the light intensity.

A solar cell can also be employed to measure irradiance. Datasheets of solar cells provide the values of the open-circuit voltage and short-circuit current at different irradiance levels to characterize them. The short-circuit current (Eq. (16)) is proportional to the irradiance and therefore it can be used to measure it. The short-circuit current is measured with a shunt resistor. Changes in temperature have a very slight effect on the short-circuit current and can be ignored [52]. Moreover, if at the same time measurements of the open-circuit voltage are carried out, the maximum output power of the solar cell can be estimated, as will be explained later.

Short-circuit current I_{sc} is the current delivered by the solar cell when its voltage is set to zero. Open-circuit voltage V_{oc} is the voltage delivered by the solar cell without load. There is a point called maximum power point P_{MPP} where the power delivered by the solar cell is maximum, which has an associated voltage V_{MPP} and current I_{MPP}. More interesting than the current and power are the current and power per unit area, current and power density, respectively [53], in order to compare the behavior of several cells with different areas. Figures 20 and 21 show current density as a function of voltage (J-V) and power density as a function of voltage (S-V), typical curves, respectively, for different irradiance levels of a solar cell.

Figures 22 and 23 show J-V and S-V curves for different temperature values. Temperature has a negligible effect on short-circuit current density. The equation for the short circuit current density is

$$J_{SC}(G) = \frac{J_{SC,ref}}{G_{ref}} G \tag{16}$$

where $J_{SC,ref}$ corresponds to the short-circuit current density at irradiance G_{ref}. Datasheets for solar cells provide the value of short-circuit current or current density $J_{SC,ref}$ at a certain

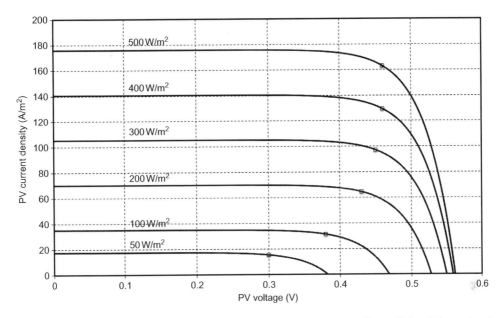

FIGURE 20 Current density versus voltage curves of a typical monocrystalline cell for different irradiance levels and a fixed ambient temperature of 25°C.

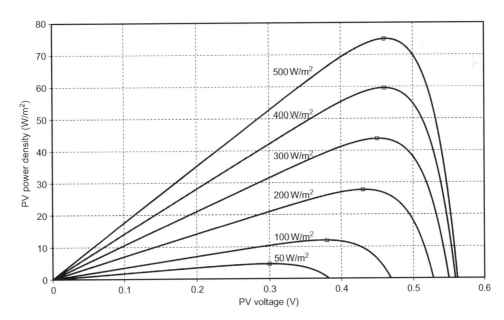

FIGURE 21 Power density versus voltage curves of a typical monocrystalline cell for different irradiance levels and a fixed ambient temperature of 25°C.

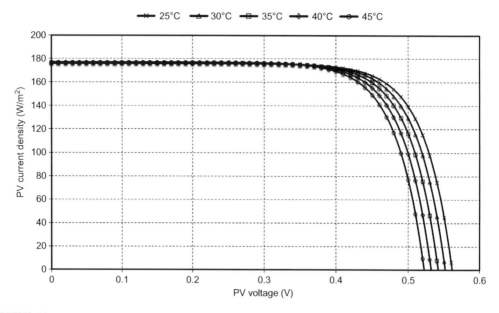

FIGURE 22   Current density versus voltage curves of a typical monocrystalline cell for different ambient temperature values and a fixed irradiance of 500 W/m^2.

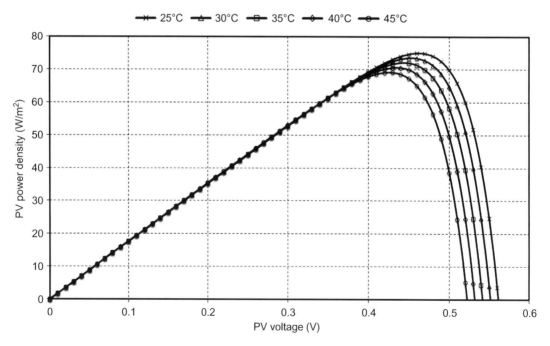

FIGURE 23   Power density versus voltage curves of a typical monocrystalline cell for different ambient temperature values and a fixed irradiance of 500 W/m^2.

irradiance G_{ref}, usually 1,000 W/m^2. The open-circuit voltage is a function of temperature and irradiance:

$$V_{OC}(T_C)(G) = V_{OC,ref} + (T_c - T_{c,ref})\frac{dV_{OC}}{dT_C} + V_t ln\frac{G}{G_{ref}} \qquad (17)$$

where $V_{OC,ref}$ corresponds to the open-circuit voltage of the solar cell at a given irradiance G_{ref} and temperature $T_{c,ref}$, T_c is the solar-cell temperature, dV_{OC}/dT_c is the voltage temperature coefficient, and V_t is the thermal voltage that is equal to kT/q (k is the Boltzmann's constant and T is the temperature expressed in Kelvin and q is the electrical charge of an electron). V_t is equal to 26 mV at 300 K. G is the incident irradiance on the solar cell.

The temperature of the solar cell is related to the ambient temperature through the nominal operating cell temperature (NOCT) parameter that is sometimes given by manufacturers in datasheets. NOCT has a value between 42°C and 48°C for silicon cells [53]:

$$T_{cell} = T_a + \frac{NOCT - 20}{800}G \qquad (18)$$

where NOCT is the cell temperature at 800 W/m^2, 20°C and a wind speed of 1 m/s in open-circuit conditions [54,55].

The following expression also models the behavior of the open-circuit voltage for solar cells that receive less than 200 W/m^2 [52]:

$$V_{OC}(T_C)(G) = \left(V_{OC,ref} + (T_c - T_{c,ref})\frac{dV_{OC}}{dT_C}\right)\left(1 + \rho_{OC}ln\frac{G}{G_{OC}}ln\frac{G}{G_{ref}}\right) \qquad (19)$$

where ρ_{OC} and G_{OC} are parameters obtained empirically. For many silicon PV modules, values of $\rho_{OC} = -0.04$ and $G_{OC} = 1,000$ W/m² are taken [52].

Fill factor (FF) is a parameter often given by manufacturers of solar cells in their datasheets. The fill factor is the ratio between the maximum power and the product of the open-circuit voltage and short-circuit current. Thus, the fill factor is given for a certain irradiance value:

$$FF = \frac{V_{MPP}I_{MPP}}{V_{OC}I_{SC}} = \frac{P_{MPP}}{V_{OC}I_{SC}} \qquad (20)$$

The equivalent circuit of a real or non-ideal solar cell is shown in Figure 24. The solar cell is modeled as a current source I_{PH} connected in parallel to diodes D_1 and D_2, which have saturation currents, I_{o1} and I_{o2}, with ideality factors 1 and 2. This sub-

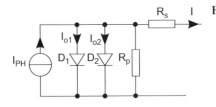

FIGURE 24   Equivalent electrical circuit of the solar cell.

circuit is a simple equivalent circuit of a solar cell, which models the ohmic losses with resistors R_s and R_p. From Kirchhoff's current law the equation for a solar cell is obtained:

$$I = I_{PH} - I_{o1}\left(e^{\frac{q(V+IR_S)}{kT}} - 1\right) - I_{o2}\left(e^{\frac{q(V+IR_S)}{2kT}} - 1\right) - \frac{V+IR_S}{R_p} \tag{21}$$

where I_{PH} is the current generated by the solar cell.

The previous equation is often given as [52]:

$$I = I_{PH} - I_0\left(e^{\frac{q(V+IR_S)}{nkT}} - 1\right) - \frac{V+IR_S}{R_p} \tag{22}$$

where n is the ideality factor. n can have a value between 1 and 2 [52] and for a silicon cell between 1.2 and 1.8 [53].

If the effects of the ohmic losses are ignored, Eq. (22) can be simplified to

$$I = I_{SC} - I_o\left(e^{\frac{q(V)}{nkT}} - 1\right) \tag{23}$$

where the short-circuit current is now the current generated by the solar cell. In the previous equations, the current terms can be substituted by current density.

PV modules are a combination of solar cells that can be connected either in parallel or series. If N_s and N_p are the number of solar cells connected in series and parallel, respectively, in the PV module, the equivalent value of the series and parallel resistance for a PV module is [55]:

$$R_{s,PVM} = \frac{N_s}{N_p} R_s \tag{24}$$

$$R_{p,PVM} = \frac{N_s}{N_p} R_p \tag{25}$$

The current converted by the PV module is multiplied by the number of solar cells connected in parallel N_p while thermal voltage V_t is multiplied by the number of solar cells connected in series N_s [55]:

$$I_{SC,PVM} = N_p I_{SC} \tag{26}$$

$$I_{0,PVM} = N_p I_0 \tag{27}$$

$$V_{t,PVM} = N_s V_t \tag{28}$$

Blätzer et al. [56] assume that since J_{MPP} and J_{SC} can be considered proportional to the irradiance, the fill factor can be determined by

$$FF = \frac{V_{MPP}}{V_{OC}} \qquad (29)$$

The efficiency of the solar cell is defined as the ratio of the output power of the solar cell at its maximum power point to the input energy received by the incident light:

$$\eta_{cell} = \frac{P_{MPP}}{GA_{cell}} \qquad (30)$$

where A_{cell} corresponds to the area of the solar cell. Thus, for a certain irradiance G corresponds an efficiency η_{cell}:

$$\eta_{cell} = \frac{P_{MPP}}{GA_{cell}} \qquad (31)$$

5.2 DC-DC Converter

A solar cell provides DC power and therefore a DC-DC converter is necessary to adapt the voltage levels of the solar cell to the energy-storage element or to the electrical load. The number of solar cells connected in series in the solar module determines the output voltage at the maximum power point V_{MPP}. In order to extract the maximum power from the solar cell, the DC-DC converter has to include a control loop to adjust the current and voltage according to the maximum power point. There are different topologies that can be employed for the DC-DC converter: buck, boost, and buck-boost converter. The DC-DC converter can be controlled with the duty cycle (pulse width modulation, PWM) or with the switching frequency (pulse frequency modulation, PFM).

The MPPT controller can maximize either the input power of the DC-DC converter that corresponds to the output power delivered by the solar cell or the output power of the DC-DC converter. If the input power of the DC-DC converter is maximized, it does not imply that the output of the DC-DC converter is also maximized since its input power and efficiency are not linearly dependent [55]. Therefore, an MPPT controller that maximizes the output power of the DC-DC converter is also a valid approach even when the power obtained from the solar cell does not correspond to P_{MPP}.

The input voltage and current of the DC-DC converter are used as input parameters by the MPPT algorithm when the input power is going to be maximized. However, output voltage and current are employed by the MPPT algorithm for the maximization of the output power. The MPPT control circuit must be able to adjust either the input or output power to its MPP during operation. Depending on the dimensions of the solar cell and the irradiance levels, a buck, a buck-boost, or a boost converter will be chosen for the power management unit of the energy harvesting solar cell supply.

There are two possible outputs for the power management unit: a battery or an electrical load. If a battery is connected and has to be charged, it can be assumed that the output voltage of the battery remains constant and the MPPT circuit has just to sense the charge current and try to maximize it. If an electrical load is connected to the power management unit, the DC-DC converter has to change its duty cycle in order to match the output resistance. MPPT algorithms traditionally track the MPP of solar cells. Thus, the algorithm gets that input voltage and current of the DC-DC converter is at V_{MPP} and I_{MPP}, respectively. The algorithm is usually implemented in a microcontroller or digitally. However, a complete analog implementation of the MPP algorithm is also possible. section S.3 of the supplemental material provides additional information about several MPPT algorithms for solar cells.

Normally, due to the intermittent behavior of light sources, a battery is employed to store the energy converted by the PV cell. Then, it can be considered that when the charge current of the battery is maximized, the power at the battery is also being maximized [57]. The MPPT algorithm to be used in this case is shown in Figure 25. Io is the output current of the DC-DC converter and δ is the duty cycle of the DC-DC converter.

The efficiency of a DC-DC converter connected to a solar cell is the ratio of the power delivered to the load to the power converted by the solar cell at MPP:

$$\eta_{DC-DC} = \frac{P_L}{P_{MPP}} \tag{32}$$

For the application under consideration it makes more sense to define the ratio as expressed by Eq. (32) instead of the classical form output to input power since the input power of the DC-DC converter depends on the MPPT circuit that controls the switching of the DC-DC converter; it is the goal of the MPPT control circuit to work at this point.

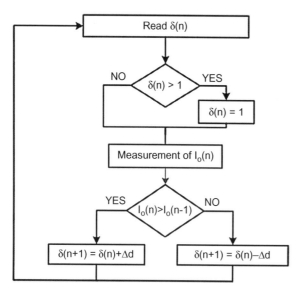

FIGURE 25   Flowchart of the MPPT algorithm that uses the output current as input parameter [58].

Therefore, the efficiency depends on two subjects: the localization of the MPP of the solar or PV cell by the MPPT circuit and the efficiency of the DC-DC converter itself and how much of the input power is delivered to the load P_L.

6. ENERGY AND POWER CONSUMPTION ISSUES

In energy harvesting applications, both power and energy are of interest. The peak power of an energy harvesting transducer alone or together with a power-management circuit is given at certain environmental conditions and with a specific output resistor that maximizes the output power. However, the power present in the environment that can potentially be harvested is usually limited and discontinuous in nature. This means that it is not just necessary to know the instantaneous power that is harvested but also its duration and its period. The mean power is calculated as

$$<P> = \frac{1}{T} \int_T p(t) \mathrm{d}t \tag{33}$$

The energy that a certain energy harvesting transducer delivers during a time duration t is equal to the mean power multiplied by the time:

$$E = <P> t \tag{34}$$

The power required by the electrical load in energy harvesting systems is also discontinuous since it operates in different consumption modes. Thus, a power management unit is necessary for adapting the voltage level of the energy harvesting to the electrical load. Moreover, an energy storage element is also needed to store the energy for later use at times when there is no available energy from the energy harvesting transducer. There are three operation modes of an energy harvesting system:

- The power consumption of the electrical load is always lower than the power provided by the environment. This case does not require an energy storage element.
- The mean power consumption of the electrical load is lower than the mean power provided by the environment. Thus, the electronic device may operate continuously but an energy storage element is required.
- The mean power provided by the environment is lower than the mean power of the electrical. Therefore, an energy storage element is necessary and the load can only operate discontinuously.

Nowadays, not only the source of an energy harvesting system is discontinuous in nature but also the electrical load. Microcontrollers, digital sensors, and RF transmitters have different power consumption modes depending on if they are active or in standby. Figure 26 shows the power consumption of an electronic load as a function of time. $<P_{load}>$ is the average power consumption of the load. P_{active} is the power consumption of the load during time interval t_{active} and P_{sleep} is the power consumption of the load in the lowest power consumption mode, and it takes place during a time interval t_{sleep}. In an energy harvesting system, the electrical load is active with a certain duty cycle that is set normally in order to be powered by the ambient source. Therefore, the mean average power of the load is low in comparison with its power peaks. Since in an energy harvesting system the load is usually in standby mode, the power consumption in this mode is what sets the mean power consumption. Thus, it has more impact on the overall mean power consumption of the load to choose components with lower power consumption in standby mode than in active mode.

The mean power consumption of the electrical load, which works in active and sleep modes, has to be equal to or lower than the mean power provided by the environment. If this is not the case, there are different ways to overcome this situation:

- Decrease the duty cycle of the load that implies increasing the time between operations if the electrical load is in active mode.
- If the main part of the mean power consumed is caused by the power consumption in sleep mode, the selection of another electrical load with a lower power consumption in sleep mode can solve the problem.
- Turn off the electrical load instead of setting it in sleep mode. However, this strategy has to take into consideration the start-up time, and therefore the power consumed during this time, for turning on the electrical load.

The duty cycle of the electrical load is defined as

$$D_{load} = \frac{t_{active}}{t_{active} + t_{sleep}} \qquad (35)$$

where t_{active} and t_{sleep} are the time the electrical load is in active and sleep mode, respectively.

Thus, the selection of the duty cycle has an important impact on the mean consumed power by the wearable system since

$$<P_{load}> = P_{active}D_{load} + P_{sleep}(1 - D) \qquad (36)$$

where P_{active} and P_{sleep} are the power consumed by the load in active mode and in sleep mode, respectively.

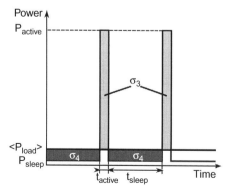

FIGURE 26   Power profile of an electrical load with different power consumption modes.

In order to choose the battery for the application, it is required to calculate the amount of energy B that must be stored. This parameter together with the maximum discharge current of the battery are the selection parameters of the battery. The supply current needed by the linear regulator or DC-DC converter for supplying the electrical load in active mode cannot be higher than the maximum discharge current of the battery.

An interesting output parameter is the required time for starting the measurements of the wearable system. It is assumed that the battery will operate all the time with its nominal operating voltage. Thin-film and solid-state batteries have flat discharge curves and therefore the battery can deliver energy to the load until it is nearly fully discharged. These batteries can also deliver high amounts of current compared to their capacity for pulse durations of 20 ms. Another advantage of thin-film batteries is their low self-discharge rate compared to supercaps [59]. The thin-film batteries can be almost fully charged at a constant voltage in less than an hour since they do not require a two-phase charge with a constant current phase followed by a constant voltage phase [59].

The calculation of the capacity of the battery is based on the work of Kansal et al. [60,61]. First of all, it is necessary to define the energy obtained from the transducer and the energy consumed by the load in a mathematical way. The harvested energy can be defined as the function of three parameters: $<P_{harvested}>$, σ_1, and σ_2 if it can be considered as periodical or quasi-periodical [20]. $<P_{harvested}>$ is the mean power delivered by the energy harvesting transducer, σ_1 corresponds to the maximum amount of energy when the power is over the mean power value, and σ_2 is defined as the maximum amount of energy obtained when the power is under the mean power value (see Figure 27):

$$<P_{harvested}> = \frac{1}{T}\int_T p_{source}(t)\mathrm{d}t \tag{37}$$

$$\sigma_1 = max_i\left\{\int_{Thighsource-i} H(p_{source}(t) - <P_{harvested}> \mathrm{d}t)\right\} \tag{38}$$

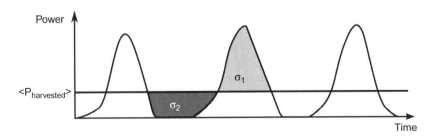

FIGURE 27 Power harvested by an energy harvesting transducer as a function of time.

where Thighsource-i is the i-th contiguous time duration for which $p_{source}(t) \geq <P_{harvested}>$ and H denotes the Heaviside step function.

$$\sigma_2 = max_i \left\{ \int_{Tlowsource-i} (<P_{harvested} - p_{source}(t)> dt) \right\} \tag{39}$$

where Tlowsource-i is the i-th contiguous time duration for which $p_{source}(t) \leq <P_{harvested}>$.

Therefore, the electronic load consumption can be defined as a function of the parameters: $<P_{load}>$, σ_3, and σ_4 (see Figure 26). σ_3 corresponds to the maximum amount of energy when the power is over the mean power of the load and σ_4 is defined as the maximum amount of energy obtained when the power is under the mean power of the load:

$$<P_{load}> = \frac{1}{T} \int_T p_{load}(t)dt \tag{40}$$

$$\sigma_3 = max_i \left\{ \int_{Thignload-i} H(p_{load}(t) - <P_{load}> dt) \right\} \tag{41}$$

where Thighload-i is the i-th contiguous time duration for which $p_{load}(t) \geq$

$$\sigma_4 = max_i \left\{ \int_{Tlowload-i} H(<P_{consumed} - p_{load}(t) -> dt) \right\} \tag{42}$$

where Tlowload-i is the i-th contiguous time duration for which $p_{load}(t) \leq <P_{load}>$. When an initial charge is required in the battery for immediately starting the load, the minimum amount of required energy is calculated by [62]:

$$B_{0min} = \sigma_2 + \sigma_3 \tag{43}$$

The energy that has to be stored in the battery is calculated as

$$B \geq B_0 + \sigma_1 + \sigma_4 \tag{44}$$

where Bo is the initial energy stored in the battery.

7. CONCLUSIONS AND FUTURE CONSIDERATIONS

Different energy harvesting sources available from the human body: either environmental sources or the human body itself have been described in this chapter. Wireless power transmission and the conversion of light into electrical energy have been evaluated as environmental sources to power wearable sensors. Moreover, thermal and mechanical energy directly harvested from the human body are also a power supply for wearable sensors.

Wireless power transmission is a promising technology that allows powering embedded sensors, circuits, and systems without wires. In recent years, several products for consumer electronics have appeared but established international standards are still missing. Several wireless technologies have been presented in this chapter, featuring different frequency bands, transmission ranges, and amounts of power. These technologies can also be used for human applications such as passive wireless monitoring. It can allow wearing passive sensors without batteries that must be replaced or, if a battery is needed, charging it without the need for wires. In these applications, the influence of the human body on the wireless system must be considered, for example, in the form of special antennas.

The conversion of light into electrical energy using solar cells is a mature technology but its integration on wearable systems presents some challenges like the integration of the solar cell in the textile and the low input power levels associated with indoor light that provides low output voltage levels.

The conversion of mechanical energy into electrical energy can be done with piezoelectric and electrodynamic transducers. Resonant transducers harvest a high amount of power when the resonance frequency matches the frequency of the mechanical vibration. The bandwidth of a single piezoelectric or electrodynamic transducer is very narrow. However, the bandwidth of an energy harvesting system can be increased using several energy harvesting transducers with different resonant frequencies or with a frequency tunable transducer. Nevertheless, non-resonant transducers also convert mechanical energy into electrical energy through the deformation or strain of the piezoelectric element without resonance. Human motion is considered as quasi-static excitation due to the low frequencies related to human activities [19].

Thermoelectric generators convert thermal into electrical energy. The main drawback for the integration of these transducers in wearable systems are the heat sinks that are necessary to maintain the temperature gradient between both sides of the thermoelectric generator [63].

The efficiency of the complete energy harvesting system is restricted by the efficiency of its components. First, the efficiency of the transducer that converts the ambient energy into electrical energy. Later, the efficiency of the power-management circuits that are used after the ambient transducer for rectifying and/or adapting the voltage levels. Furthermore, the efficiency of the DC-DC converters that are employed for adapting the voltage levels of the energy-storage devices to the electrical loads to be powered. Moreover, batteries and capacitors also have an efficiency that depends on the temperature and the current employed for charging them. Moreover, batteries and capacitors also have self-discharge rates due to leakage currents.

The harvested energy is a function of the size of the energy harvesting transducer. Microelectromechanical systems (MEMS)-based energy harvesting transducers provide high miniaturization and easy integration, which is of special interest for wearable systems, but its small size provides less electrical energy.

Wearable systems can be powered with energy harvesting generators that employ light, heat, motion, or RF energy as ambient energy. For the selection of the appropriate energy harvesting source, the utilization of a data logger can be of special interest. The data logger can store data from light sensors, accelerometers, temperature sensors, etc., over time in the real conditions under consideration. It is normally powered by batteries and contains a microcontroller, an SD card, sensors, and a GPS receiver [64]. A wifi interface offers the possibility of online transmission and processing of the available data. The SD card is capable of storing data for several days, whereas the battery will provide the required electrical energy. Figure 28 shows a block diagram of a data logger. Later, the stored data can be analyzed and it can be determined which energy harvesting transducer better suits the application.

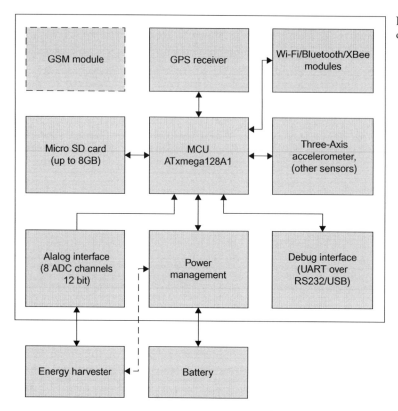

FIGURE 28 Block diagram of a data logger [64].

References

[1] J.A. Chavez, J.A. Ortega, J. Salazar, A. Turo, M.J. Garcia, SPICE model of thermoelectric elements including thermal effects, Instrum. Meas. Technol. Conf., 2000. IMTC 2000. Proc. 17th IEEE 2 (2000) 1019–1023.

[2] M. Lossec, B. Multon, H.B. Ahmed, et al., Sizing optimization with thermal and electrical matching of a thermogenerator placed on the human body, Proc. Int. Conf. Renewable Energy Eco-Design Electr. Eng. (2011).

[3] S. Lineykin, S. Ben-Yaakov, Modeling and analysis of thermoelectric modules, Appl. Power Electron. Conf. Expos., 2005. APEC 2005. Twentieth Ann. IEEE 3 (2005) 2019–2023. Available from: http://dx.doi.org/doi:10.1109/APEC.2005.1453336.

[4] M. Lossec, B. Multon, H. Ben Ahmed, Sizing optimization of a thermoelectric generator set with heatsink for harvesting human body heat, Energy Convers. Manag. 68 (2013) 260–265.

[5] V. Leonov, T. Torfs, P. Fiorini, C. Van Hoof, Thermoelectric converters of human warmth for self-powered wireless sensor nodes, IEEE Sens. J. 7 (2007) 650–657.

[6] T. Torfs, V. Leonov, C. Van Hoof, B. Gyselinckx, Body-heat powered autonomous pulse oximeter, in: Sensors, 2006. 5th IEEE Conference on, 2007, pp. 427–430.

[7] V. Leonov, T. Torfs, N. Kukhar, C.V. Hoof, R. Vullers, Small-size BiTe thermopiles and a thermoelectric generator for wearable sensor nodes, Proc. 5th Eur. Conf. Thermoelectrics (2007).

[8] Overview of thermoelectric/seebeck elements, last accesed november 2013. URL: <http://www.thermalforce.de/de/product/thermogenerator/index.php> (Last Accessed: 05.07.14).

[9] B.A. Edwards, Comparison of thermoelectric properties of arc-melted and hot-pressed half-heuslers (2007).

[10] N. Mohan, T.M. Undeland, W.P. Robbins, et al., Power Electronics: Converters, Applications and Design, Wiley, 1995.

[11] Ultralow Voltage Step-Up Converter and Power Manager ltc3108 data sheet, last accessed December 2013.

[12] Auto-Polarity, Ultralow Voltage Step-Up Converter and Power Manager. LTC3109 Data Sheet, last accessed December 2013.

[13] Armstrong oscillator, last accesed december 2013. URL: <http://en.wikipedia.org/wiki/Armstrong_oscillator> (Last Accessed: 05.07.14).

[14] J. Damaschke, Design of a low-input-voltage converter for thermoelectricgenerator, Industry Applications, IEEE Transactions on 33 (1997) 1203–1207.

[15] T. Starner, Human-powered wearable computing, IBM Syst. J. 35 (1996).

[16] F. Moll, A. Rubio, An approach to the analysis of wearable bodypowered systems, Mixed Sig. Des. Workshop (2000).

[17] N. Shenck, J. Paradiso, Energy scavenging with shoe-mounted piezoelectrics, Micro, IEEE 21 (2001) 30–42.

[18] D. Fourie, Shoe mounted PVDF piezoelectric transducer for energy harvesting, Morj Report, vol. 19, Spring 2010, 2010, pp. 66–70.

[19] Y. Liu, Active energy harvesting, Ph.D. thesis, The Pennsylvania State University, 2006.

[20] L. Mateu, Energy Harvesting from Passive Human Power, Ph.D. thesis, Universitat Politecnica de Catalunya, 2009.

[21] Piezo Film Sensors Technical Manual, Measurement Specialities, last accesed october 2013. <http://www.msiusa.com> (Last Accessed: 05.07.14).

[22] L. Mateu and F. Moll, Optimum piezoelectric bending beam structures for energy harvesting using shoe inserts, J Intelligent Mat Syst Struct, Vol. 16, pp. 835–845, 2005.

[23] M. Rossi, Acoustics and Electroacoustics, Artech House, Inc., Norwood, MA, 1988.

[24] S. Platt, S. Farritor, H. Haider, On low-frequency electric power generation with pzt ceramics, Mechatronics, IEEE/ASME Transactions on 10 (2005) 240–252.

[25] L. Mateu, F. Moll, Review of energy harvesting techniques for microelectronics, Proc. SPIE Microtechnol. New Millenium (2005) 359–373.

[26] L. Mateu, H. Zessin, P. Spies, Analytical method for selecting a rectification technique for a piezoelectric generator based on admittance measurement, Journal of Physics: Conference Series, vol. 476, IOP Publishing, 2013, pp. 012111.

[27] G. Tian, Active Energy Harvesting on Piezoelectric Materials: Experimental Demonstration and Standalone Circuit Implementation, Ph.D. thesis, The Pennsylvania State University, 2008.

[28] K. Ishida, T. Huang, K. Honda, T. Sekitani, H. Nakajima, H. Maeda, M. Takamiya, T. Someya, T. Sakurai, 100v ac power meter system-ona-film (sof) integrating 20v organic cmos digital and analog circuits with floating gate for process-variation compensation and 100v organic pmos rectifier, Solid-State Circuits Conf. Digest Tech. Papers (ISSCC), 2011 IEEE Int., IEEE (2011) 218–220.

[29] K. Ishida, T.-C. Huang, K. Honda, Y. Shinozuka, H. Fuketa, T. Yokota, U. Zschieschang, H. Klauk, G. Tortissier, T. Sekitani, et al., Insole pedometer with piezoelectric energy harvester and 2v organic digital and analog circuits, Solid-State Circuits Conf. Digest Techn. Papers (ISSCC), 2012 IEEE Int., IEEE (2012) 308–310.

[30] Y.K. Ramadass, A.P. Chandrakasan, An efficient piezoelectric energy harvesting interface circuit using a bias-flip rectifier and shared inductor, Solid-State Circuits, IEEE J. 45 (2010) 189–204.

[31] E. Lefeuvre, A. Badel, C. Richard, L. Petit, D. Guyomar, Optimization of piezoelectric electrical generators powered by random vibrations, Dans Symposium on Design, Test, Integration and Packaging (DTIP) of MEMS/MOEMS, Citeseer, 2006.

[32] G. Ottman, H. Hofmann, A. Bhatt, G. Lesieutre, Adaptive piezoelectric energy harvesting circuit for wireless remote power supply, IEEE Trans. Power Electron. 17 (2002) 669–676.

[33] G. Ottman, H. Hofmann, G. Lesieutre, Optimized piezoelectric energy harvesting circuit using step-down converter in discontinuous conduction mode, IEEE Trans. Power Electron. 18 (2003) 696–703.

[34] P. Nikitin, K.V.S. Rao, S. Lazar, An overview of near field uhf rfid, in: RFID, 2007. IEEE International Conference on, 2007, pp. 167–174.

[35] I. Mayordomo, T. Drager, P. Spies, J. Bernhard, A. Pflaum, An overview of technical challenges and advances of inductive wireless power transmission, Proc. IEEE 101 (2013) 1302–1311.

[36] Murata, Wireless Power Transmission Modules, 2013. URL: <www.murata.com/products/wireless_power>, last accessed November 2013.

[37] Powercast, Wireless Power Solutions, 2013. URL: <www.powercastco.com>, last accessed November 2013.

[38] O. Jonah, S. Georgakopoulos, M. Tentzeris, Orientation insensitive power transfer by magnetic resonance for mobile devices, Wireless Power Transfer (WPT), 2013 IEEE (2013) 5−8.

[39] K. Miwa, H. Mori, N. Kikuma, H. Hirayama, K. Sakakibara, A consideration of efficiency improvement of transmitting coil array in wireless power transfer with magnetically coupled resonance, Wireless Power Transfer (WPT), 2013 IEEE (2013) 13−16.

[40] C. Gabrieli, Compilation of the Dielectric Properties of Body Tissues at RF and Microwave Frequencies, Technical Report AL/OE-TR-19960004, Physics Department, King's College London, 1996.

[41] U.N.L. of Medicine, The Visible Human Project, 2013. URL: <www.nlm.nih.gov/research/visible/visible_-human.html>, last accessed November 2013.

[42] Farsens, Battery Free Sensor Solutions, 2013. URL: <www.farsens.com>, last accessed November 2013.

[43] G. Marrocco, Rfid antennas for the uhf remote monitoring of human subjects, Antennas and Propagation, IEEE Transactions on 55 (2007) 1862−1870.

[44] D. Ranasinghe, T. Kaufmann, Wearable RFID tags, RFID J. (2012).

[45] R. Bashirullah, Wireless implants, microwave magazine, IEEE 11 (2010) S14−S23.

[46] J.-C. Chiaon, Wireless implants for personalized medicine and chronic monitoring, IEEE Life Sci. Newsletter (2013).

[47] A. Reinders, Options for photovoltaic solar energy systems in portable products, in: proceedings of TCME 2002, Fourth International symposium, 2002, pp. −.

[48] J. Randall, N. Bharatula, N. Perera, T. von Buren, S. Ossevoort, G. Tröster, Indoor tracking using solar cell powered system: Interpolation of irradiance, Int. Conf. Ubiquitous Comput. (2004).

[49] IXOLAR High Efficiency SolarMD slmd121h09l data sheet, last accesed November 2013. URL <http://ixapps.ixys.com/Viewer.aspx?p = http%3a%2f%2fixapps.ixys.com%2fDataSheet%2fSLMD121H09L-DATA-SHEET.pdf> (Last Accessed: 05.07.14).

[50] SILONEX Inc., Datasheet NORPS-12, Last accessed November 2013. URL: <http://www.farnell.com/data-sheets/409710.pdf> (Last Accessed: 05.07.14).

[51] Vishay, Datasheet TEMT6000X01, Last accessed November 2013. URL: <http://www.vishay.com/docs/81579/temt6000.pdf> (Last Accessed: 05.07.14).

[52] A. Luque, S. Hegedus, com. Handbook of Photovoltaic Science and Engineering, Wiley, 2011

[53] M.T. Penella-López, M. Gasulla-Forner, Powering Autonomous Sensors, Springer, 2011.

[54] M.K. Fuentes, A Simplified Thermal Model for Flat-Plate Photovoltaic Arrays, Technical Report, Sandia National Labs, Albuquerque, NM (USA), 1987.

[55] N. Femia, G. Petrone, G. Spagnuolo, M. Vitelli, Power electronics and control techniques for maximum energy harvesting in photovoltaic systems, vol. 11, CRC Press, 2012.

[56] D.B. Atzner, A. Romeo, H. Zogg, A. Tiwari, Cdte/cds solar cell performance under low irradiance, in: 17-th EC PV Solar Energy Conference, Munich, Germany, 2001.

[57] D. Shmilovitz, On the control of photovoltaic maximum power point tracker via output parameters, in: Electric Power Applications, IEE Proceedings-, volume 152, IET, 2005, pp. 239−248.

[58] D. Shmilovitz, Photovoltaic maximum power point tracking employing load parameters, in: Industrial Electronics, 2005. ISIE 2005. Proceedings of the IEEE International Symposium on, volume 3, IEEE, 2005, pp. 1037−1042.

[59] T. Cantrell, SILICON UPDATE-LiOn King-A Look at" Battery-in-aChip" technology, Circuit Cellar-The Magazine Compu. Appl. (2009) 62.

[60] A. Kansal, J. Hsu, S. Zahedi, M.B. Srivastava, Power management in energy harvesting sensor networks, Technical Report TR-UCLA-NESL200603-02, Networked and Embedded Systems Laboratory, UCLA, 2006. URL: <http://nesl.ee.ucla.edu/fw/kansal/kansal_tecs.pdf> (Last Accessed: 05.07.14).

[61] A. Kansal, D. Potter, M.B. Srivastava, Performance aware tasking for environmentally powered sensor networks, SIGMETRICS '04/Performance '04: Proceedings of the joint international conference on Measurement and modeling of computer systems, ACM Press, New York, NY, USA, 2004. Available from: http://doi.acm.org/10.1145/1005686.1005714, pp. 223−234.

[62] L. Mateu, Energy Harvesting from Human Passive Power, Ph.D. thesis, Universitat Politecnica de Catalunya, 2009.

[63] S. Priya, D. Inman, Incorporated. Energy Harvesting Technologies, Springer Publishing Company, 2008

[64] DATA-LOGGER to characterize vibrations for energy harvesting systems, last accesed november2013. URL: <http://www.iis.fraunhofer.de/content/dam/iis/en/dokumente/Embedded-Communication/Data_Logger.pdf> (Last Accessed: 05.07.14).

Supplemental Material: Energy Harvesting at the Human Body

S.1. ENERGY HARVESTING FROM TEMPERATURE GRADIENT AT THE HUMAN BODY: DC-DC CONVERTER DESIGN FOR ULTRA-LOW INPUT VOLTAGES

In the implementation of a low-input step-up converter of Figure S1 a standard MOSFET T2 in parallel to T_1 is used to cope with the high ON-resistance of a typical JFET. Here, the JFET is only switching at start-up, whereas it shuts off completely afterwards, then the ON-state current is only flowing through MOSFET T2. This is a consequence of the negative voltage across C_1, which further decreases after the start-up sequence so that T1 cannot turn ON anymore. This behavior can also be noticed in the graphs of VL2 and IT1 (Figure S2), which show some measurement results of a prototype of the DC-DC converter of Figure S1.

The converter design of Figure S1 can also be built up with two MOSFET transistors with different threshold voltages. This is useful if the JFET is replaced by a MOSFET with a threshold voltage close to zero and high ON-state resistance, which is employed for starting up the converter. Consequently, the second MOSFET can be a standard one with a high threshold voltage and a lowest possible ON-resistance to be used in steady-state operation to minimize ohmic losses. This configuration is shown in Figure S3 and can be built up for example with the combination of zero-threshold transistor ALD110900 of Advanced Linear Devices [A1] and the standard MOSFET BSH105 [A2]. The chargepump (see Figure S1) consisting of the diodes D_4 and D_5 as well as the capacitors C_4 and C_5 gives a DC bias to the gate of T_2 to guarantee lowest possible input start-up voltage, if the oscillation amplitude of the secondary inductance L_2 is too low to reach the threshold voltage of T_2.

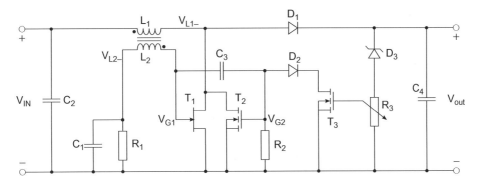

FIGURE S1 Self-oscillating boost converter with additional switching MOSFET and voltage regulation.

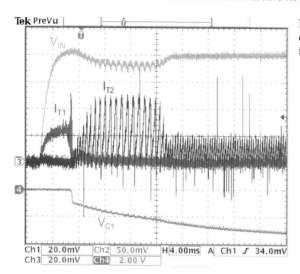

FIGURE S2 Waveforms of regulated self-oscillating boost converter with additional switching MOSFET.

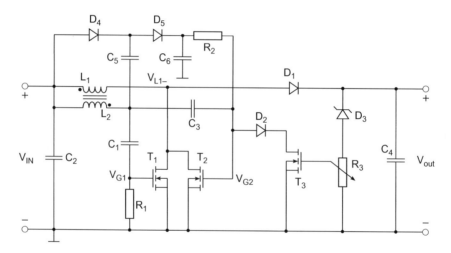

FIGURE S3 Meissner oscillator based converter using two switching MOSFETs with different threshold voltages.

If higher output voltages are desired or the forward voltage losses of diodes D_4 and D_5 are not dominant in relation to the output power, it can be more convenient to use the circuit of Figure S3 in flyback configuration, like in Figure S4. Using high-ohmic zero-threshold MOSFETs for T_1, very low start-up currents and low turn ratios of the transformer can be achieved. Because of the last mentioned characteristics, the last two mentioned circuits can be expanded to work independently from the input voltage polarity. How this can be established is described in the next sections.

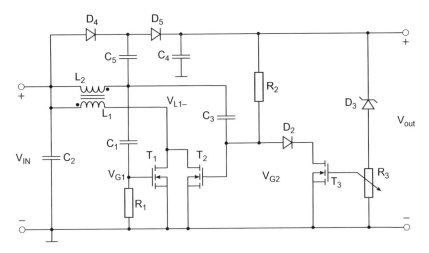

FIGURE S4   Meissner oscillator-based converter using two switching MOSFETs in a flyback configuration.

S.1.1 Bipolar DC-DC Converter Design

A block diagram of a "bipolar" DC-DC converter, i.e., a DC-DC converter working with both positive and negative input polarities is depicted in Figure S5. This implementation is useful, for example, if a TEG is connected to the inputs, which exposed to positive and negative temperature gradients results in inverse output voltages. A scenario can be, for instance, the transition between day and night, summer and winter, or indoor and outdoor use.

In the circuit of Figure S5 a polarity switch consisting of a bridge rectifier structure is employed before the DC-DC converter stage, which is controlled by the two comparators U_1 and U_2. This implementation creates a positive supply voltage independent from the polarity of the input voltage between the terminals V_{TEG1} and V_{TEG2}. It should be pointed out that this architecture does not work for alternating currents between the terminals V_{TEG1} and V_{TEG2}. Otherwise, an active or passive diode has to be connected between the output of the polarity switch and capacitor C_{in}.

The diodes D_1 to D_4 are supposed to work in a start-up phase of the bipolar converter, where the output voltage between the terminals $V_{out}+$ and V_{out-} is too low to supply the comparators U_1 and U_2, and finally the transistors T_1 to T_4 cannot be switched ON. In this phase the polarity switch consequently works as a standard bridge rectifier. To assure the lowest possible start-up voltage of the bipolar architecture the diodes D_1 to D_4 are supposed to have the lowest possible forward voltages, which is in general contrary to their leakage current. One possible choice that can be considered is the Silicon Schottky diode BAT60A [A3], which has a typical forward voltage of 0.12 V at 10 mA and a typical leakage current of 0.3 mA at 5 V reverse voltage at room temperature. To guarantee a low start-up voltage, it is additionally desired that the current below start-up DC-DC converter stage (see Figure S5) be as small as possible. For that task, for example, the described converters of Figure S3 and Figure S4 are best suited, but the architecture of Figure S5 is able to work with any kind of converter type.

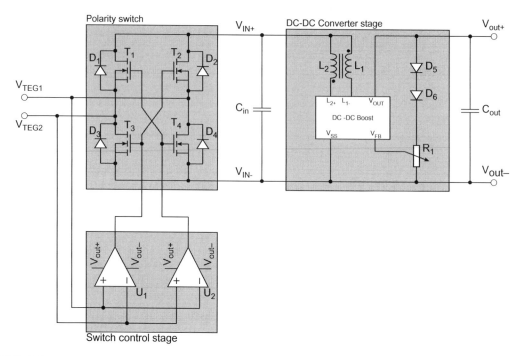

FIGURE S5   Block diagram of bipolar DC-DC converter design.

S.1.2 ASIC Design and Demonstrator

With some modifications it is possible to implement the low-input voltage DC-DC converter of Figures S1 to S4 on an application-specific integrated circuit (ASIC). This is done in UMC 180 nm technology [A4], where zero-threshold MOS transistors, even with low ON-resistances, are available. Figure S6 depicts the Fraunhofer ASIC [A5] die on a 10 Euro-Cent coin, whereas Figure S7 shows a photo of the evaluation board assembled with the ASIC in a QFM5 × 5 package.

The evaluation board with the DC-DC converter ASIC was designed especially for use in conjunction with a wireless demonstrator, which is able to deliver sensor data without employing any battery or other comparable energy storage device. A block diagram of the transmitter device is shown in Figure S8, which is supposed to work with the heat energy of the human body. The demonstrator uses a solid-state thermogenerator as its inputs $V_{TEG}+$ and V_{TEG-} and a STM110 transmitter module by the ENOcean [A6] to send the temperature difference of the TEG, input current, and voltage to a PC or laptop, for example. The data is sent via an RCM120 receiver (also by ENOcean) connected to an FTDI232B [A7] serial-to-USB converter. For measuring the output current of the TEG, simply a shunt resistor R_{Shunt} in the ground path of the TEG is used. This resistor creates a voltage drop measured by a current sense amplifier, which adapts the small voltage drop across R_{Shunt} to the first input of the ADC in the transmitter module. In the same way, U_2 measures the input voltage across the TEG and U_3 subtracts the two outputs of the temperature sensors

FIGURE S6   Photo of the Fraunhofer ASIC die on a 10 Euro-Cent coin.

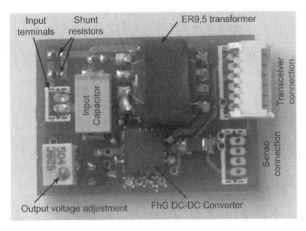

FIGURE S7   Photo of the evaluation board with the Fraunhofer converter ASIC.

so that the voltage at V_{TSENS} (and at the third input of the ADC) is proportional to the temperature gradient of the TEG. Figure S9 shows a complete setup of the demonstrator with software developed especially for illustrating the measured signals of the TEG device on a laptop.

S.1.3 Maximum-Power Point Tracking for Impedance Matching

If an energy-storage device like a battery or a capacitor is to be charged by a thermogenerator or a solar cell, for example, generally a voltage converter is used, which can be adjusted automatically to operate the energy transducer in the maximum power point.

The benefit of efficiency when using a maximum-power-tracking loop can be estimated using a diagram where the output power is plotted as a function of the output current of a

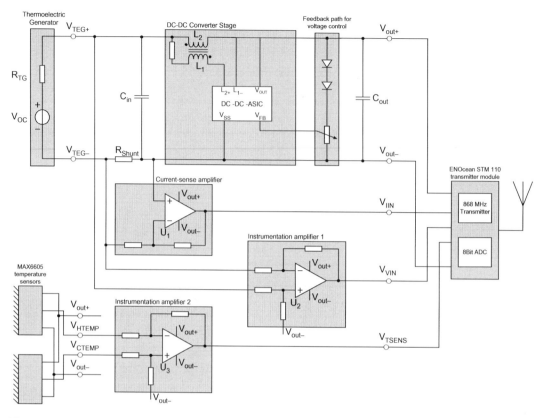

FIGURE S8   Block diagram of transmitter device.

FIGURE S9   Image of a TEG energy harvesting setup.

thermoelectric generator. For a typical TEG, for example, the 127-150-26 [A8] from the company "thermalforce.de" or the PKE-128-A-1027 [A9] from the company "Peltron GmbH," with $\alpha_m = 0.05$, the graph of the output power P_{TG} as a function of the output current I_{TG} is as shown in Figure S10 for open-circuit-voltages V_{oc} of 100, 200, 300, 400, and 500 mV. Besides these voltages, the corresponding temperature gradients ΔT of the TEG are noted.

For example, at a temperature gradient of 10 K the TEG works at its maximum power point of $P_{TG} = 6.2$ mW at an output current of $I_{TG} = 25$ mA. Now it is considered that a reduction of the applied temperature gradient from 10 K to 6 K occurs and that the output current remains at 25 mA (see Figure S10). In this case, the output power of the TEG drops to $P_{TG} = 1.2$ m, whereas the maximum power point would be at $P_{TG} = 2.5$ mW and $I_{TG} = 15$ mA. This example shows that $100 \cdot \frac{2.5 \text{ mW} - 1.2 \text{ mW}}{1.2 \text{ mW}} = 108\%$ of output power of the TEG could be gained using an adaptive method for always working in an optimum power point of an energy transducer. How that works is shown in the following.

Figure S11 illustrates a configuration (introduced at the beginning of this subsection) with a battery as the load for the voltage converter. Under this condition it can be assumed that the maximum power is supplied to the load, when the output current I_{BAT} of the converter is maximized since the output voltage V_{BAT} changes slowly. In fact, for the following analysis, V_{BAT} is assumed to be constant. As an example, a buck-boost converter is used for the voltage conversion. Therefore, the voltage conversion ratio for the configuration of Figure S11 is:

$$\frac{V_{BAT}}{V_{TG}} = \frac{D}{1 - D} \tag{45}$$

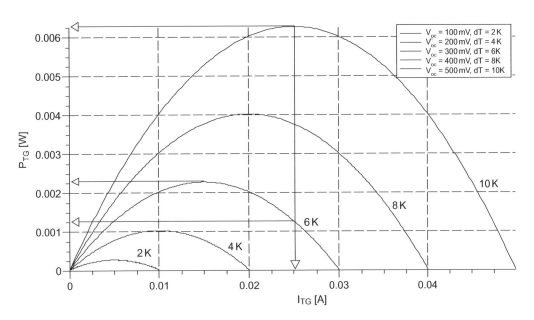

FIGURE S10 Output power vs. output current of a typical TEG for different open-circuit voltages V_{oc}.

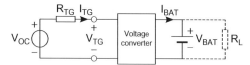

FIGURE S11   Matching a battery load to an energy transducer employing a voltage converter.

Consequently, for I_{TG} it can be calculated:

$$I_{TG} = \frac{V_{OC} - V_{TG}}{R_{TG}} = \frac{1}{R_{TG}}(V_{OC} - V_{BAT}\frac{1-D}{D}) \tag{46}$$

Finally, the output power of the TEG is determined as follows:

$$P_{TG} = V_{OC}I_{TG} - R_{TG}I_{TG}^2 =$$

$$= \frac{V_{OC}}{R_{TG}}\left(V_{OC} - V_{BAT}\frac{1-D}{D}\right) - \frac{1}{R_{TG}}\left(V_{OC} - V_{BAT}\frac{1-D}{D}\right)^2 =$$

$$= \frac{V_{OC}^2}{R_{TG}} - \frac{V_{OC}V_{BAT}}{R_{TG}} \cdot \frac{1-D}{D}$$

$$- \frac{1}{R_{TG}}\left(V_{OC}^2 - 2V_{OC}V_{BAT}\frac{1-D}{D} + V_{BAT}^2 \cdot \frac{(1-D)^2}{D^2}\right) = \tag{47}$$

$$= \frac{V_{OC}V_{BAT}}{R_{TG}} \cdot \frac{1-D}{D} + \frac{V_{BAT}^2}{R_{TG}}\left(\frac{1-D}{D}\right)^2 \cdot$$

A maximum of the output power of the TEG can be achieved by differentiating P_{TG} respect to the duty cycle D:

$$\frac{\Delta P_{TG}}{\Delta D} = \frac{V_{OC}V_{BAT}}{R_{TG}} \cdot \frac{-D-(1-D)}{D^2} -$$

$$- 2 \cdot \frac{V_{BAT}^2}{R_{TG}} \cdot \frac{1-D}{D} \cdot \frac{-D-(1-D)}{D^2} = \tag{48}$$

$$= \frac{V_{OC}V_{BAT}}{R_{TG}} \cdot \frac{-1}{D^2} + 2 \cdot \frac{V_{BAT}^2}{R_{TG}} \cdot \frac{1-D}{D^3} = 2 \cdot \frac{V_{BAT}^2}{R_{TG}} \cdot \frac{1-D}{D^3} - \frac{V_{OC}V_{BAT}}{R_{TG}D^2}$$

and equaling the resulting function to zero:

$$\frac{\Delta P_{TG}}{\Delta D} = 2 \cdot \frac{V_{BAT}^2}{R_{TG}} \cdot \frac{1-D}{D^3} - \frac{V_{OC}V_{BAT}}{R_{TG}D^2} = 0$$

$$\Rightarrow 2V_{BAT} \cdot \frac{1-D}{D} - V_{OC} = 0 \Rightarrow \frac{1}{D} - 1 = \frac{V_{OC}}{2V_{BAT}} \Rightarrow \frac{1}{D} = \frac{V_{OC}}{2V_{BAT}} + 1 \tag{49}$$

$$\Rightarrow D_{MPP} = \frac{1}{\frac{V_{OC}}{2V_{BAT}} + 1} = \frac{2V_{BAT}}{V_{OC} + 2V_{BAT}} \cdot$$

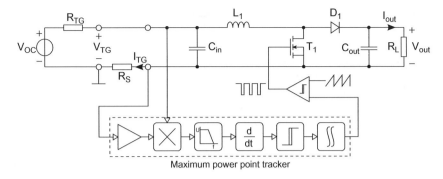

FIGURE S12 Design example of a maximum power point tracker.

It can be summarized, that for maximizing the output power of the TEG using a battery at the output of the voltage converter (Figure S11), the duty cycle D_{MPP} is only dependent on V_{OC} and V_{BAT}. In reality, the open-circuit voltage V_{OC} and the internal resistance R_{TG} cannot be measured without disconnecting the TEG from the rest of the circuit. Therefore, it is more feasible to find an electrical structure that measures the output power of the energy transducer by finding the maximum by itself. Only one general solution is shown in the following, since there is a lot of work on this topic already published in the literature. For more information, the work of Sullivan et al. [A10] or Koutroulis et al. [A11], for example, can be helpful.

A general algorithm for a "maximum power point tracker" (MPPT) can be found using Figure S10. Depending on which side of the maximum power point the regulation starts, the expressions for the algorithm are

$$Area(1): \Delta P > 0 \Rightarrow \Delta D \uparrow, \tag{50}$$

$$\Delta P < 0 \Rightarrow \Delta D \downarrow, \tag{51}$$

$$Area(2): \Delta P > 0 \Rightarrow \Delta D \downarrow, \tag{52}$$

$$\Delta P < 0 \Rightarrow \Delta D \uparrow. \tag{53}$$

From these statements it can be deduced that the starting point has to be chosen according to the implemented solution. Figure S12 shows a block diagram of a possible design employing a boost converter. Output current and voltage of the energy transducer, V_{TG} and I_{TG}, are used as an input to the MPPT and are multiplied to deliver a signal proportional to the output power of the transducer. Generally, a shunt resistor R_S transforms the output current I_{TG} of the TEG into a voltage signal, which is amplified afterwards (see Figure S12). The signal after the multiplier is fed through a low-pass filter to remove the current ripple due to the switching of the converter and to prevent the MPPT to regulate on the resulting ripple of the power signal. The output signal of the filter is then differentiated to know if the power is rising or falling. A comparator is connected to the differentiator in order to increase or decrease the duty cycle of the converter. This is realized with an integrator later, which sums up the comparator signal as a function of time.

Whether the output of the comparator is inverted or not related to its input determines which algorithm (either Eqs. (50)/(51) or Eqs. (52)/(53)) is used. In fact, it has to be decided to start the operation of the circuit either at duty cycle $D = 0$ or at $D = 1$. The proposed MPPT can only work properly when the output power of the transducer does not change faster than the reaction of the MPPT; otherwise, it could happen that the maximum power point changes to a point in the wrong area (see diagram in Figure 9(b)) and the algorithm fails to lead the duty cycle in the right direction. Therefore, this design is more practical for TEGs than for solar cells, for example, since in general the temperature gradient applied to a TEG does not change as fast as the light intensity applied to a solar cell. In the literature other solutions can be found. For example, Sullivan et al. [A10] work with the help of a flip-flop to switch between the algorithms of Eq. (50)/(51) and Eq. (52)/(53).

For the MPPT loop of Figure S12, there are both analog and digital solutions possible, whereas for a digital solution at least one analog-to-digital and one digital-to-analog converter is necessary as well as a microcontroller. For energy harvesting applications, it is in general a choice of which architecture consumes more power.

For the analog solution in Figure S12, a circuit is presented in Figure S13, where one operational amplifier is needed per stage. The other parts are standard circuits with operational amplifiers for differentiator, comparator, and integrator. Sometimes it makes sense to use an amplifier stage instead of a comparator, because its gain is lower for a possible DC component from the output of the differentiator. The output of the integrator serves as the control signal for the PWM of the converter.

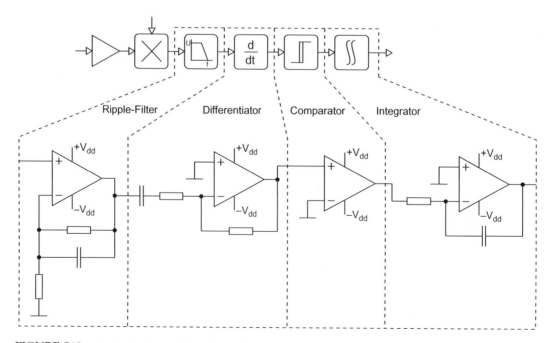

FIGURE S13   Analog implementation of a maximum power point tracker.

For the multiplier, a typical design is not provided here. For the multiplier task, commercial integrated circuits like the AD633 [A12] from Analog Devices can be used, for example. Nevertheless, especially the AD633 consumes some mA of current, which can be too much for energy harvesting applications. However, it is possible to get around the multiplier by measuring only current instead of power. For instance, if a battery at the output of the voltage converter is connected like in the previously described solution in Figure S11, the output current can be considered to be constant in relation to the output voltage. In this case, the MPPT tracks the maximum output power of the converter, which is additionally more accurate than tracking the output power of the transducer. This can be understood considering that a typical boost converter has a higher efficiency at higher input voltages, which means that the maximum output power of the transducer is at a different point than the maximum power delivered to the battery.

S.2. ENERGY HARVESTING FROM FOOT MOTION: AC-DC CONVERTER

S.2.1 AC-DC Linear Rectifiers

Han et al. [A13] employed a voltage-doubler circuit as AC-DC rectifier. Figure S14(a) shows the half-bridge rectifier implemented with a diode-pair rectifier while Figure S14 (b) shows the active version with two comparators connected to transistors $MP1$ and $MN1$.

Figure S15(a) shows a half wave rectifier with a voltage-doubler block, which is displayed in Figure S15(b) [A14]. Therefore, only one half of the piezoelectric wave is rectified and later multiplied by a factor 2. The half-wave rectifier is composed of a PMOS transistor, inverter, and comparator. When the output voltage of the piezoelectric element is higher than the voltage after the PMOS, this is turned on and the clock signal φ_1 is low. Then, capacitor C_1 of Figure S15(b) is charged with the input voltage. Moreover, the two PMOS transistors $MP2$ and $MP5$ are turned on and capacitor C_2 is charged with the output voltage of the piezoelectric element. When the output voltage of the piezoelectric element becomes lower than the voltage after the PMOS transistor $MP1$, signal φ_2 is low and the PMOS transistors connected to it in the voltage-doubler circuit are turned on. Then, the capacitor connected between nodes 10 and 11 is connected in series with the capacitor connected between node 9 and ground and so the output voltage obtained is twice the piezoelectric output voltage. For the design of synchronous rectifiers, the power consumption of the comparator must be taken into the consideration.

The comparator designed for this rectifier consumes only 165 nW. Another linear rectifier consists on using a negative voltage converter composed of two PMOS transistors and two NMOS transistors, as shown in Figure S16, followed by a diode that can be either a Schottky diode or an active PMOS transistor [A15,A16]. The transistors employed in the negative voltage converter are connected as switches and not as diodes. The MOSFETs required for this topology must have low drain-source resistance and low capacitance. The first part of the circuit converts the negative half-waves into positive ones. MOSFETs MP1

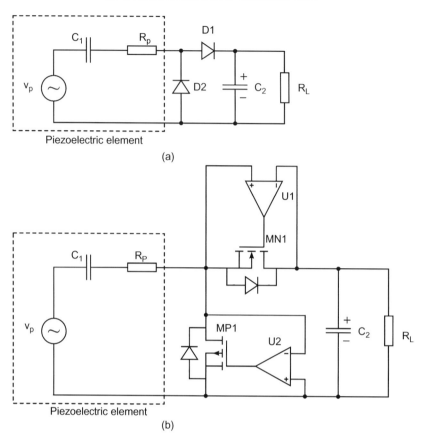

FIGURE S14   AC-DC rectifiers: (a) half-bridge diode rectifier and (b) half-bridge active rectifier [A13].

and MN2 conduct during the input positive half-wave voltage, while MP2 and MN1 conduct during the negative half-wave. Then, the diode blocks the current that could flow from the output again to the input. An improvement can be done using an active diode composed of a PMOS transistor and a comparator. If the output voltage V_{out} is higher than V_{pos}, the output of the comparator is equal to its positive supply voltage and MP3 is turned off. MP3 is controlled by the comparator output voltage during steady-state operation. However, while the output voltage does not reach the minimum supply voltage required by the comparator, the body diode of MOSFET MP3 blocks the current flowing from the output to the input.

The AC-DC rectifier requires a DC-DC converter to transfer the energy extracted to the PVDF film to the battery (see Figure 15). There is an optimum voltage level on the storage capacitor C_2 for which the power flowing from the piezoelectric transducer is maximum. Thus, the DC-DC converter must work with an appropriate duty cycle to achieve this optimum voltage. Ottman et al. [A17,A18] use a step-down converter as DC-DC converter to regulate the input voltage V_{C2} of the converter to the value that maximizes the power

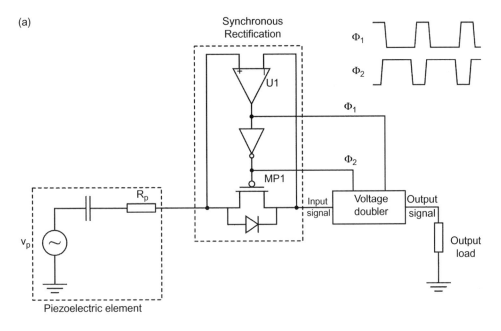

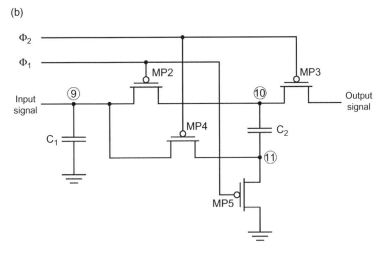

FIGURE S15 (a) Half-wave synchronous rectifier and voltage doubler. (b) Transistor-level circuit of the voltage doubler [A14].

provided by the piezoelectric element (see Eq. (12)). This concept is developed in two different ways for a sinusoidal mechanical excitation. The first implementation consists of an adaptive circuit that maximizes dynamically the current that charges the battery [A17], whereas the second implementation includes the calculation of the optimum duty cycle and the design of a circuit with a fixed duty cycle equal to the optimum [A18].

(a)

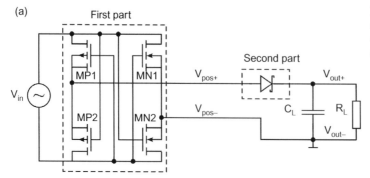

FIGURE S16 Active rectifier composed of a negative voltage converter and (a) a diode or (b) an active diode.

(b)

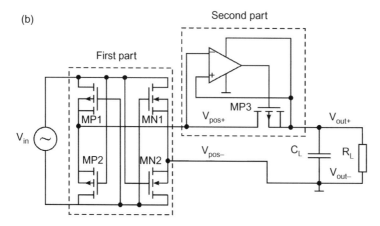

One implementation assumes that output and input power of the converter are related by a constant value. Thus, the maximization of the power delivered by the piezoelectric element is equivalent to the maximization of the output power of the converter. Since a battery is connected at the output of the DC-DC converter, it can be considered that its output voltage is constant. Then, the maximization of the output power of the DC-DC converter is equivalent to the maximization of its output current. This method is widely used in control algorithms for maximum power point trackers of, for example, solar cells [A19]. The efficiency of this converter, excluding the rectifier, is between 74% and 88% [A17]. Eq. (54) describes the algorithm of the control circuit of the DC-DC converter (Figure S17):

$$D_{i+1} = D_i + K \cdot sgn\left(\frac{\partial I}{\partial D}\right) \tag{54}$$

where D_{i+1} is the value of the duty cycle for the next iteration, D_i is the actual duty cycle and K is the multiplying coefficient of the sign function sgn that is applied to the partial derivative of output current I to duty cycle D.

In the other implementation, the optimum duty cycle is calculated and its value is fixed in the control circuit that drives the converter [A18]. Assuming that the step-down

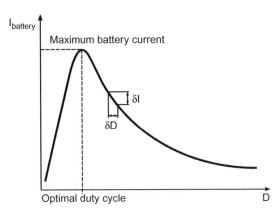

converter is working in discontinuous current conduction mode (DCM), the power delivered by the piezoelectric element is [A18]:

$$P_{in} = \frac{D^2\left(\frac{2I_p}{\pi} - \frac{2\omega C_1 V_{out}}{\pi}\right)\left(\frac{2I_p}{\pi} + \frac{D^2 V_{out}}{2Lf_s}\right)}{2Lf_s\left(\frac{2\omega C_1}{\pi} + \frac{D^2}{2Lf_s}\right)^2} \qquad (55)$$

where D is the duty cycle of the switching converter, I_p is the piezoelectric peak current, C_1 is the piezoelectric capacitance, ω is the angular frequency of the piezoelectric element, V_{out} is the output voltage of the step-down converter, L is the inductance employed in the converter, and f_s is the switching frequency of the converter.

D_{opt} is the duty cycle of the switching converter that provides the maximum power, and is calculated with $\frac{\partial P_{in}}{\partial D} = 0$,

$$D_{opt} = \sqrt{\frac{4V_{C2}\omega LC_1 f_s}{\pi(V_{C2} - V_{battery})}} \qquad (56)$$

where V_{C2} is the voltage on C_2 and $V_{battery}$ is the battery voltage. V_{C2} is equal to half of the piezoelectric open-circuit voltage V_{oc}. For large values of V_{oc}, D_{opt} remains almost constant and it is the value fixed in the control circuit.

Kong et al. propose another resistive-matching circuit, also with a switching step-down converter but removing capacitor C_2 [A20]. The optimum duty cycle $D_{1,opt}$ is calculated from the equivalent input resistance of a buck-boost converter working at discontinuous conduction mode:

$$D_{1,opt} = \sqrt{\frac{2L}{R_{in,opt}T_s}} \qquad (57)$$

where T_s is the switching period of the buck-boost converter and $R_{in,opt}$ is the resistance with a value equal to the modulus of the internal impedance of the piezoelectric transducer. The input voltage of the buck-boost converter is considered constant since the switching frequency of the oscillator that controls the gate of the transistor is much higher than the excitation frequency of the piezoelectric element.

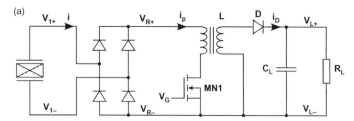

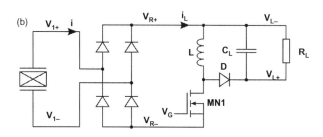

FIGURE S18   SECE energy harvesting circuits (a) flyback architecture and (b) buck-boost architecture [A22].

Another passive technique is the synchronous electric charge extraction (SECE) [A21]. Figure S18 shows the SECE circuit that includes a full-wave bridge rectifier followed by (a) flyback and (b) buck-boost converters. When a maximum voltage is detected at the rectified voltage V_R, the control circuit turns on transistor $MN1$. Then, the piezoelectric current flows through inductor L and the energy is transferred from the piezoelectric element to the inductor. When the current flowing through inductor L reaches its maximum, the rectified voltage is zero and transistor $MN1$ is turned off. From this moment, the piezoelectric transducer is in open circuit and the current starts to flow from inductor L to storage capacitor C_L and resistive load R_L through diode D [A22].

S.2.2 AC-DC Nonlinear Rectifiers

Figure S19 (a) and (b) show the parallel SSHI circuit and its associated waveforms at steady state, respectively [A23]. Current flows through the diode bridge when the voltage on the piezoelectric element is equal to the voltage V_L on the load plus the voltage drop of the diode bridge. When the piezoelectric voltage reaches its peak, switch S is closed and the piezoelectric current flows through inductor L_{res}. Thus, a resonant LC circuit is created with the piezoelectric internal capacitor and the voltage on the piezoelectric element changes its polarity in a time given by Eq. (58).

$$t_I = \pi \sqrt{L_{res} C_1} \tag{58}$$

The absolute value of the piezoelectric voltage after inversion is lower than before the inversion due to the losses in the switching circuit and inductor. The piezoelectric voltage inversion is characterized by γ, which is the ratio of the voltage after (V_{after}) to the voltage

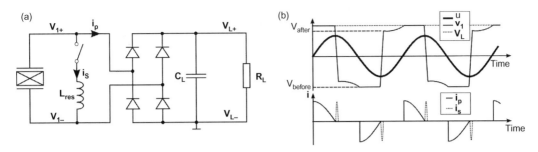

FIGURE S19   Parallel SSHI energy harvesting technique (a) circuit (b) waveforms [A23].

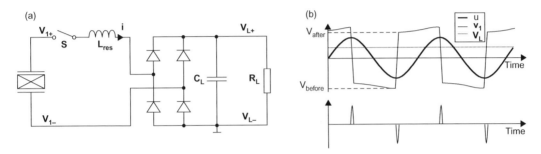

FIGURE S20   Series SSHI energy harvesting technique (a) circuit and (b) waveforms [A23].

before (V_{before}) the polarity change of the piezoelectric transducer [A23]. The value of γ affects the efficiency of the SSHI [A24].

$$\gamma = -\frac{V_{after}}{V_{before}} \tag{59}$$

Figures S20(a) and (b) show the series SSHI harvesting circuit and the associated waveforms at steady state, respectively. The inductor is connected in series with the piezoelectric element and the current flows through the inductor, the diode bridge, and the load only when the switch is closed. As in the parallel SSHI, the control circuit has to detect the peak voltage on the piezoelectric transducer and close the switch at this moment.

A modified parallel SSHI circuit is shown in Figure S21 where the number of diodes has been reduced from six to just two and the number of MOSFETs remains two, compared to the parallel SSHI converter. This SSHI converter has the same topology as a new version of the series SSHI converter presented in [A26] but operates like the parallel SSHI converter. When the piezoelectric voltage reaches its positive peak value, diode $D1$ conducts and the control signal changes from negative to positive voltage. Then, $MP1$ turns off and $MN1$ starts conducting. Since $D1$ and $MN1$ are now turned on, the piezoelectric element is connected in parallel to inductor L and the piezoelectric voltage is inverted. Once the inversion is completed, diode $D1$ does not conduct anymore and $D2$ starts conducting. During the rectification phase of the negative semi-cycle, $D2$ and $MN1$ conduct.

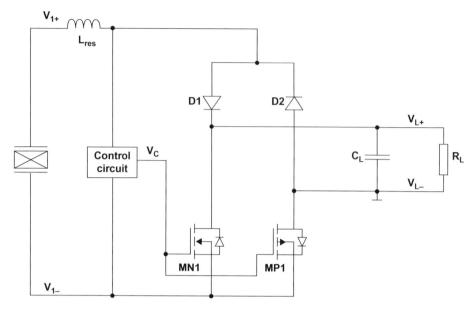

FIGURE S21   Schematic of the modified parallel SSHI converter [A25].

When the piezoelectric voltage reaches its negative peak value, diode $D2$ and $MP1$ are turned on and the piezoelectric element is connected in parallel to inductor L. Once the inversion is completed, diode $D1$ starts conducting. During the positive semi-cycle, $D1$ and $MP1$ conduct.

The series and parallel SSHI techniques provide comparable harvested power but their optimum loads are different. A DC-DC voltage converter can be used after the diode bridge to adjust the voltage after the SSHI to the value obtained with the optimum load.

S.3. ENERGY HARVESTING FROM LIGHT: MPPT ALGORITHMS

MPPT algorithms traditionally track the MPP of solar cells. Thus, the algorithm gets an input voltage and current at the DC-DC converter equal to V_{MPP} and I_{MPP}, respectively.

There are two ways to implement an MPPT algorithm: change the duty cycle of the PWM control circuit of the DC-DC converter or change the reference voltage of the control loop [A27]. In the first case, a microcontroller is necessary while in the second case an implementation done just with an analog circuit is possible.

Several algorithms to implement the MPPT are presented in this section. There are some algorithms that calculate the MPP in an indirect way since the output power of the solar cell is not measured. The simplest of these algorithms is the constant voltage (CV) algorithm where the voltage at the output of the solar cell V_{PV} is measured and compared

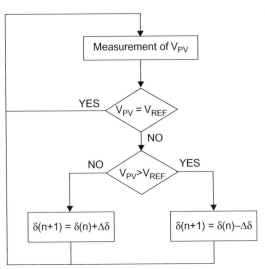

FIGURE S22 Flowchart of the constant voltage algorithm [A28].

with a reference voltage that is equal to V_{MPP} of the solar cell being employed. However, the reached MPP is not real since it does not take into consideration temperature and irradiance variations. Dolara et al. exposed that when the solar cell is in low insulation conditions, the CV technique could be more effective than either the perturb and observe (P&O) or the incremental conductance (IC) method [A28]. Figure S22 displays the flow chart of the CV method.

An evolution from the previous algorithm are the fractional open-circuit voltage and fractional short-circuit current algorithms that are based on the relationship between V_{MPP} and V_{OC} and between I_{MPP} and I_{SC}, respectively. The open-circuit voltage algorithm is based on the following relation between V_{MPP} and V_{OC} under different irradiance and temperature conditions:

$$V_{MPP} \approx k_1 V_{OC} \tag{60}$$

where k_1 is the proportionality constant.

I_{MPP} has an almost linear relation with I_{SC} for different temperature and irradiance conditions. Thus, in the short-circuit current algorithm I_{MPP} and I_{SC} are related by

$$IMPP \approx k_2 I_{SC} \tag{61}$$

where k_2 is the proportionality constant.

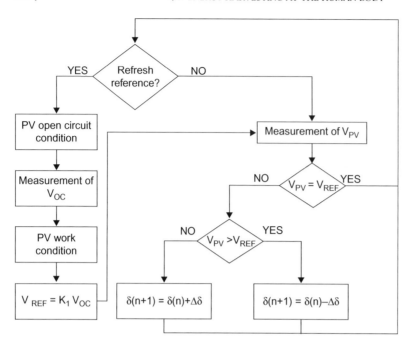

FIGURE S23 Flowchart of the fractional open circuit voltage algorithm [A28].

k_1 has a value range between 0.71 and 0.78 and k_2 between 0.78 and 0.92 depending on the solar cell under consideration [A19]. Fractional methods take periodic measurements of V_{OC} or I_{SC} to compute V_{MPP} or I_{MPP}, respectively. Both methods need to disconnect the module to perform the measurements, which causes a temporary loss of power. This problem is solved by employing an extra solar cell where the open-circuit voltage or short-circuit current is measured. Figure S23 shows the flowchart of the fractional open-circuit voltage method. Fractional methods are simple and cheap since no microcontroller is needed, just a constant voltage reference is fixed. However, the MPP is never matched since constants k_1 and k_2 do not take into consideration real temperature and irradiance conditions.

The P&O algorithm is the most used algorithm. It measures and stores the voltage and current of the solar cell to compute its power. Afterwards, the actual power point is moved with a fixed step size in order to acquire a new power point to be compared with the previous one. Depending on the difference of these two power points (positive or negative), the algorithm moves the next point of operation in the same or contrary direction as before in order to find the MPP. These algorithms allow reaching the MPP independently on the solar cell employed. Figure S24 shows a flowchart diagram of the P&O algorithm. Its main drawback is the oscillation around the MPP. However, the magnitude of the oscillation can be reduced, reducing the step size between power points. Nevertheless, if this step is too small, the dynamic response becomes slower. Another option to overcome this problem is to use a variable step-size algorithm that adapts the step size to the absolute value of the difference between the two power points under evaluation in the

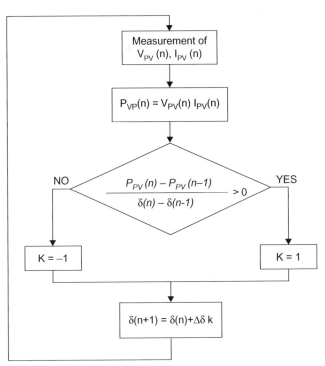

FIGURE S24 Flowchart of the P&O algorithm [A28].

algorithm. Some authors differentiate between the P&O and hill climbing (HC) algorithms [A19] and others say that P&O and incremental conductance are part of the HC technique [A27]. Either if it is considered as the same algorithm or as different algorithms, there are two ways to move the operation point of the PV cell through the MPPT control circuit of the DC-DC converter: modifying the duty cycle or modifying the reference voltage (current) to compare with the operation voltage (current) of the solar cell. If the duty cycle is modified the algorithm is always called P&O while if the reference voltage is the parameter modified it is called either P&O or HC. Nevertheless, P&O and HC algorithms do not have a good dynamic response when fast environmental changes occur, but there are improved versions that can surpass this drawback.

The IC method is based on the fact that the slope of the PV array power curve with respect to the voltage is zero at the MPP, positive to the left, and negative to the right of the MPP [A29,A30]. The condition at the MPP is expressed as

$$\frac{dP_{PV}}{dV_{PV}} = I_{PV} + V_{PV}\frac{dI_{PV}}{dV_{PV}} = 0 \tag{62}$$

Thus, rearranging terms, the previous condition is equivalent to

$$\frac{I_{PV}}{V_{PV}} = -\frac{dI_{PV}}{dV_{PV}} \tag{63}$$

Thus, the MPP can be tracked comparing the instantaneous conductance (I_{PV}/V_{PV}) with the incremental conductance ($\Delta I_{PV}/\Delta V_{PV}$):

$$\frac{dI_{PV}}{dV_{PV}} = -\frac{I_{PV}}{V_{PV}} \text{ at MPP} \tag{64}$$

$$\frac{dI_{PV}}{dV_{PV}} > -\frac{I_{PV}}{V_{PV}} \text{ left at } MPP \tag{65}$$

$$\frac{dI_{PV}}{dV_{PV}} < -\frac{I_{PV}}{V_{PV}} \text{ right of } MPP \tag{66}$$

This method also features a modified version that does not suffer from fast transients due to environmental condition changes. The IC method monitors both voltage and current of the PV as does the P&O method. However, it is not necessary to calculate the PV power. This algorithm needs to be implemented digitally. Moreover, the MPP reached is real and module independent. Figure S25 shows the flowchart for the IC algorithm.

The dP/dV or dP/dI feedback control method works in a similar way as the IC method. It computes the slope dP/dV or dP/dI and tries to drive it to zero. The derivative is computed and depending on its sign the duty ratio is incremented or decremented to reach the MPP. This technique increases the convergence speed and matches the real MPP but requires more computational power than the IC.

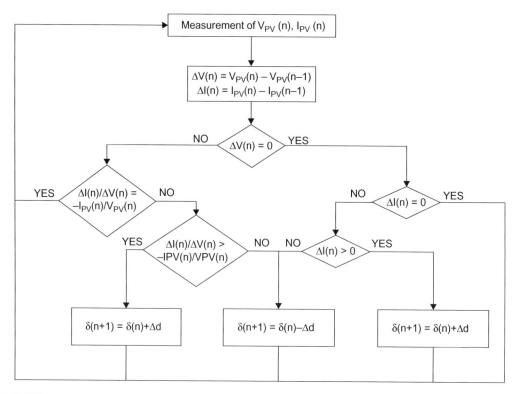

FIGURE S25   Flowchart of the IC algorithm [A19].

References for the Supplemental Material

[A1] Advanced Linear Devices, Inc., last accesed november 2013. URL: <http://www.aldinc.com> (Last Accessed: 06.07.14).

[A2] BSH105, N-channel vertical D-MOS logic level FET, NXP founded by Philips, last accessed december 2013. URL: <http://www.nxp.com/pip/BSH105.html> (Last Accessed: 06.07.14).

[A3] Infineon Technologies AG, last accesed november 2013. URL: <http://www.infineon.com> (Last Accessed: 06.07.14).

[A4] United Microelectronics Corp., last accesed november 2013. URL: <www.umc.com> (Last Accessed: 06.07.14).

[A5] Fraunhofer-Gesellschaft zur Forderung der angewandten Forschung e.V., last accesed november 2013. URL: <http://www.fraunhofer.de> (Last Accessed: 06.07.14).

[A6] enocean®, last accesed november 2013. URL: <http://www.enocean.com/en/home/> (Last Accessed: 06.07.14).

[A7] Future Technology Devices International Ltd., last accesed november 2013. URL: <http://www.ftdichip.com/FTProducts.htm> (Last Accessed: 06.07.14).

[A8] Datasheet Thermogenerator TEG-127-150-26, last accessed 2011. URL: <http://www.thermalforce.de/de/product/thermogenerator/TG127-150-26e_.pdf> (Last Accessed: 06.07.14).

[A9] Datasheet Thermogenerator PKE128A1027, last accessed 2011. URL: <http://www.peltron.de/peltrierelemente_standard.htm> (Last Accessed: 06.07.14).

[A10] C.R. Sullivan, M.J. Powers, A high-efficiency maximum power point tracker for photovoltaic arrays in a solar-powered race vehicle, Power Electron. Spec. Conf. (1993) 574–580.

[A11] E. Koutroulis, K. Kalaitzakis, N.C. Voulgaris, Development of a microcontroller-based, photovoltaic maximum power point tracking control system, IEEE Trans. Power Electron. 16 (2001) 46–54.

[A12] Low cost analog multiplier, last accessed December 2011. URL: <http://www.analog.com/static/imported-files/data_sheets/AD633.pdf> (Last Accessed: 06.07.14).

[A13] J. Han, A. von Jouanne, T. Le, K. Mayaram, T.S. Fiez, Novel power conditioning circuits for piezoelectric micropower generators, Appl. Electron. Conf. Exposition 3 (2004) (2004) 1541–1546, APEC '04. Nineteenth Annual IEEE.

[A14] T. Le, J. Han, A. Von Jouanne, K. Mayaram, T. Fiez, Piezoelectric micro-power generation interface circuits, IEEE J. solid-state circuits 41 (2006) 1411–1420.

[A15] C. Peters, D. Spreemann, M. Ortmanns, Y. Manoli, A cmos integrated voltage and power efficient ac/dc converter for energy harvesting applications, J. Micromechanics Microeng. 18 (2008) 104005.

[A16] Y. Manoli, Energy harvesting-From devices to systems, ESSCIRC, 2010 Proc., IEEE (2010) 27–36.

[A17] G. Ottman, H. Hofmann, A. Bhatt, G. Lesieutre, Adaptive piezoelectric energy harvesting circuit for wireless remote power supply, IEEE Trans. on Power Electron. 17 (2002) 669–676.

[A18] G. Ottman, H. Hofmann, G. Lesieutre, Optimized piezoelectric energy harvesting circuit using step-down converter in discontinuous conduction mode, IEEE Trans. Power Electron. 18 (2003) 696–703.

[A19] T. Esram, P. Chapman, Comparison of photovoltaic array maximum power point tracking techniques, IEEE Trans. Energy Conversion 22 (2007) 439–449.

[A20] N. Kong, D.S. Ha, A. Erturk, D.J. Inman, Resistive impedance matching circuit for piezoelectric energy harvesting, J. Intell. Mater. Syst. Struct. 21 (2010) 1293–1302.

[A21] Y. Liu, Active energy harvesting, Ph.D. thesis, The Pennsylvania State University, 2006.

[A22] E. Lefeuvre, A. Badel, C. Richard, D. Guyomar, Piezoelectric energy harvesting device optimization by synchronous electric charge extraction, J. Intell. Mater. Syst. Struct. 16 (2005) 865.

[A23] S. Priya, D. Inman, Incorporated. Energy Harvesting Technologies, Springer Publishing Company, 2008

[A24] L. Mateu, H. Zessin, P. Spies, Analytical method for selecting a rectification technique for a piezoelectric generator based on admittance measurement, Journal of Physics: Conference Series, vol. 476, IOP Publishing, 2013, pp. 012111.

[A25] L. Mateu, L. Lühmann, H. Zessin, P. Spies, Modified parallel SSHI AC-DC converter for piezoelectric energy harvesting power supplies, Telecomm. Energy Conf. (INTELEC), 2011 IEEE 33rd Int., IEEE (2011) 1–7.

[A26] M. Lallart, D. Guyomar, An optimized self-powered switching circuit for nonlinear energy harvesting with low voltage output, Smart Mater. Struct. 17 (2008) 035030.

[A27] A.K. Abdelsalam, A.M. Massoud, S. Ahmed, P.N. Enjeti, High-performance adaptive perturb and observe MPPT technique for photovoltaic-based microgrids, Power Electron., IEEE Trans. 26 (2011) 1010–1021.

[A28] A. Dolara, R. Faranda, S. Leva, Energy comparison of seven MPPT techniques for PV systems, J. Electromagn. Anal. Appl 3 (2009) 152–162.

[A29] K. Hussein, I. Muta, T. Hoshino, M. Osakada, Maximum photovoltaic power tracking: an algorithm for rapidly changing atmospheric conditions, IEE Proc. Gene., Transm Distrib. 142 (1995) 59–64.

[A30] A. Safari, S. Mekhilef, Simulation and hardware implementation of incremental conductance MPPT with direct control method using Cuk converter, Ind. Electron., IEEE Trans. 58 (2011) 1154–1161.

Introduction to RF Energy Harvesting

W.A. Serdijn, A.L.R. Mansano, and M. Stoopman

Delft University of Technology, Delft, The Netherlands

The ever-decreasing power consumption of integrated circuits provides the opportunity to use an energy harvester to power a simple wireless sensor node. This makes the sensor node truly autonomous and can significantly extend lifetime. With a limited power budget, these sensors are designed to sense, process, and wirelessly transmit information such as temperature, humidity, location, and identification.

In many applications, this power can be supplied by an RF energy harvester when other energy sources such as light, vibrations, and thermal gradients are not available.

Far-field[1] radio frequency energy harvesting (RFEH) is suitable for long-range wireless power transfer, i.e., cm range for high-frequency on-chip antennas to several meter range for off-chip antennas. This makes RFEH suitable for battery-less sensors in a WSN remotely powered by a hub (i.e., RF source). RFEH suits many applications, such as smart house, smart grid, Internet of things (IoT), and wireless body area networks (WBAN). Especially in the last few years, the WBAN application is gaining importance due to the growing importance of health care in society as health needs to be continuously monitored to identify chronic diseases or prevent illness. Examples of WBANs are a sensor array for monitoring ExG signals [1–2] and a disposable battery-less band-aid sensor [3].

In WBAN applications, sensors may require power on the order of micro-watts, depending on how they operate. For example, a temperature sensor is not required to update its momentary value very often as temperature is a slowly varying quantity in most applications. On the other hand, the peak power consumption of a duty-cycled sensor might be significantly larger than the harvested (average) power. In such a case, the energy provided by the RFEH can be stored in a capacitor or battery that periodically supplies energy to the sensor. Figure 1 presents the typical energy profile stored in the

---

[1] In the far-field region, the radiating fields dominate the non-radiating reactive fields of the transmitter. An object is in the far-field region of an antenna when the distance from the antenna is larger than $2D^2/\lambda$, where D is the largest dimension of the antenna and λ is the wavelength [4].

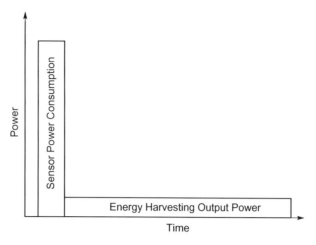

FIGURE 1   Energy profile of a sensor and RFEH.

capacitor and the energy required by the sensor. In this example, the RFEH is periodically connected to a capacitive load, thus efficient energy transfer is required during capacitor charging to minimize losses and charging time. Obviously, the energy supplied by the RFEH over time should be greater than the energy consumed by the sensor.

In this chapter, first RF energy harvesting fundamentals and practical limitations are discussed. Subsequently, an overview is given of different rectifier topologies and circuit implementations with their advantages, disadvantages, and challenges. Then, a compensation scheme is introduced in order to improve the robustness of RF energy harvesters in a practical environment. Finally, an antenna-rectifier co-design example is given to realize a high-performance RF energy harvester. The design is verified by measurements in an anechoic chamber and with energy harvesting from an ambient RF source.

1. RF ENERGY HARVESTING FUNDAMENTALS AND PRACTICAL LIMITATIONS

An RF energy harvester designer finds himself at the interface between the electromagnetic radiating fields and the electronic circuitry. In order to optimize this interface for the best possible performance, the designer needs to be equipped with knowledge from both domains. In this section, the basic understanding of antenna fundamental properties and their relationship to power density, impedance, voltage, and current is introduced.

1.1 Wave Propagation, Antenna Effective Area, and Available Power

To understand how the terminal voltage and current of a receiving antenna relate to the radiated power from a transmitting antenna, we start by deriving Friis' transmission equation. When a signal generator at the transmitter side forces a time-varying current through an antenna structure, electromagnetic radiation is produced as the radiation mechanism of any antenna is based on the acceleration of electric charge. The relative distribution of radiated power in space depends on the radiation pattern. A hypothetical isotropic radiator radiates equally well in all directions such that the power uniformly spreads out across the surface of an imaginary sphere. In reality, antennas radiate and/or receive more effectively in some directions than in others. The directivity D is used to describe the radiation intensity in a given direction compared to the radiation intensity of an isotropic antenna. The power density S [W/m^2] at a distance d from the RF source therefore equals the total radiated power by the transmitter divided by the surface area of a sphere scaled with the directivity of the transmitting antenna [4]:

$$S = \frac{P_{TX}}{4\pi d^2} D_{TX} = \frac{P_{EIRP}}{4\pi d^2} \tag{1}$$

where $P_{EIRP} = P_{TX}D_{TX}$ is defined as the equivalent isotropic radiated power (EIRP).

The amount of power collected by a receiving antenna located in the far-field region of the transmitter can be found by using the concept of antenna effective area. The antenna effective area, A_{eff}, is defined as the ratio of the available power, P_{av}, to the power density of a plane wave incident on the antenna:

$$A_{eff} = \frac{P_{av}}{S} \tag{2}$$

The available power is defined as the maximum power that can be extracted from an antenna and can be delivered to the load (i.e., power at the input of the RF energy harvester). The actual dissipated power in the antenna load only equals P_{av} in case of a lossless and perfectly impedance-matched antenna-electronics interface. For any receiving antenna, it can be proven that the maximum effective antenna area is closely related to its maximum directivity D_{RX} [4]:

$$A_{eff} = \frac{\lambda^2}{4\pi} D_{RX} \tag{3}$$

The available power to an RF energy harvester in free space using a lossless and perfectly aligned receiving antenna is then given by

$$P_{av} = A_{eff}S = \left(\frac{\lambda}{4\pi d}\right)^2 D_{RX} P_{EIRP} \tag{4}$$

This equation, known as the Friis' transmission equation, gives a fundamental limit to the available power as a function of distance, power, frequency, and antenna gain. The $(\lambda/4\pi d)^2$ term is often referred to as "free space path loss." This expression, however, suggests that free space somehow attenuates a propagating electromagnetic wave with decreasing wavelength and increasing distance. This is a misconception. Firstly, the radiated power is not lost over distance, but merely spreads out across the surface area. Secondly, the wavelength enters the equation because of the effective antenna area. A shorter wavelength corresponds to a smaller effective antenna area and therefore is less effective at capturing energy from the incoming wave. In order to capture the same power at a shorter wavelength, the physical area of the antenna needs to be increased.

1.2 Antenna-Rectifier Interface Voltage

Not only the available power is of concern in the design of RF energy harvesters; the available voltage swing is equally important. This is due to the fact that practical electronic components used for rectification such as diodes and MOS transistors inherently are voltage-controlled devices. Therefore, the first concern in the design of a highly sensitive RF energy harvester is to *activate* the rectifier by generating a sufficiently large voltage swing at the input of the rectifier.

To link the antenna terminal voltage to the available power, we can use the antenna-rectifier equivalent circuit model as depicted in Figure 2. For now, the antenna impedance is assumed to be purely real for convenience. However, it is also valid to assume that the imaginary part of the antenna impedance is absorbed into the matching network; it does not make a difference for the following analysis.

The voltage and current ratio at the antenna terminals may be modeled by using a Thévenin or Norton equivalent circuit. Here, the voltage induced by the electric field is represented by a Thévenin equivalent voltage source, V_A. The radiation resistance, R_{Rad}, relates the voltage and current ratio to the available power. The conduction loss resistance, R_{Loss}, is related to the antenna radiation efficiency η_A by

$$\eta_A = \frac{R_{Rad}}{R_{Rad} + R_{Loss}} \tag{5}$$

The total antenna series resistance amounts to $R_A = R_{Rad} + R_{Loss}$, which corresponds to the real part of the antenna input impedance. The input of the rectifier is modeled as a parallel combination of $R_{rec,p}$ and $C_{rec,p}$. In this model, the power dissipated in $R_{rec,p}$ is seen as the actual power transferred to the (ideal) rectifier DC output port. This resistance is also referred to as a loss-free resistor and becomes useful when modeling an ideal rectifier [5].

To relate V_A to P_{av}, a loss-less and conjugate impedance-matched network is assumed. In this scenario, it follows that $V_{in} = \frac{1}{2}V_A$ and $R_{in} = R_{Rad} + R_{Loss}$. Hence, the power relation for a *lossy* antenna with conjugate impedance matched interface is written as

$$\eta_A P_{av} = \frac{(V_A/2)^2}{2(R_{Rad} + R_{Loss})} \tag{6}$$

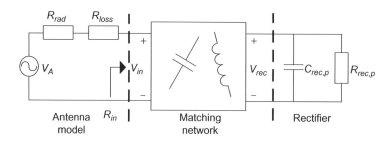

FIGURE 2   Antenna-rectifier interface with impedance matching network.

Using (6) and $R_A = R_{Rad} + R_{loss}$ we can deduce that

$$V_A = \sqrt{8\eta_A R_A P_{av}} \tag{7}$$

Equation (7) quickly reveals the expected antenna voltage. As an example, a standard 50 Ω antenna with 90% radiation efficiency and $P_{av} = -20\,\text{dBm}$ (10 μW) has an open-circuit terminal voltage of only 60 mV. This is much too low to overcome the threshold voltage of a standard CMOS transistor, which lies around 450 mV in 90 nm IC technology.

The relationship between V_A and the radiated power of the transmitter is found by substituting (4) into (7):

$$V_A = \sqrt{\frac{R_A \eta_A D_{RX} P_{EIRP}}{2}} \frac{\lambda}{\pi d} \tag{8}$$

Note that this general equation holds for *any* type of antenna. The relations between the antenna equivalent circuit elements are summarized in Table 1.

In general, the antenna voltage V_A is *not* the voltage swing at the input of the rectifier. Usually an impedance-matching network transforms the high rectifier input impedance to match with the antenna impedance. This interface impedance transformation can provide a significant passive voltage boost as it increases the voltage swing at the input of the rectifier for the same input power, hence, improving the sensitivity (i.e., power-up threshold) of the RF energy harvester.

To calculate the passive voltage boost for a given antenna-electronics interface, the equivalent circuit in Figure 2 again can be used. If the matching network is lossless and assures a conjugate matched interface, then the input power seen from R_{in} must equal the power transferred to the rectifier:

$$\frac{(V_A/2)^2}{2R_{in}} = \frac{V_{rec}^2}{2R_{rec,p}} \tag{9}$$

As $R_{in} = R_A$, the passive voltage gain, $G_{V,boost}$, is calculated as

$$G_{V,boost} = \frac{V_{rec}}{V_A} = \frac{1}{2}\sqrt{\frac{R_{rec,p}}{R_A}} \tag{10}$$

TABLE 1   Antenna Equivalent Circuit Elements

Thévenin equivalent voltage	$V_A = \sqrt{8\eta_A R_A P_{av}}$
Radiation resistance	$R_{rad} = \eta_A R_A$
Conduction loss resistance	$R_{loss} = (1 - \eta_A)R_A$
Antenna resistance	$R_A = R_{rad} + R_{loss}$
Radiation efficiency	$\eta_A = \dfrac{R_{rad}}{R_{rad} + R_{loss}}$

Note that $G_{V,boost}$ is independent of the matching network implementation and only depends on the source and load conditions. Substitution of (10) into (7) leads to the available voltage swing at the rectifier input terminals:

$$V_{rec} = \sqrt{2\eta_A R_{rec,p} P_{av}} \tag{11}$$

Intuitively, this makes sense; for a given antenna radiation efficiency and available input power, the voltage swing can *only* be increased by increasing the antenna load $R_{rec,p}$. Hence, for an interface with a given source and load resistance, one cannot design for a desired voltage boost to increase wireless range if a conjugate match is required simultaneously. *The designer therefore needs to design for the largest $R_{rec,p}$ possible and subsequently co-design the antenna impedance for conjugate matching.* This conclusion is a key point that needs to be considered during the design procedure. The rectifier input impedance needs to be designed in such a way that it maximizes the input voltage swing in order to improve the sensitivity.

To elaborate further on the principle of passive voltage boosting, a detailed equivalent antenna-rectifier circuit is depicted in Figure 3. Here, an N-stage rectifier with capacitive load is directly connected to a loop antenna. The reactive element jX_A represents the energy stored in the near field and is inductive since a loop antenna below its first anti-resonance frequency is assumed. By using a loop antenna, the rectifier capacitance can be compensated for without using other external components. The rectifier output is modeled as a Thévenin equivalent circuit. The input impedance of the rectifier is mainly capacitive, where $R_{rec,s}$ is the real part of the impedance and $X_{rec,s} = 1/(\omega C_{rec,s})$ represents the imaginary part.

For an N-stage rectifier with a capacitive load, the output voltage in steady state can be written as

$$V_{out} = V_A G_{V,boost} N \eta_V \tag{12}$$

where $G_{V,boost}$ is the passive voltage boost obtained from the LC resonating network and η_V is the voltage efficiency of a single rectifying stage. When $X_{rec,s} \gg R_{rec,s}$ and the interface is at resonance ($X_{rec,s} = X_A$), it holds that

$$G_{V,boost} = \left| \frac{V_{rec}}{V_A} \right| \approx \frac{X_{rec,s}}{R_{rec,s} + R_A} \tag{13}$$

Note that this expression is in a different form compared to (10) and suggests that an increase of $X_{rec,s}$ or decrease of $R_{rec,s}$ results in a larger passive voltage boost. This indeed

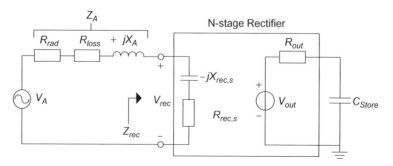

FIGURE 3   Antenna-rectifier interface equivalent circuit model with passive voltage boosting.

corresponds to an increased parallel resistance, $R_{rec,p}$, when performing a series-to-parallel impedance transformation at the frequency of interest:

$$R_{rec,p} = R_{rec,s} \left(1 + \left(\frac{X_{rec,s}}{R_{rec,s}} \right)^2 \right) \tag{14}$$

Hence, increasing $X_{rec,s}$ or decreasing $R_{rec,s}$ (thereby increasing the Q factor and parallel load resistance ($R_{rec,p}$) results in a larger available voltage swing at the rectifier input terminal.

The improvement in sensitivity can be described in terms of the interface impedance, the rectifier properties, and the rectified output voltage. When combining (12) and (13) and using the antenna parameters as given in Table 1, the minimum required available power for a desired V_{out} can be written as

$$P_{av} = \frac{(R_{rec,s} + R_A)^2}{8 \eta_A R_A} \left(\frac{V_{out}}{X_{rec,s} N \eta_V} \right)^2 \tag{15}$$

From (15), it is evident that the choice of the antenna-rectifier interface impedance plays a crucial role in the optimization of highly sensitive RF energy harvesters. The rectifier input resistance, $R_{rec,s}$, and reactance, $X_{rec,s}$, depend on the rectifier implementation and decreases with the number of stages N. Increasing the number of stages also indirectly reduces the efficiency due to the body effect when using a standard technology. These issues will be discussed further in the rectifier topology section. If, for example, the design parameters are $\eta_A = 0.8$, $X_{rec,s} = 400\ \Omega$, $N = 3$, and $\eta_V = 0.8$, the curves in Figure 4 show the minimum required available power to generate a voltage of 1.5 V across a capacitive load for three different values of $R_{rec,s}$ as a function of antenna resistance, R_A.

A low resistive antenna-rectifier interface translates into a significant improvement in sensitivity due to the large passive voltage boost, with the minimum of each curve at $R_A = R_{rec,s}$. In this particular example, the minimum required available power to generate 1.5 V for a 50 Ω interface equals −11.2 dBm (76.29 μW), while only −19.9 dBm (10.17 μW) is required for a 10 Ω interface, resulting in an 8.7 dB sensitivity improvement.

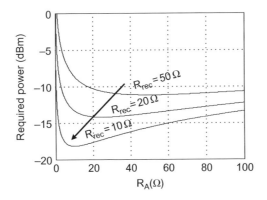

FIGURE 4 Calculated minimum required power to generate 1.5 V across a capacitive load vs. antenna resistance R_A and various $R_{rec,s}$ for $\eta_A = 0.8$, $X_{rec,s} = 400\Omega$, $N = 3$, and $\eta_V = 0.8$.

1.3 Practical Limitations

There are several practical limitations that can strongly influence the available DC power for a sensor node during harvesting. The theoretical line-of-sight available power of Eq. (4) can be extended to a more practical form:

$$P_{av,DC} = G_{RX}(\theta, \varphi) P_{EIRP} \left(\frac{\lambda}{4\pi d} \right)^n (1 - |\Gamma|^2) |cos\ \Psi|^2 \eta_{PCE} \tag{16}$$

where

- gain $G_{RX}(\theta, \varphi) = \eta_A D_{RX}(\theta, \varphi)$ is a function of the azimuth angle θ and the elevation angle φ and depends on the type of antenna being used;
- n is the propagation exponent which can range between 0.8 to 1.8 in a (highly) reflective line-of-sight environment and can get as high as 8.6 for non-line-of-sight environments [6];
- $(1 - |\Gamma|^2)$ is the interface impedance mismatch efficiency, where $\Gamma = (Z_L - Z_A{}^*)/(Z_L + Z_A)$ is the reflection coefficient;
- $|cos\ \Psi|^2$ is the polarization loss factor and depends on the angle Ψ between the transmitting and receiving antenna and their polarization; and
- η_{PCE} is the RF-DC power conversion efficiency and highly depends on input power and frequency.

2. IMPEDANCE MISMATCH, LOSSES, AND EFFICIENCY

Impedance mismatch occurs when $R_A \neq R_{rec,s}$ or $X_{rec,s} \neq X_A$. In such a scenario, the antenna is not able to deliver the full available power to the rectifier. This is a serious concern and requires the designer's full attention as impedance variations introduced by on-body antennas can strongly degrade the power conversion efficiency. In high-Q interfaces, circuit techniques may be required to automatically tune the interface to minimize impedance mismatch.

Additional power loss is introduced by the losses that occur in a practical matching network, the antenna, and the rectifier implementation. The rectifier voltage efficiency, η_V, can be optimized in an orthogonal way, and usually is on the order of 80% or higher. The power conversion efficiency, PCE, however, highly depends on the rectifier topology and the input power. The radiation efficiency, η_A, depends on the antenna size and becomes low when the area is scaled down too much. The (single-stage) rectifier efficiency, η_V, depends on the components and topology being used.

2.1 Available Components and Technology

Practical components used for rectification need a minimum voltage in order to conduct current. A typical Schottky diode has a threshold voltage of about 0.3 V, while the threshold voltage of an ordinary diode or a CMOS transistor is usually slightly higher, depending on the technology being used. The minimum required power to reach the rectifier threshold voltage with conjugate matching is given by

$$P_{power-up} = \frac{V^2_{rec,threshold}}{2\eta_A R_{rec,p}}$$

(17)

Note that $V_{rec,threshold}$ does not neccesarily equal the diode or transistor threshold voltage, it is also a function of the number of rectifying stages and the circuit topology. For a single Schottky rectifier implementation with equivalent parallel input resistance, $R_{rec,p} = 1\,k\Omega$ and $\eta_A = 0.8$, the minimum required power for rectification equals 56.25 μW (-12.5 dBm). The power conversion efficiency will likely be low around this power-up threshold as the diode is just barely forward biased. Typically, the rectifier becomes more efficient for larger input power levels.

If $V_{rec,threshold}$, for example, can be reduced to 0.1 V, the power-up threshold is lowered to -22 dBm, giving a sensitivity improvement of 9.5 dB. This can be achieved by means of circuit techniques or using more advanced components/technology with low or zero threshold voltage transistors. The latter usually comes at higher costs, so in this case circuit techniques are favorable. Some of these circuit techniques will be discussed in the next section.

2.2 Regulations and Maximum Achievable Distance

The distance an RF energy harvester can operate from a dedicated RF source is practically determined by the maximum allowed radiated power. The license-free industrial, scientific, and medical (ISM) frequency bands are often used for RF harvesting applications since they allow for high equivalent isotropic radiated power with small antenna areas. Although national restrictions may also apply, the European Radio communications Commission (ERC) limits the maximum P_{EIRP} at 868 MHz to 3.28 W and to 4 W at 2.45 GHz [7]. The U.S. Federal Communications Commission (FCC) limits the power levels at the 915 MHz and 2.45 GHz bands to 4 W EIRP [8].

As a practical example, Figure 5 shows the available DC power versus distance when transmitting the maximum allowed power in the European 868 MHz band. A line-of-sight scenario is assumed where $P_{EIRP} = 3.28$ W, $\lambda = 0.345$ m, $G_A = 1.25$, $(1 - |\Gamma|^2) = 0.9$, and

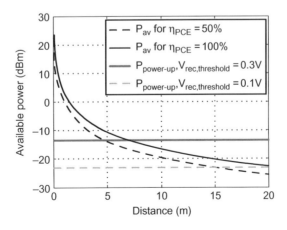

FIGURE 5 Calculated available power and power-up threshold vs. distance.

$|cos \Psi|^2 = 0.8$. The available power is shown for both a power conversion efficiency (PCE) of 50% and 100%. The power-up threshold creates an upper limit to the maximum achievable range. In this scenario, a reduction in rectifier threshold voltage from 0.3 V to 0.1 V increases the maximum achievable distance from 5 meters to 14.7 meters for $\eta_{PCE} = 50\%$. This indicates the importance of minimizing $P_{power-up}$ in case the regulations do not allow increasing the radiated power.

3. DISTRIBUTION OF HARVESTED POWER IN A REALISTIC ENVIRONMENT

The harvested power can vary significantly in a realistic environment due to the many unknown variables in the propagation channel. Figure 6 shows the measured harvested

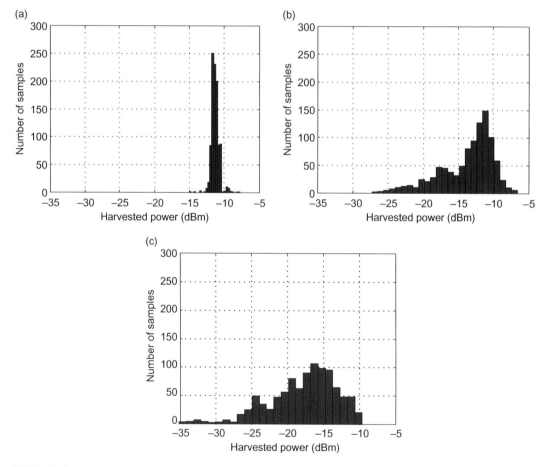

FIGURE 6   Distribution of harvested power for three scenarios: (a) line-of-sight and aligned antenna, (b) line-of-sight random angular and lateral misalignment, and (c) non-line-of-sight with random angular and lateral misalignment.

power in an office room ($\sim$24 m^2) for three different scenarios using a Powercast P2110-EVAL-02 RF energy harvesting development kit at a 3.5 meter distance from the 3 W EIRP dedicated RF source. In each scenario, 1,000 samples are taken over a period of 30 minutes.

The first figure shows a static office environment with line-of-sight and aligned antenna. The mean value is -11.26 dBm with a standard deviation of 0.57 dBm. A scenario shown in Figure 6(b) demonstrates harvested power in a case with random line-of-sight angular and lateral misalignment. Here, a mean value of -13.7 dBm was measured with a standard deviation of 3.86 dBm. The last scenario is non-line-of-sight with random misaligned antennas in a highly dynamic office room with people walking in between the transmitter and RF harvester. The mean value is -17.86 dBm with a standard deviation of 4.67 dBm. Although these measurements strongly depend on the environment and the RFEH, they demonstrate that the available power can vary significantly. These fluctuations need to be considered at system level. Moreover, these measurements also demonstrate the need for an intelligent RF source that tracks the sensor node and steers the highly directive beam towards the sensor to always provide sufficient available power to the RF harvester in any given scenario.

3.1 Ambient RF Power

Harvesting RF energy from ambient sources such as TV, GSM, and WLAN base stations sounds very promising as these signals are omnipresent in urban environments and open the possibility to realize truly autonomous wireless sensors. Unfortunately, this is only feasible for a limited number of applications as the available power generally is very low and unreliable. In [9], an experiment was conducted to measure the received signal strength and probability in an urban and suburban area between 800 and 900 MHz. In an urban area, the received signal strength was most likely around -20 to -40 dBm with a peak probability of 31% around -33 dBm. In a less densely populated environment, this peak was around -37 dBm with a probability of 27%. Similar results have been found in [10], where the ambient RF energy available from a GSM-900 cell was found to be on the order of -30 dBm at 200 m distance. These power levels are generally considered to be too low to be used for RF harvesting as they are around or below the power-up threshold.

4. CHARGE PUMP RECTIFIER TOPOLOGIES

Rectifiers are very well known as power conversion devices that convert alternating current (AC) to direct current (DC). In the beginning of the twentieth century, rectifiers were usually applied to high-voltage and high-power applications. The same principal of rectifying current and voltage can be applied to RF energy harvesting, where rectifiers convert AC power, captured by an antenna, to DC power that is subsequently delivered to a load. Generally, the rectified voltage is also boosted to achieve higher levels. Thus, the rectifier also operates as a charge pump, and for this reason, the term charge pump rectifier (CPR) is commonly found in RFEH literature.

An example of a CPR is presented in Figure 7. This is one of the simplest CPR topologies, and comprises only diodes and capacitors. In the positive half-period of a sinusoidal voltage applied across the anode-cathode junction, the diode conducts current and thus transfers charge to the capacitor connected at the cathode terminal, which stores this charge, thereby producing or increasing its DC voltage. Since the circuit has two diodes, the voltage is rectified in both semi-cycles. The diode can be seen as a switch with a voltage source (V_D) in series. The equivalent circuits in both half-periods are represented in Figure 8 (a) and (b). The diodes have losses, represented by V_D, that limit the CPR output voltage and, consequently, the power conversion efficiency (PCE). The steady-state output voltage of this topology is described by (18), where V_{IN} is the input peak voltage:

$$V_{out} = 2V_{IN} - 2V_D \tag{18}$$

As can be noticed (18), V_D is an undesired factor and has to be minimized for better power conversion.

Various techniques and topologies that have been developed to reduce V_D, and increase PCE and the sensitivity of CPRs, will be briefly discussed here. A Schottky diode is a special type of diode built on a metal-semiconductor substrate which has low V_D (0.15 V − 0.4 V) [11,12]. One approach is the replacement of conventional diodes, in the circuit of Figure 7, by Schottky diodes. The circuit topology is not different from Figure 7; the significant change is in the reduction of diode voltage drop.

The advantage of diodes is that they conduct current in one direction only, thus, no flow-back current losses are associated with the use of diodes in case the CPR receives high voltage levels. The main drawback of Schottky diodes is their manufacturing cost. Therefore, these diodes are not always available in CMOS technologies because their use adds to the costs considerably.

Due to the costs and processing disadvantages of Schottky diodes, many research groups have been looking into solutions to reduce CPR losses using standard CMOS

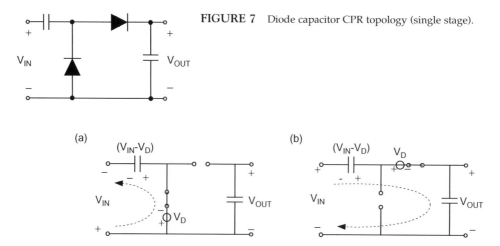

FIGURE 7   Diode capacitor CPR topology (single stage).

FIGURE 8   Equivalent circuit in negative (a) and positive (b) half-period.

processes. An integrated CMOS rectifier has a clear advantage over discrete Schottky diodes as MOS transistors can easily be scaled and have four terminals compared to two terminal diodes. This allows for a much more versatile design approach with many different topology and circuit techniques to design for the optimum rectifier input impedance.

The CPR in Figure 9(a) replaces diodes by MOS transistors connected as diodes. However, MOS transistors also present a voltage drop, which depends on the threshold voltage (V_{TH}) of the MOS transistor. For this reason, techniques to reduce the threshold voltage have been developed. A known technique is the floating gate threshold voltage compensation [13]. This technique consists of MOS transistors connected as diodes with a capacitor connected between the drain and gate terminals. Assuming the initial voltages on these capacitors are zero, before operation, the capacitors must be charged with a voltage close to V_{TH}. After that, the diode-connected transistors will have their conduction losses reduced since the capacitors work as series voltage sources, as shown in Figure 9(b), and thus drive the gates with higher voltages. The extra voltage allows the transistor channel to be inverted with lower gate-source voltages, thereby minimizing V_{TH} effects.

The advantages of this topology are good V_{TH} compensation and simple circuit implementation. On the other hand, the pre-charging of the capacitors needs to be accurate in order to avoid excessive extra potential, which can lead to negative turn-on voltages. As a consequence, losses due to flow-back current become very significant. Moreover, pre-charging requires a battery, thus, making this technique unattractive to truly autonomous solutions.

Other techniques for V_{TH} compensation are presented in Figure 10 (a)–(c)[14–16]. In Figure 10 (a) the compensation relies on the output voltages of the succeeding stages that are fed back to the compensated stage. This method does not require pre-charging of the gates, and the circuit implementation has moderate complexity. However, it requires a large amount of stages to provide enough voltage for compensation. In the technique presented in Figure 10(b) the threshold voltage of M_{N1} is reduced by a biasing circuit. M_{N2} conducts current that is limited by R1, and a gate-source voltage is produced across M_{N2}. The gate-source voltage of M_{N2} sets a lower threshold voltage to M_{N1}. For M_{P2} the threshold compensation is similar to M_{N1}. This technique also requires a large amount of stages and it needs extra components, such as resistors and capacitors, to build the compensation scheme. In both approaches, flow-back current may also increase losses at high input power. Figure 10(c) shows a technique that employs an auxiliary CPR chain for V_{TH}

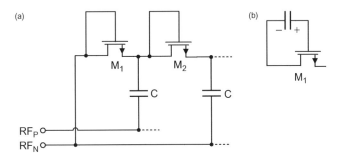

FIGURE 9　(a) Diode-connected MOS stages and (b) capacitor boosting the drain-gate voltage.

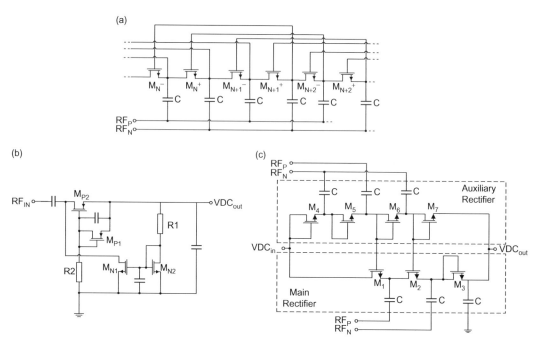

FIGURE 10   (a)–(c) static V_{TH} compensation.

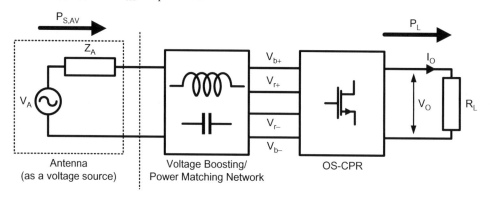

FIGURE 11   Block diagram of switching rectifier.

compensation. The compensation is very effective, but chip area is significantly increased and the received power is divided between the main and auxiliary chain.

Another technique found in the literature comprises dynamic V_{TH} compensation and voltage boosting to assist the MOS transistors switching on/off [17]. Figure 11 shows the principle diagram consisting of a boosting network and a charge pump rectifier. The CPR is connected to four branches provided by a boosting network. The signals Vb+ and Vb− are the switching signals, and Vr+ and Vr− are connected to the input of the rectifier. Figure 12 (a) and (b) present the details of the boosting network and one stage of the rectifier circuit. Note that the PMOS transistors in Figure 12(b) are set to operate as voltage-

(a)

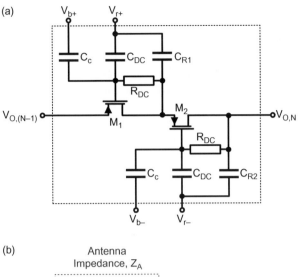

(b)

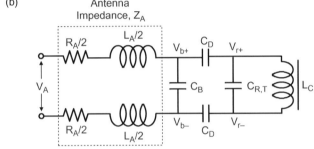

FIGURE 12 (a) Boost network schematic and (b) rectifier stage.

controlled switches to reduce voltage drops. In each stage, the output voltage is fed back to the transistor gate through R_{DC} and C_{DC}. Capacitor C_C couples the AC voltage from the boosting network to the gate of the PMOS. As a result, the PMOS gate voltage has two components, one DC component for threshold compensation, and one AC component to turn on/off the transistors in the positive/negative half-periods. The advantages of this topology are reduced number of stages, no pre-charging, and reduced flow-back losses. On the other hand, the circuit implementation is more complex and its efficiency depends on the boosting network quality factor. Considering that the other techniques require a matching network as well, the quality factor will also affect their efficiency. A comparison among the various approaches can be found in [17].

5. EFFECT OF LOAD AND SOURCE VARIATIONS

In the previous section, CPR topologies and techniques to improve efficiency/sensitivity were presented. This section discusses how CPR efficiency (PCE) and impedance (Z_{IN}) are affected by load variations and input power deviations.

Let us assume that a resistive load is connected to the output of the CPR. As the input power increases, the input voltage is expected to increase, thus the output voltage also increases. Since the load is a constant resistance, the output current also increases. The output current is reflected to the input as more charge needs to be transferred to the capacitors of the rectifier and, as a result, the input current increases. Despite an input voltage and current increase, it is hard to affirm whether the rectifier input impedance increases or decreases since the voltage-to-current ratio is highly nonlinear. Thus, Z_{IN} tendency should be determined by transient simulations or measurements in order to reach a more accurate estimation. Analytical analysis can also be performed. However, it is less accurate as mathematical reduction is usually applied to simplify the analysis.

Besides input power deviation, load deviation also affects input impedance and efficiency. The PCE and Z_{IN} as a function of load impedance are nonlinear and change with topology and V_{TH} compensation. In Figure 13 efficiency vs input power (P_{IN}) for a constant resistive load is depicted. Most of the topologies presented in this section have similar behavior to the plot of Figure 13. As P_{IN} increases, PCE increases until a maximum point, and drops as P_{IN} further increases. Changes in Z_{IN}, voltage drop losses, and flowback losses are the factors that cause the PCE to drop. In addition, the differences among technologies, topologies, and compensation schemes cause differences in peak efficiency, minimum PCE, and sensitivity.

5.1 Optimum Power Transfer Techniques

The concept of high-Q passive voltage boosting to improve sensitivity has the disadvantage that the antenna-rectifier interface becomes very sensitive to small impedance variations. These variations can be the result of process mismatch, variation in input power level, or environment changes caused by on-body antennas. This is extensively reported in various articles on printed flexible antennas [18–20]. Also, the rectifier nonlinearity presents a challenging task to the designer. A robust RF energy harvester therefore requires a

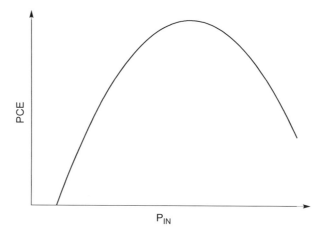

FIGURE 13   PCE variation vs. input power.

mechanism that compensates for these effects. A well-known compensation method used in various applications is maximum power point tracking (MPPT), as proposed in [21,22]. MPPT optimizes the DC-DC energy transfer *after* the rectification is performed. Another possible solution to this problem is to adjust the antenna-rectifier interface impedance using a resonance control loop as shown in Figure 14 [23,24].

The principle of the control loop is the following: after the harvester initially charges the storage capacitor, C_{STORE}, to the turn-on voltage of the loop, the slope information of V_{out} is obtained using a differentiating network. Subsequently, a sample and comparator stage compares the slope information of V_{out} with the previous sample and determines if the slope has increased or decreased. This information is fed to a finite-state machine that determines if the up-down counter should keep counting or change count direction. The output of the n-bit up-down counter is used to control an n-bit binary weighted capacitor bank at the interface. This way, the control loop continuously maximizes the slope of V_{out}, corresponding to maximum energy transfer with minimum charging time. Once the loop is calibrated, it can be turned off so that it is not loading the rectifier for very low power levels.

The core of the harvester consists of a conventional n-stage cross-connected bridge rectifier. In this structure, the output voltage and common-mode gate voltage generated during rectification provide additional biasing and effectively reduce the required turn-on voltage. Due to this V_{TH} self-cancellation, the rectifier can be activated at lower input power levels than other similar topologies. Another benefit is its symmetry, as this circuit cancels all even order harmonic currents that can be re-radiated, and therefore improves power efficiency. The required transistor width is determined by analyzing the charging time and input impedance for different input voltages when changing the transistor width.

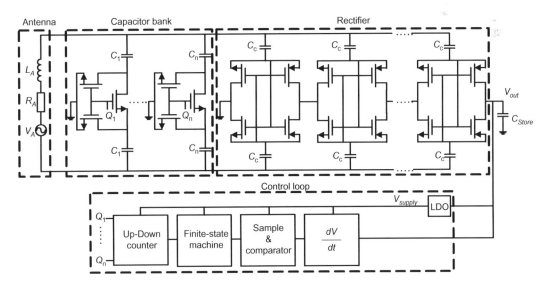

FIGURE 14 A multi-stage RF energy harvester circuit with resonance control loop.

Then, a trade-off is made between minimum charging time, maximum voltage boost, and output voltage (i.e., number of stages).

The capacitor bank consists of $(2^n - 1)$ unit capacitor switches, where each unit consists of a main switching transistor, two small biasing transistors to enhance the Q-factor, and two custom-designed metal-metal capacitors. The desired tuning range and accuracy can be set with the number of bits. A low-dropout regulator (LDO) offers a stable supply voltage for the control loop. The required bandwidth of the loop is on the order of kHz as the charging time is relatively slow. Because of this, the average power consumption of the mainly digital loop can be designed with negligible power consumption.

The resonance control loop optimization is verified with simulations and shown in Figure 15(a) for $P_{av} = -20$ dBm using a relatively small load capacitance of 10 pF in order to reduce simulation time. The binary "CountDirection" is the output signal of the finite-state machine that controls the up-down count direction. The control loop keeps counting up or down (depending on the initial direction) as long as the slope of V_{out} keeps increasing. The slope decreases when the capacitor bank has passed the optimum capacitance. In this case, "CountDirection" turns from "1" to "0" and the loop inverts the counting direction. At the optimum capacitance code, "CountDirection" oscillates between "1" and "0" at the clock frequency.

This scheme ensures that small impedance variations, which may occur in a realistic environment, will be compensated for by the control loop, making the RF energy harvester very robust while it benefits from the passive voltage boost obtained from the high-Q antenna. This is demonstrated in Figure 15(b), where the antenna inductance is varied between 80 nH $\leq L_A \leq$ 160 nH to mimic antenna environment changes. The control loop is able to cope with these large variations despite the high-Q antenna network and maximizes V_{out} in each scenario.

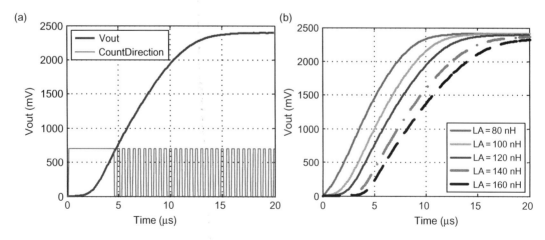

FIGURE 15   Control loop optimization: (a) simulated V_{OUT} and binary count direction signal and (b) V_{OUT} for variations in L_A.

6. ANTENNA-RECTIFIER CO-DESIGN

Co-designing the antenna with the rectifier allows optimization of the performance as the interface is no longer constrained by the traditional 50 Ω characteristic. The design methodology described in this section requires a highly inductive antenna that is conjugate matched to the rectifier input impedance. This antenna impedance can be realized using a modified folded dipole antenna as depicted in Figure 16. This compact loop antenna has additional short-circuited arms in order to fine-tune the antenna impedance.

A prototype antenna is designed and fabricated on 1.6 mm FR4 substrate. The antenna impedance is simulated in CST Microwave Studio using frequency, transient, and transmission line solvers, and all converge to approximately $11 + j398.8$ Ω at 868 MHz, thereby ensuring a high degree of confidence in the simulation procedure and accuracy.

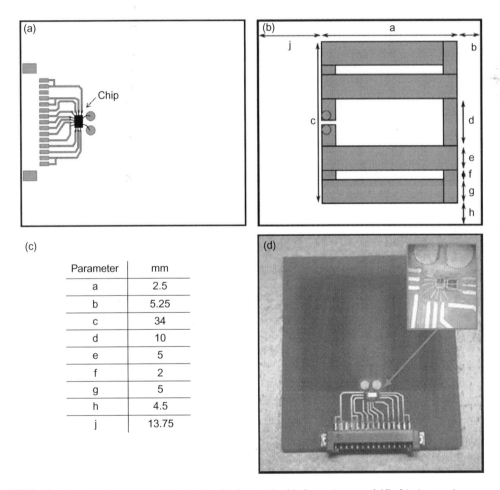

Parameter	mm
a	2.5
b	5.25
c	34
d	10
e	5
f	2
g	5
h	4.5
j	13.75

FIGURE 16   Fabricated antenna: (a) back side, (b) front side, (c) dimensions, and (d) chip integration.

The radiation efficiency and maximum directivity of this prototype antenna is 46.5% and 1.55 dBi, respectively. This corresponds to a maximum antenna gain of −1.78 dBi.

The rectifier chip is integrated on the back side of the antenna to minimize its effect on the antenna performance. The RF inputs are bond wired to vias (2 mm in diameter) that connect to the antenna feed point. The other control signals are connected to a control logic and measurement board to evaluate the performance. All off-chip connections (vias, bondwires, PCB traces, connector) are included in the antenna simulations to accurately determine the input impedance at the antenna feedpoint (Figure 17).

6.1 Measurements and Verification

At the end of the design and fabrication phase, the performance needs to be evaluated by verifiable measurements. This way, the performance can be compared properly to the state of the art. For this reason, the RF harvester first is measured in an anechoic chamber to mimic a free-space condition. The setup (Figure 18) is calibrated at 868 MHz using two

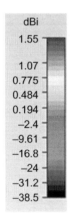
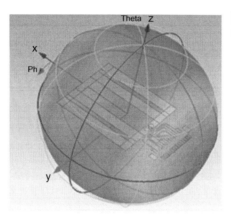

FIGURE 17   Antenna radiation pattern and directivity.

FIGURE 18   Measurement setup in anechoic chamber.

identical broadband log periodic antennas (HG824−11LP-NF) separated by 3.6 meters to ensure far-field conditions. Since the harvesting antenna dimensions ($a = \lambda/11.3$ and $c = \lambda/9.8$) are much smaller than the wavelength, the antenna performance is included in the measurements by defining the available input power as the maximum power available from an isotropic antenna ($G_A = 0$ dBi). Using this definition, the RF-DC power conversion efficiency is defined as

$$PCE = \frac{P_{load}}{P_{av,iso}} = \frac{V_{out}^2}{R_{load}P_{av,iso}} \tag{19}$$

This isotropic available power is determined by measuring the received power using the reference antenna at the harvester position and subsequently adding the reference antenna gain to obtain $P_{in} = P_{av,iso}$. The measured power is within ± 0.5 dB agreement with the theoretical available power calculated from (4). As the control loop is off-chip, its power consumption during calibration is not included in the measurements.

To demonstrate the adaptability of the resonance control loop, the measured output voltage for closed and open-loop scenarios is shown in Figure 19(a), where Rload = 1 MΩ and Pin = − 20 dBm. During a capacitance sweep (open loop), the control loop is continuously counting from the minimum to the maximum capacitance value. Clearly, an optimum capacitance exists that corresponds to maximum power. In the closed-loop scenario, the loop optimization is activated at 0.5 seconds and maximizes the power dissipation in the load. The charging time vs. P_{IN} for a 450 nF load capacitance is shown in Figure 19(b).

The measured V_{out} vs. input power is depicted in Figure 20(a) and shows an excellent sensitivity for low power levels. A capacitor can be charged to 1 V with only −26.3 dBm input power and takes approximately 2 seconds for a 450 nF capacitor. As the charging time scales linearly with the capacitor size, these graphs give a good indication of the available energy as a function of input power and time.

Figure 20 (b) shows that the power efficiency peaks around −15 dBm with a maximum of 31.5%. This includes all losses of the antenna, interface, and rectifier.

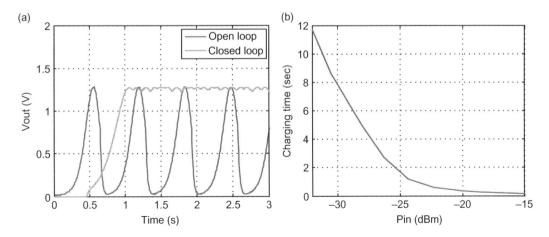

FIGURE 19 (a) V_{out} for capacitance sweep and optimizing control loop and (b) charging time vs. P_{IN}.

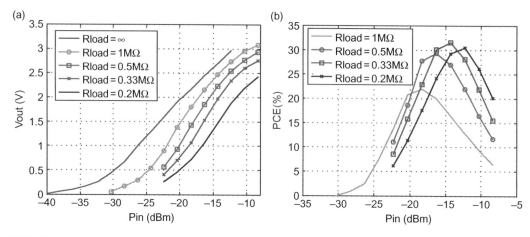

FIGURE 20   (a) V_{out} vs. P_{IN} and (b) PCE vs. P_{IN}.

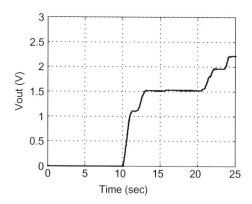

FIGURE 21   Measured ambient RF energy harvesting from a GSM-900 mobile phone at 2 meters' distance.

To verify the performance in a realistic environment, the RF harvester is tested for both a dedicated and ambient RF source. First, a dedicated 1.78 W EIRP RF source was used to measure the line-of-sight distance in an office corridor. In this experiment, 1 V could be generated from a 25-meter distance, corresponding well with −26.3 dBm sensitivity. The energy harvesting from ambient RF sources is demonstrated by measuring the output voltage when using a GSM-900 mobile phone from a 2-meter distance (Figure 21). Although the frequency and power levels varied greatly during a call (peak power levels of −4.6 dBm were measured between 886 MHz and 907 MHz), the RF harvester was able to charge a capacitor to 2.2 V. Both experiments demonstrate the feasibility of RF energy harvesting for a wide range of applications.

7. CONCLUSION

RF energy harvesting is an increasingly popular research topic that has gained a lot of attention since the introduction of RFID, autonomous wireless sensors, and wearable devices. RF energy harvesting fundamentals such as wave propagation, antenna efficiency, and antenna-rectifier interface have been presented. It has been shown that the choice of antenna-rectifier interface impedance plays a crucial role in the optimization of highly sensitive RF energy harvesters.

Practical limitations, which include impedance mismatch, losses, government regulations, and harvested-power distribution in realistic environments, affect RF energy-harvesting performance and must be taken into account in the design during the specification phase. Rectifier topologies and techniques for reducing losses have been presented. One simple scheme is a rectifier consisting of Schottky diodes and capacitors. Despite great reduction of power losses, Schottky diodes are expensive and are not available in most IC technologies. Therefore, CMOS-based solutions have been developed by many research groups to align low process cost with good performance.

Besides rectifier topologies, optimum power transfer techniques were introduced to increase sensitivity and efficiently transfer energy from antenna to rectifier. An example of an RFEH employing antenna-rectifier co-design has been shown and measurement results have been presented. A peak efficiency of 31.5% has been achieved at −15 dBm, showing state-of-the-art performance of the RF energy harvesting by employing a resonance control loop.

The fundamental RF energy harvesting properties, challenges, techniques, and design procedures covered in this section give the designer the essential knowledge to optimize and realize high-performance RF energy harvesters for wearable device applications.

Acknowledgement

The authors gratefully acknowledge the technical and financial support provided by IMEC-NL/Holst centre at Eindhoven, the Netherlands and Conselho Nacional de Desenvolvimento Científico e Tecnológico (CNPq), Brazil.

References

[1] J. Yoo, Y. Long, S. Lee, K. Yongsang, Y. Hoi-Jun, A 5.2 mW Self-configured wearable body sensor network controller and a 12 u W wirelessly powered sensor for a continuous health monitoring system, IEEE J.Solid State Circuits 45 (1) (2010) 178−188.
[2] Z. Yanqing, Z. Fan, Y. Shakhsheer, J.D. Silver, A. Klinefelter, M. Nagaraju, et al., A Batteryless 19 uW MICS ISM-band energy harvesting body sensor node SoC for ExG applications, IEEE J.Solid State Circuits 48 (1) (2013) 199−213.
[3] S. Lee, Y. Long, R. Taehwan, H. Sunjoo, Y. Hoi-Jun, A 75W Real-time scalable body area network controller and a 25W ExG sensor IC for compact sleep monitoring applications, IEEE J.Solid State Circuits 47 (1) (2012) 323−334.
[4] C.A. Balanis, Antenna theory, Analysis and Design, 2nd ed., Wiley, New York, NY, USA, 1997.
[5] S.S. Singer, A pure realization of loss-free resistor, IEEE Trans. Circuits Syst.-I 51 (8) (2004).
[6] H.J. Visser, R.J.M. Vullers, RF energy harvesting and transport for wireless sensor network applications: principles and requirements, Proc. IEEE 101 (6) (2013) 1410−1423.

[7] ERC Recommendation 70-03 Relating to the Use of Short Range Devices (SRD), (2009). ERC/REC 70-03, European Radiocommunications Committee.

[8] Federal Communications Commission (2009), Code of Federal Regulations, Title 47, Part 15.

[9] Salter, T.S., (2009). Low Power Smartdust Receiver with novel applications and improvements of an RF Power Harvesting Circuit. PhD Thesis, University of Maryland, Maryland, Washington, D.C. USA.

[10] H.J. Visser, A.C.F. Reniers, J.A.C. Theeuwes, Ambient RF energy scavenging: GSM and WLAN power density measurements, Proc. Eur. Microw. Conf. (2008) 721–724. Amsterdam, the Netherlands.

[11] http://powerelectronics.com/site-files/powerelectronics.com/files/archive/powerelectronics.com/images/SchottkyDiodes.pdf, (Last Accessed: 27.06.14).

[12] F. Kocer, M.P. Flynn, A new transponder architecture with on-chip ADC for long-range telemetry applications, IEEE J. Solid State Circuits 41 (5) (2006) 1142–1148.

[13] T. Le, K. Mayaram, T. Fiez, Efficient far-field radio frequency energy harvesting for passively powered sensor networks, IEEE J. Solid State Circuits 43 (5) (2008) 1287–1302.

[14] G. Papotto, F. Carrara, G. Palmisano, A 90-nm CMOS threshold compensated RF energy harvester, IEEE J. Solid State Circuits 46 (9) (2011) 1985–1997.

[15] H. Nakamoto, et al., A passive UHF RF identification CMOS tag IC using ferroelectric RAM in 0.35- technology, IEEE J. Solid State Circuits 42 (1) (2007) 101–110.

[16] H. Lin, K.H. Chang, S.C. Wong, Novel high positive and negative pumping circuits for low supply voltage, Proc. IEEE Int. Symp. Circuits Syst. (ISCAS) 1 (1999) 238–241.

[17] A. Mansano, S. Bagga, W. Serdijn, A high efficiency orthogonally switching passive charge pump rectifier for energy harvesters, IEEE Trans. Circuits Syst. I 60 (7) (2013) 1959–1966.

[18] D'errico R., Rosini R, Delaveaud C., A et al (2011). Final report on the antenna-human body interactions, around-the-body propagation, www.WiserBAN.eu, (Last Accessed: 27.06.14).

[19] D. Masotti, A. Costanzo, S. Adami, Design and realization of a wearable multi-frequency RF energy harvesting system, Proc. 5th Eur. Conf. Antennas Propag. (2011) 517–520.

[20] K. Koski, E. Marodi, A. Vena, et al., Characterization of Electro-textiles using wireless reflectometry for optimization of wearable UHF RFID tags, Proc. Prog. Electromagnet. Res. Symp. (2013) 1188–1192.

[21] A. Dolgov, R. Zane, Z. Popovic, Power management system for online low power RF energy harvesting optimization, IEEE Trans. Circuits Syst. I: Reg. Pap. (2010) 1802–1811.

[22] P.-H. Hsieh, T. Chiang, An RF energy harvester with 35.7%, Symp. VLSI Circuits (VLSIC)   (2013) C224–C225.

[23] M. Stoopman, W.A. Serdijn, K. Philips, A robust and large range optimally mismatched RF energy harvester with resonance control loop, IEEE Int. Symp. Circuits Syst. (ISCAS) (2012) 476–479.

[24] M. Stoopman, S. Keyrouz, H.J. Visser, K. Philips, W.A. Serdijn, A self-calibrating RF energy harvester generating 1V at −26.3 dBm, Symp. VLSI Circuits (2013).

Low-Power Integrated Circuit Design for Wearable Biopotential Sensing

Sohmyung Ha[1], Chul Kim[1], Yu M. Chi[2] and Gert Cauwenberghs[1]

[1]Department of Bioengineering at University of California, San Diego, La Jolla, CA, USA,
[2]Cognionics, Inc., San Diego, CA, USA

1. INTRODUCTION

Wearable sensor systems require long-term power autonomy with extremely limited power. Circuit techniques dictate the power consumption of the whole system. However, achieving the lowest possible power consumption must not sacrifice system performance. Systematic approaches with holistic considerations are necessary to realize intelligent low-power wearable sensors.

A solid understanding of the biopotential signals and the electrode-body interface is of primary importance for low-power, high-performance sensor design. Therefore, the first two sections review the basics of biopotentials and electrode-body interfaces.

The following considerations and requirements are essential to accomplish low-power circuits for wearable sensors:

- Low input-referred noise ($1-5\,\mu V_{rms}$)
- Low power consumption ($<100\,\mu W$) optimizing noise energy efficiency (NEF)
- Application-dependent signal gain ($1-1,000$); variable gain in some applications
- Medium to high signal dynamic range ($40-80$ dB)
- Input impedance much higher than the electrode-body interface (>10 GΩ)
- Low to medium frequency range (0.1 Hz-10 kHz); configurable bandwidth in some applications
- DC blocking; or DC coupling not to be affected by electrode offset voltages
- Motion artifact rejection
- High common-mode rejection ratio (CMRR >80 dB)

- High power supply rejection ratio (PSRR > 80 dB)
- Proper grounding
- Rejection of mains interference
- Small silicon area (<10 mm^2)
- No or few off-chip components

All these issues are inter-correlated in various ways. A deep understanding of the trade-offs and the challenges are required. The basic trade-offs and architectural design topologies are explained in the following sections. More specific design reviews for low-power amplifier and ADC are covered next. Lastly, more practical considerations for sensor designs are presented.

2. BIOPOTENTIAL SIGNALS AND THEIR CHARACTERISTICS

Biopotentials, such as EEG, ECG, EMG, etc., are generated from volume conduction of currents made by collections of electrogenic cells. EEG is the electrical potential induced from collective activities of a large number of neurons in the brain. ECG results from action potentials of cardiac muscle cells, and EMG from contractions of skeletal muscle cells. Various other biopotentials (EOG, ERG, EGG, etc.) also result from collective effects of large numbers of electrogenic cells or ionic distribution.

Almost all biopotentials, including EEG, ECG, and EMG, of which characteristics are shown in Figure 1, range over very low frequency, typically less than 1 kHz. They are very low in amplitude, ranging tens to hundreds of μV when measured by a surface electrode. Since EEG and ECG range down to less than 1 Hz, recording of these signals faces challenges in electrode offset voltage, which may reach up to 100 mV, varying slowly over

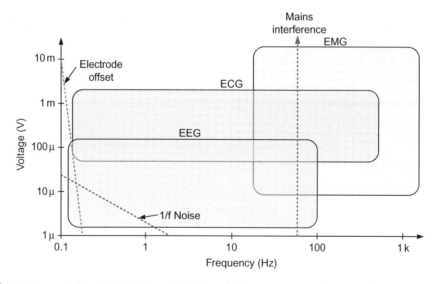

FIGURE 1   Characteristics of EEG, ECG, and EMG in relation to mains interference and noise [1–3].

time. In addition, flicker noise of electronic circuitry, which is described further in Section 3.2, falls within the EEG signal range. Also, common-mode interference from the mains and irrelevant biopotentials should be sufficiently rejected.

3. ELECTRODE-BODY INTERFACE AND ELECTRODE NOISE

3.1 Electrode-Body Interface

The coupling of biopotential signals from the body into the front-end amplifier is accomplished through electrodes. At a fundamental level, the electrode interfaces ionic currents in the body with electrical currents in the electronic instrumentation. In practice, because the electrode comprises the first stage of the signal chain, its properties can dominate the overall noise and performance of the acquisition system, making its design and selection crucially important.

Broadly speaking, there exist three classes of biopotential electrodes in the literature: wet, dry, and non-contact (Figure 2) [3,4]. All types of electrodes ideally measure the exact same biopotential signals and are largely differentiated by the presence of a gel and the resulting contact impedance to the body.

Wet electrodes are the most common type and considered the "gold standard" for both clinical and research applications. A typical wet electrode consists of a silver-silver chloride (Ag/AgCl) metal that is surrounded by a wet or solid hydrogel, containing chloride. Other kinds of metals can be used (gold is common for EEG) if the DC stability of the Ag/AgCl electrode is not necessary. The primary drawbacks with wet electrodes are its longevity and comfort. Wet electrodes degrade as the moisture content evaporates limiting its useful

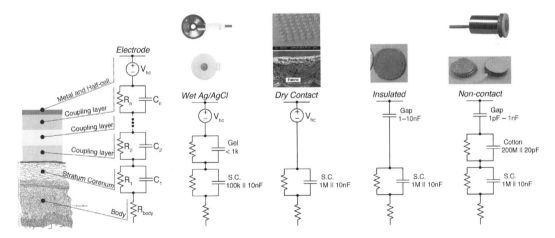

FIGURE 2 Electrical coupling of the skin-electrode interface for various electrode topologies, including wet-contact gel-based Ag/AgCl, dry-contact MEMS and metal plate, thin-film insulated metal plate, and non-contact metal plate coupling through hair or clothing such as cotton. The insets show examples of practical electrodes for each category [4].

lifetime to, at most, a few days. Many users also report skin irritation and discomfort from the gels and adhesives that contact the skin.

Dry electrodes operate without the use of an explicit wet/gel coupling media. The metal in the electrode directly contacts the skin to couple biopotential signals. In practice, however, virtually all dry electrodes still rely on some degree of moisture which is gathered from the environment or emitted from the body (e.g., sweat). Compared to the wet electrodes, the performance of a dry electrode usually increases over time as more moisture permeates the skin-electrode interface, resulting in increased coupling. On bare skin, dry electrodes normally exhibit higher contact impedances than wet electrodes − on the order of hundreds of kΩ versus tens of kΩ [5]. However, with modern high input impedance amplifiers, this is rarely an issue. As with wet electrodes, Ag/AgCl contact materials tend to show the best performance, especially in terms of drift noise, which is important for diagnostic ECG applications.

There are also dry contact electrodes with capacitive coupling between the electrode and the body instead of coupling via galvanic conduction. Capacitive contact electrodes utilize a thin dielectric layer to form an insulated contact to the body. Compared to standard dry contact electrodes, capacitive electrodes offer a galvanically isolated, chemically inert surface, maximizing user safety and electrode longevity. However, the capacitive interface precludes the measurement of true DC potentials and may result in long settling times depending on the bias resistor and the amount of coupling to the body. In other respects, the capacitive contact electrodes operate similar to dry contact electrodes.

The final type of electrodes, non-contact, can be thought of as a special case of dry electrodes. They operate not only without gel, but also through an insulation layer such as clothing, enabling signal acquisition without direct skin contact. As expected, the coupling impedance can be very high, on the order of tens of pF in parallel with hundreds of MΩ. Obtaining acceptable signals requires the use of special, very high input impedance active electrodes. Because there is no direct skin contact, movement artifacts are a major, unsolved issue, especially for ambulatory use. Non-contact electrodes are also highly sensitive to environmental conditions such as humidity and the exact insulating material. Non-contact electrodes tend to work well on natural fabrics (e.g., cotton) under high humidity where the fabric actually becomes slightly conductive, offering a galvanic path to the skin. In contrast, it is difficult to obtain acceptable signals on high insulating synthetic fabrics (e.g., polyester) due to triboelectric artifacts.

3.2 Electrode Noise

In addition to the circuit noise from the amplifier components, electrodes can be a significant noise contributor in the signal chain [4,6]. Unlike circuit noise, however, comprehensive models for electrode noise do not exist, in part because the mechanisms for electrode noise are not well understood. In general, electrode noise is strongly correlated with the contact impedance, but the actual level is significantly higher than just the thermal noise from the resistive portion of the impedance.

The aggregate sum of the electrode noise sources can be quite large, on the order of $\mu V/\sqrt{Hz}$ at 1 Hz, even for wet electrodes. This far exceeds the noise contribution of circuit components, illustrating the importance of proper electrode selection. Due to integrated current noise, both wet and dry electrodes have sharp $1/f^2$ spectra, which show up as baseline drifts in the time domain.

Non-contact electrodes can pick up additional noise from the insulating material between the metal and skin. As an example, acquiring signals through fabrics can be noisy due to the intrinsic high resistance of the fabric (>100 MΩ). This amounts to the equivalent of inserting a large resistor in series with the amplifier input and can add significant noise in the signal bandwidth.

4. LOW-POWER ANALOG CIRCUIT DESIGN TECHNIQUES FOR BIOPOTENTIAL SENSORS

Power consumption of a circuit can be estimated by accumulating the products of supply voltage V_{DD} for each block with its current. All strategies for low power design are simple combinations of reducing supply voltage and current. However, there are several considerations and limitations in pursuit of low power design without compromising the performance of the system. Trade-offs between noise, bandwidth, dynamic range, and several other should be considered, and a deep understanding of basic principles is necessary. Fundamental trade-offs and essential low-power analog design techniques for biopotential sensors are discussed in this section.

4.1 Subthreshold Weak Inversion Operation of MOS Transistors

Figure 3(a) illustrates a cross-section of fabricated NMOS and PMOS field-effect transistors in an integrated circuit CMOS process. Both have four terminals: gate (G), source (S), drain (D), and body (B). Their schematic drawing symbols are depicted in Figure 3(b). MOSFET operation can be separated into two modes according to the voltages between gate and source terminals: strong inversion and weak inversion (or subthreshold) operation, as shown in Figure 3 (c).

Counter to standard practices in analog CMOS circuit design, the weak inversion (subthreshold) region of CMOS operation has proven a favorable regime for low-power biomedical circuit design. In conventional design and particularly for high-speed applications, weak inversion operation has been considered as non-ideality in the cut-off region and its current has been labeled as leakage current. Recently, weak inversion has become increasingly important because its low power and low bandwidth characteristics are well suited for biomedical and other low-power sensor applications, owing to superior transconductance efficiency. Furthermore, transistors in deep submicron technology operating in weak inversion do not suffer from many process-dependent problems plaguing the above-threshold strong inversion region, such as gain-limiting effects of velocity saturation in electron and hole mobility [7].

Transistor model equations in weak inversion are simpler, are more transparent, and scale over a wider range than in strong inversion. The electron energy of a transistor in weak inversion is based entirely on the Boltzmann distribution, independent of process technology. The drain current through the transistor channel flows not by drift, but by

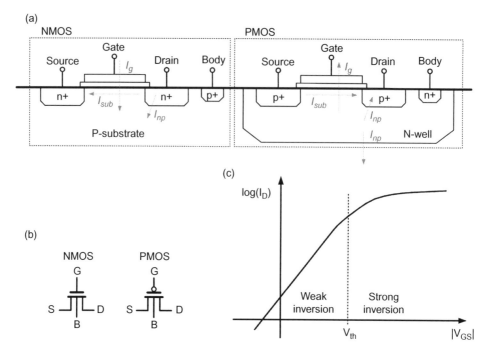

FIGURE 3 (a) Cross-section of NMOS and PMOS FETs fabricated in a CMOS process, (b) their schematic drawing symbols, and (c) logarithm of drain current I_D as a function of gate-source voltage with body tied to source, and drain voltage biased in the saturation region.

diffusion, and changes exponentially with gate voltage. Thus, weak inversion operation is particularly well suited for implementing translinear circuits and log-domain filters.

Drain current I_{DS}, transconductance g_m, and unity-gain frequency f_t in weak inversion are as follows [8]:

$$i_{DS} = i_{DS0} \frac{W}{L} e^{v_{GS}/(nV_t)}(1 - e^{-v_{DS}/V_t}) \tag{1}$$

$$g_m = \frac{I_{DS}}{nV_t} \tag{2}$$

$$f_t = \frac{I_{DS}}{2\pi nV_t(C_{gs} + C_{gd} + C_{gb})} \propto I_{DS} \tag{3}$$

Because the transconductance is linearly proportional to drain current, so is unit-gain frequency. Thus, the trade-off between current and bandwidth is very straightforward: the larger the current, the wider the bandwidth.

Thermal noise in saturation and weak inversion is proportional to drain current as follows [9]:

$$\overline{i_{n,th}^2} = 2q\overline{I_{DS}}\Delta f \tag{4}$$

where Δf is the signal bandwidth. The relative noise (inverse of the signal-to-noise ratio) is inversely proportional to drain current:

$$\frac{\overline{i_{n,th}^2}}{I_{DS}^2} = 2q\frac{\Delta f}{I_{DS}} \tag{5}$$

Therefore, signal-to-noise ratio is linearly proportional to bias current I_{DS} in weak inversion. For a majority of biomedical applications with narrow signal bandwidth, the lower currents of circuits in weak inversion still offer adequately large signal-to-noise ratio at maximum energy efficiency.

Flicker noise, also known as $1/f$ noise or pink noise, is also a significant noise source at low frequency. Random captures of carriers in traps near the Si/SiO_2 interface and some other mechanisms are known to be a main source of $1/f$ noise [10,11], which is given by

$$\overline{i_{n,f}^2} = \frac{g_m^2 K}{C_{ox} WL} \cdot \frac{1}{f}\Delta f \tag{6}$$

where K is a process-dependent constant, W and L are width and length of the MOS transistor, and C_{ox} is the gate oxide capacitance. PMOS transistors are known to have less $1/f$ noise than NMOS transistors, and therefore should be used in the input differential pair of a front-end amplifier for low-noise low-frequency applications in biosensing. Enlarging the MOS device size also decreases $1/f$ noise inversely proportional to area.

The $1/f$ noise corner frequency f_c serves as an important indication for the proportion of $1/f$ noise relative to thermal noise over the spectrum. The $1/f$ noise corner f_c is defined as the frequency at which $1/f$ noise and thermal noise are at equal magnitude, which in weak inversion is given by

$$\overline{i_{n,th}^2} = \overline{i_{n,f}^2}$$
$$2nkTg_m\Delta f = \frac{g_m^2 K}{C_{ox} WL} \cdot \frac{1}{f_c}\Delta f \tag{7}$$
$$f_c = \frac{K}{C_{ox} WL}g_m\frac{1}{2nkT}$$

The $1/f$ noise corner can vary from a few 100 Hz to a few MHz depending on quality of process fabrication. Also, it depends on bias current: the lower the bias current, the lower the $1/f$ noise corner and hence the smaller the relative $1/f$ contribution to the overall noise.

For low-noise biomedical applications such as EEG acquisition, chopper stabilization techniques are widely used to reduce $1/f$ noise further. To be effective, the chopping frequency needs to be significantly higher than the $1/f$ noise corner frequency, typically ranging between 100 Hz and 10 kHz in these applications. Other techniques such as auto-zeroing and correlated double sampling can be used to reduce $1/f$ noise as well.

Three types of currents in MOS transistors need careful consideration in biomedical circuit design, below. These three are depicted in Figure 3(a) as subthreshold current I_{sub}, gate leakage I_g, and pn-junction reverse-bias leakage I_{np} [12].

TABLE 1   Measured Performances of State-of-the-Art Instrumentation Amplifiers

	Harrison 2003 [13]	Yazicioglu 2008 [14]	Zou 2009 [15]	Verma 2010 [16]
Application	Neural	EEG	Multimodal	EEG
VDD [V]	5	3	1	1
Power [μW]	80	6.9	0.337	3.5
Bandwidth [kHz]	7.2	N/A	0.292	N/A
Low-frequency Cut-off [Hz]	0.025	N/A	0.005–3.6	N/A
Input-Referred Noise [μVrms]	1.6	0.59	2.5	1.3
	(0.025 Hz– 7.2 kHz)	(0.5–100 Hz)	(0.05–460 Hz)	(0.5–100 Hz)
NEF	4.0	4.3	3.26	N/A
CMRR [dB]	$\geq$ 83	> 120	$\geq$ 71.2	> 60
PSRR [dB]	$\geq$ 85	N/A	$\geq$ 84	N/A
Input Impedance [MΩ]	N/A	> 1000	N/A	>700
	($\sim$ 20 pF)		($\sim$ 8 pF)	
Area [mm^2]	0.16	0.45	< 0.5	0.3

4.2 Requirements for Instrumentation Amplifiers

One of the most challenging parts in the design of wearable physiological monitoring systems is the implementation of instrumentation amplifiers (IAs), which acquire biopotentials from electrodes and perform analog signal processing and conditioning.

IAs are subject to almost all the challenging design specifications as discussed in the introductory section. As a point of reference, measured performances of some state-of-the-art IAs are shown in Table 1.

4.3 Basic Instrumentation Amplifier

A classic three-opamp IA is adequate for achieving large input impedance, large CMRR, and sufficient gain. However, it consumes large power and area since it uses three amplifiers [17]. For a micropower biopotential acquisition front-end, the configurations shown in Figure 4(a) and (b) are widely used [13]. The AC-coupling input capacitors C_C block electrode offset voltages. Owing to favorable matching performance of capacitors in integrated C.

In MOS processes, the gain can be precisely controlled. A large resistor R_f, typically substituted by a pseudoresistor or a switched-capacitor circuit (Figure 10), establishes DC biasing of the voltage at the input nodes of the amplifier and performs highpass filtering together with C_f. Mismatch in capacitor values results in degradation of CMRR. A practical

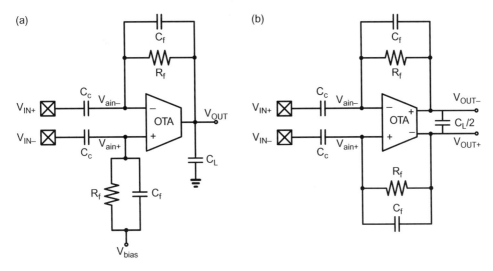

FIGURE 4 A general architecture of (a) single-ended output and (b) fully differential instrumentation amplifier [13].

CMRR that these architectures can achieve is about 60 to 70 dB. In addition, C_c dominates the input impedance. Therefore, the value of C_C needs to be set by considering CMRR and the input impedance.

The transfer function of the IA is as follows [13]:

$$A_V(s) = \frac{C_C}{C_f} \cdot \frac{1 - \frac{sC_f}{g_m}}{\left(1 + \frac{1}{sR_fC_f}\right) \cdot \left(1 + \frac{s(C_C + C_L + C_CC_L/C_f)}{g_m}\right)} \tag{8}$$

where g_m is transconductance of the OTA. Its passband gain is determined by the ratio of capacitors C_c to C_f. The high-pass cut-off frequency, f_{HP}, is given by

$$f_{HP} = \frac{1}{2\pi R_f C_f} \tag{9}$$

The lowpass cut-off frequency, f_{LP}, is controlled by the load capacitor C_L and is approximately given by

$$f_{LP} \cong \frac{g_m C_f}{2\pi C_C C_L} \tag{10}$$

when $C_L \gg C_f$ and $C_C/C_f \gg 1$. The right-half-plane zero at $g_m/2\pi C_f$ can be canceled by inserting a $1/g_m$ resistor in series with C_f. However, it can be ignored in many cases because it is located at much higher frequency than the frequency range of interest for biomedical applications.

The main noise contributors of the single-stage IA are the OTA and the feedback resistor R_f. The input-referred noise due to these sources is

$$\sqrt{\overline{v_{ni,R_f}^2}} = \sqrt{\frac{4kT}{R_f} \cdot \frac{1}{2\pi f C_{in}}}$$

$$\sqrt{\overline{v_{ni,A_{amp}}^2}} = \sqrt{\overline{v_{ni,sys}^2}} \cdot \left(\frac{C_f + C_{in} + C_C}{C_C}\right) \tag{11}$$

where C_{in} refers to parasitic capacitance at the OTA input nodes V_{IN1} and V_{IN2} [17]. The noise from R_f can be dominant at low frequencies of interest. Thus, the following criteria needs to be met in order to reduce the contribution of the noise from R_f [18]:

$$C_C \ll \frac{2}{3} \frac{f_{LP}}{f_{HP}} \text{ where } f_{LP} \approx \frac{g_{m1} C_f}{2\pi C_C C_L} \text{ and } f_{HP} = \frac{1}{2\pi R_f C_f} \tag{12}$$

In practical circuits, the noise from the OTA is typically dominant over the noise from R_f [18].

As a benchmark in the design of front-end IAs for low noise and low supply current, the noise efficiency factor (NEF) is used to compare the current-noise performance:

$$NEF = V_{rms,in} \sqrt{\frac{2I_{tot}}{\pi V_t \cdot 4kT \cdot BW}} \tag{13}$$

where $V_{rms,in}$ is the total input-referred noise, I_{tot} is the total current drain in the system, V_t is the thermal voltage, and BW is the -3-dB bandwidth of the system [19]. NEF corresponds to the normalized supply current relative to that of a single BJT with ideal current load for the same noise level, defining the theoretical limit (NEF = 1). In practice, differential IAs with input differential pairs incur twice the supply current for the same transconductance, with NEF values greater than 2. The state-of-the-art IAs typically have NEF of 2.5 to 10. It should be noted that NEF is a trade-off between bandwidth, noise, and current — not power.

4.4 Amplifier Design Techniques and Considerations

The operational transconductance amplifier (OTA) is the most important block in an IA, as shown centrally in Figure 4 (a) (single-ended) and (b) (fully differential). Among many kinds of amplifiers, the most popular two amplifiers, symmetrical OTA (or current mirror OTA) and folded cascode OTA, are shown in Figure 5 [13,16]. Both are one-stage amplifiers since both have only one high-impedance node at V_{OUT}. The gain of both amplifiers is the product of the transconductance of the input pair and the output impedance. Fully differential amplifiers (not shown) add complexity in the design with the need for common-mode feedback, but offer superior CMRR and PSRR over single-ended solutions.

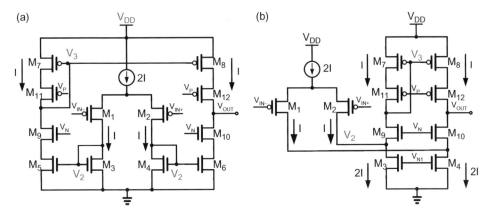

FIGURE 5   (a) Symmetrical OTA and (b) folded cascode OTA.

4.4.1 Noise and Power Perspective

Noise of the front-end amplifier for biomedical applications is critical. With given transconductances g_{m1} for M_1-M_2, g_{m3} for M_3-M_6 and g_{m7} for M_7-M_8, the input-referred thermal noise for the each amplifier in Figure 5 is as follows [13]:

$$\overline{v_{ni,sym}^2} = \frac{16kT}{3g_{m1}}\left(\frac{3}{4}n + 2\frac{g_{m3}}{g_{m1}} + \frac{g_{m7}}{g_{m1}}\right)\Delta f \tag{14}$$

$$\overline{v_{ni,folded}^2} = \frac{16kT}{3g_{m1}}\left(\frac{3}{4}n + \frac{g_{m3}}{g_{m1}} + \frac{g_{m7}}{g_{m1}}\right)\Delta f \tag{15}$$

where n is dependent on the capacitance ratio between depletion and oxide capacitance, which is generally between 1.2 and 1.8. In these equations, M_1 and M_2 are assumed in the weak inversion region while the others are in the strong inversion region. The effect from cascode MOSFETs are negligible in terms of noise. These equations clearly show that in order to obtain good noise performance, g_{m1} should be as large as possible. With a given bias current, the input pair transistors M_1-M_2 should operate in the weak inversion for higher g_m/I_D efficiency. In contrast, M_3-M_8 need to be in the strong inversion region in order to get lower g_m. Due to the approximation that g_{m1} is much larger than g_{m3} and g_{m7}, the input-referred noise is simplified by

$$\overline{v_{ni,sym}^2} \approx \overline{v_{ni,folded}^2} \approx \frac{4nkT}{g_{m1}}\Delta f = \frac{4n^2kTV_t}{I_1}\Delta f \propto \frac{1}{current} \tag{16}$$

With a proper sizing in the amplifier, there is only one factor that can be manipulated by circuit designers: current. Therefore, the trade-off between power consumption and noise is straightforward: the larger the current, the lower the noise.

For more stringent constraint in power consumption, the telescopic amplifier is superior to the other amplifiers because it has only two branches. However, output swing range is much narrower than the two amplifiers aforementioned.

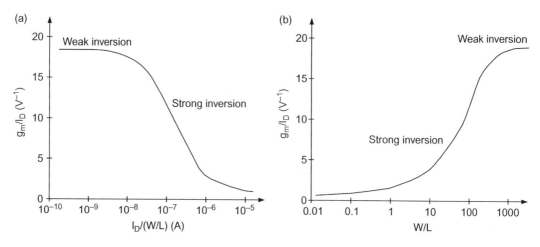

FIGURE 6   Transconductance efficiency (a) as a function of bias current with a fixed size (W/L = 100) and (b) as a function of size with a given bias current (I_D = 100 nA).

THE g_m/I_D DESIGN METHODOLOGY

The g_m/I_D design methodology is very practical for low-power and low-noise design [20,21]. The transconductance efficiency g_m/I_D indicates the efficiency of a transistor in achieving higher gain with lower current. g_m/I_D is larger in the weak inversion than in the strong inversion, as shown in Figure 6(a). It means that a transistor operates more efficiently in weak inversion. As Figure 6(b) depicts, the operation region of a transistor with a given bias current can be manipulated by sizing for desired transconductance efficiency. However, it should be noted that larger device size for higher transconductance efficiency consumes larger area, inducing larger parasitic capacitances.

4.4.2 Stability Perspective

Negative feedback is utilized in almost all amplifiers to acquire a precise gain, set independent of the open-loop gain by a feedback ratio of linear passive components, along with widened bandwidth and increased noise suppression. However, feedback in high-gain systems may be subject to possible sources of instability, which requires careful design consideration. IAs for biomedical applications do not require wide bandwidth since the frequency range of interest is rather low, typically less than 1 kHz. However, due to low-power constraints, typical currents are a few tens to hundreds of nA, and transistors are typically sized large for better matching and low noise performance, resulting in low conductances over large parasitic capacitances, which may worsen the stability.

Figure 7 shows an open-loop Bode plot for the amplifiers. Since V_{OUT} is the only high-impedance node (as mentioned above), the pole at V_{OUT} is located at very low frequency. The feedback establishes the closed loop gain indicated by the dashed horizontal line in Figure 7. The intersection of the closed loop gain of the feedback and open loop gain of the amplifier determines the unity loop-gain frequency f_K. The phase margin is given by

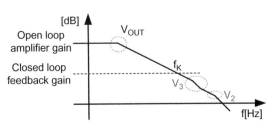

FIGURE 7   Open-loop Bode plot for the one-stage amplifiers.

$$\text{Phase Margin} = 90° - \tan^{-1}\frac{f_K}{f_{p2}} \tag{17}$$

where f_{P2} is the second pole frequency. The position of the second pole is hence critical in the phase margin. If M_3-M_8 in Figure 5 (a) and (b) are assumed to have identical overdrive voltage, V_3 node has larger gate capacitance due to the larger device size, and may generate the second pole. However, this pole is always followed by a zero due to double signal paths, so it does not affect the phase margin seriously. Furthermore, in a fully differential amplifier, the pole-zero pair at the V_3 node is not induced. The pole at V_2 may be lower than the pole-zero pair due to the large junction capacitances of M_1 and M_2, which are sized for large input-pair transconductance and low $1/f$ noise. In some cases, this may generate a doublet, causing relatively long settling time, although within acceptable range for biomedical applications.

4.5 Noise Across Sampling Capacitor

Thermal noise in a resistor can be modeled by the Nyquist-Johnson noise [22,23] as a series voltage source or a parallel current source as follows:

$$\begin{aligned}
\overline{v_{n,R}^2} &= 4kTR\Delta f \\
\overline{i_{n,R}^2} &= \frac{4kT}{R}\Delta f
\end{aligned} \tag{18}$$

Thermal noise, which is induced by random electron motions, is proportional to the absolute temperature.

An ideal capacitor induces no noise. However, due to ohmic coupling with nearby resistive elements such as switches in most practical settings, the integrated noise in a sampling capacitor is typically

$$\overline{v_{n,C}^2} = \frac{kT}{C} \tag{19}$$

It depends only on the capacitor size, other than absolute temperature. This also implies that the noise voltage is inversely proportional to root of capacitance. A 100-fold larger capacitor is required in order to decrease noise voltage 10-fold. Where kT/C noise limits the system's performance, it is difficult to overcome noise just by enlarging the capacitor size.

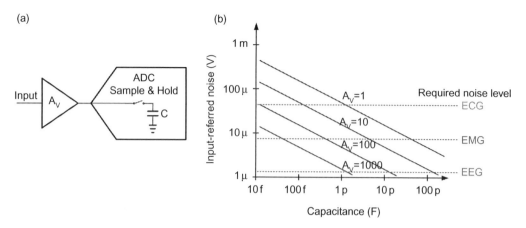

FIGURE 8   (a) A simplified biopotential architecture with an amplification stage and an ADC. (b) The input-referred integrated noise as functions of the capacitance C and the amplification gain A_V.

Figure 8(a) depicts the main functional components of a typical biopotential sensor architecture. It consists of an amplification stage and an ADC with a sample-and-hold input. Figure 8(b) shows the input-referred noise due to the kT/C noise of the sample-and-hold circuit with respect to the capacitor size and the gain A_V. The horizontal dashed lines refer to the required noise levels for sensing different types of biopotential signals. In order to avoid off-chip components, capacitor size is limited to be a few pF typically. Therefore, for the ADC with a sample-and-hold input, pre-amplification is required in order not to be affected by kT/C noise. Note though that for ADCs without sample-and-hold input such as a continuous $\Delta\Sigma$ ADC, pre-amplification may not be necessary.

Autozeroing techniques to an AC-coupled front-end could in principle be used to bypass electrode offset voltage issues and cancel $1/f$ noise [24]. However, just kT/C noise for $1-10$ pF sampling capacitance is about tens of μV as shown by the $A_V = 1$ curve in Figure 8 (b). Therefore other circuit techniques that do not rely on capacitive sampling, such as chopper stabilization, are typically used to mitigate low-frequency noise.

4.6 Chopper Stabilization Techniques

The chopper modulation technique is widespread and essential to mitigate $1/f$ noise and other low-frequency noise, such as popcorn noise, voltage offsets, and drifts. It is particularly used for sensitive acquisition of relatively weak biopotentials such as EEG, which requires very low input-referred noise, less than 1-2 μV_{rms}.

The principles of the chopper modulation technique for amplifiers, which have been extensively studied [24−27], are illustrated in Figure 9. The low-frequency band-limited input signal V_{in} is modulated in front of the amplifier by a square-wave chopping signal. The resulting waveform is V_a: now the signal is lifted to the chopping frequency f_{ch}, and the low-frequency aggressors do not fall within the signal band. After amplification and demodulation with the same chopping signal in V_b, the amplified input signal is shifted back to DC and the aggressors are now moved to f_{ch}. All the undesired aggressors and the

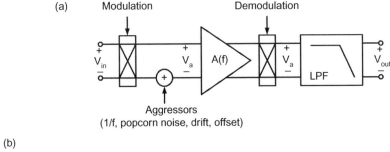

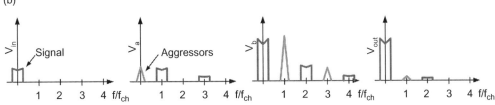

FIGURE 9   (a) Block diagram and (b) frequency-domain illustration of the principle of the chopper technique [3].

harmonics are filtered out through the low-pass filter, and the desired input signal is restored at the output V_{out}.

There are mainly two challenges involved in the design of a chopper-stabilized amplifier: residual offset and output ripple [28].

The residual offset is mainly caused by non-idealities of the input chopper modulator. The mismatch of the clock-feedthrough and the charge injection in the input chopper generates spikes, which are demodulated at the output chopper into a residual output offset. In order to minimize the offset, at first, careful design and layout is required. A continuous [14,16,28,29] or digital [30] DC servo-loop can reduce the residual offset, and mitigate the signal distortion problem that is caused by the finite bandwidth of the amplifier. Also, filtering techniques [31−33] can be applied to resolve the issue.

Output ripple is induced by the input offset of the amplifier, and can saturate the output of the amplifier because the offset is also amplified. The ripple can be reduced by a continuous ripple-reduction loop [28] and a digital foreground calibration [30].

4.7 Pseudoresistors for Sub-Hz High-Pass Cut-Off

The high-pass cut-off frequency needs to be well below 1 Hz for typical biomedical signals, including ECG and EEG. The high-pass cut-off frequency of the typical IA in Figure 4 is determined by the time constant formed by its coupling capacitance C_c and feedback resistance R_f. Realizing sub-Hz time constants with on-chip capacitors and poly-resistors consumes an impractically large area for practical implementation in a chip. Using a 1−10 pF capacitor and a PMOS-based (MOS-bipolar) pseudoresistor is the dominant method. An alternative approach to implement high resistances is based on switched-capacitor logic.

The most prevalent solutions are PMOS-based MOS-bipolar pseudoresistors as shown in Figure 10(a−f). The most basic topology among these is a PMOS whose gate and body

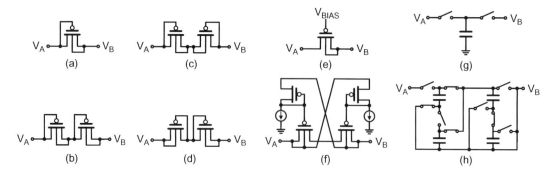

FIGURE 10 On-chip pseudoresistor implementations: (a) single MOS-bipolar pseudoresistor; (b) pseudoresistor with two MOS-bipolar elements in series for twice higher resistance and greater voltage range; (c) symmetrical version with outwardly connected gates; (d) symmetrical version with inwardly connected gates; (e) voltage-biased pseudoresistor for resistance tunability; (f) balanced tunable pseudoresistor with wider linear range; (g) switched-capacitor implementation; and (h) switched-capacitor implementation with 10-times larger effective resistance than (g) [3].

terminals are connected as in Figure 10(a) [34,13]. The gate-connected drain is recommended to be placed in a leakage-sensitive side such as the floating inputs of an OTA. The resistance between the two nodes V_A and V_B is at its maximum when they are equal. When $V_A > V_B$, the forward pn-junction current flows. When $V_A < V_B$, the through current is dominated by substrate leakage current. The two currents are unbalanced, and so is the resistance. The pseudoresistors of two PMOSs with different configurations are shown in Figure 10(b−d) [35,36]. A configuration with the PMOS gate connected to a bias voltage in Figure 10(e) results in a controllable resistance by gate voltage [36−38].

However, the resistance of the PMOS-based pseudoresistors in Figure 10(a−e) drops drastically when the voltage across moves away from zero, inducing signal-dependent distortion while limiting the voltage dynamic range [39]. The pseudoresistor in Figure 10 (f) has balanced resistance with wider linear range up to a few hundred mV. An even wider linear range can be achieved by using an auxiliary amplifier [40].

Switched-capacitors can be also used to implement high resistance on-chip as shown in Figure 10(g). In this topology, switching frequency f_s controls the resistance precisely by

$$R = \frac{1}{f_s C}$$

where C is capacitance of the capacitor in the middle. In order to obtain resistance as large as 1 TΩ, both f_s and C should be made as small as $C = 1$ fF and $f_S = 1$ kHz. However, a 1 fF is too small for manufacturability, and a 1-kHz clock is also not adequate for biomedical applications because of its close proximity to signal frequencies of interest. The switched-capacitor pseudoresistor in Figure 10(h) realizes a 10-fold resistance increase by charge sharing in the switched-capacitor circuits [16]. This topology mitigates the manufacturability and interference issues.

4.8 CMRR Enhancement Techniques

Common-mode interference is a difficult challenge for biomedical signal sensing systems. The major source of the interference comes from electric power lines, which are

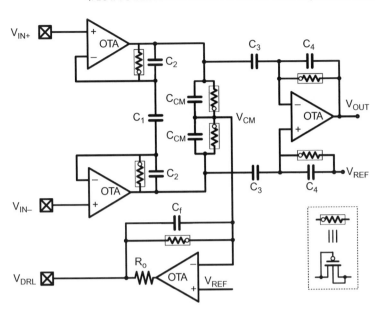

FIGURE 11 An instrumentation amplifier with DRL.

electrically coupled to the human body. High CMRR is required in the system to reject the common-mode interference in order to ensure high signal quality.

Accurate component matching between differential signal lines and between the channels is the most fundamental requirement to accomplish high CMRR. Good matching involves techniques from careful layout to smart architectural design choices. For example, the regulated cascode current mirror [41] in the current sources and the fixed-V_{DS} input pair can enhance CMRR because the techniques decrease mismatch between the differential signals over wide dynamic range [33]. Furthermore, the mains interference can be rejected by a mixed-signal notch filter [42].

The most effective and popular techniques to enhance CMRR are active grounding by the driven right leg (DRL) technique and input impedance-boosting techniques.

4.8.1 DRL Technique

The DRL technique feeds the amplified input common-mode voltage into the body through an additional electrode, which has been placed on the right leg for ECG measurements. This negative feedback reduces the impedance in the feedback loop, attenuating the common-mode interference voltage at the sensor inputs [43,44] by factor of the feedback loop gain.

Figure 11 depicts an example implementation of a capacitive DRL circuit interfacing with a 3-OTA IA, requiring one additional OTA. V_{CM} represents the common-mode voltage sensed as the average of the two electrode inputs. The DRL circuit drives this common-mode voltage to a virtual ground through high-gain feedback to the body. R_o is typically on the order of MΩ in order to limit current flowing to the body for safety.

From two electrode inputs, V_{IN+} and V_{IN-}, to the DRL electrode V_{DRL}, the gain of the negative feedback loop $A_{V,DRL}$ can be simplified by

$$A_{V,DRL} = -\frac{C_f}{2C_{CM}} \qquad\qquad 20)$$

By the feedback, the electrode impedance and the common-mode voltage are reduced by factor $1 + A_V$. In order to obtain large gain in the DRL circuit, an open-loop amplifier can be employed [45]. However, the feedback by the DRL circuit requires careful design for stability. Typically, large capacitance up to a few nF for C_f [46] is required to ensure stability due to variation in electrode impedances. A digitally assisted DRL circuit has the capability to have larger gain at the mains frequency for higher rejection and lower gain elsewhere for stability [47]. In dry electrode applications, common-mode feedback to one of the differential inputs in the front-end increases CMRR, and ensures its stability independent of electrode impedance variations [30].

4.8.2 Input Impedance-Boosting Techniques

Variations and mismatch in electrode impedances also degrade CMRR, reduce signal amplitude, and make the system more susceptible to movement artifacts. Thus, the input impedance of the biopotential sensor should be much higher than the impedance of the electrode and the interface between the body and the electrode. In many cases, the input impedance of biopotential sensors is limited by the parasitic switched-capacitor resistance of the input chopper [14,29,33] or by the AC-coupled input capacitors [13,15,16]. A positive feedback can bootstrap the AC-coupled input capacitors to boost the input impedance [30,48,49], achieving input impedance on the order of $G\Omega$. In order to further boost the input impedance to $T\Omega$ levels, a unity-gain amplifier with active shielding can be used to bootstrap capacitance of the input transistor and all other parasitic capacitance [50].

5. LOW-POWER DESIGN FOR ADCS

Digitization of recorded and processed analog signals is necessary for further digital signal processing and digital RF communication. The tight power and low noise constraints demand ultra-low power ADCs at low frequency range (1−10 kHz) without sacrificing noise performance, while requiring no or very little static current drain and scalable power consumption with respect to sampling rate for multi-modal recording applications. Both successive-approximation-register (SAR) ADC and oversampling $\Delta\Sigma$ ADC are superior architectures for achieving the specifications with the lowest power dissipation.

SAR ADC is the dominant architecture for low-power medium-resolution (8−10 bits) biomedical applications due to its simple architecture involving few analog circuits and its low power consumption at low frequency without static power consumption [16−17,45,51−56]. A typical SAR ADC consists of SAR logic, a clocked-comparator, and a capacitor input DAC as shown in Figure 12. It performs a binary-search-based successive approximation of the sampled input by controlling the switching logics of the input capacitor DAC while producing a one-bit quantization result per each clock from MSB to LSB [57−58].

There are many techniques and architectures to minimize the power consumption of an SAR ADC. Using main and sub binary weighted DAC arrays with a series attenuation capacitor can reduce the total size of the capacitor array; leading to reduced power

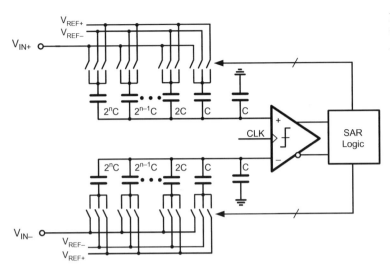

FIGURE 12 A general architecture of differential SAR ADC.

consumption in the ADC driver and also in the capacitor DAC [59]. Also, the folded capacitor DAC architecture with divided reference voltages reduces the size of capacitor DAC, resulting in further power savings [52]. Using charge-recycling switching methods results in further saving of switching power consumption in the capacitor DAC [60].

SAR ADCs are generally considered to be most energy efficient for medium-precision low-sampling-rate digitization. However, most micropower SAR ADCs operate at signal levels substantially (3–4 orders of magnitude) greater than the typical signal level of physiological signals (Figure 1). They require a significant amplification before analog-to-digital conversion for sub-μV resolution. Furthermore, sampling at Nyquist frequency demands anti-aliasing filtering. The cost of amplification and anti-aliasing filtering are often not accounted for in ADC energy metrics. Most critically, sampling of biopotentials at μV resolution is problematic due to kT/C sampling noise on capacitors; several tens of μV for pF-range size limits on capacitors in integrated circuits.

$\Delta\Sigma$ ADCs are an alternative solution with the following strengths [61–65]:

- Resolution and sampling rate can be dynamically reconfigured, with sampling rate proportional to power consumption, so they are adequate for multimodal biopotential sensor applications.
- They include only a few simple analog components.
- They are suited for low-power and low-voltage operation.
- They can achieve high resolution such as 12–16 bits easily without complex circuit and layout techniques.
- For continuous-time $\Delta\Sigma$ topologies, there is no kT/C sampling noise.

A Gm-C incremental $\Delta\Sigma$ ADC with widely configurable resolution and sampling rate is shown in Figure 13 [61]. A transconductance (G_m) cell converts the differential input voltage signal to a current, approximately linear over the voltage range of typical biopotentials. The difference between this current and a feedback current is integrated and the resulting voltage is compared for three-level quantization of the feedback current,

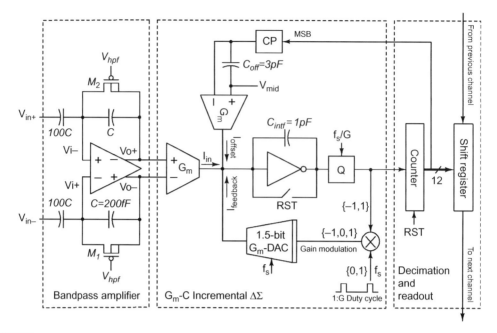

FIGURE 13 A biopotential acquisition system utilizing G_m-C incremental $\Delta\Sigma$ ADC [61].

implementing a continuous-time first-order $\Delta\Sigma$ modulator. A continuous-time oversampling ADC avoids the need for anti-aliasing filter and sample-and-hold circuits preceding the ADC. In addition, its duty cycle control feature in the feedback offers precise digital gain programmability from 1 to 4,096.

Another example incremental $\Delta\Sigma$ ADC for non-invasive biopotential recording is given in Figure 14 [62]. It receives unbuffered biopotential signals and performs amplification, signal conditioning, and digitization using only a single OTA.

Another alternative solution for achieving high resolution for low-power biomedical applications is the hybrid architecture of SAR and $\Delta\Sigma$ conversions, as shown in Figure 15 [63]. It performs a successive approximation conversion in the first phase and $\Delta\Sigma$ conversion in the next phase with the residue from the previous phase. Thus, it can achieve more resolution than conventional SAR ADCs, and much shorter conversion time than typical $\Delta\Sigma$ ADCs.

In addition, there are other alternative ADC architectures such as an asynchronous level-crossing ADC [66] and a bio-inspired ADC with successive integrate-and-fire operation [67].

6. LOW-POWER DIGITAL CIRCUIT DESIGN TECHNIQUES

Integration of digital circuits enables versatile programmable functionality in wireless sensor applications. Digital post-processing reduces the amount of data significantly,

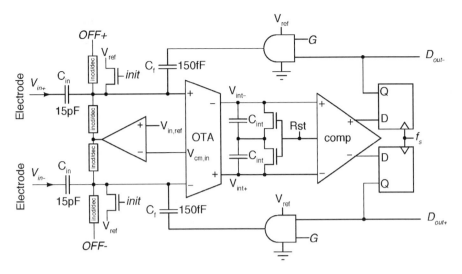

FIGURE 14 An incremental $\Delta\Sigma$ ADC that directly interfaces biopotential signals [62].

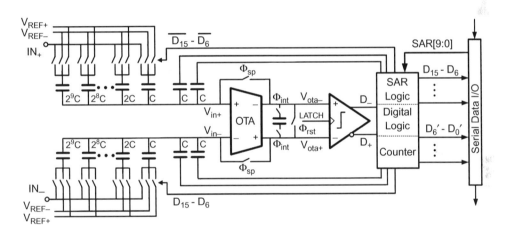

FIGURE 15 A SAR-$\Delta\Sigma$ hybrid architecture [63].

relaxing the power consumption in data communications. For example, digital pattern recognition embedded in wearable sensors is capable of detecting epileptic seizures without communicating all the data over a wireless link, leading to much power savings for the overall system [18,68]. In addition, various digital feedback and control signals may offer more adaptive functionality and better performance to the sensor operation.

Power dissipation in digital circuits is related to four factors — activity factor α relating the probability of a switching event, capacitance of the circuit C, clock frequency f_{CLK}, and power supply voltage V_{DD}:

$$P_{avg} = \alpha C V_{DD}^2 f_{CLK} \tag{21}$$

The strategies to reduce power dissipation in digital circuits are simple: minimize each of the four factors.

6.1 Minimum Energy Design Methodology

Lowering power supply voltage V_{DD} reduces the current drain in digital circuits drastically, and slows down the operation speed. To execute a given operation, operation of the digital circuit with lower V_{DD} normally consumes less active energy, but drains larger leakage current due to longer operation time. Therefore, an optimized V_{DD}, which is called the minimum energy point (MEP), exists where the power consumption for an operation is minimized considering both the active and leakage energy consumption [69–70]. The active energy consumption is as follows:

$$E_{ACTIVE} = C_{eff} V_{DD}^2 \tag{22}$$

where C_{eff} is the average effective switched capacitance per the given operation. The leakage energy per the operation is given by

$$E_{LEAK} = \beta C_{eff} L_{DP} V_{DD}^2 e^{-\frac{V_{DD}}{n\phi_t}} \tag{23}$$

where β is a constant, and L_{DP} is the logic depth of the critical path. The total energy consumption is the summation of the two energy components:

$$E_{TOTAL} = C_{eff} V_{DD}^2 \left(1 + \beta L_{DP} e^{-\frac{V_{DD}}{n\phi_t}}\right) \tag{24}$$

Figure 16 illustrates the location of the MEP as a balance between the active and leakage energies where the total energy is at the minimum. The MEP can vary by temperature and workload.

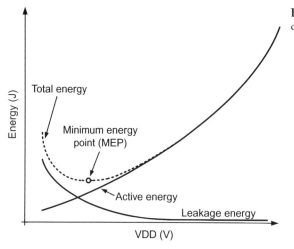

FIGURE 16   Energy consumption of a digital circuit as a function of V_{DD}.

Several other techniques have been developed for optimizing the balance between high-performance and low-power digital circuits. We briefly outline a few of the prevailing techniques under continued development, each targeting one or more of the factors determining the power-performance trade-off:

- V_{DD}: dynamic voltage scaling, voltage domain design, etc.
- Activity factor α: disabling of unused logics, clock gating, etc.
- Capacitance C: advances in CMOS technology, gate sizing, floorplanning for reducing wire lengths, etc.
- Clock frequency f_{CLK}: dynamic frequency scaling, etc.
- Architectural methodologies: parallel processing, pipelining, retiming, unfolding/folding, etc.
- Leakage power: exploiting diversity in multiple threshold voltage devices, power gating, etc.

The details of these and other low-power digital circuit and architectural techniques are beyond the scope of this chapter and can be found in other books and papers, e.g., [68, 71−75].

7. ARCHITECTURAL DESIGN FOR LOW-POWER BIOPOTENTIAL ACQUISITION

System-level architectural design may lead to significant advances in minimizing power consumption without compromising performance. Block-level design approaches should be considered before optimization within each block. This section reviews architectural design choices for low-power biopotential acquisition with some examples.

7.1 Architectural Design Strategies

An intelligent choice for functional block allocation in a system can result in significant improvements in power consumption and performance of the whole system. A typical architecture of wearable biopotential sensors comprises an analog front-end, a digital signal processor, an ADC between the two, an RF transmitter, and a power harvesting and management unit as shown in Figure 17.

Preprocessing inside of the sensor is more favorable than transmitting all raw data out, particularly in high-dimensional or high-bandwidth sensor applications where the amount of useful information may be significantly lower than the data rate requirements of the raw signals. Transmitting the processed data instead of all raw data reduces power

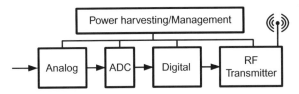

FIGURE 17 General architecture for wireless biopotential acquisition.

consumption in the RF transmitter considerably. In the aforementioned example of seizure detection in EEG sensing, 18 electrodes produce 200-Hz 12-bit data resulting in a total data rate of 43.2 kbps. Employing digital processing for feature extraction and seizure detection reduces the data bandwidth to 2 kbps with a 10-fold reduction in power dissipation [18,68].

Equally importantly, energy-efficient analog preprocessing can lead to significant reduction of data bandwidth and power in the ADC and the digital block. A low-power ADC and digital blocks can substitute a high-performance ADC and digital signal processor. For example, typical EEG/ECoG-based brain-computer interface applications do not need raw data, but spectral characteristics of the recordings. Thus, extracting spectral power of the required frequency bands in analog front-end can reduce power dissipation and complexity in the other blocks [76]. There are many other kinds of analog preprocessing for low-power high-performance biosignal processing such as the QRS detection in ECG [77].

Manipulating functional positioning within the analog block is also very important to maximize the performance while minimizing the power consumption. There are various block-level designs in the analog domain for each application-specific requirement. Separating functions into each block can optimize each function respectively, and maximize programmability on gain and bandwidth [17]. Figure 18(a) shows such an architecture comprising an instrumentation amplifier, a variable-gain/bandwidth amplifier, and an ADC driver, which is connected to an ADC. In contrast to the separation of all functions into each block, the other extreme alternative is an architecture combining all of analog signal conditioning function and analog-to-digital conversion into one block, as shown in Figure 18(b). This architecture removes unnecessary power consumption in the VGA/filter stage and the ADC driver. In addition, it achieves digital controllability in gain and bandwidth by changing duty cycle of integration clock and over-sampling ratio [62].

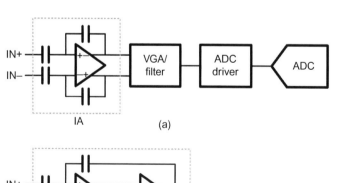

FIGURE 18 Alternative signal acquisition architectures. (a) Standard modular approach with isolated blocks for amplification, filtering, and ADC. (b) Hybrid architecture combining all functions into a single analog block including embedded ADC [62].

7.2 Power Domain Design

How the power domains are designed in a system may contribute more to power savings in system performance than improvements by circuit techniques. In most wearable sensors, power is harvested in various ways or supplied from a battery. In any case, the amount of the power is very limited. Therefore, power should be managed wisely. The first step of power management is to design power domains in the system.

Having only one lowest possible power supply in the system is most preferable and does not involve complicated power management schemes and level shifts. However, having different kinds of power levels can be more beneficial in order to reduce the power consumption further. Each block may require different supply voltage, V_{DD}. A block with narrower voltage dynamic range can operate with a lower V_{DD}, resulting in lower power. Higher V_{DD} is necessary for blocks requiring a wide dynamic range. In addition, higher V_{DD} is favorable for better performance in some applications such as a switch, which requires lowest possible turn-on resistance. [78]. In a system with multi-power domains, more careful design is required at the interfaces where different power levels meet. Many techniques such as level shifting, AC-coupling, etc., can be utilized at the interface.

There are several ways to design multi-power domains. 1) A multiple of power levels can be generated directly from the power-harvesting circuit [79]. This method meets different V_{DD} requirements of different blocks and optimizes the power consumptions. However, it requires more complex design in the energy-harvesting block. 2) Multiple power levels can be generated by regulators from a high V_{DD}, which is scavenged or supplied from a battery. This approach can decrease current consumption in some blocks, but the headroom is still wasteful. 3) Contrary to method 2), some higher voltages can be generated from a lower voltage using a power management unit (PMU) such as a charge pump. It can be optimized in terms of power consumption. However, it may induce more complexity because of adding a PMU block [80].

8. PRACTICAL CONSIDERATIONS

Outside of controlled laboratory conditions, the largest noise sources will likely be electromagnetic interference (EMI) and movement artifacts. The most common symptom of EMI is seen as 50/60 Hz power line pickup. Triboelectric charging during subject movement is also a large, albeit less understood, source of interference. As a subject moves, the body to environment potential changes due to charge generation as the body strikes the ground. This can couple into the system just like EMI and is often mistaken for movement artifacts due to electrode-skin displacements.

EMI can be reduced or eliminated by a few simple techniques. Common-mode interference is easily attenuated through the use of a DRL. The technique is well understood and operates by actively biasing the body potential towards a fixed circuit reference potential through the use of negative feedback. This has the effect of reducing the magnitude of common mode interference seen by the amplifiers. Differential pickup of EMI is mitigated through the use of active electrodes where an amplifier is placed in close proximity to the electrode. Alternatively, the use of shielded lead wires is also high effective at accomplishing the same goal — minimizing the total area of high impedance traces that are susceptible to external electric fields.

Movement artifacts are a difficult challenge due to the lack of quantified metrics and clear design methodology. Any physical displacement between the electrode and the skin will necessarily generate noise, sometimes many orders of magnitude larger than the actual signal. Movement artifacts can be eliminated by increasing the physical coupling pressure between the electrode and skin, but may conflict with the need for subject comfort and wearability. Reducing movement artifacts highly relies on mechanical and industrial design, and solutions are dependent on the specific end application. In addition, electrode–tissue impedance measurement and signal processing techniques can be used to quantify and suppress movement artifacts [3, 48, 81–86].

9. CONCLUSION

This chapter reviewed principles of micropower mixed-signal analog and digital integrated circuit design for non-invasive biopotential sensing in wearable clinical and ambulatory physiological monitoring systems. A systematic approach to g_m-C circuit design using MOS transistors operating in subthreshold and weak inversion offers high energy and noise efficiency where resolution and bandwidth scale approximately linearly with power consumption. The methodology was illustrated with practical examples of such high-efficiency g_m-C MOS circuits for front-end amplification, bandpass filtering, and analog-to-digital conversion. A critical factor in the performance of wearable integrated bioinstrumentation is the electrode-body interface, and practical guidelines of front-end circuit design were given.

Glossary

EEG electroencephalography, electroencephalogram
ECG (EKG) electrocardiography, electrocardiogram
EMG electromyography, electromyogram
FET field-effect transistor
MOS metal–oxide–semiconductor
MOSFET metal–oxide–semiconductor field-effect transistor
NMOS n-channel MOSFET
PMOS p-channel MOSFET
CMOS complementary metal–oxide–semiconductor
BJT bipolar junction transistor
Opamp operational amplifier
OTA operational transconductance amplifier
ADC analog-to-digital converter
DAC digital-to-analog converter
MSB most significant bit
LSB least significant bit
SAR successive approximation register
ΔΣ sigma-delta

References

[1] J.G. Webster, Medical Instrumentation: Application and Design, forth ed., John Wiley & Sons, New York, 2010.

[2] R.F. Yazicioglu, C. van Hoof, R. Puers, Biopotential Readout Circuits for Portable Acquisition Systems, Springer, The Netherlands, 2009.

[3] S. Ha, C. Kim, Y.M. Chi, A. Akinin, C. Maier, A. Ueno, et al., Integrated circuits and electrode interfaces for noninvasive physiological monitoring, IEEE Trans. Biomed. Eng. 61 (2014) 1522−1537.

[4] Y.M. Chi, T.-P. Jung, G. Cauwenberghs, Dry-Contact and Noncontact Biopotential Electrodes: Methodological Review, IEEE Rev. Biomed. Eng. 3 (2010) 106−119.

[5] A. Baba, M. Burke, Measurement of the electrical properties of ungelled ECG electrodes, Int. J. Biol. Biomed. Eng. 2 (2008) 89−97.

[6] E. Huigen, A. Peper, C.A. Grimbergen, Investigation into the origin of the noise of surface electrodes, Med. Biol. Eng. Comput. 40 (2002) 332−338.

[7] C. Mead, Introduction to VLSI Systems, Addison Wesley, 1979.

[8] R. Sarpeshkar, Ultra Low Power Bioelectronics: Fundamentals, Biomedical Applications, and Bio-Inspired Systems, Cambridge University Press, Cambridge, 2010.

[9] R. Sarpeshkar, T. Delbruck, C.A. Mead, White noise in MOS transistors and resistors, IEEE Circuits Devices Mag. 9 (1993) 23−29.

[10] Mcworther, A. L. (1955). $1/f$ noise and related surface effects in germanium, Sc. D. Thesis, Cambridge: Massachusetts Institute of Technology.

[11] A. Van der Ziel, Unified presentation of $1/f$ noise in electron devices: fundamental $1/f$ noise sources, Proc. IEEE 76 (1988) 233−258.

[12] K. Roy, S. Mukhopadhyay, H. Mahmoodi-Meimand, Leakage current mechanisms and leakage reduction techniques in deep-submicrometer CMOS circuits, Proc. IEEE 91 (2003) 305−327.

[13] R.R. Harrison, C. Charles, A low-power low-noise CMOS amplifier for neural recording applications, IEEE J. Solid State Circuits 38 (2003) 958−965.

[14] R.F. Yazicioglu, P. Merken, R. Puers, C. van Hoof, A 200 μW Eight-Channel EEG Acquisition ASIC for Ambulatory EEG Systems, IEEE J. Solid State Circuits 43 (2008) 3025−3038.

[15] X.D. Zou, X.Y. Xu, L.B. Yao, Y. Lian, A 1-V 450-nW Fully Integrated Programmable Biomedical Sensor Interface Chip, IEEE J. Solid State Circuits 44 (2009) 1067−1077.

[16] N. Verma, A. Shoeb, J. Bohorquez, J. Dawson, J. Guttag, A.P. Chandrakasan, A Micro-Power EEG Acquisition SoC With Integrated Feature Extraction Processor for a Chronic Seizure Detection System, IEEE J. Solid State Circuits 45 (2010) 804−816.

[17] M.J. Burke, D.T. Gleeson, A micropower dry-electrode ECG preamplifier, IEEE Trans. Biomed. Eng. 47 (2000) 155−162.

[18] R.R. Harrison, The design of integrated circuits to observe brain activity, Proc. IEEE 96 (2008) 1203−1216.

[19] M.S.J. Steyaert, W.M.C. Sansen, Z.Y. Chang, A Micropower Low-Noise Monolithic Instrumentation Amplifier for Medical Purposes, IEEE J. Solid State Circuits 22 (1987) 1163−1168.

[20] C.C. Enz, F. Krummenacher, E.A. Vittoz, An analytical MOS transistor model valid in all regions of operation and dedicated to low-voltage and low-current applications, Analog Integr. Circuits Signal Process. 8 (1995) 83−114.

[21] F. Silveira, D. Flandre, P.G.A. Jespers, A g_m/I_D based methodology for the design of CMOS analog circuits and its application to the synthesis of a silicon-on-insulator micropower OTA, IEEE J. Solid State Circuits 31 (1996) 1314−1319.

[22] H. Nyquist, Thermal agitation of electric charge in conductors, Phys. Rev. 32 (1928) 110−113.

[23] J.B. Johnson, Thermal agitation of electricity in conductors, Phys. Rev. 32 (1928) 97−109.

[24] C.C. Enz, G.C. Temes, Circuit techniques for reducing the effects of op-amp imperfections: Autozeroing, correlated double sampling, and chopper stabilization, Proc. IEEE 84 (1996) 1584−1614.

[25] K.C. Hsieh, P.R. Gray, D. Senderowicz, D.G. Messerschmitt, A Low-Noise Chopper-Stabilized Differential Switched-Capacitor Filtering Technique, IEEE J. Solid State Circuits 16 (1981) 708−715.

[26] C.C. Enz, E.A. Vittoz, F. Krummenacher, A CMOS chopper amplifier, IEEE J. Solid State Circuits 22 (1987) 335−342.

[27] C. Menolfi, Q.T. Huang, A low-noise CMOS instrumentation amplifier for thermoelectric infrared detectors, IEEE J. Solid State Circuits 32 (1997) 968−976.

[28] R. Wu, K.A.A. Makinwa, J.H. Huijsing, A Chopper Current-Feedback Instrumentation Amplifier With a 1 mHz 1/f Noise Corner and an AC-Coupled Ripple Reduction Loop, IEEE J. Solid State Circuits 44 (2009) 3232−3243.

[29] T. Denison, K. Consoer, W. Santa, A.T. Avestruz, J. Cooley, A. Kelly, A 2 μW 100 nV/$\sqrt{\text{Hz}}$ chopper-stabilized instrumentation amplifier for chronic measurement of neural field potentials, IEEE J. Solid State Circuits 42 (2007) 2934−2945.

[30] J.W. Xu, R.F. Yazicioglu, B. Grundlehner, P. Harpe, K.A.A. Makinwa, C. Van Hoof, A 160 mu W 8-Channel Active Electrode System for EEG Monitoring, IEEE Trans. Biomed. Circuits Syst. 5 (2011) 555−567.

[31] C. Menolfi, Q.T. Huang, A fully integrated, untrimmed CMOS instrumentation amplifier with submicrovolt offset, IEEE J. Solid State Circuits 34 (1999) 415−420.

[32] R. Burt, J. Zhang, A micropower chopper-stabilized operational amplifier using a SC notch filter with synchronous integration inside the continuous-time signal path, IEEE J. Solid State Circuits 41 (2006) 2729−2736.

[33] R.F. Yazicioglu, P. Merken, R. Puers, C. Van Hoof, A 60 μW 60 nV/$\sqrt{\text{Hz}}$ Readout Front-End for Portable Biopotential Acquisition Systems, IEEE J. Solid State Circuits 42 (2007) 1100−1110.

[34] T. Delbruck, C.A. Mead, Adaptive photoreceptor with wide dynamic range, Proc. 1994 IEEE Int. Symp. Circuits Syst. 1994 (4) (1994) 339−342.

[35] H. Wu, Y.P. Xu, A 1 V 2.3/spl mu/W Biomedical Signal Acquisition IC, 2006 IEEE Int. Solid State Circuits Conf. Dig. Tech. Papers (2006) 119−128.

[36] W. Wattanapanitch, M. Fee, R. Sarpeshkar, An Energy-Efficient Micropower Neural Recording Amplifier, IEEE Trans. Biomed. Circuits Syst. 1 (2007) 136−147.

[37] R.H. Olsson, D.L. Buhl, A.M. Sirota, G. Buzsaki, K.D. Wise, Band-tunable and multiplexed integrated circuits for simultaneous recording and stimulation with microelectrode arrays, IEEE Trans. Biomed. Eng. 52 (2005) 1303−1311.

[38] M.S. Chae, Z. Yang, M.R. Yuce, L. Hoang, W.T. Liu, A 128-Channel 6 mW Wireless Neural Recording IC With Spike Feature Extraction and UWB Transmitter, IEEE Trans. Neural Syst. Rehabil. Eng. 17 (2009) 312−321.

[39] X. Zou, X. Xu, L. Yao, Y. Lian, A 1-V 450-nW Fully Integrated Programmable Biomedical Sensor Interface Chip, IEEE J. Solid State Circuits 44 (2009) 1067−1077.

[40] M.T. Shiue, K.W. Yao, C.S.A. Gong, Tunable high resistance voltage-controlled pseudo-resistor with wide input voltage swing capability, Electron. Lett. 47 (2011) 377−378.

[41] E. Sackinger, W. Guggenbuhl, A High-Swing, High-Impedance MOS Cascode Circuit, IEEE J. Solid State Circuits 25 (1990) 289−298.

[42] J.L. Bohorquez, M. Yip, A.P. Chandrakasan, J.L. Dawson, A Biomedical Sensor Interface With a *sinc* Filter and Interference Cancellation, IEEE J. Solid State Circuits 46 (2011) 746−756.

[43] B.B. Winter, J.G. Webster, Reduction of Interference Due to Common-Mode Voltage in Biopotential Amplifiers, IEEE Trans. Biomed. Eng. 30 (1983) 58−62.

[44] B.B. Winter, J.G. Webster, Driven-Right-Leg Circuit-Design, IEEE Trans. Biomed. Eng. 30 (1983) 62−66.

[45] L. Fay, V. Misra, R. Sarpeshkar, A Micropower Electrocardiogram Amplifier, IEEE Trans. Biomed. Circuits Syst. 3 (2009) 312−320.

[46] T. Degen, H. Jackel, Enhancing interference rejection of preamplified electrodes by automated gain adaption, IEEE Trans. Biomed. Eng. 51 (2004) 2031−2039.

[47] M.A. Haberman, E.M. Spinelli, A Multichannel EEG Acquisition Scheme Based on Single Ended Amplifiers and Digital DRL, IEEE Trans. Biomed. Circuits Syst. 6 (2012) 614−618.

[48] N. Van Helleputte, S. Kim, H. Kim, J.P. Kim, C. Van Hoof, R.F. Yazicioglu, A 160 μA Biopotential Acquisition IC With Fully Integrated IA and Motion Artifact Suppression, IEEE Trans. Biomed. Circuits Syst. 6 (2012) 552−561.

[49] Q.W. Fan, F. Sebastiano, J.H. Huijsing, K.A.A. Makinwa, A 1.8 mu W 60 nV/$\sqrt{\text{Hz}}$ Capacitively-Coupled Chopper Instrumentation Amplifier in 65 nm CMOS for Wireless Sensor Nodes, IEEE J. Solid State Circuits 46 (2011) 1534−1543.

[50] Y.M. Chi, C. Maier, G. Cauwenberghs, Ultra-High Input Impedance, Low Noise Integrated Amplifier for Noncontact Biopotential Sensing, IEEE J. Emerg. Sel. Top. Circuits Syst. 1 (2011) 526−535.

[51] N. Verma, A.P. Chandrakasan, An ultra low energy 12-bit rate-resolution scalable SAR ADC for wireless sensor nodes, IEEE J. Solid State Circuits 42 (2007) 1196–1205.

[52] L. Yan, J. Yoo, B. Kim, H.J. Yoo, A 0.5-μV_{rms} 12-μW Wirelessly Powered Patch-Type Healthcare Sensor for Wearable Body Sensor Network, IEEE J. Solid State Circuits 45 (2010) 2356–2365.

[53] S. Lee, L. Yan, T. Roh, S. Hong, H.J. Yoo, A 75 μW Real-Time Scalable Body Area Network Controller and a 25 mu W ExG Sensor IC for Compact Sleep Monitoring Applications, IEEE J. Solid State Circuits 47 (2012) 323–334.

[54] J. Yoo, L. Yan, D. El-Damak, M.A. Bin Altaf, A.H. Shoeb, A.P. Chandrakasan, An 8-Channel Scalable EEG Acquisition SoC With Patient-Specific Seizure Classification and Recording Processor, IEEE J. Solid State Circuits 48 (2013) 214–228.

[55] M. Khayatzadeh, X. Zhang, J. Tan, W.S. Liew, Y. Lian, A 0.7-V 17.4-uW 3-Lead Wireless ECG SoC, IEEE Trans. Biomed. Circuits Syst. 7 (2013) 583–592.

[56] Y.-J. Min, H.-K. Kim, Y.-R. Kang, G.-S. Kim, J. Park, S.-W. Kim, Design of Wavelet-Based ECG Detector for Implantable Cardiac Pacemakers, IEEE Trans. Biomed. Circuits Syst. 7 (2013) 426–436.

[57] J.L. Mccreary, P.R. Gray, All-MOS Charge Redistribution Analog-to-Digital Conversion Techniques-Part I, IEEE J. Solid State Circuits 10 (1975) 371–379.

[58] R.E. Suarez, P.R. Gray, D.A. Hodges, All-MOS Charge Redistribution Analog-to-Digital Conversion Techniques-Part II, IEEE J. Solid State Circuits 10 (1975) 379–385.

[59] A. Agnes, E. Bonizzoni, P. Malcovati, F. Maloberti, A 9.4-ENOB 1 V 3.8 μW 100 kS/s SAR ADC with Time-Domain Comparator, 2008 IEEE Int. Solid State Circuits Conf. Dig. Tech. Papers (2008) 246–610.

[60] B.P. Ginsburg, A.P. Chandrakasan, An energy-efficient charge recycling approach for a SAR converter with capacitive DAC, Proc. 2005 IEEE Int. Symp. Circuits Syst. (2005) 184–187.

[61] M. Mollazadeh, K. Murari, G. Cauwenberghs, N. Thakor, Micropower CMOS Integrated Low-Noise Amplification, Filtering, and Digitization of Multimodal Neuropotentials, IEEE Trans. Biomed. Circuits Syst. 3 (2009) 1–10.

[62] Y.M. Chi, G. Cauwenberghs, Micropower integrated bioamplifier and auto-ranging ADC for wireless and implantable medical instrumentation, 2010 Proc. Eur. Solid State Circuits Conf. (2010) 334–337.

[63] S. Ha, J. Park, Y.M. Chi, J. Viventi, J. Rogers, G. Cauwenberghs, 85 dB dynamic range 1.2 mW 156 kS/s biopotential recording IC for high-density ECoG flexible active electrode array, 2013 Proc. Eur. Solid State Circuits Conf. (2013) 141–144.

[64] J. Garcia, S. Rodriguez, A. Rusu, A Low-Power CT Incremental 3rd Order Sigma Delta ADC for Biosensor Applications, IEEE Trans. Circuits Syst. I Regul. Papers 60 (2013) 25–36.

[65] J.R. Custodio, J. Goes, N. Paulino, J.P. Oliveira, E. Bruun, A 1.2-V 165-mu W 0.29-mm(2) Multibit Sigma-Delta ADC for Hearing Aids Using Nonlinear DACs and With Over 91 dB Dynamic-Range, IEEE Trans. Biomed. Circuits Syst. 7 (2013) 376–385.

[66] L. Yongjia, Z. Duan, W.A. Serdijn, A Sub-Microwatt Asynchronous Level-Crossing ADC for Biomedical Applications, IEEE Trans. Biomed. Circuits Syst. 7 (2013) 149–157.

[67] H.Y. Yang, R. Sarpeshkar, A Bio-Inspired Ultra-Energy-Efficient Analog-to-Digital Converter for Biomedical Applications, IEEE Trans. Circuits Syst. I Regul. Papers 53 (2006) 2349–2356.

[68] A.P. Chandrakasan, R.W. Brodersen, Minimizing power-consumption in digital CMOS circuits, Proc. IEEE 83 (1995) 498–523.

[69] A. Wang, A. Chandrakasan, A 180-mV subthreshold FFT processor using a minimum energy design methodology, IEEE J. Solid State Circuits 40 (2005) 310–319.

[70] Y.K. Ramadass, A.P. Chandrakasan, Minimum energy tracking loop with embedded DC-DC converter enabling ultra-low-voltage operation down to 250 mV in 65 nm CMOS, IEEE J. Solid State Circuits 43 (2008) 256–265.

[71] A.P. Chandrakasan, S. Sheng, R.W. Brodersen, Low-power CMOS digital design, IEEE J. Solid State Circuits 27 (1992) 473–484.

[72] J.M. Rabaey, A. Chandrakasan, B. Nikolic, Digital Integrated Circuits: A Design Perspective, second ed., Prentice Hall, 2003.

[73] C. Piguet, Low-Power Electronics Design, first ed., CRC Press, 2005.

[74] N. Weste, D. Harris, CMOS VLSI Design: A Circuits and Systems Perspective, forth ed., Addison-Wesley, 2010.

[75] P.R. Panda, A. Shrivastava, B.V.N. Silpa, K. Gummidipudi, Power Efficient System Design, first ed., Springer, 2010.

[76] F. Zhang, A. Mishra, A.G. Richardson, B. Otis, A Low-Power ECoG/EEG Processing IC With Integrated Multiband Energy Extractor, IEEE Trans. Circuits Syst. I Regul. Papers 58 (2011) 2069−2082.

[77] Y.-J. Min, H.-K. Kim, Y.-R. Kang, G.-S. Kim, J. Park, S.-W. Kim, Design of Wavelet-Based ECG Detector for Implantable Cardiac Pacemakers, IEEE Trans. Biomed. Circuits Syst. 7 (2013) 426−436.

[78] D. Park, S. Cho, Design Techniques for a Low-Voltage VCO With Wide Tuning Range and Low Sensitivity to Environmental Variations, IEEE Trans. Microw. Theory Tech. 57 (2009) 767−774.

[79] K. Chen, Y.-K. Lo, W. Liu, A 37.6 mm^2 1024-channel high-compliance-voltage SoC for epiretinal prostheses, 2013 IEEE Int. Solid State Circuits Conf. Dig. Tech. Papers (2013) 294−295.

[80] D. Han, Y. Zheng, R. Rajkumar, G. Dawe, M. Je, A 0.45 V 100-channel neural-recording IC with sub-μW/channel consumption in 0.18 μm CMOS, Trans. Biomed. Circuits Syst. 7 (2013).

[81] S. Kim, R.F. Yazicioglu, T. Torfs, B. Dilpreet, P. Julien, C. Van Hoof, A 2.4 μA continuous-time electrode-skin impedance measurement circuit for motion artifact monitoring in ECG acquisition systems, Proc. Symp. VLSI Circuits Digest Techn. Papers (2010) 219−220.

[82] R. Yazicioglu, S. Kim, T. Torfs, H. Kim, C. Van Hoof, A 30 μW analog signal processor ASIC for portable biopotential signal monitoring, IEEE J. Solid-State Circuits 46 (2011) 209−223.

[83] D. Buxi, S. Kim, N. van Helleputte, M. Altini, J. Wijsman, R.F. Yazicioglu, et al., Correlation between electrode-tissue impedance and motion artifact in biopotential recordings, IEEE Sens. J. 12 (2012) 3373−3383.

[84] A. Griffiths, A. Das, B. Fernandes, P. Gaydecki, A portable system for acquiring and removing motion artefact from ECG signals, J. Phys. Conf. Ser. 76 (2007) 012038.

[85] S. Kim, H. Kim, N. Van Helleputte, C. Van Hoof, R.F. Yazicioglu, Real time digitally assisted analog motion artifact reduction in ambulatory ECG monitoring system, Proc. 34th Annu. Int. Conf. IEEE Eng. Med. Biol. Soc (2012) 2096−2099.

[86] H. Kim, S. Kim, N. Van Helleputte, T. Berset, G. Di, I. Romero, et al., Motion artifact removal using cascade adaptive filtering for ambulatory ECG monitoring system, Proc. 2012 IEEE Biomed. Circuits Syst. Conf (2012) 160−163.

Wearable Algorithms: An Overview of a Truly Multi-Disciplinary Problem

Guangwei Chen, Esther Rodriguez-Villegas, and Alexander J. Casson

Imperial College, London, UK

1. INTRODUCTION

Wearable sensors are quickly emerging as next-generation devices for the ubiquitous monitoring of the human body. Illustrated in Figure 1, these are highly miniaturized sensor nodes that connect to the body and record one (or potentially more) physiological parameters before wirelessly transmitting the recorded signals to a base station such as a smartphone, PC, or other computer installation.

For end users there are a number of features that successful wearable sensors must include: it is essential that they are easy-to-use, socially acceptable, and long lasting. The power consumption of the sensor node is a critical factor in realizing all of these features as the current draw of the sensor sets the physical size of the battery required, which determines the device size and operating lifetime, which in turn affects the ease of use.

To illustrate current trends, Table 1 shows the 2013 performance of ten state-of-the-art wearable units for monitoring the human EEG (electroencephalogram [1]). It can be seen that a number of high-quality, highly miniaturized units are now available commercially and that these can easily offer over 8 hours of recording time. Twenty-four hour recording periods are starting to be offered by research stage units. This level of power performance is likely sufficient for performing any one EEG recording experiment.

However, the power level still falls far short of creating simple *pick up and use* devices. Substantial improvements in system power consumptions are required to realize units that

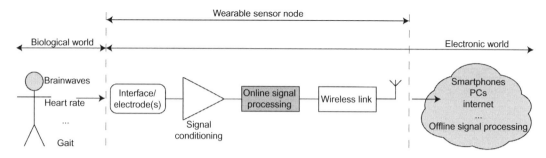

FIGURE 1   Wearable sensor nodes connect the biological world to the electronic world and consist of an interface/electrode(s), amplification and signal conditioning, and wireless transmission of the collected physiological data. Adding online signal processing has a critical role in realizing next-generation devices that have increased functionality and longer operational lifetimes.

can be reliably re-used session after session without having to worry that the device will stop working due to the battery. This is a major source of frustration for users, and limited battery life is the major obstacle to the widespread deployment of wearable sensor systems today. Strategies for maximizing the operational lifetime in next-generation devices are therefore essential.

This chapter will explore how low-power signal analysis algorithms can be integrated into wearable sensors in order to improve the operational lifetime. We will demonstrate that while there are many examples of low-power electronics available in the literature, and similarly many examples of automated processing algorithms, creating successful algorithms for use in a wearable device is not a matter of just connecting the two together. Instead, new *wearable algorithms* are emerging at the interface of these disciplines, and these rely on the close fusion of the application requirements, the sensor node design, the signal-processing design, and the electronic design in order to realize the lowest possible levels of power consumption and to maximize the battery lifetime.

Our objective is to provide practical insights into the creation of these new algorithms. In section 2 we consider the detailed power performance of a current sensor node based on Texas Instrument's popular MSP430 processing chip [14]. This study allows us to demonstrate the design trade-offs present, and the benefits of on-sensor node signal processing in terms of both maximizing operational lifetime and in increasing the range of battery technologies that are suitable for use with a particular senor. Based on this, in section 3 we consider the theory behind wearable algorithms and establish the key objectives that successful algorithms and hardware implementations must meet. This allows us to explore current and emerging techniques for realizing wearable algorithms using very low-power consumption dedicated circuitry in Section 4. We conclude by summarizing the 2013 state-of-the-art and motivate future developments as we move towards realizing truly wearable algorithms for our wearable sensor nodes.

TABLE 1 Approximate specifications of 2013 state-of-the-art low channel count EEG systems for wearable brainwave monitoring. Many devices come in different models and configurations; only one potential configuration is reported here. Physical sizes are as given by the manufacturer and are not directly comparable: some are for the recorder unit alone while others are for the complete recorder plus electrodes system.

Device	Actiwave [2]	Emotiv [3]	B-Alert [4]	Neurosky [5]	Sleep zeo [6,7]	Enobio [8]	Cognionics [9]	Quasar [10]	Mindo [11]	IMEC [12,13]
Channels	4	14	4	1	1	8	16	12	4	8
Sampling frequency [Hz]	128	128	256	512	128	500	500	240	512	1000
Resolution [bits]	8	14	16	12	12	24	24	16	16	16
Size [mm]	37 × 27 × 8.5	–	127 × 57 × 25	225 × 115 × 165	–	225 × 115 × 165	46 × 56 × –	–	165 × 145 × 50	165 × 145 × 50
Weight [g]	8.5	116	110	90	24	65	75	500	100	100
Battery life [hours]	13	12	8	8	8 (1 night)	8	4	24	20	20
Wireless?	No	Yes	Yes	Yes	Yes	Yes	Yes	Yes	Yes	Yes
Dry electrodes?	No	No	No	Yes	Yes	Yes	Yes	Yes	Yes	Yes
Status	Commercial									Research

2. WHY DO WEARABLE SENSORS NEED ALGORITHMS?

We begin by investigating the practical challenges in the design of miniature wearable devices that use standard off-the-shelf components. Our goal is to minimize the device volume and maximize the device-operating lifetime, while under the constraint of having restricted hardware resources available. We will see how this motivates the use of online, real-time, signal processing as part of the device design and in turn, in section 3, how this leads to wearable algorithms.

The key to the success of a wearable device is the minimization of its size and weight as these directly affect the device's discreteness and comfort. As surface mount components are nowadays small compared to batteries, leaving aside the application-specific interface/electrode(s), the size of the batteries dominates the overall volume of the sensor node. The essential starting point in node design is therefore a consideration of suitable battery technologies, sizes, and performances (section 2.1). We can then consider the hardware platform used to collect the physiological data (section 2.2) and the wireless transmitter used (section 2.3). The design decisions made at this stage have a large impact on the node operating lifetime, and the presence of any online signal processing, as we will see in a practical design example (section 2.4).

2.1 Battery Selection

To guide our investigation, Table 2 summarizes the specifications of four off-the-shelf primary batteries from three different size groups that are potentially suitable for powering wearable sensor nodes. This shows five battery specifications that are critical to consider for low-power wireless design. The physical size, which as discussed above dominates the device volume, and the energy storage capacity, typically expressed in mA-hours, are the well-known parameters. However, all batteries also have an internal

TABLE 2   Specifications of four non-rechargeable disposable batteries potentially suitable for wearable sensor nodes with varying physical sizes and battery technologies. Three classes of battery size are considered: the cylindrical cell (CYC), button cell (BC), and coin cell (CC). (Rechargeable lithium polymer (LiPo) batteries can have higher energy densities than the chemistries listed here, but the minimum physical sizes available are also generally bigger.)

ID	Group	Name	Type	Nominal Voltage [V]	Max Continuous Current [mA]	Nominal Current [mA]	Nominal Capacity [mAh]	Size (Diameter × Height) [mm × mm]
B1	CYC	Xeno XL-050F (1/2 AA) [15]	LiSOCl$_2$	3.5	50	1	1200	14.5 × 25.2
B2	BC	Duracell DA675 (Size 675) [16]	Zn(OH)$_4$	1.4	16	2	600	11.6 × 5.4
B3	BC	Duracell DA13 (Size 13) [17]	Zn(OH)$_4$	1.4	6	0.9	290	7.9 × 5.4
B4	CC	Renata CR2430 [18]	LiMnO$_2$	3	4	0.5	285	24.5 × 3.0

resistance, and this means that the energy stored cannot be optimally discharged into all possible loads. This leads to further important battery parameters, which are discussed below.

2.1.1 Supplied Voltage

Firstly, the supplied battery voltage must meet the operation requirements of the electronic circuitry used in the wearable device. Importantly, the supplied voltage is not the same as the battery nominal voltage as there will be an internal voltage drop in the battery due to the internal resistance, and this drop will vary depending on the current draw. Most off-the-shelf low-power microcontrollers and transceivers today require somewhere between 1.8 V and 3.6 V, and batteries that supply a voltage outside this range must be used with a DC-to-DC voltage converter or stacked in series to increase the delivered voltage. However, these techniques reduce the battery lifetime due to either the extra power consumption from the additional circuitry or due to the increased internal impedance.

2.1.2 Maximum Continuous Current

Secondly, the maximum continuous current (also referred to as the maximum average current supply), $I_{avg(max)}$, limits the average current draw from the battery. In theory, a 200 mAh battery can provide 1 mA for 200 hours, or 200 mA for 1 hour. In practice, each battery actually has a maximum supported current draw and if more current than this is drawn the effective capacity will not be the full 200 mAh reported value. The average current required by the system must be smaller than the corresponding $I_{avg(max)}$ for the specific battery. Otherwise, more than one battery will have to be used in parallel, possibly together with diodes to prevent any non-rechargeable batteries from inadvertently charging. Again, the battery must be able to provide this current without a significant drop in its voltage supply because of the internal impedance.

2.1.3 Maximum Pulse Current Capability

Thirdly, for short periods a battery can provide more than $I_{avg(max)}$, up to a maximum pulse value $I_{pulse(max)}$. $I_{pulse(max)}$ must be large enough to guarantee that variations in the supply voltage, which will drop if the current consumption increases, will not exceed the operating range of any circuit in the wearable sensor. The value of $I_{pulse(max)}$ is a function of how long the pulse must be provided (the hold time); Figure 2 shows typical maximum values of pulse current with various pulse widths for the batteries listed in Table 2. There is a clear decrease in the maximum pulse current that can be provided as the duration of the required pulse increases. This is of particular importance when selecting the radio transceiver block and protocol as this generally determines both the peak current draw and how long it is required for.

2.1.4 Effective Capacity and Lifetime

Combining the above effects allows the effective battery capacity (C_{eff}) to be found. This is a critical parameter for long-term monitoring applications and it will generally be smaller than the nominal battery capacity, unless the system average current consumption (I_{sys}) matches the manufacturer's recommended value (also referred to as the nominal

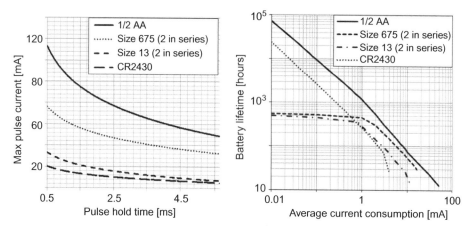

FIGURE 2 Measured discharge characteristics of the small-size batteries from Table 2. Zinc-air batteries are tested in a combination of two in series with no air management. 2.1 V is considered as the discharge voltage. Left: Battery maximum supported pulse current vs. pulse hold time. Right: Battery lifetime vs. average current consumption.

current or the standard discharge current). The value of C_{eff} is hence a function of I_{sys} and it will be lower than its nominal value when I_{sys} is larger than the nominal current.

Based on this, in order for a certain battery to be a viable option for a specific system it must meet the system requirements not only in terms of size and supply voltage, but also in terms of the average and maximum dynamic currents that the system needs to operate. Once the necessary conditions are met the lifetime of the battery, $LT(I_{sys})$, can be determined as

$$LT(I_{sys}) = \frac{C_{eff}(I_{sys})}{I_{sys}} \tag{1}$$

Figure 2 also shows a plot of the measured battery lifetime $LT(I_{sys})$ versus the average current consumption I_{sys} for our battery examples. From this it is clear how, although the CYC battery used requires more volume, it also provides much better performance in all other aspects. Comparatively the zinc-air BC batteries have similar performance only if the effective average current drain is higher than 1 mA. This is due to their high self-discharge characteristic when exposed to air, and for applications with ultra-low effective power consumption (i.e., $I_{sys} < 1$mA) zinc-air batteries should be used with air management so that only a limited (but necessary) amount of air goes into the batteries. Alternatively, in cases when the continuous current is not a limiting factor, the CC battery may be preferable since it has a superior combined performance in terms of self-discharge rate and pulse current capability.

2.2 Hardware Platform

The hardware platform in a wearable sensor node is responsible for collecting the physiological data (data acquisition) and for packaging and passing this data to the wireless transmitter. It must also control the transmitter operation (wireless transmission control), and there are a number of design factors that need to be studied.

2.2.1 Data Acquisition

As shown in Figure 1, a general wearable sensor may contain signal conditioning circuitry (typically an amplifier and an analog-to-digital converter (ADC)), and some optional signal-processing blocks to reduce the amount of data for wireless transmission. In the signal-conditioning block there is at least one anti-aliasing filter that restricts the bandwidth of the input signal before passing it to the ADC. For wearable applications seeking to minimize size, the anti-aliasing filter may be realized as a first-order RC low-pass filter. This is a physically very small circuit, but its use will come at the cost of requiring a higher ADC sampling frequency to ensure that the analog input signal is correctly represented in the digital domain. In turn, this implies that the amount of data passed through the system is increased, leading to an increase in power consumption due to the data multiplication. If desired, and as considered in section 2.4, downsampling of the data can be performed in the digital domain to obtain data compression. It is to avoid similar data multiplication that successive approximation register (SAR) ADCs tend to be preferred in wearable applications, rather than Σ-Δ ones, which are intrinsically based upon oversampling. In addition, SAR converters can operate with a lower peak dynamic current.

2.2.2 Wireless Transmission Controller

After acquisition, the physiological data is temporarily stored in a buffer and packetized to be sent to the wireless transmitter. During this process the controller decides on the structure of the transmission protocol, including the total size of a packet frame (L_{frame}) and the size of the packet header. Both limit the amount of physiological data (L_{data}) that can be carried in each over-the-air data packet and mean that the full over-the-air rate available to the wireless transmitter cannot be used to transmit useful physiological data. The utilization of the protocol ($\eta_{protocol}$) for the data can be expressed as

$$\eta_{protocol} = \frac{L_{data}}{L_{frame} + T_{latency} \cdot R_{air}} \tag{2}$$

where R_{air} is the over-the-air data rate and $T_{latency}$ is the total latency introduced by the data acquisition, packet packaging, and transceiver interfacing.

Note that although longer packet sizes increase the data throughput and relieve the communication overhead, they also increase the transmission time, exponentially increasing the risk of interruption from undesired radio frequency (RF) interferences. This error rate will be explored in more detail below. Further, to avoid extra transceiver control complexity and overhead, the length of each data packet should be designed to match the size of the buffer present in the transceiver being used. This prevents the buffer from overflowing, and here we consider using this optimal packet size only.

2.3 Wireless Transmitter

Finally, the packetized physiological data is passed to the wireless transmitter, which sends it out to the sensor node base station. There are three factors that dominate the power consumption at this stage: the quality of the packet transmission, the hardware overhead, and the over-the-air data rate.

2.3.1 Quality of the Packet Transmission

Unstable transmission quality can cause unnecessary retransmission overhead or even packet loss. The successful transmission rate (η_{tx}) for one packet can be calculated as

$$\eta_{tx} = (1 - PER) \cdot \varphi_{QOL} = (1 - BER)^N \cdot \varphi_{QOL} \tag{3}$$

where PER is the packet error rate, BER is the bit error rate, N is the bit length of a packet, and φ_{QOL}, the quality of link (QOL) factor, which estimates the probability of the RF channel being clear throughout the transmission process. It decays with longer transmission times.

From (3) it can be seen how a long packet size is not advisable. Also, the QOL of radio transceivers is affected by the transmission range and the transmission power. Long transmission ranges decrease the strength of the RF signal and hence lower the QOL. Generally, the transmission RF power can be tuned to trade-off QOL and power consumption.

In addition to this, to compensate for the BER and improve the link quality, chip manufacturers have introduced different features into transceivers so as to strengthen their error tolerance capability. Forward error correction (FEC) and automatic acknowledgement (auto-ack) are two examples that will be used in the test cases considered below. However, the use of these can come with extra cost, the largest one being FEC, which sacrifices half of the available bandwidth to create redundancy.

2.3.2 Hardware Overhead

In order to allow the wireless transmitter and receiver to recognize each other, and to synchronize, the transceiver appends a preamble signal (with size $L_{preamble}$) and its own transmission identification data (with size L_{txid}) to the front of every packet automatically. Once the packet reaches the receiver these extra bits are examined and removed by the packet-handling hardware. Again, this results in not all of the over-the-air data rate being available for useful data transmission. The hardware efficiency η_{hw} is a key factor in this and can be calculated as

$$\eta_{hw} = \frac{L_{data}}{L_{frame} + L_{preamble} + L_{txid} + (T_{cal} + T_{switch}) \cdot R_{air}} \tag{4}$$

where the last term in the denominator accounts for a certain overhead generated by the time required to calibrate the PLL module of the RF synthesizer before transmission: T_{cal} is necessary in battery powered systems in order to avoid the frequency drift caused by variations in the supply voltage, and T_{switch} is the time taken by the transitions between different low-power states.

2.3.3 Air Data Rate vs. Effective Data Rate

Given all of these factors the effective data transmission rate can be calculated, and this must be sufficient for the wanted application. Moreover, for the same amount of data, the faster the data rate the lower the effective power consumption as the transmitter can be turned off for more of the time. However, the overheads discussed in previous sections decrease the effective data rate and must not be ignored in power management. The effective bandwidth (R_{eff}) of the transmission process can be estimated as

$$R_{eff} = \eta_{protocol} \cdot \eta_{tx} \cdot \eta_{hw} \cdot R_{air} \tag{5}$$

2.4 Practical Example and the Impact of Data Compression

We now take the design constraints from sections 2.1, 2.2, and 2.3 and apply them to a real sensor platform to demonstrate how they impact the node design and performance. Further, we also demonstrate the practical impact of on-sensor node signal processing for improving the operating lifetime. Our architecture is shown in Figure 3, and is based upon the popular MSP430 microcontroller to be representative of many current sensor nodes.

2.4.1 Node Design

Our system has a single channel that starts at the high-pass filter, normally used to eliminate out-of-band low frequency components such as electrode drift. This is followed by an amplifier that conditions the normally very weak signals for the subsequent blocks, a low-pass anti-aliasing filter (first-order RC filter made using discrete surface mount components) and a 10-bit SAR ADC, which is part of the MSP430 chip (MSP430F2274).

Here, we investigate two variants on this basic system: one using a Texas Instruments CC2500 transceiver as the transmitter stage and one using a Nordic RF24L01 + transceiver. Both of these operate in the 2.4 GHz band and are connected to the MSP430 by the serial peripheral interface (SPI) with maximum 10 MHz and 8 MHz clocks, respectively. The CC2500 has an option to enable FEC, while the RF24L01 + provides hardware support for auto-ack.

We use Texas Instrument's SimpliciTI protocol stack with both transceivers, although it was originally designed for the CC2500 only. SimpliciTI is a low-power, lightweight wireless network protocol dedicated for battery-operated devices that require long battery life. From [19] the header of a non-encrypted SimpliciTI packet frame contains 96 bits. Because the RF24L01 + has half the buffer size of the CC2500 (256 bits), one field in the header is used to store the byte length of the user data, and this has been modified for the RF24L01 + . In the user data field we add a time stamp for the recording, 20 bits in length, to each packet as an additional header. Given the memory constraints of the used MSP430 the rest of the space in a packet allows the CC2500 to transmit 34 samples (340 bits),

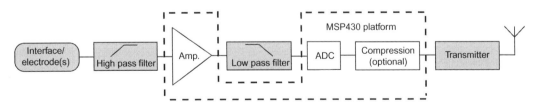

FIGURE 3 Example MSP430-based wireless sensor node used in the design examples presented here. The platform can be contrasted with the general node architecture presented in Figure 1. In our examples we use both a Texas Instruments CC2500 and a Nordic RF24L01 + as suitable low-power transmitter stages.

TABLE 3 Measured current consumption of the hardware platform and its constituent parts, excluding the wireless transmitter. Note that the CC2500 system has a 10 MHz clock while the RF24L01+ uses an 8 MHz clock.

Block	Peak Current [mA]	Duty Cycle	Effective Current [mA]
Sensor	0.27	100%	0.27
Amplifier	0.36	100%	0.36
10-bit SAR ADC	1.10	19.8% (45 µs per sample)	0.22
Compression by downsampling (optional)	2.50 (CC2500 system) 2.10 (RF24L01 + system)	57.5% (130 µs per sample) 71.9% (163 µs per sample)	1.44 1.51
Total	4.23 (CC2500 system) 3.83 (RF24L01 + system)	–	2.29 2.36

whereas the RF24L01 + can transmit 18 samples (180 bits). In total, the packet size is 456 bits for the CC2500 and 294 bits for the RF24L01 + .

To maximize the data throughput, minimize latency, and allow the transmitters to be duty cycled to reduce the average power, the air data rates of the transceivers are set to their maximum values: 500 kbits/s for the CC2500 and 2 Mbits/s for the RF24L01+ . As recommended in [20], the preamble signal of the CC2500 is set to 96 bits. The RF24L01+ has 8 bits fixed preamble length [21] and both the CC2500 and RF24L01+ have extra identification data appended by hardware with lengths of 16 and 19 bits. Overall, assuming 100% of QOL, the effective data rate can be calculated from Eqs. (2) to (5) as 106 kbits/s and 338 kbits/s, respectively.

2.4.2 Optional Data Compression

To cover the physiological range we assume a bandwidth of up to 1 kHz. However, due to the use of a first-order passive RC anti-aliasing filter we sample at 4 kHz. Given this oversampling we again investigate two variants on the system: one where we transmit all of the collected data and one where we first downsample the data by a factor of 2. This lets us explore the impact of even modest data compression on our node lifetime.

The downsampling is implemented using a tenth-order Kaiser window digital low-pass filter (FIR) running on the MSP430. Our MSP430 model contains no hardware multiplier to reduce the computational burden of this processing.

2.4.3 Power Performance Results

For the core system, excluding the wireless transmitter, Table 3 shows the current consumption of each block present. The typical current profiles of the two transceivers are illustrated in Figure 4, where label *A* marks the section corresponding to the transmission overhead, *B* marks the actual wireless transmission times, and *C* indicates the optional auto-ack reception.

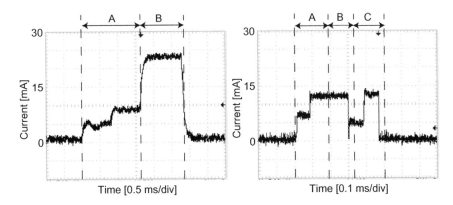

FIGURE 4 Measured current consumption profiles of the two transceivers with $V_{dd} = 2.4$ V. Left: CC2500, with no FEC, packet size with overhead is 568 bits, 34 samples. Right: RF24L01 +, with auto-ack, packet size with overhead is 321 bits, 18 samples. Section *A* is the overall transmission overhead, *B* is the actual wireless transmission time, and *C* the optional auto-ack reception.

TABLE 4 Current consumption of the sensor nodes with compression enabled and disabled. Only some of the miniature batteries listed in Table 2 are capable of powering the system for 24 hours.

Transmitter	Configuration	Effective Current [mA]	Peak Current [mA]	Batteries Passed
CC2500	Compression off, FEC off	15.8	24.3	B1
	Compression off, FEC on	18.5	24.3	B1
	Compression on, FEC off	7.4	25.6	B1, B2
	Compression on, FEC on	13.5	25.6	B1, B2
RF24L01 +	Compression off, auto-ack off	4.5	14.3	B1, B2, B3
	Compression off, auto-ack on	5.6	15.2	B1, B2, B3
	Compression on, auto-ack off	3.5	15.1	B1, B2, B3, B4
	Compression on, auto-ack on	3.8	16.1	B1, B2, B3

Combined with the wireless transmitter, the two systems are compared in Table 4, which shows the current consumption of the systems when tested over 24 hours using the different batteries given in Table 2. It can be seen how some configurations only pass the 24-hour test with certain batteries. Not all miniature batteries are suitable for long-term monitoring! This has direct implications on the kind (and number) of batteries that may be required in a specific design, and consequently on the device size. For 24-hour monitoring applications with 22 kbits/s data rate, the smallest wireless configuration can be achieved by using one lithium manganese dioxide CR2430 3V battery.

Turning on the data compression leads to substantial reductions in the effective current drawn by the system. Reductions by 20 to 30% are achieved, and in the CC2500 case without FEC the reduction is 53%. In all of the cases this reduction allows more of the battery technologies from Table 2 to be used, giving greater freedom in the system design and in the optimization of the device size. In the best case the effective current consumption of the entire system goes down to 3.5 mA, equivalent to a net power consumption of 480 nJ/ bit. It is important to highlight, however, that only reductions in the effective current are achieved by the data compression. There are no substantial differences in the peak currents, and in some cases these may now become the limiting factors in the system design.

2.5 Summary

Designing a wearable sensor node involves a careful set of trade-offs between the electronic components used, the battery technology selected, and the implementation of any real-time signal processing. This section has demonstrated that in battery selection, which dominates the end physical size of the device, both the average and peak current draws have to be taken into consideration. Our presented numbers can be used as a realistic guide for system designers when distributing their power budget and when estimating the size and kind of battery required to operate their device for a certain length of time. Further, quantitative measured results have shown for two different transmitters the potential benefits of onboard signal processing for increasing the operational lifetime of the device. The challenge now is to realize more advanced signal processing to extend the operating lifetime even further.

3. WHAT ARE WEARABLE ALGORITHMS?

Section 2 demonstrated that by simply downsampling data by a factor of two reductions in the total system, power consumption of up to 53% could be achieved. Moreover, this was done with the MSP430 active for up to 72% of the time. Clearly, if more data reduction could be provided, or if the signal-processing platform (our MSP430) could be turned off more of the time to reduce its effective power, even greater increases in operational lifetimes could be provided.

The challenge, of course, is in realizing accurate data reduction algorithms that can operate within the limited power budgets available. *Wearable algorithms* is the name given to the emerging signal-processing approaches attempting to do this, and they differ from conventional algorithmic approaches in three important respects. In this section we explore these in detail and establish the theory behind, and requirements of, wearable algorithms.

3.1 Power–Lifetime Trade-Off

The example given in section 2 showed one case where online data compression can be used to increase the operational lifetime of a sensor node and to allow more battery technologies and hence physical sizes. We now consider the more general case and put bounds on the performance required in order to provide power beneficial signal processing, following the analysis originally introduced in [1,22,23].

Considering the example system given in Figure 1, the power consumption of the entire system can be approximated by

$$P_{system} = NP_{sc} + P_{alg} + CP_t \tag{6}$$

where P_{sc} is the power consumption of the front-end amplifier and any other signal conditioning such as the ADC and N is the number of simultaneous recording channels present with one front-end per channel. P_{alg} is then the power budget available for implementing the signal-processing algorithm, while P_t is the power consumption of the transmitter. C is the ratio between the size of the raw physiological data and the size of the data actually sent from the transmitter. If the signal processing passes all of the collected data to the transmitter $C = 1$, and as more data reduction is provided, this number decreases.

Transmitters are commonly specified in terms of the energy per bit (J) required to transmit data effectively, and in this case P_t can be approximated as

$$P_t = Jf_sRN \tag{7}$$

where f_s is the sampling frequency and R is the resolution of the ADC, which together define the total number of bits of physiological data collected.

As a result, if the inequality

$$P_{alg} < Jf_sRN(1 - C) \tag{8}$$

is satisfied, a system with data reduction will consume less power than one that doesn't have data reduction present. Taking typical values [1,22], $f_s = 200$ Hz, $R = 12$ bits, $J = 5$ nJ/bit and $N = 8$ channels, gives a maximum possible power budget (when $C = 0$ so complete data reduction with no actual data transmitted) of 96 μW.

In reality C will be somewhat larger than zero, bringing this budget down. Also, to account for the approximations, and to ensure decent improvements in lifetime are provided for the effort expended, the typical power target may be reduced by a factor of 10 to approximately 1 to 10 μW. This is very low indeed, and occurs for a multi-channel wearable sensor node, as opposed to the single-channel system from section 2, as both more front-ends are required and there is substantially more physiological data to be transmitted.

As we established in section 2, minimum power consumption is not the only design criteria: more energy can potentially be provided if physically larger batteries are used, but at the cost of a physically larger device. For a battery of volume V and energy density D operating over a lifetime T, the system power budget available is

$$P_{system} = \frac{VD}{T} \tag{9}$$

Combining (9) with (6) and (8) gives a three-way trade-off between the amount of data reduction achieved, the power budget available to implement the signal-processing algorithm, and the operational lifetime that is then possible. This trade-off is plotted in Figure 5, using the same values as before, and using $P_{sc} = 25$ μW. The normalized lifetime T_n

$$T_n = \frac{T}{VD} \tag{10}$$

is plotted in place of T so that the curve is independent of battery technology.

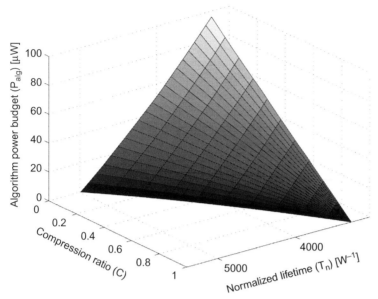

FIGURE 5 Online signal processing embedded in wearable sensors can be used to increase the operational lifetime of the device if the algorithm means that not all of the raw physiological data needs to be transmitted, and if the algorithm can operate within the power budget shown.

This shows that if a 50% data reduction is provided by an embedded signal-processing algorithm within a 10 μW power budget, the operational lifetime of the eight-channel wearable sensor node can be increased by 15%. If 80% data reduction was achieved ($C = 0.2$) the lifetime would be increased by 28%.

Inevitably these figures are approximations, and typical values need to be mapped to the wearable sensing situation under consideration, but in all cases the power budgets available are very low. In 2010 authors in the *IEEE Signal Processing* magazine posed the question: *"What does ultra low power consumption mean?"* and came to the conclusion that it is where the *"power source lasts longer than the useful life of the product"* [24]. This is exactly what is required for maximizing the operating lifetimes of our wearable sensors. However, to realize such low-power signal processing, huge advances in power performance are still required. The aim of wearable algorithms is to bridge this gap and to bring algorithm power consumptions down into the needed microWatt and sub-microWatt levels.

3.2 *Big Data* Performance Testing

Classic signal-processing algorithms are assessed in terms of the *performance* obtained and the *cost* of getting this performance. There are many different metrics that can be used to quantify these, but human physiological signals are highly variable and inevitably algorithms are not perfect leading to a trade-off between the *performance* and *cost*. For example, when attempting to detect events such as falls, a number of correct detections will be made along with a number of false detections. Different algorithms, and different versions of the same algorithm, can provide different trade-offs, and these can be plotted as a curve as shown in Figure 6.

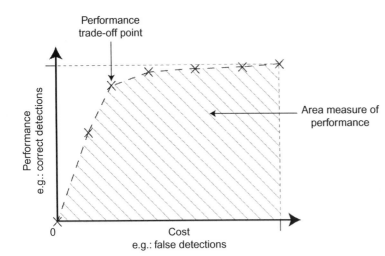

FIGURE 6 Algorithm performance is inevitably a trade-off between *performance* and *cost*. In this case, the number of correct detections of an event and the number of false, incorrect, detections that an event has occurred. Often algorithms can be used with different detection thresholds to allow different points in this trade-off to be used. In this case, the area under the performance curve can be used as a measure of the overall success of the algorithm.

Unfortunately, ensuring that this performance testing is accurate and representative of the actual underlying algorithm performance is a major challenge [25,26]. Here we illustrate one example situation to demonstrate this and show that it is very easy to inadvertently skew algorithm results so that they appear better or worse than they actually are.

In event detection algorithms there are three main metrics that are used to assess the operation. (Often the same metrics as discussed here are used, but under different names.) The sensitivity is the main *performance* metric and shows how many of the actual events that are present are correctly detected:

$$Sensitivity = \frac{TP}{TP + FN} \times 100\% \tag{11}$$

where TP is the number of correct detections (true positives) and FN is the number of real events that are missed and not detected (false negatives).

The specificity and the selectivity are then different *cost* metrics. The specificity shows how many non-events that should not be detected are indeed not detected:

$$Specificity = \frac{TN}{TN + FP} \times 100\% \tag{12}$$

where FP is the number of incorrect, false, detections of an event (false positives) and TN is the number of non-detections (true negatives). Often it is not easy to define what a true negative actually is. The selectivity shows what fraction of all of the detections made are in fact correct:

$$Selectivity = \frac{TP}{TP + FP} \times 100\% \tag{13}$$

However, Eqs. (11) to (13) do not account for the fact that in the emerging *Big Data* era the available test data is made up of multiple different recordings, potentially from different test sites and with different durations and numbers of events present. To actually calculate a figure to plot on trade-off curves such as Figure 6 we need to take the results from each recording and combine the results together. This can be done in a number of ways.

Say there are M records available for testing, recorded from different people or from the same person at different points in time. Let each record be indexed by i. The *arithmetic mean sensitivity* can be found by calculating the sensitivity in each individual record and then averaging these values:

$$Arithmetic\ mean\ sensitivity = \frac{1}{M}\sum_{i=1}^{M}\frac{TP_i}{TP_i + FN_i} \times 100\% \tag{14}$$

Alternatively, the number of correct detections and the number of missed detections can be summed separately to give the *total sensitivity*:

$$Total\ sensitivity = \frac{\sum_{i=1}^{M} TP_i}{\sum_{i=1}^{M}(TP_i + FN_i)} \times 100\% \tag{15}$$

This treats all of the records as if they were one long record concatenated together.

The performance of an example EEG spike detection algorithm [25,26] using these two metrics is shown in Figure 7. It can be seen that these two approaches give quite different pictures of the performance, particularly in the 20 to 40% *cost* region. This occurs because one of the records available for testing contains many more actual events than the other records. Nevertheless, both reporting approaches are mathematically and conceptually correct, so which is more suitable for the performance evaluation?

This is just one possible example, and there are many other factors that can potentially skew reported performance results and that need to be taken into account (see, e.g., [27−29]). Of all of the challenges facing wearable algorithms, it is possible that performance metrics is the key one where there is still substantial need for exploration,

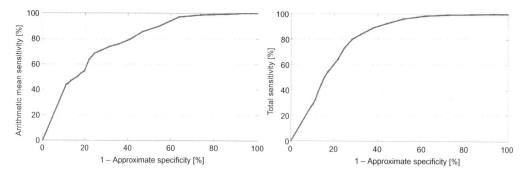

FIGURE 7 Different methods for reporting the results from the same algorithm can give very different pictures of the apparent level of algorithm performance. Both are mathematically and conceptually correct, so which is more suitable for the performance evaluation? Left: Arithmetic mean averaging method (14). Right: Total sensitivity averaging method (15).

improvement, and better understanding of how our algorithms actually operate over time and over multiple people. The availability of *Big Data* is driving this: it is no longer feasible or acceptable to test algorithms using data from just one subject, or to simply report the performance for each individual subject separately. Of course, in turn, wearable sensors are also driving *Big Data*: the aim of wearable sensors is prolonged physiological monitoring intrinsically giving us much more data to use in our algorithms.

3.3 Performance–Power Trade-Off

Finally, section 3.1 established that wearable algorithms need to operate with the lowest levels of power consumption, ideally into the sub-microWatt range. However, absent from the performance metrics discussion in section 3.2 was any consideration of the power consumption! True wearable algorithms are assessed in terms of the three-way trade-off between *performance*, *cost*, and *power consumption*.

Inevitably this leads to difficult decisions for the system designer: is it preferable to maximize *performance*, or to minimize *cost*, or to minimize *power consumption*? Is an algorithm with very low-power consumption, but comparatively low algorithm performance, a better choice than a higher power, higher performance algorithm? These choices are driven by the application that the wearable sensor is to be used in and there are many design options available. This leads to the key hallmark that differentiates wearable algorithms from previous approaches. Designs for wearable algorithms must span four levels: the human monitoring application design, the signal-processing design, the performance-testing design, and the circuit design, simultaneously. There are interactions between all of these different levels, making wearable algorithms a truly multi-disciplinary problem.

3.4 Summary

Wearable algorithms are a new discipline distinguished by the requirement for very low-power hardware implementations, *Big Data* performance testing, and power consumption aware performance testing. The algorithm design must also span four levels: the human monitoring application design, the signal-processing design, the performance testing design, and the circuit design, and exploit the new design trade-offs that are present when considering all of these levels at the same time. Inevitably this creates a very large design space to be explored, and this space is far from being fully mapped out. Nevertheless, there are emerging techniques that can be used to help realize wearable algorithms, and these form the subject of section 4.

4. WEARABLE ALGORITHMS: STATE-OF-THE-ART AND EMERGING TECHNIQUES

The signal processing applied in the design example in section 2 was a relatively straight-forward downsampling of the physiological data, based on a tenth-order FIR digital filter. The MSP430 used did not have a hardware multiplier present and as such, while it was duty cycled to save power, the MSP430 was still on for up to 72% of the time. Even with this relatively modest duty cycling of the signal-processing platform, total power reductions of up to 53% were achieved. Using more sophisticated signal processing approaches, and coupling these with advanced hardware implementations, can offer even greater improvements in node lifetime. Potential avenues for achieving this are explored here to highlight the main developments.

4.1 Making the Signal Processing Algorithm

4.1.1 Procedure

Online signal processing for wearable sensors can take two forms:

- Application agnostic compression applied equally to all of the physiological data, essentially like making a zip file in real-time.
- *Intelligent* signal processing where analysis of the current data drives the operation of the senor node.

As we will see in section 4.3 most current wearable algorithms focus on this second option; the core stages required are shown in Figure 8. The raw input signal (y) is first passed to a feature extraction stage which *emphasizes* the points in the signal of interest: signal processing is applied such that interesting sections of the input signal are amplified relative to the non-interesting sections. For example, in ECG heart beat detection the aim would be to highlight the time at which the QRS complex occurs.

These features are then normalized to correct for the fact that physiological signals vary widely between different people and in the same person over time. This may be due to different underlying disorders present, changes in the signal due to age (as happens with the EEG [30]), or due to changes in the interface/electrodes(s) contact over time. Normalization aims to correct for these changes (to some extent) to allow reliable and robust operation of the algorithm, even if a subject-dependent classifier is used.

The final step is generating an output and using the input physiological signal to actually make a decision. Generally this takes the form of a classification engine that might provide a binary answer: Is a heartbeat present in this section of data? Is the subject awake or asleep? The output could also be multi-class: in this section of data is the subject awake, in light sleep (stages 1 and 2), deep sleep (stages 3 and 4), or in REM sleep? This output can then be used to control the operation of the wearable sensor node to maximize its operational lifetime. For example, in sleep monitoring applications the node might go into

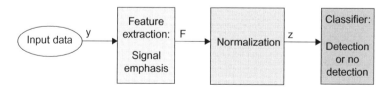

FIGURE 8   The three principal stages of a low computational complexity algorithm: feature extraction, normalization, and classification.

low-power mode whenever it detects that the human subject is awake. Alternatively, the sampling frequency of an ECG may be varied between a high rate during the heartbeat (QRS complex) and a low rate in between beats [31].

4.1.2 Feature Extraction

There are many possible features that could be chosen for use as the signal processing basis, and this forms one of the main design decisions impacting algorithm performance. Many of the algorithms considered in section 4.3 are based upon frequency domain information. For example, sleep onset in the EEG is known to be characterized by a reduction in the presence of 8 to 13 Hz (alpha) activity with this being replaced by 4 to 8 Hz (theta) activity [32]. The feature extraction in this case may therefore be a Fourier transform to follow the changes in these frequency bands. Alternatively, time-frequency transforms such as the continuous wavelet transform or discrete wavelet transform could be used [33,34].

For use in wearable sensor nodes the number of different features used is ideally minimized to reduce the power consumption and the number of circuit stages required. Hence it is important to make an optimal feature choice. A recent study that investigated 63 different features for highlighting seizure activity in the EEG [35] found that discrete wavelet transform-based features obtained the best performance (with an area under the performance curve of 83%), while fractal dimension and bounded variation features offered little over chance performance (53%). However, this doesn't rule out the possibility that these other features if used with a different classification approach would get better performance, or that combinations of more than one feature could again lead to better performances.

Further, all features are also not equal in terms of the power required to calculate them, and this must weigh the choices made. Indeed when corrected for run time, [35] prefers a time-domain feature known as the line-length as the single best feature for highlighting seizure activity in scalp EEG.

4.1.3 Classification Engines

The simplest decision-making scheme is a *threshold* where the normalized input feature is simply compared to a fixed detection threshold:

$$
\begin{aligned}
&\text{If feature} > \text{threshold} \\
&\quad \text{Make detection} \\
&\text{Else} \\
&\quad \text{No detection.}
\end{aligned}
\tag{16}
$$

This threshold can easily be varied to produce performance trade-off curves as shown in Figure 6 and has a very low computational complexity. Such an approach has been used in systems such as [35,36] to obtain low complexity performance.

More recently, machine-learning approaches have been used to automatically determine the best detection parameters, and this is particularly useful when dealing with multiple features and the need to select optimal separating planes between classes. Support vector machines (SVM) [37] are quickly becoming the most popular choice of machine learning approach as they achieve high classification accuracies and have recently been implemented at the circuit level with low-power consumptions [38–40].

4.2 The Hardware Platform: Analog Vs. Digital; Generic Vs. Custom

4.2.1 Analog Signal Processing

The physiological world that wearable sensors are attempting to monitor is intrinsically analog. Signals such as the EEG, ECG, or glucose concentration vary continuously and in continuous time. In contrast, the world of smartphones, PCs, and the Internet is digital with data represented by strings of binary data (either 0 or 1). These numbers can only represent finite, quantized input values and are taken at discrete time points at a particular sampling frequency. At some point a conversion between the two domains must take place, and this leads to a design choice over which domain to use when implementing wearable algorithms.

In 1990 Eric Vittoz published work on the fundamental power consumption limits of analog and digital processing, which has been since expanded on several times [41–43]. In the analog domain, the basic building block is taken as the integrator, modeled as an ideal transconductor charging and discharging a capacitor at frequency f. (A transconductor is an analog circuit block that outputs a current directly proportional to the input voltage.) The minimum power consumption required for this is found as [43]

$$P_{min} = 8kT \cdot f \cdot SNR \qquad (17)$$

where k is Boltzmann's constant, T the temperature, and SNR the signal-to-noise ratio.

Similarly, in the digital domain the minimum required power consumption can be found by considering how many elementary operations are required, and the power consumption per elementary operation, E_{tr}. For a single pole digital filter this is estimated as [43]

$$P_{min} = 50B^2 \cdot f \cdot E_{tr} \qquad (18)$$

where B is the digital word length in bits. In general, E_{tr} scales with the size of the CMOS technology used for the circuits, down to a fundamental noise governed limit.

These power equations are intended as fundamental limits and so based upon assumptions, in particular that the analog circuits are limited only by the noise floor. Several further bounds have also been derived since these, including ones that assume process variations and the matching of transistors further limits the analog circuits, and ones that offer more general modeling approaches [44,45]. Nevertheless, the underlying trends that (17) and (18) show are illustrative and widely accepted. They are visualized in Figure 9.

This shows that while the fundamental digital limit is well below the analog one, the practical limit can be a lot higher. Moreover, the analog limit is a strong function of the SNR ratio, and hence dynamic range, while for values over 6 bits the digital limit is a much weaker function. The result is that analog processing is generally accepted to be superior for low dynamic range applications. There are numerous sources that note the important role of analog processing systems in such applications, both presently and into

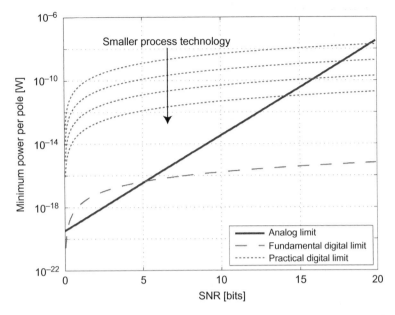

FIGURE 9   The fundamental limits of power consumption for analog and digital filters from [43]. For low SNR (signal-to-noise ratio), and hence low dynamic range, applications an analog approach may be preferable.

the future (see, e.g., [24,45−47]). As a result, several of the signal-processing algorithms considered in section 4.3 incorporate some form of analog signal processing.

Beyond these general trends, however, determining the precise cross-over point for switching between analog and digital processing is very difficult for a particular circuit topology. In fact, even if this was known, the dynamic range of many physiological signals is highly debatable. The EEG systems considered in Table 1 used everywhere between 8 and 24 bits, while classical pen writer-based systems had a range of approximately 7 bits [48]. A key challenge is that once processing is started in the analog domain, additional ADCs to take the results into the digital domain want to be avoided, which means that *all* of the processing must be analog.

4.2.2 *Fully Custom Hardware*

The next choice is on the general design approach: to use generic off-the-shelf components that are easily available commercially or to fabricate custom-designed microchips that implement just the algorithm operations of interest and which can be highly optimized for the wanted application, or a mixture of the two. The general trade-off in performance is illustrated in Figure 10.

In general, the fully custom microchip is the preferred option as it allows complete customization, and so there are no possible excess blocks that are present but not 100% necessary. By implementing everything on the same silicon chip significant miniaturization can

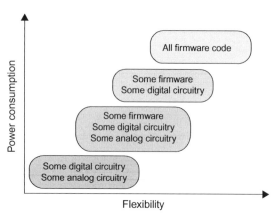

FIGURE 10 In general, off-the-shelf hardware with algorithms implemented in firmware code offer the greatest flexibility in the system design, but also the highest power consumption. As more dedicated digital and then analog hardware circuits are added, lower power consumptions can be achieved at the cost of reduced flexibility. This reduced flexibility is also a key motivation for the attention given to performance assessment in section 3.

also be achieved compared to buying multiple chips and needing a larger PCB to connect everything together.

It is important to highlight that while this is the route to best performance it is inflexible, time consuming, and expensive. It is also not *guaranteed* to deliver the best performance, and careful design is required to minimize power consumption. For example, Figure 9 showed that the power consumption of digital processing is heavily dependent on the technology used to fabricate the microchip: the smaller the technology the lower the power consumption. However, for prototype runs and academic use the smallest processing nodes can be prohibitively expensive, potentially making off-the-shelf-devices that can use such nodes (due to having large fabrication runs) a better design choice.

This said, at present all of the algorithms investigated in section 4.3 use some form of custom microchip and there is little doubt that, while it presents significant design challenges, it is the best approach for realizing wearable algorithms today.

4.3 Towards Wearable Algorithms: Examples from the Literature

To investigate the state-of-the-art we systematically reviewed papers published in *IEEE Transactions* and other journals since 2010 that implement some form of algorithm for use in wearable sensors. The performances of these algorithms are summarized in Table 5 and discussed below. (Note that there are also algorithms for implantable sensors, e.g., [49,50], which are not considered here. Some of the designs reported here are for use in both implanted and body surface recordings.)

Inevitably it is very difficult to capture all of the information and to directly compare different algorithms that are used for different purposes, with different algorithmic approaches, and assessed using different performance metrics, but Table 5 does provide key insights into the main approaches that are currently being used and the current state-of-the-art.

Broadly there are three categories of on-chip algorithm implementation present: firstly, highly optimized, but generic, on-chip processors that can be used to implement any

TABLE 5 Summary of current algorithms implemented in low-power hardware for wearable systems. Many papers report more than one operating point or setup and only one representative case is summarized here. (−) indicates that the information was not reported or was not clear for the case used. Unless trivial, to avoid extrapolations, performances are as given by the authors and have not been reprocessed to use consistent units, although this does make direct comparisons difficult. Note that some algorithms are single channel while others can analyze more than one channel of data at the same time.

Paper	Aim	Features	Classifier	Algorithm Performance	Circuit Basis	Power Performance
GENERIC PROCESSORS						
[51]	ECG heart beat detection	Frequency information (CWT)	Maxima detection and threshold	Sensitivity 99.65% Selectivity 99.79%	Custom CoolFlux processor	12.8 pJ/cycle, 1 MHz clock, 0.4−1.2 V
GENERIC PROCESSORS WITH ACCELERATORS						
[36]	EEG seizure detection	Frequency information (FFT)	Threshold	−	Custom ARM Cortex M3 processor	0.99 μW, 0.8 V
[52]	EEG band power extraction	Frequency information (FIR filter)	−	−	Custom MSP430 core with FFT and CORDIC accelerators	19.3 μJ/512 samples , 0.7 V
	ECG heart beat detection	Frequency information (IIR filter)	Adaptive threshold	−		16.4 μJ/heart beat, 0.7 V
[38]	EEG seizure detection	Frequency information (FIR filter)	SVM	−	Custom MSP430 core with SVM accelerator	273 μJ/classification 0.55−1.2 V
	ECG arrhythmia detection	Time domain morphology	SVM	−		124 μJ/classification 0.55−1.2 V
[39]	ECG arrhythmia detection	Time domain morphology	SVM	−	Custom Tensilica processor with added instructions and SVM	10.24 μJ/classification 0.4 V
FULLY HARDWARE ELECTRONICS						
[53]	EEG application agnostic compression	Compressive sensing		∼10 dB SNDR, x10 data compression	Custom digital circuits	1.9 μW, 0.6 V
[54]	ECG heart beat detection	Frequency information (DWT)	Maximum-likelihood type	Error rate 0.196%	Standard cell digital circuits	13.6 μW, 3 V

(Continued)

TABLE 5 (Continued)

Paper	Aim	Features	Classifier	Algorithm Performance	Circuit Basis	Power Performance
[55]	ECG heart beat detection	Frequency information (DWT)	Maximum-likelihood type	—	Custom digital circuits	0.88 pJ/sample, 20 kHz clock, 0.32 V
[40]	EEG seizure detection	Frequency information (FIR filter)	SVM	Detection rate 82.7% False rate 4.5%	Custom digital circuits	2.03 µJ/classification 128 classifications/s
	EEG blink detection			Detection rate 84.4%		1 V
[56]	ECG artifact removal	Time domain electrical impedance tomography	LMS adaptive filter	~10 dB increase in Signal-to-Artifact power	Off-the-shelf MSP430 with analog co-processing	—
[31]	ECG adaptive sampling frequency	Frequency information (Band-pass filter)	R-peak search algorithm	x7 data compression	Off-the-shelf MSP430 with switched capacitor analog	30 µW, 2 V (Gives x4 power reduction of full system)
[57]	EEG band power extraction	Frequency information (Band-pass filter)	—	—	Switched capacitor analog processing	3.12 µW, 1.2 V
[58]	EEG band power extraction	Frequency information (CWT)	—	—	Continuous time analog processing	60 pW, 1 V

algorithm using software; secondly, generic processors that are combined with application-specific features or accelerators to decrease the power consumption of the key signal processing stages; and finally, fully hardware-based algorithms.

4.3.1 Generic Processors

Our sensor node in section 2 was based on the MSP430 as an easy-to-use, easily available low-power platform. However, in active mode the power consumption was 5.5 mW (at 10 MHz clock, 2.2 V supply). Thus, to realize microWatt levels of power consumption this MSP430 must be powered down for very large amounts of time, limiting the signal processing that can be provided.

There have been a number of recent publications that implement platforms compatible with the MSP430 instruction set allowing algorithm code to be directly re-used, but with a reduced power consumption. The platform described in [59] presented such an architecture consuming 175 µW in active mode (at 25 MHz clock, 0.4 V supply). Note that this was partly achieved by the use of an advanced 65 nm CMOS technology, and for very low-power use the off/leakage current is substantial: 1.7 µW. No specific algorithm was presented for use on this processor and so it is not included in Table 5.

In contrast [51] presented a high-performance general processor based on the *CoolFlux* platform rather than the MSP430 instruction set. This was used to implement an ECG heart beat detection algorithm that consumed only 13 µW (single-channel analysis only). In active mode the processor core power consumption is 1.45 mW, with a high 100 MHz clock.

Analog signal processing is also possible on generic platforms, with [60] presenting a programmable analog chip operating from 2.4 V. This demonstrated an analog FIR filter that could be used for feature extractions and consumed 7 µW at 1 MHz, although again no full algorithm was presented.

4.3.2 Generic Processors with Accelerators

As seen in Figure 8, the main stages in a wearable algorithm are the feature extraction and classification. To improve power performance these stages can therefore be implemented in dedicated hardware while the rest of the algorithm is still implemented in software on a processing core. An algorithm for seizure detection based upon frequency band changes calculated using the FFT was presented in [36]. By providing dedicated hardware for calculating the FFT, while running the rest of the algorithm on a customized ARM Cortex M3 processor, this achieved an 18 times reduction in the power consumption with the total power being less than 1 µW. No measure of the algorithm detection performance was given, however.

Alternatively, [38] used a customized MSP430 core for calculating the features and then an SVM accelerator for reducing the energy impact of the classification stage. The rationale being that the features are often specific to an application, while the classification engine (the SVM) could be re-used in many different situations. This acceleration reduced the classification energy costs by up to 144 times compared to a software-only implementation, although for EEG processing the energy per classification is 273 µJ. Thus, if a fast update rate is required, the total power consumption may still be quite high. Similarly, [39] built upon this to also accelerate some of the feature generations.

4.3.3 *Fully Hardware Electronics and Design Trends*

Most of the works considered in Table 5 do not use any specific software platform and instead achieve very low-power consumption by using only dedicated and highly customized hardware circuits. While there are many different approaches to realizing low-power fully custom electronics, nevertheless a number of common design trends are seen.

Firstly, very low supply voltages (typically in the 0.5–1 V range, and down to 0.32 V in [55]) are widespread. This directly reduces the power consumption (as $P = VI$), but has a big impact on the operation speed of a circuit [55]. Physiological signals are normally low frequency, typically below 1 kHz, and so this trade-off is often acceptable provided that a large number of operations are not required. In addition, often multiple power zones are used with different supply voltages and clock speeds provided to different parts of a single chip depending on the processing required at a particular time. For example, [36] had 18 different voltage domains, while [52] had 15. Dynamic voltage scaling (allowing speed scaling), clock gating (where the clock is disconnected to prevent unnecessary switching), and power gating (where entire sections are turned off to reduce leakage currents) form different aspects of these multiple power zones.

Secondly, very few of the considered circuits are based purely on conventional architectures and a recurring theme is the use of new and simplified topologies. For example, [54] introduces a small modification to the classification process with a small impact on classification performance, but which more than halves the number of circuit blocks required to implement it. A new digital filter topology was introduced in [40], while [55] only used integer coefficients in the filter stages to simplify the multiplications required. The objective with these approaches is to minimize the system complexity and to minimize the transistor count. Doing this gives fewer transistors to power, fewer sources of leakage current, and hence power savings.

The third trend is for the use of analog signal processing, which is used in four of the algorithms considered and is a very powerful method for minimizing the transistor count. A hybrid approach was used in [31] and [56], whereby analog signal processing was combined with an off-the-shelf MSP430 for motion artifact removal and heartbeat detection, respectively. Both [57] and [58] use fully analog approaches for calculating frequency information.

All but three of the entries in Table 5 make use of frequency information as part of the signal-processing algorithm. It is by far the most common basis for the feature extraction stage, and as a result highlights the potential for the use of analog signal processing in this role. In particular, a recent publication [58] presented a continuous wavelet transform (CWT) circuit for performing time-frequency analysis on scalp EEG signals while consuming a nominal power of only 60 pW. This CWT is only a feature extraction stage, but the picoWatt power level is far below any of the other circuit blocks considered in this chapter. It is achieved by the use of very low processing currents and a fully analog signal processing approach. It highlights the potential key role of analog signal processing will have in future systems, and that there is real opportunity to create truly wearable algorithms where *the power source lasts longer than the useful life of the product*.

5. CONCLUSIONS

Wearable algorithms are an emerging truly multi-disciplinary problem where to achieve the lowest levels of power consumption innovations are required on multiple fronts: in the human-monitoring application design, in the signal-processing design, in the performance-testing design, and in the circuit design. This presents a large, four-dimensional, multi-disciplinary design space that has not yet been fully explored by a long way. Many challenges and opportunities are present, and while innovative design at all of the four levels in isolation will be beneficial, for future systems it is critical to exploit the multi-disciplinary factors present and the interactions between the different levels.

This chapter has presented a practical overview of the state-of-the-art in wearable algorithms with the aim of maximizing the operational lifetime of wearable sensor nodes: the detailed design decisions required in a typical system (section 2), the fundamental trade-offs faced by wearable algorithms (section 3), and the current low-power circuit techniques employed (section 4). Each one of these topics could easily occupy an entire book by themselves, but this would not allow the inter-disciplinary links to be drawn out, as we have attempted to do here and which we believe is essential for realizing truly wearable systems.

To conclude, we would reiterate that our aim here has been to maximize operational lifetime as this is the current major obstacle to the large-scale deployment of wearable sensors. However, wearable algorithms do not stop with increased battery life. We believe that there are at least eight essential benefits to using very low-power signal processing embedded in wearable sensor nodes:

- Reduced system power consumption
- Increased device functionality, such as alarm generation
- Reliable and robust operation in the presence of unreliable wireless links
- Minimized system latency
- Reduction in the amount of data to be analyzed offline
- Enabling of closed-loop recording—stimulation devices
- Better quality recordings, e.g., with motion artifact removal
- Real-time data reduction for improved privacy

The challenge remains in realizing accurate algorithms that can operate within the power budgets available. As we have presented here, substantial progress has been made toward realizing these goals in recent years, and our examples should aid the designers of next-generation systems. Nevertheless, there is much progress still to be made in this rapidly evolving field, and much scope for substantially better *intelligent* systems in the future.

References

[1] A.J. Casson, D.C. Yates, S.J. Smith, J.S. Duncan, E. Rodriguez-Villegas, Wearable electroencephalography, IEEE Eng. Med. Biol. Mag. 29 (no. 3) (2010) 44–56.
[2] camNtech Actiwave (2013) Home page. [Online]. Available: <http://www.camntech.com/>.

[3] Emotiv EEG systems (2013) Home page. [Online]. Available: <http://www.emotiv.com/>.

[4] Advanced Brain Monitoring (2013) Home page, B-Alert X4. [Oneline]. Available: <http://advancedbrain monitoring.com/>.

[5] NeuroSky (2013) Home page, MindWave. [Online]. Available: <http://www.neurosky.com/>.

[6] Sleep Zeo (2013) Home page. [Online]. Available: <http://www.myzeo.com/sleep/>.

[7] J.R. Shambroom, S.E. Fabregas, J. Johnstone, Validation of an automated wireless system to monitor sleep in healthy adults, J. Sleep Res. 21 (no. 2) (2012) 221–230.

[8] Neuroelectrics (2013) Home page, Enobio. [Online]. Available: http://neuroelectrics.com/.

[9] Cognionics (2013) Home page, Mini data acquisition system. [Online]. Available: <http://www.cognionics.com/>.

[10] Quasar USA (2013) Home page, DSI 10/20. [Online]. Available: <http://www.quasarusa.com/>.

[11] Mindo (2013) Home page, 4H Earphone. [Online]. Available: <http://www.mindo.com.tw/>.

[12] IMEC (2013) Holst centre and Panasonic present wireless low-power active-electrode EEG headset. [Online]. Available: <http://www.imec.be/>.

[13] S. Patki, B. Grundlehner, A. Verwegen, S. Mitra, J. Xu, A. Matsumoto, et al., Wireless EEG system with real time impedance monitoring and active electrodes, Proc. IEEE Biomed. Circuits Syst. Conf. (2012) 108–111, Hsinchu.

[14] Texas Instruments (2013) Home page, MSP430 microcontroller. [Online]. Available: <http://www.msp430.com/>.

[15] Xeno Energy (2010) Thionyl Chloride Lithium battery XL-050F specifications. [Online]. Available: <http://www.xenousa.com/pdf/XL-050F.pdf>.

[16] Duracell (2010) Zinc air battery DA675 datasheet. [Online]. Available: <http://www.farnell.com/datasheets/6247.pdf>.

[17] Duracell (2010) Zinc air battery DA13 datasheet. Available: <http://www.farnell.com/datasheets/6248.pdf>.

[18] Renata (2010) 3V Lithium battery CR2430 datasheet. [Online]. Available: <http://www.renata.com/fileadmin/downloads/productsheets/lithium/3V_lithium/CR2430_v07.pdf>.

[19] Texas Instruments (2010) SimpliciTI: Simple modular RF network specification. [Online]. Available: <http://focus.ti.com/docs/toolsw/folders/print/simpliciti.html>.

[20] Texas Instruments (2010) CC2500 low-cost low-power 2.4 GHz RF transceiver datasheet. [Online]. Available: <http://focus.ti.com/lit/ds/symlink/cc2500.pdf>.

[21] Nordic Semiconductor (2010) nRF24L01 + single chip 2.4 GHz transceiver product specification. [Online]. Available: <http://www.nordicsemi.com/eng/content/download/2726/34069/file/nRF24L01P_Product_Specification_1_0.pdf>.

[22] D.C. Yates, E. Rodriguez-Villegas, A key power trade-off in wireless EEG headset design, Proc. 3rd Int. IEEE Neural Eng. Conf. (2007) 453–456, Kohala coast, Hawaii.

[23] A.J. Casson, E. Rodriguez-Villegas, Generic vs custom; Analog vs digital: On the implementation of an online EEG signal processing algorithm, Proc. 30th Int. IEEE Eng. Med. Biol. Soc. Conf. (2008) 5876–5880, Vancouver.

[24] G. Frantz, J. Henkel, J. Rabaey, T. Schneider, M. Wolf, U. Batur, Ultra-low power signal processing [DSP Forum], IEEE Signal Process. Mag. 27 (no. 2) (2010) 149–154.

[25] A.J. Casson, E. Luna, E. Rodriguez-Villegas, Performance metrics for the accurate characterisation of interictal spike detection algorithms, J. Neurosci. Methods 177 (no. 2) (2009) 479–487.

[26] A.J. Casson, E. Rodriguez-Villegas, Interfacing biology and circuits: quantification and performance metrics, in: Iniewski (Ed.), Integrated Bio-Microsystems, Wiley, 2011, pp. 1–32.

[27] A. Temko, E. Thomas, W. Marnane, G. Lightbody, G.B. Boylan, Performance assessment for EEG-based neonatal seizure detectors, Clin. Neurophysiol. vol. 122 (no. 3) (2011) 474–482.

[28] R. Akbani, S. Kwek, N. Japkowicz, Applying Support Vector Machines to imbalanced datasets, Proc. 15th Eur. Conf. Mach. Learn. (2004) 39–50, Pisa.

[29] L. Logesparan, A.J. Casson, E. Rodriguez-Villegas, Assessing the impact of signal normalization: Preliminary results on epileptic seizure detection, Proc. IEEE Eng. Med. Biol. Conf. (2011) 1439–1442, Boston.

[30] P.E.M. Smith, S.J. Wallace, Clinicians Guide to Epilepsy, Arnold, London, 2001.

[31] R.F. Yazicioglu, K. Sunyoung, T. Torfs, K. Hyejung, C. Van Hoof, A 30 μW analog signal processor ASIC for portable biopotential signal monitoring, IEEE J. Solid State Circuits 46 (no. 1) (2011) 209–223.

[32] A. Rechtschaffen, A. Kales (Eds.), A Manual of Standardized Terminology, Techniques and Scoring System for Sleep Stages of Human Subjects, Public Health Service, U.S. Government Printing Office, Washington DC, 1968.

[33] S. Mallat, A Wavelet Tour of Signal Processing: The Sparse Way, Third ed., Academic Press, San Diego, 2008.

[34] P.S. Addison, J. Walker, R.C. Guido, Time–frequency analysis of biosignals, IEEE Eng. Med. Biol. Mag. 28 (no. 5) (2009) 14–29.

[35] L. Logesparan, A.J. Casson, E. Rodriguez-Villegas, Optimal features for online seizure detection, Med. Biol. Eng. Comput. 50 (no, 7) (2012) 659–669.

[36] S.R. Sridhara, M. DiRenzo, S. Lingam, S.-J. Lee, R. Blazquez, J. Maxey, et al., Microwatt embedded processor platform for medical System-on-Chip applications, IEEE J. Solid State Circuits 46 (no. 4) (2011) 721–730.

[37] A. Statnikov, C.F. Aliferis, D.P. Hardin, I. Guyon, A gentle Introduction to Support Vector Machines in Biomedicine: Theory and Methods, World Scientific, Singapore, 2011.

[38] K.H. Lee, N. Verma, A low-power processor with configurable embedded machine-learning accelerators for high-order and adaptive analysis of medical-sensor signals, IEEE J. Solid State Circuits 48 (no. 7) (2013) 1625–1637.

[39] M. Shoaib, N.K. Jha, N. Verma, Algorithm-driven architectural design space exploration of domain-specific medical-sensor processors, IEEE Trans. Very Large Scale Integration (VLSI) Syst. 21 (no. 10) (2013) 1849–1862.

[40] J. Yoo, Y. Long, D. El-Damak, M.A.B. Altaf, A.H. Shoeb, A.P. Chandrakasan, An 8-channel scalable EEG acquisition SoC with patient-specific seizure classification and recording processor, IEEE J. Solid State Circuits 48 (no. 1) (2013) 214–228.

[41] E.A. Vittoz, Future of analog in the VLSI environment, Proc. IEEE Int. Symp. Circuits Syst. vol. 2 (1990) 1372–1375, New Orleans.

[42] E.A. Vittoz, Low-power design: Ways to approach the limits, Proc. IEEE Int. Solid State Circuits Conf. (1994) 14–18, San Francisco.

[43] C.C. Enz, E.A. Vittoz, CMOS low-power analog circuit design, Tutorial IEEE Int. Symp. Circuits Syst. (1996) 79–133, Atlanta.

[44] P. Kinget, M. Steyaert, Impact of transistor mismatch on the speed-accuracy-power trade-off of analog CMOS circuits, Proc. IEEE Custom Integr. Circuits Conf. (1996) 333–336, San Diego.

[45] R. Sarpeshkar, Analog versus digital: Extrapolating from electronics to neurobiology,", Neural Comput. 10 (no. 7) (1998) 1601–1638.

[46] S.A.P. Haddad, W.A. Serdijn, Ultra Low-Power Biomedical Signal Processing: An Analog Wavelet Filter Approach for Pacemakers, Springer, Dordrecht, 2009.

[47] L. Tarassenko, Interview with Lionel Tarassenko, Electron. Lett. 47 (no. 26) (2011) s29.

[48] G.L. Krauss, R.S. Fisher, The Johns Hopkins Atlas of Digital EEG: An Interactive Training Guide, Johns Hopkins University Press, Baltimore, 2006.

[49] B. Gosselin, M. Sawan, An ultra low-power CMOS automatic action potential detector, IEEE Trans. Neural Syst. Rehabil. Eng. 17 (no. 4) (2009) 346–353.

[50] M.T. Salam, M. Sawan, D.K. Nguyen, A novel low-power-implantable epileptic seizure-onset detector, IEEE Trans. Biomed. Circuits Syst. 5 (no. 6) (2011) 568–578.

[51] J. Hulzink, M. Konijnenburg, M. Ashouei, A. Breeschoten, T. Berset, J. Huisken, et al., An ultra low energy biomedical signal processing system operating at near-threshold, IEEE Trans. Biomed. Circuits Syst. 5 (no. 6) (2011) 546–554.

[52] J. Kwong, A.P. Chandrakasan, An energy-efficient biomedical signal processing platform, IEEE J. Solid State Circuits 46 (no. 7) (2011) 1742–1753.

[53] F. Chen, A.P. Chandrakasan, V.M. Stojanovic, Design and analysis of a hardware-efficient compressed sensing architecture for data compression in wireless sensors, IEEE J. Solid State Circuits 47 (no. 3) (2012) 744–756.

[54] Y.-J. Min, H.-K. Kim, Y.-R. Kang, G.-S. Kim, J. Park, S.-W. Kim, Design of wavelet-based ECG detector for implantable cardiac pacemakers, IEEE Trans. Biomed. Circuits Syst. 7 (no. 4) (2013) 426–436.

[55] O.C. Akgun, J.N. Rodrigues, Y. Leblebici, V. Owall, High-level energy estimation in the sub-V_T domain: Simulation and measurement of a cardiac event detector, IEEE Trans. Biomed. Circuits Syst. 6 (no. 1) (2012) 15–27.

[56] N. Van Helleputte, S. Kim, H. Kim, J.P. Kim, C. Van Hoof, R.F. Yazicioglu, A 160 μA biopotential acquisition IC with fully integrated IA and motion artifact suppression, IEEE Trans. Biomed. Circuits Syst. vol. 6 (no. 6) (2012) 552−561.

[57] F. Zhang, A. Mishra, A.G. Richardson, B. Otis, A low-power ECoG/EEG processing IC with integrated multi-band energy extractor, IEEE Trans. Circuits Syst. I 58 (no. 9) (2011) 2069−2082.

[58] A.J. Casson, E. Rodriguez-Villegas, A 60 pW g_mC Continuous Wavelet Transform circuit for portable EEG systems, IEEE J. Solid State Circuits 46 (no. 6) (2011) 1406−1415.

[59] D. Bol, J. De Vos, C. Hocquet, F. Botman, F. Durvaux, S. Boyd, et al., SleepWalker: A 25-MHz 0.4-V sub-mm^2 7 μW/MHz microcontroller in 65-nm LP/GP CMOS for low-carbon wireless sensor nodes, IEEE J. Solid State Circuits 48 (no. 1) (2013) 20−32.

[60] C.R. Schlottmann, S. Shapero, S. Nease, P. Hasler, A digitally enhanced dynamically reconfigurable analog platform for low-power signal processing, IEEE J. Solid State Circuits 47 (no. 9) (2012) 2174−2184.

Mining Techniques for Body Sensor Network Data Repository

Vitali Loseu[1], Jian Wu[2], and Roozbeh Jafari[2]

[1]MPI Lab, Samsung Research America, [2]University of Texas at Dallas, Richardson, Texas, USA

1. INTRODUCTION

Body sensor network (BSN) research is maturing and increasingly moving from preliminary sensor, node, and network evaluation to investigation of more partial applications ranging from fall and posture detection [1,2], telemedicine, rehabilitation, and sports training [3,4]. These systems are composed of lightweight wearable sensors that capture different modalities of data from the human body. This data may include inertial measurements, electrocardiogram (ECG) readings, electromyogram (EMG) readings of the muscle activities, skin conductance level, blood pressure, and several other modalities. The diversity of applications means that the data collection setup has to be uniquely tailored for each application and the type of sensor nodes. This step can either be performed manually or considered to be a parametric optimization problem subject to the application constraints. Manual sensor selection and placement is usually based on prior knowledge of sensor efficiency in detecting a given movement or physiological phenomenon, or it is based on prior knowledge of a particular sensor-placement combination for avoiding a given type of artifact.

For example, positioning a peripheral oxygen saturation sensor on the ear is effective in reducing motion artifacts [5]. Because the ear is located close to the human vestibular system, it is exposed to limited random movements. Similarly, movements related to food intake (e.g., eating) are likely to be better observed with sensors on lower and upper arms [6]. Sensor placement and selection can also be done automatically. While this approach is more general, in a sense it can adapt to the changing application conditions, but it also requires additional resources. This can mean redundant sensor placement, increased sensory and processing load, or even increased communication load [7]. The automated optimization criteria for sensor placement can generally be classified into short-term and long-term strategies, where the former means optimization of the system for the quality of recognition, and the latter corresponds to increasing the sensor network lifetime. The short-term strategy approach can be demonstrated by maximizing the probability of detecting a target [8] or improving hazard detection by drivers on the road [9]. An example of the long-term strategy can be demonstrated by the approach that tries to optimize the network lifetime while meeting minimum sensing constraints [10]. All of the above automatic sensor management approaches only determine the initial sensor placement. They do not address imprecision of sensor placement or sensor displacement during the experiments.

Sensors of different types are often deployed by researchers, and often the experiments are highly controlled, i.e., the conditions of sensors are carefully monitored. If the sensors are to be used in natural settings, it is often challenging to ensure ideal laboratory settings, control, and monitoring would be available. This is especially true for the BSNs because it is often difficult to ensure sensor placements remain intact after the original deployment. This problem is often addressed by continuous feedback and calibration using additional sensor modalities. For example, sensor fusion of two dual-axis accelerometers and three dual-axis gyros is performed to properly estimate roll, pitch, and yaw of movement despite possible orientation changes and sensor drift [11]. A similar sensor fusion algorithm is designed for a body-worn sensor network that consists of a tri-axis accelerometer, tri-axis magnetometer, and a tri-axis gyroscope [12]. Direction cosine matrix (DCM) in the case of inertial sensors and Kalman filters are often popular choices for calibration and error correction due to the ability to both estimate the desired value and the noise in the system. Both of the above examples rely on the wearable sensor for both estimation and correction; however, that is not always desirable because the setup becomes less seamless and clunky to wear. Additionally, while a Kalman filter or a DCM-based approach are able to track the orientation change of the sensor nodes, the performance of these algorithms degrades drastically if the displacement affects the nature of the biomechanical observations.

For example, if a sensor node is to get displaced from just above the elbow to just below the elbow, the biomechanical model of the observed movement completely changes and can no longer be described by the previous estimations. This problem can be addressed via utilizing a non-wearable infrastructure. An example of such an approach can be found in an application for indoor map building, where inertial sensors are fused with environment-referenced sensors, such as ultrasonic, optical, magnetic, or RF sensors [13]. The approach works very well due to the availability of a great variety of information. It also moves some of the power consumption away from the wearable device and to a static infrastructure, where the power consumption is not as much of an issue. However, the static infrastructure is also a significant limiting factor in terms of the practical deployment and utilization, and adds costs. Context-aware data fusion is another way of tackling the original problem. Instead of directly using additional sensor modalities as input, a context-aware data fusion relies on establishing a context of a particular action to extract additional information ([14,15]).

BSN platforms are desirable because they provide a relatively inexpensive way to collect realistic and, more importantly, quantitative data about the subjects in their natural environment. Furthermore, sensor nodes can be inserted into common wearable items such as belt buckles, shoes, or pendants, which promotes the adoption by end-users. Ease-of-use in BSN nodes allows them to be quickly deployed, and data collected and processed. A problem that has not received sufficient research attention is storing and tracking the collected data. The data collected from these wearable systems are especially valuable for medical applications. The ability to search and compare BSN observations can potentially shed light on diseases such as Parkinson's disease [16]. Parkinson's disease is a neurological disorder, but many of its symptoms, such as slow automatic movements (e.g., blinking), inability to finish some movements, impaired balance while walking, muscle rigidity, and tremor, severely affect human movements and can be observed with the help of wearable inertial sensors [17]. Wearable nodes can enable monitoring and detection of the onset of a disease. Based on the above properties of BSNs, we envision that in the near future, large repositories of BSN data will be available. The information in such repositories can be utilized to address the problems associated with data collection, including error reduction and variations due to the subject's unique movement execution. This can be achieved by extracting and comparing structural properties of the data. Comparison of the signals relies on the idea that similar movements have inherently similar structure, while not being exactly the same. This idea is important, because it suggests that while observations may not match in their entirety, due to the data collection artifacts and variations in individual subject performance, they still have a distinguishable structural similarity.

The real challenge with this type of approach comes with implementing all the logic on the sensor nodes to avoid a heavy communication energy cost. BSN sensor nodes are highly constrained in terms of memory, processing power, and battery lifetime. This means that not all collected data can be stored on the wearable device, communicated wirelessly for an indefinite amount of time, or processed with complicated and possibly slow computational approaches on the device itself. At the same time, the wearable devices have a potential to produce very large data sets over time. This suggests that the data representation approach needs to significantly reduce the complexity of the data, while maintaining the characteristic properties of the signal, which in our case is the signal structure. This task is further complicated by the possibility of errors in the signal and inter-subject variability in movement performance. This problem can be solved by applying limited processing that exclusively focuses on identifying unique (in terms of a given application) structural blocks and transitions between them. In the context of the activity recognition application, this approach can be further simplified to detecting transitions in the signal that uniquely characterize each movement. For this step to be successful, it is essential for the system to extract the properties of the signal capable of capturing such characteristic transitions. While in other system components (such as communication), redundancy may be acceptable and even desirable, the resource and time constraints of the BSNs demand that the considered set of signal properties be minimal. With these requirements in mind, the chapter reviews existing mining techniques and presents a novel data-mining model for large BSN data repositories based on a pilot application on human movement monitoring.

2. MACHINE LEARNING APPROACHES TO DATA MINING

The goal of data mining is identifying relevant objects. The relevance of an object may be defined by some features or parameters and its similarity to other objects. This task is trivial in a well-structured and indexed database. However, when the data is not trivially structured, defining features and measures of similarity are not obvious. Possible techniques addressing this issue can be generally partitioned into two phases. Information retrieval is the first phase, where important information for a given application is extracted from possibly noisy data. Object summarization is the second phase, where features of interest are defined in the context of relevant information extracted during the first phase. The first phase combines information theory with the properties of the specific object type. The second phase tries to identify the best way to store and parse the metadata extracted during the first phase to efficiently mine the data and identify relevant objects. Before looking into the details of BSN data mining, it is important to look into a set of basic machine learning techniques that are often used for data mining.

2.1 Mining Techniques

The most simple classification rule based on a set of instances is called 1R or 1-Rule. In this approach, the system selects one attribute of the collected sensor readings and makes a classification decision based on it. While this a very simple approach, it tends to work reasonably well for some applications [18]. The rule selection can be described as follows: For each possible attribute, the system can count how often each value of that attribute appears in any given class and make an attribute-class assignment based on the value appearing most often. It can calculate the error of all of the attributes based on the cross validation set and select the attribute with the least error. This algorithm faces two major issues: First, it may not be able to account for the values that are missing in the training set. Second, when an attribute has a large range of values, it is prone to over-fitting (or may detect trends specific to the training data that are not desired for detection and mining).

Statistical modeling is a potential solution to the problem. Instead of selecting only one attribute, the system can select all of the attributes, assuming that they are independent. Such an approach is known as Naive Bayes and can perform well when the assumptions hold [19]. The approach has two major problems. First, it assumes that each of the attributes is independent, which is likely not the case for many real problems. This problem has been extensively studied in [20,21], where the authors propose a semi-Bayes approach to modeling the actual data dependencies, correct data bias and manage attribute weights. The second problem is the assumption that the attributes have normal distribution, which may not hold in many practical applications.

Another way to address the issue of different attributes having diverse ranges is based on divide-and-conquer. Typically, this suggests creation of a tree-like structure, where each node corresponds to a specific attribute [22]. This way an attribute does not correspond to a whole level of the tree, meaning that at the same level different branches may use different attributes. This approach works in a top-down manner, where, at each level of the tree, it seeks the best attribute to split the remaining data. The difficulty of this approach lies in selecting proper attributes at levels in the tree. It often uses feature selection algorithms such as information gain [23], mutual information [24], and utility-based solutions such as Bayesian information criterion (BIC) [25]. Shortcomings of these approaches are defined by their respective assumptions. For example, information gain tends to work very well when attributes have very few possible values, suitable for binary attributes [26]. Information-gain performance decreases as the number of possible values increases due to the nature of the entropy calculations.

Previously described approaches work best with nominal attributes; however, the idea can be extended to the numerical attributes as well. The most simple and relatively effective approach is known as linear regression, where the idea is to represent class values as linear combinations of the attributes and their respective weights [27]. The idea is to calculate proper weights during the training process and apply the classifiers on the validation data. While this approach often works very well, there is a serious drawback: it assumes that the data can be modeled in a linear fashion, which may not be a valid assumption. This problem can be addressed with the help of logistic regression and then evaluated with log-likelihood maximization [28]. A major problem with this approach is probabilities not adding up to 1 when the logistic regression is applied to multiple classes.

In instance-based learning, the training trials themselves are used to evaluate unknown samples. It is done with the help of a distance function defined for the data of interest. For classification purposes, the system measures the distance from an unknown trial to the training sample and selects the one with the shortest distance. A simple example of this learning time is the 1-nearest-neighbor (1-NN) approach. However, this approach considers each attribute equally just like Naive Bayes. Additionally, a specific classification can be heavily affected by the outliers that do not represent the class well. These problems can be partially addressed by a k-NN approach, where instead of finding the nearest sample in the training data, the system looks for a consensus among k nearest neighbors [29]. However, k-NN approaches are very slow.

Clustering algorithms are applied when there is no predetermined class to be detected, but rather the observed instances are split into natural groups. During the clustering step, the instances are combined together, based on strong resemblance, to form groups that can act as classes during the detection process. There are many approaches for clustering implementation, but they mainly focus on bringing the similar instances together, while separating the dissimilar instances. One of the most commonly used clustering approaches is known as k-means [30]. It takes the training instances and the number of desirable clusters as an input and groups instances together based on their proximity. In the context of the Euclidean distance, the k-means approach iteratively minimizes the total squared distance from each instance to the cluster centers. It generally has two weaknesses. First, the best number of clusters is not always obvious, while a bad choice can result in improper grouping. Second, the iterative approach heavily depends on the initial selection of the cluster centers. Different random selections can result in significantly different clustering.

2.2 Structural Recognition in BSN

In the context of BSN data, the idea of structural data representation and recognition is explored in [31]. This approach has a major weakness. The comparison evaluation is based on the value of Levenstein distance (or edits distance) [32]. Edit distance calculation assigns the same weight to deletion, insertion, and substitution operations. It is not a problem when the compared strings have similar size. However, BSNs can observe the same movement at different speeds, which may mean that the speed of movement execution can dominate the edit distance value. It is possible to manually manage the weights of each one of the three edit distance operations; however, that would generate a heuristic approach [33]. Another way to deal with this issue is to normalize the length of each primitive in motion transcripts [34]. While this approach might work in some specific applications, in general it is very hard to predict how to scale parts of movements depending on the overall execution speed. A possible solution to this problem is to identify significant transitions in the motion transcripts and base the comparison on variations in these transitions. In the field of speech processing, a similar function is often performed by n-gram features. N-grams are substrings of length N. They were first introduced by Shannon [35] as a means of analyzing vulnerability of ciphers but since have been extensively used in the field of speech and text recognition.

N-grams [36] proved to be useful for structural parameter extraction when used for spoken language recognition [37]. N-grams can be used to capture phoneme, in the case of spoken language, and grammatical constructs, in the case of written language, to identify bodies of speech or text. Similarly, n-grams can be used to analyze text summaries [38] or translation quality [36] with respect to co-occurrence statistics. While good at recognizing major structural differences, n-grams can also be used in the case of fine grain spelling error correction [39]. In addition to maintaining structural information of the considered string, n-grams can significantly reduce the amount of information that needs to be stored and verified. Instead of storing a large body of text, the system can identify important transitions and improve both the memory usage and the execution speed of the search. These important n-grams can be better organized with a suffix tree [40], which would increase the speed of identifying language constructs [41]. In fact, suffix trees are often used to index large amounts of data in the natural language processing and other fields. For example, in the field of molecular biology, DNA sequences can be indexed with the help of suffix trees [42]. In [43] the authors discuss an efficient query algorithm on a large compressed body of text using suffix trees. The general effectiveness of the suffix trees is discussed in the work trying to identify local patterns in an event sequence database [44].

At first glance, the above examples have little in common with data collected from BSNs. Suffix tree approaches normally index a uni-dimensional data set, while BSNs normally have a set of multiple sensors with multiple dimensions of sensing. This problem can be resolved by combining all data readings and representing them with uni-dimensional primitives [31]. While this simple approach seems to resolve the issue, it fails to recognize that each one of the sensing dimensions (or individual orthogonal sensing axis of sensors such as accelerometer) can observe variations such as changing speed and amplitude of the signal. In a text data set, the variations are one dimensional, just like the data itself; this is not the case in multidimensional sensor readings of BSNs. Furthermore, it is not clear how variations occurring in multiple sensing dimensions should be handled in the context of a one-dimensional primitive. It is possible that different combinations of signal variations may hinder the structural consistency of the combined primitive representation. Classical mining techniques are not well suited for BSN data due to their inability to capture structural and relational properties of the signals and not providing a compressed data representation. Therefore, novel data-mining methodologies are required to address these shortcomings.

3. MINING BSN DATA

3.1 BSN System

The BSN system in this chapter consists of a set of wearable nodes placed on the human body to collect inertial activities from human movements and a computer that maintains the BSN repository and facilitates data organization and mining. The wearable nodes are connected to the computer via wireless radios. The system overview is shown in Figure 1 and the details on the operation of the system are described in the following sections and summarized in section 7.

3.2 Wearable Sensor Hardware

Each wearable sensor has a tri-axial accelerometer (providing x-, y-, and z-axis of acceleration) and a bi-axial gyroscope (providing x and y axis of angular velocity) that is sampled at 50 Hz. This sampling frequency is high enough to provide acceptable resolution of the movements, and has been previously suggested by several investigations on physical movement monitoring applications [45,46]. Furthermore, it satisfies the Nyquist criterion

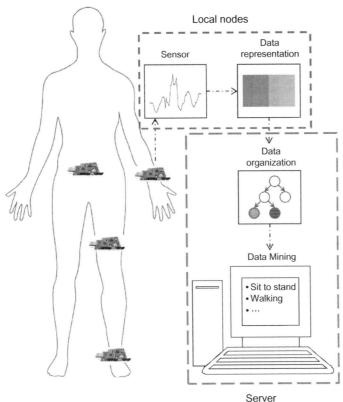

FIGURE 1 BSN mining system overview.

TABLE 1   Pilot Application Movements

No.	Description
1	Stand to Sit
2	Sit to Stand
3	Stand to Sit to Stand
4	Kneeling, right leg first
5	Turn counter clockwise 90 degrees
6	Look back clockwise
7	Move forward (1 step)
8	Move to the left (1 step)
9	Move to the right (1 step)
10	Jumping

[47]. After collecting the data, each node sends its readings to the base-station, which forwards all of the received data to the PC via a USB connection for further processing.

3.3 Pilot Application

To illustrate data mining in a large BSN repository, the proposed approach is applied to a classification problem, where the classification is performed by mining a database of inertial sensor data. Furthermore, as the mining approach is designed for a large data set, the aim is to make it as fast as possible. The database in this case consists of data of ten movements from three subjects. The details of the experimental movements can be found in Table 1. Every subject repeated each movement ten times. Each subject was asked to wear nine sensor nodes positioned on both ankles, both thighs, both wrists and upper arms, and one on the belt (as shown in Figure 2).

3.4 Desirable Solution Properties

A data-mining technique for the BSN data repositories should take raw sensor readings as an input and perform a computationally efficient search in the repository to identify signals similar to the input. Due to a potentially large size of the repository, the approach needs to be fast yet reliable. Sensor readings can be viewed as observations made by the system. Upon receiving an observation as the input, the search approach should be able to identify a movement class to which the observation belongs, so that it can be stored in the appropriate place in the repository. It should be able to compare signals of two movements and find possible similarities. Furthermore, it should be able to identify similar portions of the signals, which can be useful to identify abnormality in movements. Finally, it needs to identify movements that contain certain instances of the signal (e.g., identify all movements where the torso moves forward).

FIGURE 2 TelosB sensor node with a custom sensor board.

1.267 in.

2.580 in.

4. DATA REPRESENTATION

A physical movement can normally be represented as a sequence of shorter motions. Capturing the structure of movement involves capturing these shorter motions and timing relationships between them. This can be done by identifying motion primitives. Following the idea introduced in [21] a clustering technique is used for primitive generation. Features are extracted from the signal and clustered. The clustering outcome is dependent on the perspective that the features can provide. This adds flexibility to the data mining because different feature sets can characterize the signal from different perspectives.

4.1 Primitive Construction

Before applying a clustering technique, it is necessary to decide what data set the clustering is applied to. One way to handle this issue is to combine all axes of sensory data to define primitives. This approach is not most suitable because, when the multidimensional data is merged into a uni-dimensional primitive, combining variations of each of the sensing axes could modify the structure of the combined primitives. An example of such an alternation can be a slight delay of the movement. This alternation does not modify the structure of the individual sensory axis signals, but, since alternations of all of the axes are independent of each other, aligning them with respect to time can significantly change the structure of the combined primitives. Figure 3 demonstrates two trials of the same movement where a slight variation in one of the axes, which does not alter the signal structure for that axis, introduces changes to final primitives. While the signals in the second trial have the same structure as the signals in the first trial, as demonstrated by the individual transcripts, their timing is inconsistent with the signals in the first trial. The bottom part of the figure demonstrates the combined transcript generated from two individual transcripts. Large vertical blocks correspond to the parts of the combined transcript of the second trial that do not have a corresponding counterpart in the first trial. This suggests, if the primitive transitions are not aligned in the original signals, the time aligned combination of these signals may not be consistent between both trials. Noise in the inertial data will introduce yet another source of error.

To avoid the issue with alignment, the reading of each sensing axis can be performed separately. Primitives are created for each one of the axes and each sensing axis is treated

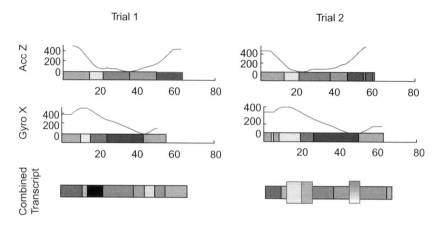

FIGURE 3 Signal alignment issue with respect to time.

as a separate classifier. This approach has an additional benefit of increasing the flexibility of the system. The system does not require all of the sensing axis used in one experiment to be present in other experiments, which means that it does not dictate a particular hardware configuration.

4.1.1 Data Clustering

Clustering is a very effective method of grouping similar data points and distinguishing between different data points. When trying to cluster BSN data, a clustering approach is normally applied to feature vectors extracted from the original signals. There are a variety of features that can be extracted from inertial data. Different approaches rely on first and second derivatives, signal mean, amplitude, variance, standard deviation, peak detection, morphological features, and more. Ideally, a small and simple feature set should be identified that would produce good results. The resultant clusters should be able to identify enough transitions in the signal so that each movement of interest would be characterized with a unique subset of such transitions. A primitive generation experiment concluded that first and second derivatives are sufficient to describe the structure of this data set. To minimize the effect of the inter-subject differences in features, the system normalizes features with respect to each subject using a standard score (or z-score) [48]. In general, the proposed approach is independent of the feature selection and only requires that selected features would represent the structure of the input signal.

There is a wide array of clustering techniques that includes hierarchical, partitional, conceptual, and density-based approaches. For BSN data mining, two clustering approaches were considered. The first approach was k-Means clustering [49]. k-Means is a hierarchical approach that attempts to partition the data in a way that every point is assigned to a cluster with the closest mean, or cluster center. The second approach was expectation maximization in Gaussian mixture models (GMM) [50]. Both approaches try to identify the centers of natural clusters of the data instead of artificially selecting points in the training set as cluster centers. GMM clustering computes the probability that any given point is assigned to every individual cluster and makes an assignment that maximizes its likelihood of such assignment. Both clustering approaches are computationally simple and identify cluster centers of the data set without any prior knowledge of the data. Both approaches start with random cluster centers and re-evaluate them after each round of assignment. Once the cluster centers stay constant within a predefined threshold, both algorithms assume to have converged to the natural cluster centers of the data and return the result.

A parameter to determine during unsupervised clustering is the number of clusters k that produces the best results. To find the best solution, k was varied from 2 to the length of the shortest observation in the training set, while evaluating parameters of both k-Means and GMM. In the case of k-Means, the decision was made based on cluster silhouette [51]. Silhouette is calculated based on the tightness of each cluster and its separation from other clusters. For every point i, the silhouette is defined as

$$s(i) = \frac{b(i) - a(i)}{\max(a(i), b(i))} \tag{1}$$

where $a(i)$ is the average distance of point i to all other points in its cluster, $b_j(i)$ is the average distance of point i to all the points in cluster j, and $b(i) = min(b_j(i))$, $\forall j$.

Silhouette $s(i)$ describes how well the point i is mixed with the similar data points and is separated from the different data points. As a result, the quality of a clustering model with k clusters and d training points can be evaluated as

$$Quality(k) = \frac{\sum_{i:1}^{d} s(i)}{d} \tag{2}$$

The larger the average silhouette value, the better is the model. Therefore, the best value of k can be selected by finding the largest $Quality(k)$ [51].

Expectation maximization (EM) was used [52] to find the best mixing parameters for the GMM. The mixing parameters such as the mean and covariance matrices depend on the number of clusters k. Once the GMM parameters are selected, there are multiple ways to evaluate the quality of clustering that include log likelihood, Akaike's information criterion (AIC) [53], and Bayesian information criterion (BIC) [54]. Table 2 demonstrates the difference between quality estimation models for a GMM with k clusters, maximum likelihood of the estimated model L, and n points in the training set.

Log likelihood reports the likelihood of the model, and AIC and BIC attempt to penalize the system for increasing the number of clusters. For the collected inertial dataset the penalty of the BIC was harsh and led to an extremely small number of clusters. As a result AIC was selected as the GMM evaluation tool.

TABLE 2   Quality Estimation of GMM-Based Clustering Using EM

Log Likelihood	AIC	BIC
$ln(L)$	$-ln(L) + 2 * k$	$-ln(L) + k * ln(n)$

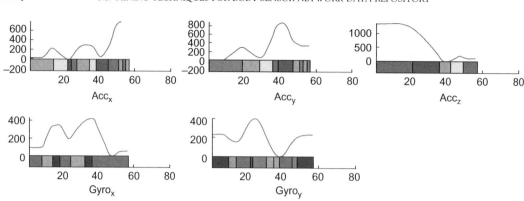

FIGURE 4   Sample transcripts for three-axis accelerometer and two-axis gyro.

4.2 Motion Transcripts

Each movement can be described as a series of primitives. When an unlabeled movement needs to be classified, the system can extract features from each of its data points, and based on the clustering technique, assign motion primitives to them. Motion transcripts are sequences of primitives over a certain alphabet assigned to movement trials. Since the data from different wireless nodes in a BSN are not comparable, the system has to make sure to differentiate between individual nodes by using a unique alphabet for each one. Figure 4 demonstrates a sample transcript generated by the ankle node for a "Lie to Sit" movement. Each one of the sensing axes uses a separate alphabet, so while they are displayed with the same color, the primitives of different transcripts are not related.

5. COMPARISON METRIC

Once the BSN data is converted to motion transcripts, the data mining requires an efficient way to classify and search the transcripts. Section 2 introduced edit distance, a common approach to comparing strings. However, edit distance does not perform well when the input data has noise and varies in length. Additionally, the edit-distance calculation is very slow with order of O(n2), where n is the length of the string. While it may be an acceptable solution for a small data set, its speed performance is not at all acceptable for a large data repository potentially containing terabytes of data. As an alternative, BSN data mining can use the notion of n-grams that track transitions in motion primitives in linear time with respect to the trial length. The goal of the n-grams is to identify common important transitions between movement primitives in string transcripts. However, the task of identifying n-grams that represent important transitions is not simple since overlapping n-grams may be extracted to improve the quality of the recognition. This means that potentially there is a very large number of n-grams that can be selected from any given transcript.

5.1 N-gram Selection

The objective of this operation is to identify a small number of *n*-grams that can uniquely characterize the movement of interest and provide means of distinguishing the movement from others in the repository. There are a variety of ways to select proper *n*-grams, once all *n*-grams are extracted from the training data. Information gain (IG) has proven to be effective in the field of natural language processing [55]. Information gain becomes complicated to compute and less effective when each evaluated feature can have a larger range of values. However, in our experiments, each *n*-gram has two possible values. A specific *n*-gram can be present in a motion trial and the value of "1" is assigned to it, or the *n*-gram can be absent with a value of "0" assigned. While IG proved effective on the described data set, the proposed approach is not dependent on this particular *n*-gram selection technique and can be modified based on the specific user demands.

IG can assess the effectiveness of a feature by tracking changes in the entropy after consideration of that feature. IG of a feature f on the collection of movements *m* is defined as

$$Gain(m, f) = H(m) - H(m|f) \tag{3}$$

where $H(m)$ defines entropy of the movement set and $H(m|f)$ defines conditional entropy of the movement set with respect to feature *f*. A slightly modified approach is used here, because when the system is looking for a target movement all other movements can be treated the same way. It is possible that a feature might be good at identifying one movement while being unable to differentiate between the rest of the movements. That feature would have a bad general information gain, but if the information gain is computed with respect to each movement, suitable features can be identified for each movement. Practically, this means that while computing information gain of a feature with respect to a particular movement m_i, the movement set is split into subsets of $\{m_i\}$ and $\{'not'\ m_i\}$ or $\{m - m_i\}$. In this case, $H(m_i|f)$ can be different for each m_i and need to be calculated individually. This means that the information can be redefined as

$$Gain(m_i, f) = H(m_i) - H(m_i|f) \tag{4}$$

where $H(m_i)$ represents the amount of expected information that set *m* carries itself with respect to movement m_i. Conditional entropy $H(m_i|f)$ defines the expected amount of information the set *m* carries with respect to feature *f* and movement m_i.

Once all *n*-grams have an IG assigned to them for each movement, the list of IGs can be sorted to select *t* *n*-grams that have the best IG. This is a very simple approach because it does not consider correlation between features, meaning that some of the features can be redundant. However, even this simple approach can generate good results [56]. Information gain performance can also suffer from movement or subject-specific signal variations. For example, if a subject has a consistent way of performing a movement, which differs from other subjects, the information gain may select the subject-specific transitions as characteristic for the whole movement. In reality these transitions will not be observed from any other subject and represent an over-fitting. The over-fitting problem can be addressed by disqualifying *n*-grams that do not appear in enough training trials before the information gain is applied. This step makes selecting a clustering technique that creates a sufficient number of clusters even more important. If the number of primitives is low, selecting only the most frequent n-grams, before the information gain is applied, is likely to result in almost identical n-gram subsets selected for every movement.

6. CLASSIFIER

Once a set of good n-grams is selected, an approach for fast movement classification and search is necessary. This approach also should not rely on the knowledge of the complete structure of the data and be able to finish classification and search based on partial information. These properties are exhibited by suffix trees [57]; more specifically, the Patricia tree was used. Patricia trees are used to represent sets of string by splitting them into substrings and assigning substrings to the edges. This idea fits naturally with n-grams that are substrings. Once all of the n-grams are selected for each movement, they are combined and assigned to the edges of a Patricia tree. The paths from the root to all leaves correspond to all possible permutations of the combined n-gram set. This idea is illustrated in Figure 5, where a sample Patricia tree is generated for six movements. The path "BBB," "AEE," and "EBB" corresponds to "Sit to Stand" movements.

Once the Patricia tree is created, each leaf of the Patricia tree corresponds to a subset of the movements. During testing, the n-grams of the test trial are used to traverse the tree and return the corresponding movement set. The result may be an empty set or it may contain one or more movements. Specifically, if not enough n-grams are present to traverse the tree to a leaf, the algorithm returns all movements assigned to the leafs of the subtree rooted at the node where the traversal terminated. For example, in Figure 5 if the traversal would terminate at Node$_2$, then the set containing {Sit to Stand, Stand to Sit, Sit to Lie to Sit, Bend and Grasp} is reported as the answer. If the traversal would terminate at Node$_5$, then only the

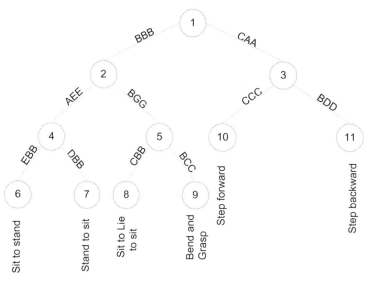

FIGURE 5   A sample Patricia tree for six movements.

set containing {Sit to Lie to Sit, Bend and Grasp} is reported. Finally, if the traversal terminates at a leaf $Node_1$, then the system reports only Step Backward as the answer.

7. DATA-MINING MODEL

Based on the constructs defined earlier, a data-mining approach is proposed for BSN data. The approach has two distinct parts: training and query processing.

7.1 Training

During the training phase of the execution, the system acquires parameters that can be used during the query processing. The training starts with selecting a portion of the available data trials for training. First and second derivatives are then extracted from each one of the trials for every sensing axis. Features are then normalized with respect to each subject using a standard score (or z-score) [48] in order to remove intersubject variations of the same movements. Then normalized features are used to define data clusters as described in section 4.1. Once the data clusters are defined, primitives are extracted for the data points in each training trial and then combined to define motion transcripts as described in section 4.2. The next step is to extract n-grams from each one of the transcripts generated for the training samples. Since the number of n-grams is very large, the system then selects a small number of t n-grams using the information gain as described in section 5.1. Finally, the system constructs a Patricia tree with selected n-grams on the edges and movement classes on the leafs as described in section 6. The overall process is demonstrated in Figure 6. The parameters defined during the training are data clusters for each sensing axis, n-grams selected with respect to the IG criteria, and the Patricia trees for classification. Clusters are represented by the cluster center coordinates, while important features selected for each sensing axis of each mote are then combined and stored.

7.2 Query Processing

When the system needs to search for a query or a movement, it receives input in the form of the sensor readings. First and second derivatives are extracted from the sensor readings of each of the sensing axes. Based on these features and clusters, defined during the training of the system, each data point of the trial is labeled with a primitive. Primitives are combined with respect to timing alignment into motion transcripts. The system then traverses the transcript of the trial and verifies if it contains important n-grams selected during the training. Using this information, the system traverses the Patricia tree defined for this sensing axis during the training and returns the set of movements assigned to the leaf where the traversal terminated. If the traversal terminated at node p that is not a leaf, the algorithm returns all the movements assigned to the leafs of the subtree rooted at node p. This allows the system to avoid introducing bias for the axis that observes the same signal for two different movements. Since all these operations are defined in terms of the individual sensing axis, an approach is required to combine the local decisions. A simple voting scheme is

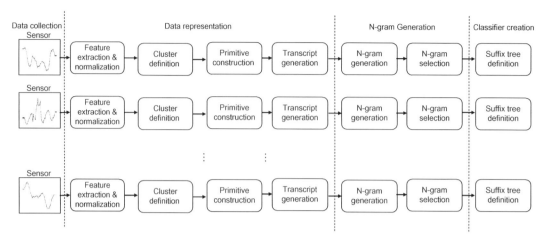

FIGURE 6   System training flow.

used, which performs well in the context of this dataset. However, this method can be improved by treating each sensory axis as an individual classifier. To determine the final decision, the individual classifiers can be combined in an intelligent way such as AdaBoost [58]. The flow of the query processing is demonstrated in Figure 7.

Because the system initially processes each sensing axis individually, it is possible to query only for a subset of axes available in the system. This can be useful when a specific sensor is not available to all users. For example, one user can use a three-dimensional gyroscope, while another may use only a two-dimensional gyroscope. Additionally, since the system uses a voting scheme, it is possible to make classification decisions based on the local view of only a subset of nodes.

8. EXPERIMENTAL RESULTS

To verify the performance of the proposed data-mining approach, it was applied to a pilot application discussed in section 3.3. The verification step was split into two phases. The first phase involved locating the correct place in the repository to store the signal for an unknown trial. To achieve this, the available data was split into two equal sets. The first half of the data was used to train the system, while the second half of the data was used to verify the classification accuracy. The second phase was used to create a representative signal template for each movement, searching the entire repository for the trials consistent with the template.

Initially the data-mining algorithm was trained on the entire data set. During the training, the algorithm selects n-gram sets, or templates, specific to each movement. The individual templates are then used to search for relevant trials in the entire repository. While evaluating the results, the results of classification when k-Means clustering is used are compared to the results when GMM clustering is used. The accuracy of the approach is considered with respect to the length of the n-gram n and number of features selected t. Finally, the accuracy trade-off is demonstrated with respect to n and t. For the second part

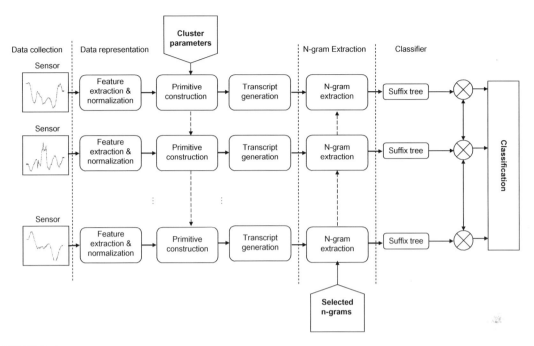

FIGURE 7   Query processing flow.

of the analysis, the average frequency of the template appearance in the repository is assessed. The computed templates are then considered in the context of individual trials. If a trial contains enough *n*-grams from a given template, it is accepted as the movement represented by that template. Otherwise, the trial is rejected. The accuracy of trial classification is reported for each of the movement templates.

8.1 *k*-Means or GMM?

The movement transcript generated based on the *k*-Means and GMM clustering was used to compare two clustering approaches for a 3-gram with the number of selected *n*-grams varying from 1 to 6 per sensing axis as shown in Table 3, and with {1..6}-gram with only 1 *n*-gram selected from each sensing axis as shown in Table 4. Both tables indicate that an increase in *t* or *n* would increase the precision and recall for both approaches until the over-fitting point is reached. It is also clear that the GMM approach outperforms the *k*-Means approach with respect to varying both *n* and *k*, and therefore is a better candidate for the example application.

8.2 Classification Accuracy

To evaluate classification accuracy, the precision and recall of movement classification were evaluated using the *n*-gram size of $n = 3$ and number of features $t = \{1, 2...5\}$ with the

TABLE 3   {3-7}-Gram Average Performance of K-Means vs. GMM with 1 *n*-Gram Selected

n	GMM		k-Means	
	Precision	Recall	Precision	Recall
3	.93	.98	.75	.86
4	.99	1	.76	.84
5	.99	1	.83	.89
6	1	1	.89	.93
7	.95	.99	.91	.93

TABLE 4   3-Gram Average Performance of K-Means vs. GMM with {1...6} *n*-Grams Selected

t	GMM		k-Means	
	Precision	Recall	Precision	Recall
1	.93	.97	.75	.86
2	.96	.98	.88	.94
3	.96	.98	.89	.95
4	.95	.96	.90	.94
5	.94	.96	.91	.96
6	.94	.96	.85	.92

GMM clustering model. Half of the dataset was used to train the system, while the other half was used to test it. The results of the classification are demonstrated in Table 5. The table contains the F-score defined as $(2 \times P \times R)/(P + R)$, where P is the classification precision and R is the classification recall. This table confirms that adding more *n*-grams would improve both average precision and average recall until an over-fitting point is reached. Note that individual values for movements sometimes decrease when an additional feature is selected. This is due to the fact that the data set has a considerable amount of noise, and while an *n*-gram improves the overall classification accuracy, it may cause confusion in classification of some trials where it appears as noise and not an important transition. Table 5 displays the number of *n*-grams extracted from each sensing axis, so the total number of the *n*-grams extracted by a sensor node should be multiplied by 5. However, even after that, the classification accuracy is fairly high for the number of features considered.

8.3 Parameter Trade-offs

The classification accuracy in the proposed data-mining approach depends on two parameters: the length of the substring *n* and the number of *n*-grams *t* selected for

TABLE 5 Classification Precision with Respect to the Number of Features Selected

Movement	Features per Sensing Axis				
	1	2	3	4	5
Stand to Sit	1	1	.958	.958	.857
Sit to Stand	1	1	.958	.958	.925
Stand to Sit to Stand	1	1	1	1	.958
Kneeling, right leg first	.80	.857	.907	.907	.958
Turn 90 degrees	.925	1	.958	.958	.958
Look back clockwise	.958	.925	1	.958	.958
Move forward	1	1	1	1	1
Move to the left	.958	1	1	1	1
Move to the right	.857	.868	.907	.83	.868
Jumping	1	1	1	1	1
Average	.95	.965	.969	.957	.958

TABLE 6 Precision with Respect to n-Gram Size and Number of n-Grams Selected

n	Features per Sensing Axis				
	1	2	3	4	5
3	.95	.97	.97	.96	.95
5	1	.97	.97	.99	.98
7	.99	.99	.99	.98	1
9	.98	.97	.97	.97	.96

classification. As t is increasing, so does the accuracy until the over-fitting point is reached. After the over-fitting point is reached, the accuracy of the approach will no longer improve with additional features. It is clear that a large n inherently is able to capture more structural information. However, since a moving window is used for n-gram extraction, a single erroneous primitive affects more n-grams for larger values of n, which is equivalent to experiencing the over-fitting sooner. The algorithm is expected to converge to the best accuracy faster for large values of n, but it also means that the over-fitting point will happen faster. Table 6 demonstrates the F-score of accuracy vs. the number of n-grams t for different values of n.

From the tables, it is clear that higher values of n are desirable before over-fitting, which means that n should be determined based on the expected amount of noise in the original

TABLE 7 Average Template Evaluation Normalized with Respect to the Template Size

T_i	Movement									
	1	2	3	4	5	6	7	8	9	10
1	**.51**	.36	.4	.37	.32	.29	.37	.35	.34	.22
2	.33	**.49**	.34	.37	.33	.27	.42	.38	.41	.29
3	.39	.41	**.54**	.42	.30	.29	.37	.34	.36	.33
4	.22	.23	.27	**.57**	.37	.21	.37	.31	.36	.27
5	.25	.28	.24	.36	**.55**	.28	.35	.37	.4	.31
6	.42	.40	.4	.44	.46	**.56**	.46	.46	.47	.26
7	.23	.25	.21	.4	.35	.24	**.66**	.42	.5	.26
8	.31	.3	.25	.33	.33	.26	.43	**.61**	.40	.23
9	.27	.27	.23	.39	.35	.26	.48	.41	**.59**	.25
10	.23	.25	.23	.27	.24	.18	.26	.26	.27	**.57**

signal. For a lower amount of noise, a higher value of n would work better, while when the amount of noise is larger, lower values of n will provide a safer solution with less risk of over-fitting. In this example, the precision is improving from $n = 3$ to $n = 5$, it is fairly stationary from $n = 5$ to $n = 7$, and finally, at $n = 9$ has decreasing performance. The fact that larger n-grams take more time to locate in the training trials should also be considered. The system can evaluate multiple possibilities during the training and generate the curves to identify the best operational point from the perspective of accuracy and speed trade-offs.

8.4 Movement Template Evaluation

Based on the parameters selected in section 8.3, T_i is defined as the combination sets of 3-grams selected for each sensing axis during the training process for movement M_i. Here the entire data set is used to train the templates. Once the training is complete and the templates are generated, the average quality measure is evaluated for each T_i. It is done by checking how often the n-grams of each T_i appear in movement trials of every movement. Intuitively, n-grams of T_i should appear more often in M_i than any other movement in order for the template to be effective. Table 7 demonstrates the results of this evaluation normalized with respect to the size of each template, meaning that on average, 51% of T_1 appear in trials of M_1, while only 36% of T_1 appear in trials of M_2.

Two observations can be made based on the results in Table 7. First, it is clear that n-grams of T_i appear most often in the M_i itself. This observation is in line with the expectation for a good template. While n-grams of the T_i, on average, appear 10% more often in the respective M_i, they also appear in trials of other movements with a sizable number of occurrences. These results suggest that a closer look on a per-trial basis is required to evaluate the template quality. The intuitive approach to this problem is to search for trials that have the

TABLE 8 Template Evaluation for Individual Trials Normalized with Respect to the Number of Trials of Each Movement

T_i	Movement									
	1	2	3	4	5	6	7	8	9	10
1	**.48**	0	0	0	0	0	0	0	0	0
2	0	**.41**	0	0	0	0	0	0	0	0
3	0	0	**.78**	.03	0	0	0	0	0	0
4	0	0	0	**.96**	0	0	0	0	0	0
5	0	0	0	0	**.93**	0	0	0	0	0
6	0	0	0	.18	.18	**.89**	.14	.14	.22	0
7	0	0	0	.03	0	0	**1**	0	.44	0
8	0	0	0	0	0	0	.07	**.96**	0	0
9	0	0	0	0	0	0	.22	0	**1**	0
10	0	0	0	0	0	0	0	0	0	**.96**

entire template present. However, in a realistic system with noisy trials, this solution is not practical, since it's unlikely that many trials will be perfect. Table 7 confirms this: the highest average values for each T_i are close to 50% and not 100%. This problem is solved by introducing a variable α, which defines the proportion of the number of n-grams from a T_i that need to be present in a trial for it to be classified as M_i. If the α value is too low, it is likely that some trials will be erroneously identified as M_i, increasing the number of false positives and decreasing the precision of the template. However, if the α value is too high, some trials of M_i will not be identified, increasing the amount of false negative errors and decreasing the recall of the template. Additionally, lower values of α can speed up the computation, which is desirable in our problem due to a potentially large number of movements in the repository. To achieve a balance between the precision and recall, as well as to promote faster data mining, the value of $\alpha = .5$ is selected. The value of 50% is also suggested by Table 7, where the average presence of more than 50% n-grams from a template T_i identifies the movement M_i associated with that template. Table 8 demonstrates the normalized number of trials of each movement identified as M_i by each of the templates T_i, meaning, for example, that template T_1 selects 78% of trials belonging to M_1.

With $\alpha = .5$, each T_i correctly identifies substantially more trials of its own movement than any other movement. This low value of α also defines a search speed increase of up to 50% since only 50% of the trials in templates need to be located. However, higher false negative and false positive errors are observed for certain templates, which suggests that a static value of α is inappropriate. Templates T_1 and T_2 do well at identifying only trials of their respective movements. However, less than half of the appropriate trials are identified. This suggests that the value of $\alpha = .5$ is too high for these templates. At the same time, T_6 and T_7 have a much better rate of recognizing trials of their respective movements, but they also falsely identify trials of other movements. This suggests that the value of $\alpha = .5$ is too low for these templates. Defining a movement-specific value of α_i, based on the training set for each template can decrease the amount of errors in the system.

9. CONCLUSION AND RECOMMENDATIONS

This chapter introduced a data-mining approach for a repository of movement data acquired by wearable sensors that can capture structural properties of the observed movements. This mining algorithm has the potential to significantly simplify the deployment and management of large data repositories since it removes the need for extensive per-subject training. The chapter demonstrated that the proposed approach has suitable performance on the pilot application and explored the trade-offs between the length of the extracted n-grams and the required number of features for the best classification results. While the results are promising, they can be improved in two ways. First, the performance of each component of the data-mining algorithm can be analyzed in detail and the implementation details may be modified for enhancing the performance and reducing the computational complexity. Second, in order to group similar movements together, the approach relies on the fact that similar sensor placement will be used.

It is a strong assumption, granted that for the inertial MEMs sensors even a small misplacement may result in changes in the sensor output. Furthermore, it is likely that during longer periods of sensor deployment, sensors can become loose or misplaced. Advanced data-mining techniques may help with slight orientation changes or displacement; however, the change may become significant with time, and the data for a given subject may become less similar to the original training data from the repository. There are two possible directions that can be taken to address this issue. The proposed data-mining approach needs to be evaluated with respect to the sensor misplacements. This step will clarify the question of "how much displacement is too much?" and will lay the foundation for tracking and compensation of the sensor position and orientation. Tracking of the sensor displacement is challenging without an additional sensory perspective because all of the sensors considered so far collect data in the local frame of reference. Additional information can be extracted from supplementary modalities, e.g., cameras or other types of wearable sensors. These additional modalities can be used to track changes in orientation, and an intelligent algorithm can be developed to continuously calibrate the motion sensors. For example, a camera available at home can be used to calibrate motion sensors every time a user passes by. These periodic calibrations will guarantee consistent system performance.

The pilot application introduced in this chapter demonstrates the behavior of the proposed mining approach based on the example of inertial wearable sensors. The same idea will apply to other types of wearable sensors; for example, ECG or GSR sensors can be analyzed in a similar fashion. To do that, the reader will need to follow the conceptual steps outlined in this chapter. First, extract features from the raw sensor signal that highlight the structural properties of the data for a given application. Then use the extracted features to create signal transcripts. Finally, execute the proposed mining approach with signal transcripts as an input. Note that other applications can have limitations that were not considered in this chapter.

Acknowledgment

This work was supported in part by the National Science Foundation, under grant CNS-1150079. Any opinions, findings, conclusions, or recommendations expressed in this material are those of the authors and do not necessarily reflect the views of the funding organizations.

References

[1] B. Lo, J. Wang, G. Yang, From imaging networks to behavior profiling: ubiquitous sensing for managed homecare of the elderly, in: Adjunct Proceedings of the 3rd International Conference on Pervasive Computing, 2005.

[2] A. Volmer, N. Kruger, R. Orglmeister, Posture and Motion Detection Using Acceleration Data for Context Aware Sensing in Personal Healthcare Systems, in: World Congress on Medical Physics and Biomedical Engineering, September 7–12, 2009, Munich, Germany. Springer, 2009, pp. 71–74.

[3] D. Brunelli, E. Farella, L. Rocchi, M. Dozza, L. Chiari, L. Benini, Bio-feedback system for rehabilitation based on a wireless body area network, March 2006, pp. 5, pp. 531–536.

[4] M. Lapinski, E. Berkson, T. Gill, M. Reinold, J. Paradiso, A Distributed Wearable, Wireless Sensor System for Evaluating Professional Baseball Pitchers and Batters, in: 2009 International Symposium on Wearable Computers. IEEE, 2009, pp. 131–138.

[5] B. Lo, L. Atallah, O. Aziz, M. El ElHew, A. Darzi, G.-Z. Yang, Real-time pervasive monitoring for postoperative care, in: 4th International Workshop on Wearable and Implantable Body Sensor Networks (BSN 2007). Springer, 2007, pp. 122–127.

[6] O. Amft, H. Junker, G. Troster, Detection of eating and drinking arm gestures using inertial body-worn sensors, in: Wearable Computers, 2005. Proceedings. Ninth IEEE International Symposium on. IEEE, 2005, pp. 160–163.

[7] N. Xiong, P. Svensson, Multi-sensor management for information fusion: issues and approaches, Inf. Fusion 3 (2) (2002) 163–186.

[8] D.E. Penny, The automatic management of multi-sensor systems, in: Proceedings of the International Conference on Information Fusion, 1998, pp. 748–755.

[9] W.K. Krebs, J.S. McCarley, T. Kozek, G. Miller, M.J. Sinai, F. Werblin, An evaluation of a sensor fusion system to improve drivers' nighttime detection of road hazards, in: Proceedings of the Human Factors and Ergonomics Society Annual Meeting, vol. 43, no. 23. SAGE Publications, 1999, pp. 1333–1337.

[10] M.A. Perillo, W.B. Heinzelman, Optimal sensor management under energy and reliability constraints, in: Wireless Communications and Networking, 2003. WCNC 2003. 2003 IEEE, vol. 3. IEEE, 2003, pp. 1621–1626.

[11] J. Vaganay, M.-J. Aldon, A. Fournier, Mobile robot attitude estimation by fusion of inertial data, in: Robotics and Automation, 1993. Proceedings., 1993 IEEE International Conference on. IEEE, 1993, pp. 277–282.

[12] K.Y. Lim, F.Y.K. Goh, W. Dong, K.D. Nguyen, I.-M. Chen, S.H. Yeo, et al., A wearable, self-calibrating, wireless sensor network for body motion processing, in: Robotics and Automation, 2008. ICRA 2008. IEEE International Conference on. IEEE, 2008, pp. 1017–1022.

[13] E.M. Foxlin, Generalized architecture for simultaneous localization, auto-calibration, and map-building, in: Intelligent Robots and Systems, 2002. IEEE/RSJ International Conference on, vol. 1. IEEE, 2002, pp. 527–533.

[14] A. Padovitz, S.W. Loke, A. Zaslavsky, B. Burg, C. Bartolini, An approach to data fusion for context awareness, Modeling and Using Context, Springer, 2005, pp. 353–367.

[15] M. Nakamura, J. Nakamura, G. Lopez, M. Shuzo, I. Yamada, Collaborative processing of wearable and ambient sensor system for blood pressure monitoring, Sensors 11 (7) (2011) 6760–6770.

[16] V. Shnayder, B. Chen, K. Lorincz, T. Fulford-Jones, M. Welsh, Sensor networks for medical care, in: SenSys 05: Proceedings of the 3rd international conference on Embedded networked sensor systems. Citeseer, 2005, pp. 314–314.

[17] D. Gelb, E. Oliver, S. Gilman, Diagnostic criteria for Parkinson disease, Arch. Neurol. 56 (1) (1999) 33–39.

[18] I. Witten, E. Frank, Data Mining: Practical Machine Learning Tools and Techniques, Morgan Kaufmann Pub, 2005.

[19] E. Frank, M. Hall, B. Pfahringer, U. of Waikato. Dept. of Computer Science, Locally weighted naive Bayes. Citeseer, 2003.

[20] D. Lewis, Naive (bayes) at forty: The independence assumption in information retrieval, Machine Learning: ECML-98, pp. 4–15, 1998.

[21] I. Kononenko, Semi-naïve Bayesian classifier, Machine LearningEWSL-91, Springer, 1991, pp. 206–219.

[22] J. Quinlan, Induction of decision trees, Mach. Learn. 1 (1) (1986) 81–106.

[23] J. Kent, Information gain and a general measure of correlation, Biometrika 70 (1) (1983) 163–173.

[24] I. Sethi, G. Sarvarayudu, Hierarchical classifier design using mutual information, Pattern Analysis and Machine Intelligence, IEEE Transactions on, no. 4, 1982, pp. 441–445.

[25] W. Chou, W. Reichl, Decision tree state tying based on penalized Bayesian information criterion, in icassp. IEEE, 1999, pp. 345–348.

[26] M. Rogati, Y. Yang, High-performing feature selection for text classification, Proceedings of the Eleventh International Conference on Information and Knowledge Management, ACM, 2002, pp. 659–661.

[27] I. Naseem, R. Togneri, M. Bennamoun, Linear regression for face recognition, IEEE transactions on pattern analysis and machine intelligence, 2010, pp. 2106–2112.

[28] A. Genkin, D. Lewis, D. Madigan, Large-scale bayesian logistic regression for text categorization, Technometrics 49 (3) (2007) 291–304.

[29] S. Rasheed, D. Stashuk, M. Kamel, Adaptive fuzzy k-nn classifier for emg signal decomposition, Med. Eng. Phys. 28 (7) (2006) 694–709.

[30] J. Hartigan, M. Wong, A k-means clustering algorithm, J. R. Stat. Soc. C 28 (1) (1979) 100–108.

[31] H. Ghasemzadeh, V. Loseu, R. Jafari, Collaborative signal processing for action recognition in body sensor networks: a distributed classification algorithm using motion transcripts, Proceedings of the 9th ACM/IEEE International Conference on Information Processing in Sensor Networks, ACM, 2010, pp. 244–255.

[32] V. Levenshteiti, Binary codes capable of correcting deletions, insertions, and reversals, Soviet Physics-Doklady 10 (8) (1966).

[33] S. Kurtz, Approximate string searching under weighted edit distance, in: Proc. of Third South American Workshop on String Processing. Citeseer, 1996, pp. 156–170.

[34] A. Marzal, E. Vidal, Computation of normalized edit distance and applications, IEEE Trans. Pattern Anal. Mach. Intell. (1993) 926–932.

[35] C. Shannon, Communication theory of secrecy systems, MD Comput. 15 (1) (1998) 57–64.

[36] G. Doddington, Automatic evaluation of machine translation quality using n-gram co-occurrence statistics, Proceedings of the Second International Conference on Human Language Technology Research, Morgan Kaufmann Publishers Inc., 2002. p. 145.

[37] A. Adami, H. Hermansky, Segmentation of speech for speaker and language recognition, in: Eighth European Conference on Speech Communication and Technology, 2003.

[38] C. Lin, E. Hovy, Automatic evaluation of summaries using n-gram co-occurrence statistics, in: Proceedings of hlt-naacl, vol. 2003, 2003.

[39] K. Kukich, Techniques for automatically correcting words in text, ACM Comput. Surv. (CSUR) 24 (4) (1992) 439.

[40] U. Manber, G. Myers, Suffix arrays: a new method for online string searches, Proceedings of the First Annual ACM-SIAM Symposium on Discrete Algorithms, Society for Industrial and Applied Mathematics, 1990, pp. 319–327.

[41] F. Pereira, Y. Singer, N. Tishby, Beyond word n-grams, in: Proceedings of the Third Workshop on Very Large Corpora, 1995, pp. 95–106.

[42] S. Burkhardt, A. Crauser, P. Ferragina, H. Lenhof, E. Rivals, M. Vingron, Q-gram based database searching using a suffix array (QUASAR), Proceedings of the Third Annual International Conference on Computational Molecular Biology, ACM, 1999. p. 83.

[43] K. Sadakane, Compressed text databases with efficient query algorithms based on the compressed suffix array, Algorithms Comput. 295–321.

[44] X. Jin, L. Wang, Y. Lu, C. Shi, Indexing and mining of the local patterns in sequence database, Intell. Data Eng. Automated Learn. IDEAL (2002) 39–52.

[45] N. Ravi, N. Dandekar, P. Mysore, M. Littman, Activity recognition from accelerometer data, in: Proceedings of the National Conference on Artificial Intelligence, vol. 20, no. 3, 2005, p. 1541.

[46] R. Mayagoitia, A. Nene, P. Veltink, Accelerometer and rate gyroscope measurement of kinematics: an inexpensive alternative to optical motion analysis systems, J. Biomech. 35 (4) (2002) 537–542.

[47] N. Stergiou, Innovative Analyses of Human Movement, Human Kinetics Publishers, 2004.

[48] G. Milligan, M. Cooper, A study of standardization of variables in cluster analysis, J. Classif. 5 (2) (1988) 181–204.

[49] J. MacQueen, et al., Some methods for classification and analysis of multivariate observations, in: Proceedings of the Fifth Berkeley Symposium on Mathematical Statistics and Probability, vol. 1, no. 281–297. California, USA, 1967, p. 14.

[50] H. Friedman, J. Rubin, On some invariant criteria for grouping data, J. Am. Stat. Assoc. (1967) 1159–1178.

[51] P. Rousseeuw, Silhouettes: a graphical aid to the interpretation and validation of cluster analysis, J. Comput. Appl. Math. 20 (1987) 53–65.

[52] G. McLachlan, T. Krishnan, The EM Algorithm and Extensions, Wiley, New York, 1997.

[53] K. Yamaoka, T. Nakagawa, T. Uno, Application of Akaike's information criterion (AIC) in the evaluation of linear pharmacokinetic equations, J. Pharmacokinet. Pharmacodyn. 6 (2) (1978) 165–175.

[54] G. Schwarz, Estimating the dimension of a model, Ann. Stat. (1978) 461–464.

[55] Y. Zhang, S. Vogel, Measuring confidence intervals for the machine translation evaluation metrics, Proceedings of TMI, vol. 2004, pp. 85–94, 2004.

[56] M. Masud, L. Khan, B. Thuraisingham, A scalable multi-level feature extraction technique to detect malicious executables, Inf. Syst. Front. 10 (1) (2008) 33–45.

[57] S. Inenaga, H. Bannai, A. Shinohara, M. Takeda, S. Arikawa, Discovering best variable length don't care patterns, in: Discovery Science, Springer, pp. 169–216.

[58] P. Viola, M. Jones, Fast and robust classification using asymmetric adaboost and a detector cascade, Adv. Neural Inf. Process. Syst. 2 (2002) 1311–1318.

Modeling Physical Activity Behavior Change

Edmund Seto[1] and Ruzena Bajcsy[2]

[1]School of Public Health, University of Washington, Seattle, Washington, USA, [2]Electrical Engineering and Computer Science, University of California, Berkeley, California, USA

1. INTRODUCTION – PHYSICAL ACTIVITY MONITORING CAPABILITIES

During the last twenty or so years, several groups, including ours, have used body sensors (a network of wireless accelerometers and other sensors that measure physiological parameters) on people during their physical activities, such as walking, falling, exercising, and the like. Most of these experiments were performed on small samples of the population, mostly on college students over relatively short time periods, and the scientific literature describing these studies is extremely scattered, distributed between engineering publications (*IEEE Transactions on Biomedical Engineering, IEEE Transactions on Systems, Man, and Cybernetics, Journal of Biomechanics*, etc.) and rehabilitation publications (*Occupational Therapy, Sports Medicine*, etc.). These small-scale studies of body sensors have rarely moved outside of the experimental laboratory to real-world practice.

Despite this, body sensors offer new possibilities to conduct increasingly detailed physical activity assessment. Beyond the simple waist-worn pedometer or accelerometer, using a variety of wearable sensors and different sensor modalities communicating with one another can provide a much richer perspective on a person's physical activity (e.g., accelerometry at different body locations to assess posture and to recognize activity types; heart rate, respiration, and galvanic skin response sensors to assess fitness levels as exercise occurs; and GPS and other types of sensors to provide time-location measurements that offer contextual information for when and where exercise occurs).

Moreover, wireless technologies and integration of sensors with mobile phones are providing opportunities to communicate the results of these assessments in more creative, interactive, and meaningful ways so as to be more effective in motivating individuals to continue their progress toward adopting healthy levels of physical activity. There remain many challenges in integrating diverse sensor data, as well as challenges related to what to do with the information derived from these data. How should this information be fed back to the person, or communicated to a person's social contacts (personal as well as health professional) so as to best help him continue exercising? Indeed, Clarke and Skiba [1] argue that there remains a "dearth of longitudinal studies to guide long-term training program designs." This argument can be made for most of the current studies in observing human physical activities, and hence provide the proper guidance for intervention.

The challenge of how to provide information feedback is not simply a technology problem, but a human-centric technology problem. Technology is a way to collect data that ultimately needs to meaningfully inform a person's ongoing process of adopting new behavior. Adopting new behavior is inherently a cognitive and psychological process that is more subjective and not as easily assessed as the more objectively measured changes in physical activity behavior. A number of theories have been proposed that govern how individuals adopt new behaviors. Indeed, these behavior-change theories — one in particular being the transtheoretical model [2] — have been applied to many health problems, including smoking cessation [3], problems with addiction [4], diet [5], and physical activity [6]. The challenge is how to integrate wireless body sensors with these theories.

The goal of this chapter is to examine these challenges. Specifically, we will consider:

- The need to utilize sensors to personalize behavior-change interventions to specific individuals
- The need to accurately quantify and model physical activities using body sensors in the context of interventions
- The transtheoretical model of behavior change and how it may be integrated with models of physical activity

1.1 Sensors can Inform Personalized Behavior-Change Interventions

In 2008, the U.S. Department of Health and Human Services issued its Physical Activity Guidelines for Americans report [7] that provides specific science-based guidelines for people of various ages to improve their health through physical activity. It recommends that children and adolescents obtain at least 60 minutes of physical activity per day, adults obtain at least 150 minutes of moderate or 75 minutes of vigorous intensity activity, and older adults try to obtain as much physical activity as younger adults. While these recommendations are based on scientific evidence of health benefits from population studies, on an individual level, a person may find such recommendations meaningless in terms of motivation to increase physical activity if he or she already meets the recommendation or feels that there is no hope of ever meeting the recommendation.

In some cases, such group standards can do more harm than good. Individuals who do not meet healthy guidelines may be ostracized by society [8], and this may lead to negative emotions or ambivalence toward healthy behavior. This suggests that while national population-based guidelines may serve as crude goals for individuals, much greater attention needs to be placed on "personalized" behavioral interventions. Within intervention programs, people need to be thought of as individuals. The ability for one person to be more physically active than another relates to many factors, including their personal motivation and barriers toward exercise. Thus, personalized intervention plans are more likely to be successful compared to intervention programs that are simply based on blindly encouraging people to take action [6]. Indeed, recent "calls to action" within the cardiovascular health community recognize the importance of behavior-change interventions, and specifically the need to combine individual-level with population-based health promotion strategies [9].

Related to the concept of personalized behavioral interventions are "adaptive" interventions. Collins et al. [10] provide a conceptual framework in which interventions are not "fixed" — every subject receiving the same format and dose of intervention — but instead, the characteristics of the intervention are tailored to individuals. Moreover, this allows interventions to change in time for a person based on their individual-level characteristics.

Body sensors can play an important role in collecting measures of physical activity to better tailor interventions to the individual. Specifically, using body sensors over time can help monitor levels of physical activity at baseline pre-intervention, changes in physical activity that occur during an intervention, and the degree to which individuals maintain improvements in physical activity post-intervention. These measures may be useful feedback to the individual, or the people working with the individual, to encourage them to change their behavior. The use of sensors is increasingly being recognized as an important departure from traditional behavioral assessments that are based on only a few, infrequent, unclear concepts — often questionnaires or qualitative interviews — rather than continuous clearly defined objective measures [11]. Indeed, sensors may play an important role in moving away from a one-size-fits-all approach to interventions to a more patient/subject-centric form of adaptive behavioral intervention.

2. PHYSICAL ACTIVITY BODY SENSOR TECHNOLOGY

Let us now consider the necessary sensor hardware and software components that enable body sensor networks for behavior-change interventions.

Firstly, there are many currently available wireless body sensors, such as those that measure physical/kinematic and dynamic activities: accelerometers, force and position sensors; and physiological measures, such as skin conductivity, perspiration, breathing intensity, EMG, EEG, heartbeat, blood pressure, etc. Furthermore, we can have environmental measures, such as cameras, audio, pollution sensors, GPS, ultrasound, and social interactions of people using phones or the Internet.

The second issue relates to how these sensors will be used on the body reliably, including both technical as well as usability considerations. Considerations include the number of sensors and their placement on the body. Because sensors on the body move during physical activities, one needs to compensate for this movement, and realize that the location of each sensor may not be reliably identified. For real-world practicality we need to know the minimum number of sensors necessary for reliable measurement of specific activities. Also, if one uses existing sensors, e.g., those on a cellphone that a person might normally carry, what is the guarantee that the user will keep the sensor on the body in the correct (assumed) orientation? Furthermore, with all populations, and the elderly specifically, there may be problems with remembering to recharge the batteries for sensors periodically. Hence, the signal may not be reliably delivered, and it is not always obvious how to detect the reasons for missing data.

Third, as described in the previous section, we would like personalized subject-specific data on activities, but personalization leads to technical challenges. We need to understand the limits of individual motion. Different people have varied kinematic and dynamic ranges of motions. This implies that we need to calibrate the reachable space for each individual. This potentially requires more accurate measurements that need to be performed in more controlled conditions. This is a practical challenge since not every subject can be forced to come into a laboratory or be bothered with occasional re-calibration procedures.

Fourth, there are costs to consider. To be practical, activity monitoring at home or in general outside of research/experimental settings demands inexpensive sensors equipped with easy-to-use interfaces. We have found that using readily available low-cost sensors like those in common mobile phones, or commercial camera sensors for the home like the Microsoft Kinect, can affect the quality of assessment. The price one pays is in coarse resolution and inaccurate data. Thus, there is a need to quantify error and to infer the correct parameters for the physical activity modeling to inform proper intervention. Moreover, general-purpose technologies like the aforementioned phone and Kinect technologies may not be appropriate or accessible to all individuals (e.g., the really young, elderly, disabled, or impoverished).

Fifth, data from body sensors needs to be processed to be intelligible. The first challenge in signal processing is in identifying and filtering out outliers. The second challenge is that data is a composition of unreliable devices (including the hardware and software) and the unpredictable behavior of the user (including non-wear, non-compliance with usage instructions, and purposeful misuse of the sensors). In order to deal with unreliable devices one needs multiple measurements of the same phenomenon in order to detect and understand the reasons for the unreliable behavior. For example, if one detects missing data from the cell phone, one can try and understand this in the context of detecting possibly the loss of battery power on the phone or the capacity of the internal memory card if data is being stored locally, or whether the phone was traveling to a remote area without cell tower coverage if the data is being stored remotely. In these cases, other internal sensors on the phone can be useful to detect whether the technical side of the body sensor network is at fault.

In contrast, the unpredictable behavior of users is a harder challenge to understand. In part it can be verified by asking the user questions, such as, Have you exercised today? and What times during the day did you exercise?, but these answers might contradict the data, causing additional challenges. It is known from psychology that people sometimes have varied perceptions of their achievements/activities. There may be real memory recall problems. There may also be shame and embarrassment related to the aforementioned ethical issues around physical activity that lead to purposeful misreporting or misuse of the devices. The body sensors may be perfectly accurate, but because they are annoyingly so, an embarrassed user may throw it in the garbage! Finally, for more socially connected body sensors, users may not want to transmit data about their activities because of lack of trust and/or privacy concerns. Moreover, issues of trust may vary depending on if the data is being sent to a healthcare professional, a family member, friends, or the public at large.

3. MODELING PHYSICAL ACTIVITY

If we are able to address the above challenges there remains the challenge of integrating the data through the use of conceptual and/or numerical models for the physical activity of interest. While review of all such models is beyond the scope of this chapter, we highlight two such models as examples. We note that in calibrating models to sensor data, the Holy Grail is to find the "right" parameters that fit the data the best and most generally for the given population, and if these can account for individual-level factors, all the better.

Consider the aforementioned work of Clarke and Skiba [1], which describes and critically evaluates two performance models: critical power (CP) and the banister impulse-response (IR) model. They justify the inclusion of these two models in the curricula of both applied exercise physiology professionals (e.g., clinical exercise physiologists, personal trainers, and coaches) and research-focused exercise physiologists. We believe these models are equally applicable to the design of exercise protocols for different personalized rehabilitation and physical exercise programs because they are based on performance models, which in turn allow for the appropriate feedback and interventions. In theory, having such models and subsequently protocols that can monitor and supervise the progress of exercise can help lead to changes in the behavior of subjects.

Modeling, of course, depends on understanding the process, selection of parameters, how to extract them from the measurements, parameter fitting and optimization and simulation and sensitivity analysis. Indeed, the performance modeling literature features each of these aspects (some cited in [12,13]).

3.1 Critical Power (CP) Model

The CP model describes the capacity of an individual to sustain particular work rates as a function of time (t). In this way, the model represents the relationship between exercise intensity and duration for an individual. The CP is defined as the power that can be sustained without fatigue for a long time. Then we can further define power P = work/time and finally W' as the finite amount of energy that is available for work above the critical power:

$$W' = (P - CP)t \tag{1}$$

Morton [14] presents four assumptions of the CP model:

- Power output is a function of two energy sources: aerobic and anaerobic.
- Aerobic energy is unlimited in capacity (one can exercise at the intensity or below of CP for infinite duration), but it is limited in the rate at which it can be converted into work.
- Anaerobic energy is unlimited in the rate of conversion (i.e., maximal power output or speed is infinite) but is limited in capacity.
- Exhaustion occurs when W' is depleted.

This model is useful for describing power duration for maximal exercise lasting from 2 to 30 minutes. As such, the CP model serves as a useful tool for devising pacing and tactical strategies in relatively short athletic competitions. But it can also provide a basis for individualized workout intensities during training, exercise, and rehabilitation procedures.

Furthermore, the model is amenable to body sensor measurements. For example, exercise above or below CP results in differences in maximal oxygen consumption.

There are limitations to the CP model, especially when we need to account for intermittent exercises and need to optimize interval workout prescription. Morton and Billat [15] extended the 2-parameter CP model as follows:

$$t = n(t_w + t_r) + \{W' - n[(P_w - CP)t_w - (CP - P_r)t_r]\}/(p_w - CP) \tag{2}$$

where t is the total endurance time, n is the number of intervals, t_w and t_r are the duration of the work and recovery phases in each interval, respectively, and P_w and P_r are power outputs during the work and rest phases, respectively.

3.2 Banister Impulse Response (IR) Model

In contrast to the use of CP for modeling individual exercise sessions or interval workouts, there is a need to model the effects of exercise over time, i.e., training. The IR model relates performance ability at a specific time to the cumulative effects of prior training loads. It describes the individual exercise dose-response relationship and handles the nonlinear time dependence and individuality in a single framework.

The model is basically a first-order differential equation describing a person's change in performance as a function of time. The solution of performance $p(t)$ is equal to the initial performance level $p(0)$ plus positive training effect (PTE = "fitness") minus the negative training effect (NTE = "fatigue"). Both PTE and NTE are expected solutions to first-order differential equation exponentials weighted with some coefficients that are specific to the activity task and the subject. The performance is a nonlinear function of training. Again, body sensors can play a useful role in monitoring changes in performance over time and successive training sessions to provide empirical data to fit such models.

In summary, both CP and IR models can be useful for guiding workout design and long-term planning, respectively. Performance models can serve as an avenue to guide professional reasoning and systematize the trial-and-error adjustments normally featured in exercise protocols. Moreover, these models also can help in identifying what feedback and intervention should be administered (e.g., Is the individual meeting the expectations of CP for a specific workout session? Is the individual training in a way that is most effective?).

Unfortunately, these models assume that individuals are motivated to optimize their exercises. For individuals just embarking on healthy lifestyle changes, or when a healthcare provider is prescribing behavioral intervention, such an assumption cannot be guaranteed. Hence, in the following, we discuss the issue of behavior-change modeling.

4. BEHAVIOR-CHANGE THEORIES RELEVANT TO PHYSICAL ACTIVITY INTERVENTIONS

Although many theories exist for explaining the factors that influence individuals, and the cognitive and psychological changes that individuals go through as they adopt new behaviors, perhaps one of the more prevalent and integrative models is Prochaska's transtheoretical model (TTM) [16]. The TTM is a conceptual (i.e., not inherently mathematical) model based on core constructs that describe the steps and ways that individuals adopt and maintain healthy behavior. The constructs describe how individuals move through six "stages" of change, and are influenced by ten "processes" of change [17], as well as some underlying assumptions about human behavior [18]. We define these briefly below.

The stages of change are ordered approximately in time. Individuals start at one of the stages, the earliest stage being "precontemplation," and through intervention, progress to the last stage of "termination."

The "stages of change" are:

1. Precontemplation: The individual is uninformed and not interested in the healthy behavior at all. He has no intention of adopting the healthy behavior at this time or in the near future.
2. Contemplation: The individual is aware of the problem, and has some intent to make a change in the near future (i.e., next 6 months). But perhaps at this stage the cons (e.g., too busy, too costly, not sure it will make a difference) outweigh the pros of adopting the healthy behavior.
3. Preparation: The individual is ready to make a change (i.e., in the next month). Perhaps they have created a plan, set goals, consulted with a trainer or have signed up for an intervention program. At this point, in his mind, the pros have outweighed the cons.
4. Action: This is when healthy behavior is observable. Changes in the person's activities can be measured and quantified during this stage.
5. Maintenance: This is when an individual has transitioned from trying a healthy behavior, and doing it occasionally, to a state in which he does it routinely and naturally. During this stage, he may be tempted to "give up" and revert to their unhealthy behavior.
6. Termination: This is the last stage in the successful adoption of behavior change. The person is completely convinced that the behavior is healthy and good for him, and has no interest in reverting back to old unhealthy habits.

While the stages of change describe the temporal sequence of cognitive states that an individual goes through, there are a number of processes of change that individuals go through to govern the transitions between the stages of change. These "processes of change" are described as follows:

1. Consciousness raising: This relates to awareness of the problem and the potential for new behavior to address the problem.
2. Dramatic relief: This relates to the use of emotional experiences to convince a person to adopt a healthy lifestyle.
3. Self-reevaluation: This is relates to how an individual sees himself (self-image) with or without the unhealthy behavior.
4. Environmental reevaluation: This relates to how an individual sees himself affecting others around him because of his unhealthy behavior.
5. Self-liberation: This relates to how an individual believes he can change himself (will power).

6. Social liberation: This relates to the empowerment of disenfranchised populations who may benefit from advocacy, provision of services, and healthy alternatives and options.

7. Counterconditioning: This relates to coping or alternative behaviors to replace unhealthy behaviors.

8. Stimulus control: This relates to removing stimuli that prompt unhealthy behavior, replacing them with stimuli or reminders that promote healthy behavior.

9. Contingency management: This relates to the use of rewards and penalties for good and bad behaviors, respectively.

10. Helping relationships: This relates to having peer, family, or professional social support to encourage healthy change.

While the processes of change affect transitions between the stages of change, they are mediated by changes in cognitive and psychological variables. First of all, the processes affect how an individual weighs the "pros versus cons" of living a healthy lifestyle, i.e., decisional balance [19]. Also, the processes of change may alter a person's "self-efficacy" − their confidence in making healthy behavior change [20]. Finally, the processes may affect one's ability to cope with "temptation" to be unhealthy or resort to previous poor habits.

4.1 A Framework for Quantitative Modeling of Physical Activity Behavior-Change Interventions

Behavior change is inherently a difficult process due to the opposite of many of the aforementioned core constructs. Individuals may not progress linearly through the stages, but may move forward as well as backward through the stages as they change cognitively and physically. As previously mentioned, forward progression through the stages can be the result of a specific intervention program of some sort aimed at helping individuals change their behavior. In our research, we have been working with health coaches who help individuals change behavior through a process called "motivational interviewing," which involves exploring a person's readiness to change through communication and listen strategies, employing empathy, and collaborating to set goals for behavior change [21]. It is a method for achieving personalized adaptive behavioral interventions that we described at the outset of this chapter.

As an example, consider Figure 1, which illustrates a physical activity intervention program in which a health coach interacts with a subject using motivational interviewing techniques over some period of time (e.g., a 6-month period). During this period, there is a feedback cycle in which the health coach provides motivational interviewing to the subject who is conditioned by various inherent characteristics (e.g., gender, age, race, culture, environment, etc.), which in turn affects his exercise (measured by sensors) and cognition (self-efficacy, decisional-balance, and temptation measured by questionnaires). The results of changes in exercise and cognition affect the type of motivational interviewing that the coach performs with the subject. Hence, this sort of feedback enables the adaptive nature of this exercise intervention.

Also consider what data from this adaptive intervention may look like from a study design and a time series perspective (Figure 2). As the subject is recruited into the

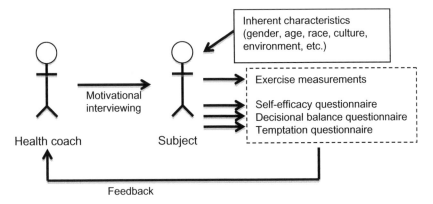

FIGURE 1   An adaptive personalized intervention for physical activity involving a health coach providing motivational interviewing to a subject based on feedback from body sensors measuring exercise, and questionnaires assessing self-efficacy, decisional-balance, and temptation over time. Note that the subject's response to motivational interviewing may depend upon various inherent individual-level characteristics.

intervention study, they are given a baseline assessment before any intervention activities begin. The goal of the baseline assessment is to determine their cognitive state toward physical activity (i.e., to figure out where they are in the stages of change). Also, sensors may be worn during this baseline assessment period to determine the individual's "normal" levels of physical activity and capabilities. It may take a few repeated measurements over a few weeks to obtain reliable baseline estimates. Let's assume that the health coach finds that the individual has no interest in physical activity or in changing his levels of physical activity. The subject would be in the precontemplative stage. As the intervention begins, the health coach will start to provide feedback to the subject based on observations of body sensor data as well as the results of ongoing cognitive assessments using questionnaires. The use of body sensors to continually evaluate physical activity levels is novel and potentially more accurate compared to the bulk of prior transtheoretical model interventions that have only used self-reported exercise questionnaires (e.g., the 7-day Physical Activity Recall, PARQ) [6].

Based on the stage of change in which the person is at, a health coach may conduct motivational interviewing to help the person identify one or more processes of change that could be used to help the individual transition to the next stage of change. Questionnaires may be used to assess the stage of change to which the individual belongs. Additionally, cognitive assessments may be used to assess changes in the mediators (self-efficacy, pros versus cons, and temptations). Specific questionnaires and assessments are presented later. The interviewing provided by the coach could be tailored to the results of the assessments. Over time, through such personalized and stage-specific intervention, the coach may help the subject establish an exercise plan, which would place the subject in the preparation stage. And when the subject begins to exercise, the various improvements in physical activity would be measureable using the body sensors. Ultimately, the study would assess changes in physical activity during the intervention and at follow-up at the end of the intervention compared to the levels of physical activity at baseline.

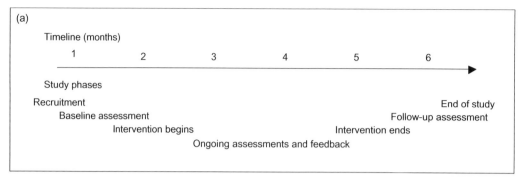

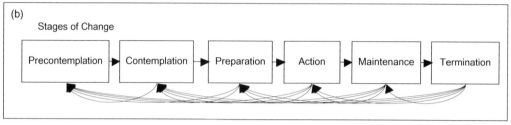

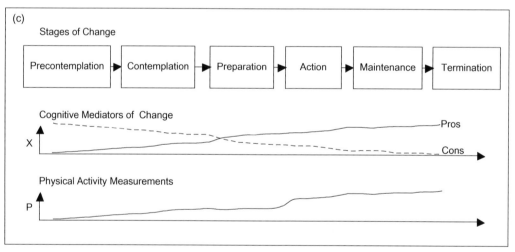

FIGURE 2 (a) A hypothetical behavior-change study design over a 6-month period, with assessments of both physical activity and cognitive aspects of behavior change done at baseline, throughout the intervention period, and in the follow-up period after intervention. Health coaches may provide feedback using motivational interviewing techniques throughout the intervention period. (b) At baseline in the study, Health Coaches may find subjects to belong to any of the stages of change. And over the course of the study, the coaches may find that subjects transition forward through successive stages, or may lapse into prior stages. (c) Ideally, making effective use of sensor data and Motivational Interviewing techniques, the health coach should assist the subject's transition continuously through the stages, improving their cognitive mediators of change X (i.e., having the "pros" outweigh the "cons" for performing physical activity regularly), and observing improvements in the level of physical activity P over time compared to the individual's baseline.

4.2 Assessing Stages of Change

In order to monitor a person's progression through the stages of change we need some way to assess his stage. Various algorithms have been developed for exercise-related stages of change, including Marcus and Simkin [22], who created the Stage of Exercise Behavior Change (SEBC) questionnaire to assign people into stages. A related approach called the Stage of Exercise Scale (SOES) developed by Cardinal et al. [23–25] asks individuals to place themselves on a 5-rung ladder, which determines their stage. The validity of these questionnaires has typically been evaluated if exercise increases as individuals progress through their assigned stages. This correlation between physical activity and stage change over time is consistent with the framework we have described in Figure 2.

Aside from assessment of stage of change, other questionnaire-based assessments of mediating factors have been developed. For instance, assessments have been developed for decisional balance [17], self-efficacy [26], and temptations [27]. These questionnaires may be used to estimate the probability of transitioning between stages based on each of these factors, respectively.

In all cases, results of such questionnaires, as is common in psychological studies, are variables that are either binary or ordinal ratings. Moreover, despite having undergone various validation studies, when applied at the individual person level, they should be viewed as only indicators, subject to error, rather than absolute assessments [28].

As a concrete example, consider the following questionnaire adapted from Marcus, Rossi et al. [17] aimed at assessing stages of change:

Please respond with how much you agree with the following statements:
I am currently physically active.
0 = Strongly Disagree, 1 = Disagree, 2 = Neither Disagree or Agree, 3 = Agree, 4 = Strongly Agree

I intend to become more physically active in the next 6 months.
0 = Strongly Disagree, 1 = Disagree, 2 = Neither Disagree or Agree, 3 = Agree, 4 = Strongly Agree

I currently engage in regular physical activity.
0 = Strongly Disagree, 1 = Disagree, 2 = Neither Disagree or Agree, 3 = Agree, 4 = Strongly Agree

I have been regularly physically active for the past 6 months.
0 = Strongly Disagree, 1 = Disagree, 2 = Neither Disagree or Agree, 3 = Agree, 4 = Strongly Agree

Note that the questions lead to ordinal, scored responses. Asked over large numbers of people, we would generally find the scored responses to each question fit normal distributions. Thresholds may be specified such that those scoring low in questions 1 and 2 would indicate subjects in the precontemplation stage. Those scoring low in question 1 and high in question 2 would indicate subjects in contemplation. Those scoring high in question 1 and low in question 3 would indicate subjects in preparation. Those scoring high in questions 1 and 3, but low in question 4 would indicate subjects in the action stage. And those scoring high in question 4 would indicate those in maintenance.

Although these assessments have traditionally been conducted fairly infrequently over the course of an intervention, new technologies may play an emerging role in more frequent (e.g., daily) assessments. Approaches like Ecological Momentary Assessment [29], in which subjects are asked in-the-moment during random or fixed-interval times about their feelings and attitudes, can be conducted on more pervasive technologies, such as personal computers, phones, and tablets.

4.3 Opportunities to Use Behavior-Change Models to Guide Intervention Programs

Over the last several years, progress has been made toward applying engineering control theory approaches to the challenges of understanding behavior-change interventions. As we have alluded to in Figure 2, data from behavioral interventions can be viewed as time series. Moreover, particularly for adaptive interventions, in which intervention characteristics change over time, more formal modeling of the feedback process (Figure 1) can benefit from more generic studies of feedback for control systems. Notable literature includes the work of Rivera et al. [30], who applied ordinary differential equations to an adaptive behavioral intervention, and Böckenholt [28], who examined the use of a Markov model in cases where there are stages (i.e., states) associated with the behavior-change process.

Because the stages of change are essentially states, they are amenable to implementation via a Markov model, whereby a person belongs to a given state at any given point in time, transitions between states are governed by probabilistic processes, and future state depends only on current state. Böckenholt [28] combines two main components to end up with a continuous-time latent Markov (CT-LM) model for behavior change. The first component considers that binary or ordinal responses to typical behavioral surveys depend upon a latent (not directly observable) state to which a person belongs. Second, changes in responses over time are modeled by a stochastic process that governs transitions between states. Detailed derivations of the model are presented in Böckenholt [28], which are discussed briefly below.

Based on prior work by Samejima [31,32], Böckenholt describes the probabilistic nature of binary or ordinal responses to questionnaires as being a function of the latent state in which a person resides:

$$Pr(y = j|s) = \Phi\left(\frac{\tau_{j+1} - \mu_s}{\sigma_s}\right) - \Phi\left(\frac{\tau_j - \mu_s}{\sigma_s}\right) \tag{3}$$

where the probability that a variable y (i.e., a question in a questionnaire) takes on a response j is conditional on the characteristics of the latent state s, and is related to the difference of two normal cumulative distribution functions Φ defined by the mean μ_s and standard deviation σ_s, for a set of lower and upper thresholds τ_j and τ_{j+1}, respectively.

Due to the challenges of collecting and processing body sensor data that we described earlier, physical activity data are also subject to error. Thus, they may also be amenable to such a probabilistic treatment as questionnaire response data, in which physical activity levels determined from sensor measurements are ranked into ordinal levels.

Next, because the responses are dependent upon the state s to which an individual belongs, we need to define switching probabilities between states. A person can belong to only one state at any time. In the CT-LM model, states are discrete, defined by the above parameters μ_s and σ_s. Switching between states follows a Markov process and depends upon time within a given state:

$$\alpha(d) = \theta[1 - \exp(-\eta d)] \tag{4}$$

where $(0 < \theta < 1)$ and $(\eta > 0)$, such that the probability of switching $\alpha(d)$ is small if the duration d of the time interval spent in a state has been short. These are called transition probabilities.

If the average time that a person spends in a state s_l is $1/\omega_{s_l}$, then ω_{s_l} is the termination rate of the state per unit of time, with transition rates between specific states l and h denoted as $\omega_{s_l s_h}$:

$$\omega_{s_l} = \sum_{h, h \neq l} \omega_{s_l s_h} \tag{5}$$

And state-switching generally following these probabilities:

$$\tau_{s_l s_h} = \omega_{s_l s_h} / \omega_{s_l} \tag{6}$$

Assuming that responses and physical activity are tracked over time longitudinally for a number of subjects, data are available to parameterize the CT-LM model. Böckenholt combines the above relationships into the CT-LM model of the probability of observing a particular response pattern over time from the i-th subject at times $t = t_0^i, t_1^i, \ldots, t_T^i$:

$$Pr(y_{t_0^i}, y_{t_1^i}, \ldots, y_{t_T^i}) = \sum_{s_u=1}^{S} \sum_{s_v=1}^{S} \cdots \sum_{s_r=1}^{S} \sum_{s_w=1}^{S} \pi_{s_u}(t_0^i) P_{y_{t_0^i} \mid s_u} \pi_{s_u s_v}(t_0^i, t_1^i) P_{y_{t_1^i} \mid s_v} \cdots \pi_{s_r s_w}(t_{T-1}^i, t_T^i) P_{y_{t_T^i} \mid s_w} \tag{7}$$

where $\pi_{s_u s_v}(t_0^i, t_1^i)$ is the probability of switching from state s_u to s_v between times t_0^i and t_1^i, $\pi_{s_u}(t_0^i)$ is the probability of being in state s_u at time t_0^i, and $P_{y_{t^i} \mid s}$ is the probability of observing the response at time t given that the i-th person is in state s.

This leads to a log-likelihood formulation from which parameters of interest (u, σ, τ, π) may be estimated using maximum likelihood estimation techniques given sufficient data observations from subjects' responses tracked over time as they transition between states.

Within this framework, there are two remaining challenges. The first is how individual-level factors are considered. In the absence of individual-level covariates in the maximum likelihood estimation, population average parameters based on the provided data will be estimated. These can be useful if the data provided are already for a specific population group. However, for larger and more general population data, if individual-level covariates are provided, e.g., related to the transition probabilities or duration of time within a state, parameters related to these individual-level covariates may also be estimated.

The second question is how feedback with the health coach is implemented. We may assume that the health coach has no state. Consistent with the concept of motivational interviewing, the coach responds simply to the state of the subject. Hence, the motivational feedback K that a health coach provides to their subject — the processes of change that they prescribe to the subject — is modeled as a function f:

$$K(t^i) = f[E(s(t^i))] \tag{8}$$

where K depends upon the expectation of the subject's state s, and results in a vector of binary (yes/no) outcomes as to whether the subject should or should not work on each of the 10 processes of change as a specific intervention strategy to encourage movement to subsequent stages of change. Evidence from the successes of stage-matched intervention studies in which individuals are matched to specific processes of change based on the stage they belong to may help define function f. Moreover, if this function is parameterized, it may be included into the transition probabilities above and subject to maximum likelihood estimation.

While the calibration of such models to real-world body sensor data has not been fully realized, the conceptual and methodological underpinnings as described by these studies will soon allow for more widespread application of numerical modeling to sensor-enabled behavioral intervention studies. For instance, assuming behavioral intervention studies lead to data from which parameters of such models can be estimated, it would be possible to estimate the expected state to which an individual belongs with greater certainty, thereby informing greater personalization of health coaching.

5. CONCLUSION

In this chapter we have explored the motivation for personalizing behavior-change interventions using body sensors. We have discussed the opportunities and challenges of using body sensors to inform exercise models. We have also presented the need to think beyond performance-oriented exercise models to those based on psychological and cognitive aspects of behavior change.

Still there remain numerous issues to be explored that are both practical and ethical in nature. For one, we have only explored a limited set of feedback options. One may be feedback just to the subject himself. Another type is feedback between a subject and a health coach. There are many other scenarios including feedback between peers undergoing interventions as well as feedback to family and friends who may provide social support. As we mentioned in the introduction of this chapter, new wireless communication capabilities have enabled all these forms of feedback. However, we have not addressed the issues of the format for such feedback (temporal frequency, symbolic/textual/verbal, computer-automated, or humanistic). With all of these forms of feedback, accurate and timely data from body sensors will play an important role in tailoring the messages to encourage behavior change. However, there are trade-offs between the breadth and depth of sensing needed to improve behavior-change programs and what can be tolerated in terms of risk to personal privacy and usability. As body sensor systems are increasingly moved from the research world into the practical world, we will need to continue to explore such practical and ethical issues.

References

[1] D.C. Clarke, P.F. Skiba, Rationale and resources for teaching the mathematical modeling of athletic training and performance, Adv. Physiol. Educ. 37 (2) (2013) 134–152.

[2] J.O. Prochaska, C.C. DiClemente, Stages and processes of self-change of smoking: toward an integrative model of change, J. Consult. Clin. Psychol. 51 (3) (1983) 390–395.

[3] L. Spencer, et al., Applying the transtheoretical model to tobacco cessation and prevention: a review of literature, Am. J. Health. Promot. 17 (1) (2002) 7–71.

[4] J.P. Migneault, T.B. Adams, J.P. Read, Application of the Transtheoretical Model to substance abuse: historical development and future directions, Drug. Alcohol. Rev. 24 (5) (2005) 437–448.

[5] L. Spencer, et al., The transtheoretical model as applied to dietary behaviour and outcomes, Nutr. Res. Rev. 20 (1) (2007) 46–73.

[6] L. Spencer, et al., Applying the transtheoretical model to exercise: a systematic and comprehensive review of the literature, Health Promot. Pract. 7 (4) (2006) 428–443.

[7] US Department of Health and Human Services, Physical Activity Guidelines for Americans, 2008.

[8] S.J. Salvy, et al., Influence of peers and friends on children's and adolescents' eating and activity behaviors, Physiol. Behav. 106 (3) (2012) 369–378.

[9] B. Spring, et al., Better population health through behavior change in adults: a call to action, Circulation 128 (19) (2013) 2169–2176.

[10] L.M. Collins, S.A. Murphy, K.L. Bierman, A conceptual framework for adaptive preventive interventions, Prev. Sci. 5 (3) (2004) 185–196.

[11] W.J. Nilsen, M. Pavel, Moving Behavioral Theories into the 21st Century: Technological Advancements for Improving Quality of Life, Pulse, IEEE 4 (5) (2013) 25–28.

[12] J.R. Fitz-Clarke, R.H. Morton, E.W. Banister, Optimizing athletic performance by influence curves, J. Appl. Physiol. (1985) 71 (3) (1991) 1151–1158.

[13] P. Hellard, et al., Assessing the limitations of the Banister model in monitoring training, J. Sports. Sci. 24 (5) (2006) 509–520.

[14] R.H. Morton, The critical power and related whole-body bioenergetic models, Eur. J. Appl. Physiol. 96 (4) (2006) 339–354.

[15] R.H. Morton, L.V. Billat, The critical power model for intermittent exercise, Eur. J. Appl. Physiol. 91 (2-3) (2004) 303–307.

[16] J.O. Prochaska, Systems of Psychotherapy: a Transtheoretical Analysis. The Dorsey Series in Psychology, Dorsey Press, Homewood, Ill, 1979. xv, 407 p.

[17] B.H. Marcus, et al., The stages and processes of exercise adoption and maintenance in a worksite sample, Health. Psychol. 11 (6) (1992) 386–395.

[18] Prochaska, Johnson, and Lee, Chapter 4: The Transtheoretical Model of Behavior Change, in The Handbook of Health Behavior Change. 2009, The Springer Publishing Company: New York.

[19] I.L. Janis, L. Mann, Decision Making: A Psychological Analysis of Conflict, Choice, and Commitment, Free Press, 1977.

[20] A. Bandura, Self-Efficacy in Changing Societies, Cambridge University Press, Cambridge; New York, 1995. xv, 334 p.

[21] S. Rollnick, W.R. Miller, C.C. Butler, Motivational Interviewing in Health Care: Helping Patients Change Behavior, Guilford Publications, 2012.

[22] B.H. Marcus, L.R. Simkin, The stages of exercise behavior, J. Sports Med. Phys. Fitness 33 (1) (1993) 83–88.

[23] B.J. Cardinal, The stages of exercise scale and stages of exercise behavior in female adults, J. Sports Med. Phys. Fitness 35 (2) (1995) 87–92.

[24] B.J. Cardinal, Construct validity of stages of change for exercise behavior, Am. J. Health. Promot. 12 (1) (1997) 68–74.

[25] B.J. Cardinal, M.L. Sachs, Prospective analysis of stage-of-exercise movement following mail-delivered, self-instructional exercise packets, Am. J. Health Promotion 9 (6) (1995) 430–432.

[26] B.H. Marcus, et al., Self-Efficacy, Decision-Making, and Stages of Change: An Integrative Model of Physical Exercise, J. Appl. Soc. Psychol. 24 (6) (1994) 489–508.

[27] H.A. Hausenblas, et al., A missing piece of the transtheoretical model applied to exercise: Development and validation of the temptation to not exercise scale, Psychol. Health 16 (4) (2001) 381–390.

[28] U. Bockenholt, A latent markov model for the analysis of longitudinal data collected in continuous time: states, durations, and transitions, Psychol. Methods 10 (1) (2005) 65–83.

[29] A.A. Stone, S. Shiffman, Ecological momentary assessment (EMA) in behavioral medicine, Ann. Behav. Med. (1994).

[30] D.E. Rivera, M.D. Pew, L.M. Collins, Using engineering control principles to inform the design of adaptive interventions: A conceptual introduction. Drug and Alcohol Dependence, 88 (2007) S31–S40.

[31] F. Samejima, Estimation of latent ability using a response pattern of graded scores, Psychometrika Monogr. Suppl. (1969).

[32] F. Samejima, Graded Response Model, in Handbook of Modern Item Response Theory, Springer, 1997. p. 85–100.

6.1

Human Body Communication for a High Data Rate Sensor Network

Jung-Hwan Hwang and Chang-Hee Hyoung

Electronics and Telecommunications Research, Daejeon, Korea

1. CAPACITIVE-COUPLING COMMUNICATION THROUGH HUMAN BODY

Human body communication (HBC) was first proposed by Zimmerman [1–3] as a novel communication technology to exchange data between electronic devices in body area networks (BAN). Its application is not limited to data transmission but extended to power transmission [4], in which electronic devices receive power required for operation simultaneously with data. Unlike wired and wireless methods, HBC uses a capacitive-coupling transmission channel, in which both transmitter and receiver are capacitively coupled by an electric field passing through the human body. The transmitter then modulates the electric field according to data to be transmitted; the receiver detects the modulated electric field and recovers the transmitted data [2]. Such capacitive coupling is possible because the human body is composed of various tissues having a high dielectric constant. The transmitter and receiver both use an electrode instead of an antenna. The electrode is attached to the human body and forms the modulating electric field at the transmitter or detects the modulated electric field at the receiver.

Compared to HBC, a wired connection has advantages of a higher data transfer rate and easiness of establishing a connection. User's behavior, however, is constrained by cables that are easily tangled. A wireless connection removes such a constraint, but has an inherent limit in terms of power consumption because it modulates a baseband signal onto the carrier frequency in order to transmit data through the air. Wireless devices should have their own energy source to operate, and a wireless solution consumes more power to transmit and receive the information. In a wireless connection, the transmitted signal experiences a path loss caused by various factors, such as free space loss, refraction, diffraction, and absorption over the channel. Such a propagation environment degrades the quality of the received signal. In addition, a wireless connection is more complex to configure and establish.

HBC has the advantages of both wired and wireless connection methods. HBC uses a capacitive-coupling transmission channel inside the human body; hence, it does not require any cable to connect devices and is less sensitive to the propagation environment outside the human body. A user can configure a communication network simply by touching devices, and it provides an intuitive service, in which a complicated procedure for a network setup is not required. A physical layer (PHY) for body area networks has been recently standardized by the IEEE 802.15 working group [5]. In addition to NB (narrow band) and UWB (ultra-wide band) wireless communications, HBC is included in the standard as a scheme for short-range communications. In the IEEE standard, data to be transmitted through the human body is spread over the selected frequency domain using a group of digital codes without a continuous frequency modulation; hence, HBC has lower circuit complexity and lower power consumption. A comparison of the HBC PHY in [5] with other PHYs for wireless communications is shown in Table 1.

TABLE 1   HBC and Other Wireless PHYs

Market Name	HBC	ZigBee	Bluetooth	Wi-Fi
Physical Layer Standard	802.15.6	802.15.4	802.15.1	802.11n
Frequency Band	5~50 MHz	900 MHz 2.4 GHz	2.4 GHz	2.4 / 5 GHz
Modulation	Frequency Selective Digital Transmission	Direct Sequence Spread Spectrum	Adaptive Frequency Hopping Spread Spectrum	Orthogonal Frequency Division Multiplexing
Data Rate	2 Mbps	250 kbps	1 Mbps	150 Mbps
Range	<3 m	100 m	10 m	100 m
Power	Very low	Low	Moderate	High
Complexity	Very simple	Simple	Complex	Very complex
Set-up time	<100 ms	30 ms	Few seconds	Few seconds
Ease of use	Easy	Easy	Normal	Hard
Application Focus	Body area network	Sensor networks, industrial control	Cable replacement	Local area networking

2. CHANNEL PROPERTIES OF HUMAN BODY

Figure 1 shows the basic principle of HBC. Each transmitter and receiver has a signal electrode that is attached to the human body. In HBC, an electric field is formed between a transmitter and receiver for a capacitive coupling between them. For this, the signal source of the transmitter creates a voltage difference between the signal electrodes and simultaneously at the load of the receiver. Also, another voltage difference is formed between the ground planes of the transmitter and receiver, which makes the sum of the voltage differences on the closed loop equal to the voltage by the signal source, where the closed loop is formed by the signal source at the transmitter, the load at the receiver, and the coupling inside the human body and in the air, respectively, with one coupling between the signal electrodes through the human body and another coupling between the ground planes through the air. Finally, the voltage difference between the signal electrodes causes a small amount of the current inside the human body due to the electric field formed inside the human body. The human body is composed of various tissues that collectively have the properties of a lossy dielectric material (i.e., each tissue has a dielectric constant and conductivity depending on the signal's frequency [6]). The current inside the human body is composed of two types of components: the displacement and the conductive currents, which are respectively related to the dielectric constant and the conductivity. The tissues of the human body have a relatively large dielectric constant [6], especially at the low-frequency band ($< 100\,\mathrm{MHz}$) used by HBC [7–11]. Therefore, the displacement current mainly contributes to the transmission of the electrical signal through the human body. In HBC, the electrical signal at the low-frequency band is transmitted over a short distance within the human body through a material having a high dielectric constant. Thereby, an HBC channel has a unique property: the transmission of the electrical signal is strongly affected by the structure of the electrodes and by the electromagnetic environment outside the human body.

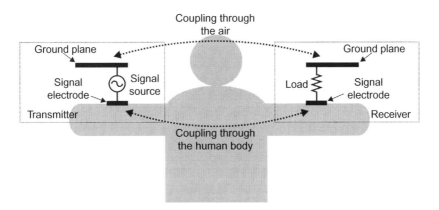

FIGURE 1   Basic principle of human body communication.

3. EFFECTS OF ELECTRODE'S STRUCTURE

In HBC, a transmitter capacitively coupled with a receiver modulates an electric field and the receiver detects the modulated electric field. This is equivalent to generating and detecting a voltage difference: a voltage difference is generated on the surface of the human body by a transmitter and then detected by a receiver [11]. The generation of the voltage difference and its detection are made possible with an electrode attached to the human body [12–15]. An electrode can have an adhesive material on its surface like a disposable ECG electrode or a metal surface for a simpler implementation. In the case of a metal surface, it can be held in contact with the human body with an attachment aid like a rubber band or be touched by hand when necessary. The electrode should have small contact resistance to minimize signal loss. HBC uses two types of electrodes: a signal electrode and a ground electrode as shown in Figure 1. Each signal electrode of the transmitter and receiver is connected to a signal source in the transmitter or a load in the receiver, while each ground electrode is connected to the ground plane of the transmitter or receiver. The structure of the electrodes affects the transmission of the electrical signal through the human body. In particular, the attachment of the ground electrode onto the body affects the channel's signal loss as it increases or decreases the signal loss that the electrical signal experiences as it is transmitted through the human body [12–14,16,17].

In two studies ([13,14]), the signal loss was shown to decrease when the ground electrode of the transmitter was attached to the body, but the measured results in another study [17] showed that the signal loss increases with the attachment of the ground electrode. Also, the decrease of the signal loss caused by the attachment of the ground electrode itself decreases gradually as the transmission distance between the transmitter and receiver increases [14]. The transmitter can be located at the wrist or the finger according to the HBC application, and the location of the transmitter also affects the operation of the ground electrode. Figure 2 shows the change in signal loss at 30 MHz caused by the ground electrode when the transmitter is located at the wrist and the finger [17]. As shown

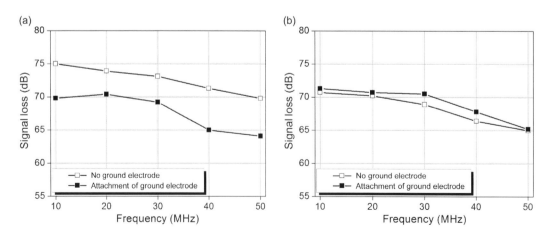

FIGURE 2   Measured signal loss at 30 MHz: (a) transmitter at wrist and (b) transmitter at finger.

in Figure 2(a), the measured signal loss decreases as the ground electrode is attached to the body when the transmitter is located at the wrist. However, as shown in Figure 2(b), the ground electrode has the opposite effect on the signal loss when the transmitter is located at the finger; the signal loss increases as the ground electrode is attached to the body. Such various effects shown by the ground electrode on the signal loss can be analyzed using the distribution of the electric fields inside the human body.

Figure 3 shows the distribution of the electric fields near the transmitter, in which the electric field was simulated using a rectangular parallelepiped human model. The simulation model has a cross-section of a different size depending on the position to separately model the wrist and finger. As shown in Figures 3(a) and 3(b), the electric fields inside the human body are distributed in different directions according to the position due to the field coupling from the signal electrode to the ground plane. When $x < -2$ cm, the fields inside the body are distributed in a direction opposite to that of the receiver (i.e., the fields direct the ground plane) because the fields are strongly coupled with the ground plane. However, as the distance from the transmitter increases, the coupling with the ground plane becomes weak; hence, the fields inside the body are distributed in the direction of the receiver, as shown when $x > 0$ cm. The degree of the field coupling with the ground plane is dependent on the location of the transmitter. When the transmitter is located at the finger, the volume of the body near the transmitter is small because the finger is much thinner and shorter than the wrist. This means that the volume of the dielectric material on the coupling path to the ground plane is small. In this case, the coupling with the ground plane becomes so weak that the fields are distributed easily in the opposite

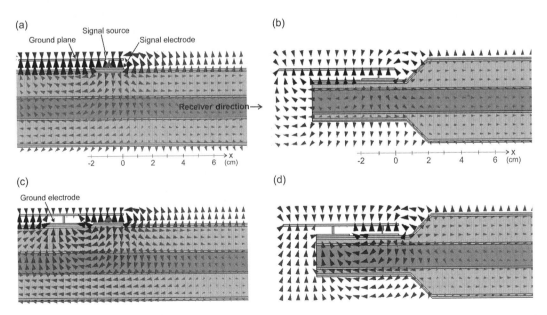

FIGURE 3 Distribution of electric fields at 30 MHz. Cases of no ground electrode: (a) transmitter at wrist and (b) transmitter at finger. Cases of attachment of the ground electrode: (c) transmitter at wrist and (d) transmitter at finger.

direction of the ground plane, i.e., in the direction of the receiver. When the ground electrode is attached on a larger part of the human body, the distribution of the electric fields takes the form shown in Figures 3(c) and 3(d).

In comparison with the distribution shown in Figures 3(a) and 3(b), more fields are distributed in the direction opposite to the receiver because the ground plane is in contact with the body through the ground electrode and, accordingly, the coupling with the ground plane becomes stronger. Thereby, the net electric field in the direction of the receiver is formed inside the human body while the field strength depends on the canceling effect between the fields of different directions. The increase of the field coupling causes an increase in the voltage difference between the signal and ground electrode and an increase in the output current of the transmitter (i.e., the output current of the signal source). The output current increases from 1.8 mA to 8.9 mA for the wrist and from 1.4 mA to 5.0 mA for the finger. When the transmitter is located at the wrist, the increase in the output current is larger because the coupling with the ground plane through the ground electrode is stronger due to the large volume of the dielectric material on the coupling path to the ground plane. The large increase in the output current caused by the attachment of the ground electrode increases the net electric field in the direction of the receiver even with the canceling effect, and the increase of the net electric field increases the voltage difference on the surface of the human body. This decreases the signal loss, as shown in Figure 2(a). However, the attachment of the ground electrode does not cause such a large increase in the output current when the transmitter is located at the finger. Therefore, the net electric field in the direction of the receiver decreases due to the canceling effect and the signal loss increases accordingly, as shown in Figure 2(b).

Considering these results, the ground electrode of the transmitter affects the signal loss through two effects. The first is the effect of the increase in the output current and the second is the increase in the field coupling of the electric fields with the ground plane. When the ground electrode of the transmitter is attached to the human body, the output current increases. When the output current is very large, there is a corresponding increase in the net electric field in the direction of the receiver and, consequently, a decrease in the signal loss [13,14], as shown in Figure 2(a). However, the ground electrode also increases the field coupling with the ground plane, causing an increase in the electric field of the opposite direction and its accompanying canceling effect. Thus, when the increase of the output current is small, the net electric field in the direction of the receiver decreases and the signal loss increases accordingly [17], as shown in Figure 2(b). Also, the canceling effect lowers the decreasing effect of the signal loss due to the attachment of the ground electrode as the transmission distance increases [14].

3.1 Effects of the Electromagnetic Environment Outside the Human Body

The human body is composed of many tissues which have a high dielectric constant, in the low-frequency bands [6], especially bands under 100 MHz, which can be easily coupled with the human body due to their short effective wavelengths inside the body [18–20]. Outside the human body, various types of electronic devices emit electromagnetic waves in a low-frequency band [17]. The emitted electromagnetic waves are coupled with the human body due to its antenna function, and the coupled waves then generate an interference signal inside the body, so the receiver receives an interference signal caused by the electromagnetic environment outside the human body along with a data signal transmitted from the transmitter. To illustrate this effect, the interference signal received at the receiver was measured in various electromagnetic environments, as shown in Figure 4. A subject was exposed to a general electromagnetic environment like that shown in Figure 4(b), while the receiver's signal-electrode was attached to the arm of the subject. The interference signal caused by the human body's antenna function was then measured at the signal electrode using the measurement setup shown in Figure 4(a).

Figure 5 shows examples of interference signals measured inside buildings and on a subway, in which power of the interference signal is presented in the frequency domain. The interference signal does not have significant power at some measurement sites inside buildings, but otherwise it has high power over a wide-band region, as shown in Figure 5 (a). Unlike the building environments, the interference signal has high power at most measurement sites inside the subway, and the high power is distributed under 30 MHz due to the high-voltage pantograph collector used on the subway system. The interference signal in HBC has a dynamic property such that its power and frequency change depending on the location; however, the interference signal is distributed mainly in the low-frequency band used by HBC. Any electronic device emitting an electromagnetic wave in the low-frequency band can be a source of interference. Therefore, the transmitter for HBC also is a significant source of interference; an electrical signal transmitted at one user's transmitter for HBC is emitted outside the human body in the form of an electromagnetic wave due to the antenna function of the human body, and the emitted electromagnetic wave is then

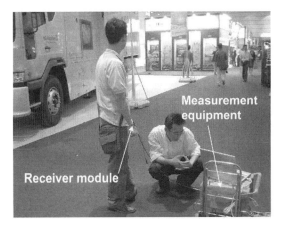

FIGURE 4 Measurement of interference signal in the electromagnetic environment: (a) measurement setup and (b) measured environment.

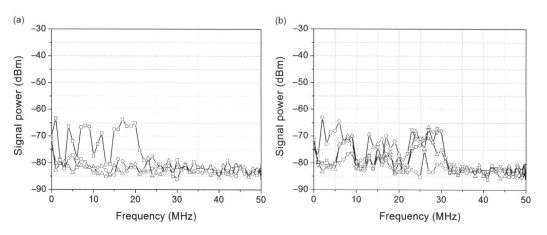

FIGURE 5   Examples of measured interference signals: (a) inside buildings and (b) on a subway.

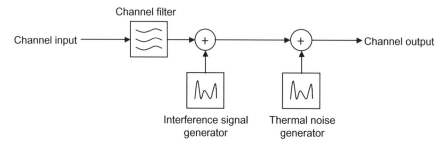

FIGURE 6   Structure of channel model for human body communication.

coupled with the human body of another user nearby. In this case, the power of the interference signal is dependent on the number of HBC users and the distances between each user. This is especially similar to that of the receiving signal for HBC when users are in close proximity to each other [18]. Therefore, the interference signal in HBC can have high power in the same frequency band occupied by the electrical signal transmitted through the human body, causing severe degradation of the bit error rate (BER) performance, a metric indicating the number of errors in the received data stream.

3.2 Channel Model for Human Body Communication

A channel model is required to design a transmission scheme and an analog front-end for HBC. The IEEE standard for HBC has been published [5], and its PHY structure is based on the channel model presented in an earlier study [21]. The channel model is composed of a channel filter and an interference signal generator[1], as shown in Figure 6. The channel filter represents the signal loss experienced by the electrical signal as it is

[1] The noise in [22] is identical to the interference signal in this book.

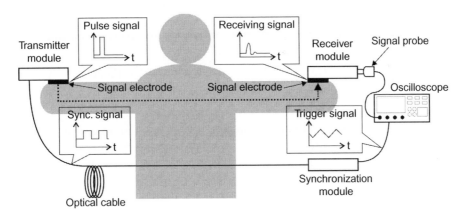

FIGURE 7 Ground-isolated measurement setup.

transmitted through the human body. An electrical signal to be transmitted through the human body is filtered by the channel filter and the interference signal generated from the interference signal generator is then added to the filtered signal. After the thermal noise is added, the channel's output signal is obtained.

Numerous researchers have studied signal loss by the human body [22–25]. For an accurate measurement of signal loss, it is important to isolate the grounds of the transmitter and the receiver from the earth ground, as the coupling between the ground planes strongly affects the signal loss. However, in previous research, the measurement equipment was connected to the earth ground [22,23], or abnormally large ground planes were used to measure the signal loss [24,25]. In order to measure the signal loss while maintaining an isolation condition between the ground planes, the measurement setup shown in Figure 7 is used [26]. The transmitter module, powered by a battery, generates the pulse signal, and the generated pulse signal is transmitted through the human body. The transmitted pulse is received at the receiver module and measured using the oscilloscope. The pulse signal has a very short width and hence has a high-frequency component to obtain the signal loss over a wide frequency band. The transmitter module simultaneously generates a signal synchronized with the pulse signal, and this synchronization signal is optically transmitted to the synchronization module using the optical cable. The synchronization module restores the synchronization signal and then triggers the oscilloscope whenever the pulse signal is generated at the transmitter module; hence, the receiving signal at the receiver module can be measured with the transmitter and receiver modules synchronized. Also, the isolation condition between the ground planes is maintained during the measurement, as the two modules are optically, not electrically, connected to each other. The pulse signal and the receiving signals are transformed into the frequency domain using the Fourier transform, and the signal loss is then computed by subtracting the receiving signal from the pulse signal in the frequency domain. The impulse response of the channel filter in Figure 6 is obtained like Eq. (1) after the transformation of the computed signal loss into the time domain [21]:

$$h(t) = h_R(t) \cdot C_h \tag{1}$$

Here, $h_R(t)$ is a reference impulse response, and C_h is a coefficient related to sizes of the ground planes and distances between the transmitter and receiver, as follows:

$$C_h = (0.0422G_T - 0.184) \cdot (0.0078G_R + 0.782) \cdot \left(\frac{120.49}{d_{body} + d_{body} \left(\frac{d_{air}}{d_{body}} \right)^5} \right)^2 \tag{2}$$

In the above equation, G_T and G_R are the ground plane's size in cm^2 at the transmitter and receiver, respectively. Additionally, d_{air} and d_{body} are the distances related to the coupling between the transmitter and receiver through two mediums, as shown in Figure 1; d_{air} is the distance in cm between the transmitter and receiver through air; and d_{body} is that between the transmitter and receiver through the human body. C_h is valid only when $10\,\text{cm}^2 \le G_T$, $G_R \le 270\,\text{cm}^2$ and $10\,\text{cm} \le d_{air}$, $d_{body} \le 200\,\text{cm}$. The reference impulse response is an impulse response when the coefficient C_h is equal to 1; it is expressed as follows:

$$h_R(t) = A_v \cdot A \cdot e^{\left(\frac{-(t-t_r)}{t_0} \right)} \cdot sin \left(\frac{\pi(t - t_r - x_c)}{w} \right) \tag{3}$$

Each HBC user has different physical parameters related to the channel; these account for various body size and component ratios of body tissues, such as fat and muscle. The signal loss in HBC is affected by these physical parameters; hence, each user has a different amount of signal loss. A_v in Eq. (3) is a random variable used to represent this type of variation in the signal loss; it follows the Gaussian distribution with a mean of 1 and a variance of 0.16^2. A, t_r, t_0, x_c, and w are the constants related to a shape of the reference impulse response [21].

At the transmitter, the electrical signal before being transmitted through the human body is filtered by mask filtering to remove harmonic components and possible interference in other frequency bands [5], thus occupying a narrow band of about 6 MHz. The interference signal is modeled only over the occupying band and can therefore be approximated with the additive white Gaussian noise, which has a frequency-independent power. The measured interference signal has the Gaussian distribution; this is reasonable because multiple electronic devices, including HBC transmitters, emit electromagnetic waves, and these waves amount to an interference signal at the receiver. After the summation of independent random variables, it follows the Gaussian distribution according to the central limit theorem. The interference signal has a different variance according to a location where HBC is used; hence, the interference signal generator from [21] is modeled with maximum variance of the interference signal, for which the measured maximum variance was found to be 2.55×10^{-5}.

4. TRANSMISSION SCHEME OF HUMAN BODY COMMUNICATION

The IEEE 802.15 working group for body area networks recently published the standard for PHY using HBC [5]. The transmitter is composed of signal-generation blocks to generate a preamble, a start-frame delimiter (SFD)/rate indicator (RI), a header, a physical layer service data unit (PSDU), and a pilot signal. The signals generated in each block are added at multiplexers (MUXs) and then sent to a transmit filter to achieve compliance with the spectral mask defined in [5]. The transmitter uses the frequency selective digital transmission (FSDT) scheme; unlike general wireless communication, a baseband signal, which results from the spreading of a data signal in the frequency domain using the Walsh code and the frequency selective code (FSC), is transmitted through a transmission channel (i.e., the human body) without modulation to transform a baseband signal into a passband signal in the IF or RF band. The transmission of the baseband signal is possible because the human body supports signal transmission at a low-frequency band where the baseband signal is distributed [7–11]. The baseband signal after the transmit filter is transmitted through the human body to a receiver. BER performance at a receiver is affected by the signal loss of the channel and by the interference signal, which is generated by the electromagnetic environment surrounding an HBC user [7]. To prevent BER degradation by an interference signal, the transmitter spreads the data signal in the frequency domain before transmitting it through the human body. The data signal is spread using the frequency-selective (FS) spreader before the transmit filter, as shown in Figure 8, and the resulting baseband signal is transmitted through the human body. Serial input data is converted to a 4-bit symbol by a serial-to-parallel (S2P) converter, and each symbol is then mapped into one of the 16-chip Walsh codes according to the mapping table shown in Figure 8. Each chip in the Walsh code is then mapped into one of the two FSCs whose chip sequence starts with "0" or "1," and it repeats "0" and "1" in the length of one FSC,

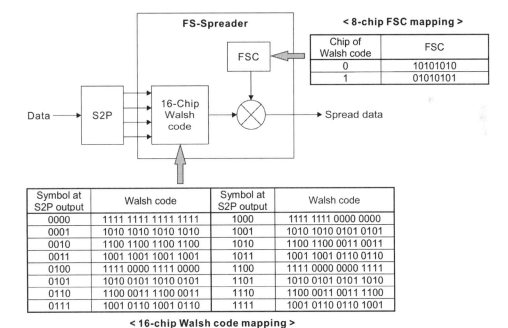

FIGURE 8 Structure of frequency-selective spreader.

as shown in the mapping table in Figure 8. This mapping table shows when each chip in the Walsh code is mapped into 8-chip length FSC. The length of the FSC is controlled according to a data rate of the serial input data to keep the chip rate at the output of the FS spreader, i.e., 42 Mcps [5]. Each Walsh code has a different fundamental frequency [8], making it possible to distribute the data signal after the spreading in a specific frequency band (i.e., to be frequency-selective) to avoid the frequency band where the interference signal is mainly distributed. Hence, the FSDT scheme has good tolerance to interference due to its frequency-selective feature along with the processing gain provided by the spreading. HBC PHY in [5], however, has not been implemented yet. Instead, an HBC modem having a similar PHY has been implemented, in which the Walsh codes are used to generate a frequency-selective data signal without FSC [8]. This section explains the transmission scheme of HBC using the HBC modem originally presented in [8].

4.1 Walsh Code

It is necessary to understand the characteristics of the spread code to understand the transmission scheme comprehensively.

The Walsh (−Hadamard) code is unique in that each non-zero code word has a Hamming weight of exactly 2^{n-1}, which implies that the distance of the code is also 2^{n-1}. In standard coding theory notation, this means that the Walsh−Hadamard code is a $[2^n, n, 2^n/2]_2$ code. The Hadamard code can be seen as a slightly improved version of the Walsh−Hadamard code, as it achieves the same block length and minimum distance with a message length of $n + 1$; i.e., it can transmit one more bit of information per code word, but this improvement comes at the expense of a slightly more complicated construction.

A Walsh (−Hadamard) code is obtained by selecting as code words the rows of a Hadamard matrix. A Hadamard matrix M_n is an $n \times n$ matrix (n is an even integer) of 1s and 0s with the property that all rows differ from other rows by exactly $n/2$ positions. One row of the matrix contains all zeros. The other rows contain $n/2$ zeros and $n/2$ ones.

For n = 2, the Hadamard matrix is

$$M_2 = \begin{bmatrix} 0 & 0 \\ 0 & 1 \end{bmatrix} \tag{4}$$

Furthermore, from M_n, the Haramard matrix M_{2n} is generated according to the relationship

$$M_{2n} = \begin{bmatrix} M_n & M_n \\ M_n & \overline{M_n} \end{bmatrix} \tag{5}$$

where $\overline{M_n}$ denotes the complement (0s replated by 1s and vice versa) of M_n. For example, the M_4 matrix is created as follows:

$$M_4 = \begin{bmatrix} 0 & 0 & 0 & 0 \\ 0 & 1 & 0 & 1 \\ 0 & 0 & 1 & 1 \\ 0 & 1 & 1 & 0 \end{bmatrix} \tag{6}$$

Each code has a fundamental frequency despite the fact that it has many frequency components. In the case of the M_4 matrix, the first row has no transitions, and the second row has three transitions and has the highest fundamental frequency component. The third and fourth rows have one and two transitions, respectively. If the matrix is rearranged by the number of transitions, the higher the index, the higher the number of fundamental frequency Walsh codes there will be. This means that the greatest power of each Walsh code is in its fundamental frequency.

4.2 Frequency-Selective Digital Transmission

Data transmitted in a digital waveform can be spread over the selected frequency domain using Walsh codes. A 64-chip Walsh code is used for HBC. The 64-chip Walsh code is rearranged according to the number of transitions; hence, the index of each code corresponds to the number of transitions. For example, a code with index number 48 of the 64-chip Walsh code, W48, has 48 transitions. It is divided into four groups with 16 codes using the index of each code. The fourth subgroup of the 64-chip Walsh code has 16 codes, the highest number of transitions and thus, the highest fundamental frequencies. If it is assumed that the system uses a clock of 32 MHz, the maximum fundamental frequency is 16 MHz. The occupied frequency band of subgroup 4 of the 64-chip Walsh code in such a case is depicted in Figure 9, which shows the simplified spectrum of the fourth subgroup, the received signal that passed through the human body and the normalized noise power from the interference signal caused by an antenna function of the human body. If the codes of this group are used in order to spread a signal, the highest signal power to be transmitted exists in the fourth subgroup mainly located in the 12 to 16 MHz frequency band.

Also, as shown in Figure 9, the normalized noise power with a 5 MHz bandwidth decreases in proportion to the frequency. An earlier study [27] also reported that there is a

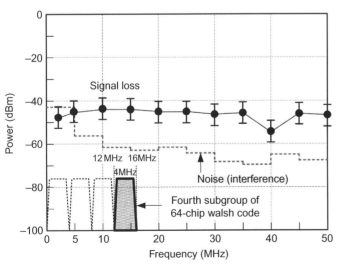

FIGURE 9 The occupied bandwidth of the 64-chip Walsh code, signal loss, and the normalized noise (interference) power.

narrow window between 10 MHz and 20 MHz in which the interference signal's power is relatively low. In the frequency response of the human body channel under study, the signal loss in the human body channel gradually increases in proportion to the frequency up to 40 MHz. Above 40 MHz, the signal loss increases more rapidly in proportion to the frequency.

Though a conventional spread system has a wider bandwidth that resembles white noise, the transmitted signal generated with the subgroup of the 64-chip Walsh code resembles the signal modulated with carrier frequency. The output spectrum appears as the baseband signal with the 4 MHz bandwidth but modulated with a carrier frequency of 14 MHz. With a group of Walsh codes, the information is spread into the selected frequency without continuous frequency modulation using the carrier frequency. The spread spectrum has the advantage of guaranteeing the quality of communication by increasing resistance to the interference of the human body coming from various appliances. The HBC modem in work [8] makes use of 16 codes out of the 64-chip Walsh codes that have the highest fundamental frequencies, i.e., the fourth subgroup (W48 ~ W63) in Figure 9. Figure 10 shows a block diagram of the HBC modem and the operation of the frequency-selective digital transmission. The serial input data becomes a 4-bit symbol by a serial-to-parallel (S2P) block. This 4-bit symbol becomes the index of the 16 codes out of the 64-chip Walsh codes. If the source data rate is 2 Mbps, the symbol rate is 0.5 Msps (symbols per second) after the serial-to-parallel converter. The final chip rate after the frequency-selective spreader is then 32 Mcps (chips per second). The FSDT scheme has the advantage of a high data rate and a simple architecture. A previous system [28] makes use of Manchester coding, which always has a transition in the middle of each bit period; hence, an error in transition will directly result in a bit error. However, in the FSDT scheme, the minimum distance of the 64-chip Walsh code is 32; therefore, the receiver can still decode even if 15 bits are lost. The spread baseband signal has only two consecutive identical chips, 00 or 11. As a result, its fundamental frequency ranges from 12 to 16 MHz with a clock frequency of 32 MHz. Through the use of the 16 spread codes with the highest fundamental frequencies, the baseband appears to be modulated with a carrier frequency of 14 MHz. The FSDT scheme can be implemented without a digital-to-analog converter (DAC) at the transmitter, an analog-to-digital converter (ADC) and circuit block related to the radio frequency (RF). This ensures extremely low power consumption and implementation in a small size.

As shown in Figure 10, the HBC modem is composed of four main blocks, the interface block (HBC IF), the transmitter block (HBC TX), the receiver block (HBC RX), and the analog front-end block (HBC AFE). The HBC IF block is the interface block between a microcontroller unit (MCU) and the HBC modem. In an HBC IF, there are register files in which to store the control information for all sub-blocks, a buffer to store the traffic data to be transmitted or received, an interrupt controller, and a serial interface block. The HBC TX is the transmitter block, which includes a scrambler, a serial-to-parallel block, and a frequency-selective spreader. The HBC TX output is connected to a signal electrode directly. The HBC RX is the receiver block, and it includes a synchronization block to search the start of a frame, a frequency-selective de-spreader, a parallel-to-serial block, and a descrambler. The HBC AFE is the analog front-end block, which has a noise reduction filter, amplifier, a clock recovery, and data-retiming block.

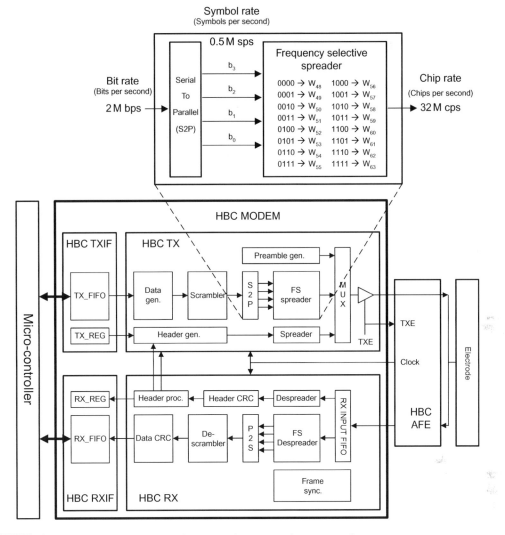

FIGURE 10 Block diagram of HBC modem using frequency selective spreader.

To prevent a loss of synchronization due to clock drift, an optional "pilot" sequence can be inserted with data in the PSDU in the IEEE standard for HBC PHY [5]. The pilot signal is inserted periodically, interleaved with a block of split data, according to the value of a predetermined insertion period. There are three pilot insertion intervals according to the information data rate. Another approach is to use CDR in order to recover a clock synchronous to the input data stream; hence, the recovered clock retimes the incoming data. A receiver implemented using CDR does not need a pilot signal, which improves the data throughput performance. In addition, it can recover the transmitted data without an ADC requiring a fast sampling frequency, which reduces the power consumption. The IEEE standard for HBC PHY also has no pilot insertion mode.

The frame structure for the FSDT scheme implemented using CDR is composed of a downlink sub-frame (or packet) and an uplink sub-frame (or packet), as shown in Figure 11 [29]. The length of a frame is 10 ms, and three sub-frame ratios are supported. The sub-frame ratios of the downlink versus the uplink are 8:2, 5:5, and 2:8. A downlink sub-frame consists of a lock time of 2 μs, a preamble of 128 bits, and a header of 64 bits. The data is shown in Figure 11. The dummy signal, which has a repeated pattern of 0 and 1 with the highest chip rate of 32 M chip per second during 2 μs, is used for the lock time. The dummy signal is used to give a CDR circuit an approximate reference frequency so that it reduces the time to align the phase to within 2 μs. An uplink sub-frame consists of a lock time of 2 μs, a header of 64 bits, and the data. There is a guard time of 2 μs for every sub-frame, which is the transition time between a receive mode and a transmit mode. From 250 kbps to 2 Mbps, four variable data rates are provided.

5. ANALOG FRONT-END FOR HUMAN BODY COMMUNICATION

The analog front-end in Figure 10 is required to receive a data signal transmitted through the human body and recover the transmitted data signal. This section presents an

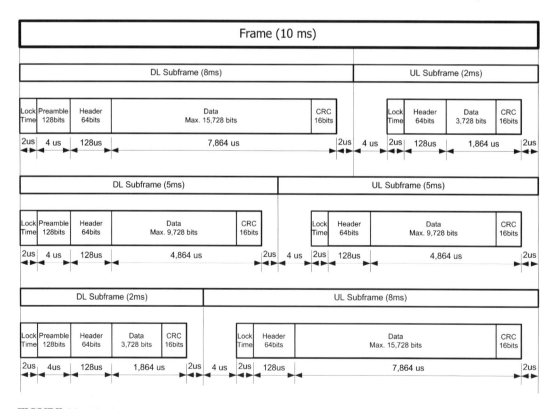

FIGURE 11   The frame structure for HBC using the FSDT scheme.

analog front-end for HBC [11] with a simple structure and which is compatible with the HBC modem of earlier work [8].

5.1 Received Signal

Design of the analog front-end requires analysis of the signal propagation through the human body. The channel model presented in the previous section is usable for performance analysis, but not suitable for design of an analog front-end because performances of an analog front-end, such as a cut-off frequency of a filter or hysteresis of a comparator, cannot be accurately modeled in simulation using the channel model. As a result, designing an analog front-end based on signal measurement is more practical. The size of the ground plane affects the signal loss in the channel. A larger ground plane enhances the electric coupling such that the size of the ground plane for the receiver is determined depending on the application, e.g., a mobile phone. In this experiment, the receiver electrode was designed with an electrode size of 5 mm $\times$ 5 mm considering typical button size of a handheld phone, and a ground plane of 50 mm $\times$ 100 mm, which is a typical phone size. Adopting a single electrode is more user-friendly than employing two electrodes for the signal and the ground. A transmitter was located on one hand and powered using a USB cable by a battery-operated mobile terminal, i.e., an ultra-mobile PC. A transmitted signal generated by a field-programmable gate array (FPGA) passed through the human body and was transferred to the receiver electrode located on the other hand. The transmitted signal consisted of a dummy pattern of 2 µs and a Manchester-coded preamble of 4 µs. The dummy signal had a repeated pattern of 0 and 1 with a 32 M chip rate during 2 µs and hence provided the clock and data recovery (CDR) circuit an approximate reference frequency to reduce the time to align the phase within 2 µs. The Manchester-coded preamble had a 4 µs length with 128 bits.

The transmission distance was approximately 150 cm. The receiver electrode was connected to the active probe to maximize the isolation between the receiver and the earth ground. The measurement setup and measured signal, which has a strong low frequency mainly around 50 kHz for 110 mV_{pp}, are shown in Figure 12. Figure 12 also shows a zoomed-in graph of Point A, which is the start point of a packet.

The measured signal was band-pass filtered by combining a high-pass filter (cut-off frequency of 8 MHz) and a low-pass filter (cut-off frequency of 30 MHz) using the mathematical function of an oscilloscope. The band-passed signal had an amplitude of about 10 mV_{pp}.

5.2 Design of an Analog Front-End

An analog front-end is designed based on the received signal measured on the human body. The signal loss in the channel (L_P) is expressed in dB. It can be calculated as follows:

$$L_P = 20\log\left(\frac{V_{RX}}{V_{TX}}\right) \qquad (7)$$

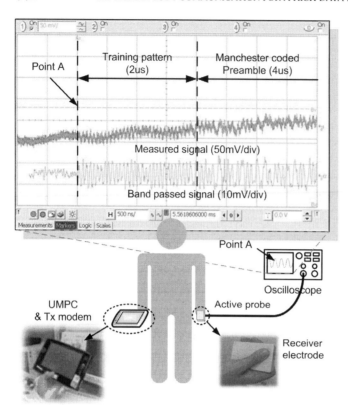

FIGURE 12 Measurement setup and the measured signal passing through a human body.

Here, V_{RX} is the received voltage and V_{TX} is the output voltage of the transmitter. If the output voltage of the transmitter is 3.3 V_{pp} and there is signal loss of approximately 60 dB through the human body [27,30], the receiver must have sensitivity of at least 3.3 mV_{pp}. As introduced in [28], the receiver separates the desired signal from all other signals induced on the channel, amplifies it to a level suitable for further processing, and determines its binary state using a comparator. A CDR circuit can be used to align the binary data to the clock. Figure 13 shows a block diagram of the analog front-end.

The first step is to define the gain of the amplifier. The gain of the receiver is determined based on the relationship between the minimum drive voltage of the comparator and the lowest amplitude of a signal that can be received. Figure 14 shows a simplified diagram of an HBC system that has a transmitter with a tri-state output buffer along with the channel (i.e., the human body) and a receiver front-end with an amplifier and comparator. High-pass filters implemented by passive and active components are omitted because they have little effect on the signal amplitude in the pass band. Because the human body is exposed to various electromagnetic environments, the received signal not only has strong low-frequency interference, but also high-frequency interference with relatively low amplitude. Positive feedback should be used to provide hysteresis and increase immunity to high-frequency noise. Maximum signal loss (L_P, dB), based on various body positions, of -70 dB [27] has been reported.

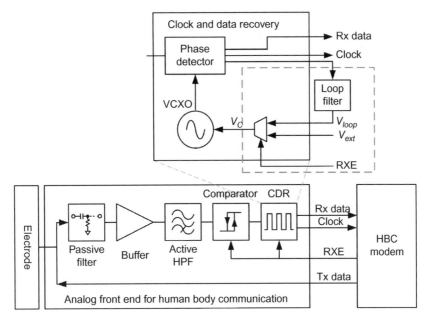

FIGURE 13   Block diagram of the analog front-end using the FSDT scheme.

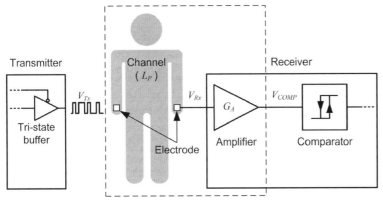

FIGURE 14   Simplified diagram of an HBC system.

L_P : Signal propagation loss in channel [dB]

G_A : Voltage gain of amplifier [dB]

To define the amplifier gain (G_A), the channel and amplifier are regarded as a single block, as depicted in the dotted box in Figure 14. The gain of the block is related to the output voltage of the transmitter (V_{TX}) and the input drive voltage of the comparator (V_{COMP}) as follows:

$$G_A = 20\log\left(\frac{V_{COMP}}{V_{TX}}\right) - L_P \tag{8}$$

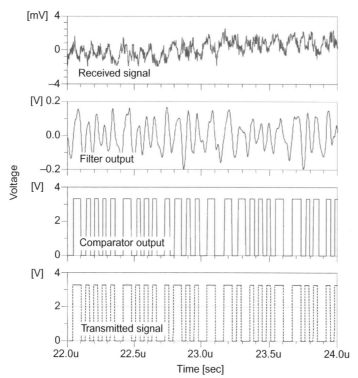

FIGURE 15 The Simulated signals from the electrode to the comparator based on measured data.

If the signal experiences a maximum path loss (L_P) of -80 dB and the comparator can be driven by an input signal of 33 mV_{pp}, the amplifier should have a gain (G_A) of 40 dB. The receiver has a sensitivity of 330 μV_{pp}. The maximum signal loss, taken into account considering various body positions, is -70 dB [15], with loss variation of -5 dB and a receiver margin of -5 dB.

The next step is to define the active filter specifications (Figure 13). A Bessel filter with a maximally flat group delay is used as it preserves the wave shape in the passband. Because the spread baseband in the FSDT scheme is directly transmitted through a human body, its spectrum occupies a broad band. In order to find the characteristics of the filter type, a transient analysis was performed based on the measured data using the Agilent Advanced Design System. Based on the simulation, a fourth-order active high-pass filter was designed with an 8 MHz cut-off frequency. The utilization of hysteresis on the comparator, together with the 200 MHz gain bandwidth product of the operational amplifier IC, can remove the high-frequency noise in the received signal, thus allowing use of a high-pass filter instead of a band-pass filter that was used in previously described experiments.

The induced noise of the 50 Hz or 60 Hz alternating current, caused by electric power transmission and fluorescent lighting, could be tens of volts on the human body. This noise is likely to saturate the active circuits with the supply voltage of 3.3 V used in the analog front-end and thus should be reduced to a point as low as the signal as the receiver can accommodate without saturation. A passive second-order high-pass filter realized using a series capacitor and two resistors connected between the supply voltage and ground offers half of the supply voltage, which provides the maximum swing range of the received signal without saturation. Figure 15 shows a simulated signal from the electrode to the comparator based on measured data. The processed signal has 4.6 ns of root-mean-square (RMS) eye jitter. This originates from noise, the induced interference, inter-symbol interference, and a band-limited channel during signal transmission, and from the physical limitations of the slew rate of the circuit and the hysteresis of the comparator. The received signal is strongly amplified, filtered, and then switched rapidly between two levels, at which point the data are finally recovered from the processed signal with jitter using a CDR circuit. A transmitted signal consisting of a spread signal using a group of Walsh codes and a Manchester-coded preamble is represented by a unipolar non-return-to-zero (NRZ) level with 0 and 3.3 V and with only two consecutive identical chips. This enables the proposed scheme to be used in asynchronous communication. The timing information is extracted and the chip stream regenerated using a CDR circuit from the received data while receiving a signal within a 2-chip delay. A voltage-controlled oscillator (VCO) is a type of oscillator whose oscillator frequency is controlled by a voltage input. The applied input voltage, named as a control voltage (V_C), determines the instantaneous oscillation frequency. The CDR is used as a local oscillator during transmission to generate a nominal frequency or a center frequency of a VCO by setting the control voltage (V_C) to a specific level, generally at the center of the supply voltage (V_{ext} in Figure 13). Within a frame, the master and slave should exchange data once with each other. This technique can periodically initialize a frequency offset between the clocks of the master and slave within a few ppm and prevent clock frequency drift. In HBC, data transfers cannot be deliberately started and stopped and thus a fast lock time is needed. The traditional frequency lock time of the phase lock loop used in a CDR is generally a few hundreds of microseconds. The slave operates as a receiver during a downlink and as a transmitter during an uplink time slot. In order to track the transmitted signal as receiving data at the downlink time slot, the output voltage of the loop filter (V_{loop}) is used as the control voltage (V_C) of VCXO by RXE in Figure 13. When no data is received at the downlink as well as at the uplink, the external voltage (V_{ext}) is used as VC to use CDR as a local oscillator. The clocks of the master and slave only exhibit a phase difference within a limited time, i.e., a frame length without buffer memory. In addition, dummy patterns of 2 μs composed of periodic data at 32 Mcps are used for training the CDR. The dummy pattern and oscillation mode during transmission can settle the control voltage to within 2 μs.

5.3 Operation of CDR

The receiver implemented using CDR has the advantage of higher data throughput performance compared to the scheme using pilot signal. The information in the frame can be used to control the CDR in order to synchronize between the HBC devices as well as to reduce the power consumption. Figure 16(a) shows the control signals related to the sub-frame ratio and the length of the received data. A control signal denoted as Rx_valid expresses the sub-frame ratio of the downlink versus the uplink and is set to a "logic high" setting during the downlink. A signal denoted as RXE expresses the length of the received data. When a master transmits data, a slave starts to receive the data, after it has passed through the human body. This point is marked as ① in Figure 16(a). The CDR of the slave in the receiving mode locks in the phase and frequency to the received signal that is represented as unipolar NRZ with the clock information of the transmitter. The control voltage of VCXO varies according to the frequency of the received signal, which is nearly identical to the frequency of the transmitter. This is the period marked as ② in Figure 16(a). If the received data is less than the allowable maximum size, the Rx data ends before the end of Rx_valid, as shown in Figure 16(a). This point is marked as ③ in Figure 16(a). In general, at the end of the received data, a lack of level transitions in the received signal for a considerable number of clock cycles should be noted. As long as the received signal is not present, the CDR is kept idle by the loss-of-signal (LOS) control signal, which is generated by the input stage of the CDR. Most CDRs have a function that

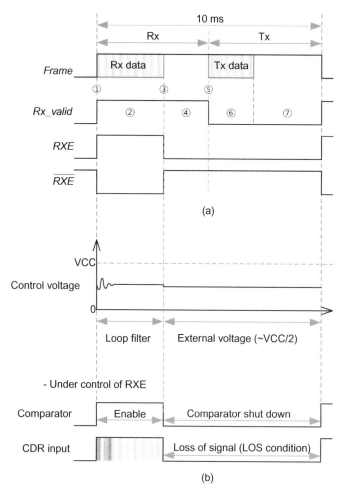

FIGURE 16   The operation of (a) the receiver enable signal (RXE) and (b) the control voltage of VCXO.

sets the control voltage of VCXO under the LOS condition. The CDR used for the FSDT scheme has a LOSIN terminal to control the control voltage. The LOS output signal is normally "logic low" and is set to "logic high" after 256 consecutive clock periods with no transition of the received signal. The LOS signal can be used to either flag external alarm circuits and/or drive the CDR's LOSIN input. When LOSIN is set to "logic high," the VCXO control voltage is switched to an internal voltage, usually half of the supply voltage, to generate a nominal frequency. The LOS signal is reset to "logic low" as soon as there are received signal transitions.

In HBC, however, random noise received from the human body is processed with the receiver chain, finally becoming binary digital data with an arbitrary pulse width. As a result, between the end of the received data and the end of the sub-frame of the downlink or the start of the sub-frame of the uplink (marked as ④ in Figure 16(a)), the phase of the incoming signal has a random difference from the phase of the local oscillator of the CDR. The processed random noise causes the CDR to lose its lock.

When the slave transmits the data marked as ⑤ in Figure 16(a), the transmitted signal leaks to the receive path of the slave. The feedback loop is formed: i) the CDR provides the clock frequency of the transmitted signal; ii) the transmitted signal leaks through an electrode to the receive path; iii) the leakage signal of the transmitted signal is processed and a comparator makes it a binary waveform with the frequency of the transmitter; iv) this signal is then connected to the CDR; and v) finally, the feedback loop is formed and the control voltage of VCXO converges to a constant value, generally half of the supply voltage, which generates the nominal frequency of VCXO. The same problem occurs when the transmitted signal is less than the allowable maximum period, marked as ⑦ in Figure 16(a).

In HBC, the user rapidly establishes connections to networks by touching the devices. A fast lock-in time of CDR is essential to characterize HBC in comparison with general wireless communication. In order to reduce the lock-in time, the frequencies of the master and the slave should be synchronized or maintained at a nominal frequency, after which the acquisition of the phase lock and the tracking of the acquired phase timing are processed. A simple solution to synchronize or maintain the frequencies of the master and the slave is to maximize the size of the data for all sub-frames. This can be accomplished by adding a dummy signal to the transmitted data for both the master and the slave. The dummy signal has a waveform with the repeated pattern of 0 and 1, with the highest chip rate, as used in the lock time. By adding the dummy signal, there is no period without transitions of the received signal even when the size of the data is less than the maximum value. As a result, no random binary signal induced on the human body is injected into the CDR. However, this method consumes additional power due to the extra pattern.

Another approach is to manage the control voltage of VCXO, as shown at the bottom of Figure 16(b). The signal of RXE depicted in Figure 16 (b) is set to "logic high" while in the presence of received data, and the signal of RXE is the complementary signal of RXE. The RXE signal is a control signal that is used to switch the control voltage of VCXO between the output of the loop filter and the external voltage. While in the presence of received data, VCXO is controlled by the output voltage of the loop filter. The control voltage of VCXO is set to the external voltage throughout all periods of the frame except in the presence of the received signal. The external voltage is generally half of the supply to generate the nominal frequency, as shown in the middle of Figure 16(b). In addition, the signal of RXE or RXE can serve as a control signal to enable or disable a comparator. According to the polarity at which the comparator needs to be enabled, one of the two signals is used. If the comparator is shut down, a CDR that is positioned after the comparator receives no transition of the signal, as shown in Figure 15. The operation of the comparator by RXE is shown in Figure 16(b). As a result, the output signal, denoted as the CDR input in Figure 16(b), intuitively describes the condition of a loss-of-signal (LOS) for the CDR. Under the LOS condition, the control voltage of VCXO in the CDR generates the nominal frequency and the CDR then becomes ready to play the role of the local oscillator during the transmitting period.

6. PERFORMANCE OF THE ANALOG FRONT-END

The performance of the analog front-end using the FSDT scheme is summarized in Table 2 [11]. The transceiver for the FSDT scheme has high sensitivity of −78 dBm and a wide dynamic range of 82 dB. The FSDT transceiver does not need complex circuitry such as analog blocks for a transmitter, a switch circuit for duplexing, a mixer for frequency translation, a low-pass filter for reducing high-frequency noise, and an ADC with a fast sampling frequency. It can reduce the power consumption as well as the degree of circuit complexity. Although the transceiver offers these advantages and thus can be implemented using off-the-shelf components, much power is dissipated.

The FSDT scheme also provides lower data rates compared to two earlier examples [7,27]. The data rate can be improved by modifying the spreading scheme while maintaining the structure of the analog front-end. If the transmission scheme uses a half-length of the Walsh code, which has a spreading factor that is twice as great, it will have the same chip rate at the output of the transmitter. Hence, it is not necessary to modify any circuit

TABLE 2   Performance Summary of the Various Analog Front-ends for HBC

Technology	FSDT '12 [11]	FSK '09 [7]	Wideband Digital Transmission '09 [27]
Frequency Band	8 ~ 22 MHz	30 ~ 120 MHz	1 ~ 30 MHz
Modulation	No	FSK	No
Interference Rejection Technique	Frequency selective Walsh spread	Adaptive Frequency Hopping	Input clamping
Data Rate	2 Mbps	10 Mbps	8.5 Mbps
Sensitivity	250 μV_{pp} (−74 dBm)	503 μV_{pp} (−65 dBm)	350 μV_{pp}
Dynamic Range	>82 dB	NA	NA
BER	<10^{-6} @ 250 μV_{pp}	<10^{-5} @ 710 μV_{pp}	<10^{-3} @ 450 μV_{pp}
Power Consumption	194.7 mW	4.6 mW	2.75 mW

in the analog front-end. Theoretically, the data rate is doubled and the signal-to-noise ratio is degraded by 3 dB. As shown in Table 2, the analog front-end using the FSDT scheme provides better BER performance at a low input power. The FSDT scheme can thus improve the data rate and maintain the target BER while reducing the processing gain in the spread spectrum signal.

7. COMMERCIALIZATION OF HUMAN BODY COMMUNICATION AND ITS CHALLENGES

For the commercialization of HBC, various transmission techniques and application models have been developed, and HBC standardization progressed simultaneously. In comparison with general wireless communication for health-monitoring sensors, for which a frequency of several hundreds of MHz (e.g., the Medical Implant Communication Service (MICS) band [5]) is typically used, HBC uses a very low frequency of less than 100 MHz, as the human body as a transmission medium supports signal transmission in a low-frequency band. This feature gives HBC a performance-related advantage: supporting high-data-rate transmission while maintaining a low level of power consumption. As HBC uses a low frequency for signal transmission, propagation delay does not occur significantly during transmission [26]; therefore, the data rate can be increased without compensation for the propagation delay. It has been presented that HBC can support a high data rate of up to 10 Mbps [29,31,32]. Also, a baseband-transmission technique [5,8–11], in which a baseband signal is transmitted through the human body without analog modulation to transform the baseband signal into a passband signal, has been proposed for HBC, allowing the power consumption of the modem and the analog front-end to be reduced. Due to the low power consumption, HBC has been applied to data communication for a capsule-type endoscope [33], in which it captures high-quality images of the inside of the bowel and transmits the captured images to a receiver on the surface of the body using HBC. Several companies have released a prototype module or a system solution using HBC [34–36]. The prototype modules and system solutions can be applied to various types of data transmission, including transmission of the identification information, but a healthcare service using wearable sensors has been also considered as one of the major applications of the modules and solutions. At the IEEE 802.15 working group, HBC PHY using the FSDT scheme has been adopted as a standard for short-range communication on the surface of the human body [5]. In addition, an interface for HBC is being standardized by the IEC TC47 standard committee [37]. The interface standard defines the electrode specifications, including the size and materials, as well as the operating conditions required to secure normal operation of the interface.

Several challenges, however, remain before the usefulness of HBC can be improved and, consequently, its potential for wearable sensors can be realized. The channel model in earlier research [21] can be applied only to signal transmission between on-body sensors, but it should be extended to signal transmission between on-body and in-body sensors, or between in-body sensors only. To do this, a modeling technique to model the human body, which is composed of various tissues, and a measurement technique to measure the channel properties inside the human body, should be studied. HBC PHY in [5] supports only a single transmission channel and cannot therefore be applied to communication between a hub and multiple nodes in a sensor network. For HBC to support multinode communication, a multichannel technique should be introduced along with an avoidance technique to deal with adjacent channel interference.

References

[1] Zimmerman, T. G. (1971). Personal area networks (PAN): near-field intra-body communication. M.S. thesis, MIT Media Laboratory, Cambridge, MA.

[2] T.G. Zimmerman, J.R. Smith, J.A. Paradiso, D. Allport, N. Gershenfeld, Applying electric field sensing to human-computer-interfaces, Comput. Human Interface Conf. (1995) 280−287.

[3] T.G. Zimmerman, Personal area networks: near-field intrabody communication, IBM Syst. J. 35 (3-4) (1996) 609−617.

[4] E.R. Post, M. Reynolds, M. Gray, J. Paradiso, N. Gershenfeld, Intrabody buses for data and power, Int. Symp. Wearable Comput. (1997) 52−55.

[5] IEEE Standard Assoiacion (2012). IEEE Standard for local and metropolitan area networks − Part 15.6: Wireless Body Area Networks.

[6] IFAC-CNR website: <http://niremf.ifac.cnr.it/tissprop>, (Last Accessed: 27.06.14).

[7] N.J. Cho, Y. Jerald, S.J. Song, et al., A 60 kb/s-10 Mb/s adaptive frequency hopping transceiver for interference-resilient body channel communication, IEEE J. Solid Stat Circuits 44 (3) (2009) 708−717.

[8] H.I. Park, I.G. Lim, S.W. Kang, W.W. Kim, Human body communication system with FSBT, IEEE 14th Int. Symp. Consum. Electron. (2010).

[9] T.W. Kang, J.H. Hwang, C.H. Hyoung, et al., Required transmitter power for frequency selective digital transmission on the effect of the human body channel, Int. Conf. Info. Commun. Technol. (2010) 17−19.

[10] T.W. Kang, J.H. Hwang, C.H. Hyoung, et al., Performance evaluation of human body communication system for IEEE 802.15 on the effect of human body channel, IEEE 15th Int. Symp. Consum. Electron. (2011).

[11] C.H. Hyoung, S.W. Kang, S.O. Park, Y.T. Kim, Transceiver for human body communication using frequency selective digital transmission, ETRI J. 34 (2) (2012) 216−225.

[12] K. Fujii, K. Ito, S. Tajima, Signal propagation of wearable computer using human body as transmission channel, IEEE Antennas Propagation Soc. Int. Symp. (2002) 512−515.

[13] K. Fujii, K. Ito, S. Tajima, A study on the receiving signal level in relation with the location of electrodes for wearable devices using human body as a transmission channel, IEEE Antennas Propagation Soc. Int. Symp. 3 (2003) 1071−1074.

[14] J.B. Sung, J.H. Hwang, C.H. Hyoung, et al., Effects of ground electrode on signal transmission of human body communication using human body as transmission medium, IEEE Antennas Propagation Soc. Int. Symp. (2006) 491−494.

[15] M.S. Wegmueller, A. Kuhn, J. Froehlich, et al., An attempt to model the human body as a communiation channel, IEEE Trans. Biomed. Eng. 54 (10) (2007) 1851−1857.

[16] J.A. Ruiz, S. Shimamoto, A study on the transmission characteristics of the human body towards broadband inter-body communications, IEEE 9th Int. Symp. Consum. Electron. (2005) 99−104.

[17] J.H. Hwang, H.J. Myoung, T.W. Kang, et al., Reverse effect of ground electrode on the signal loss of human body communication, IEEE Antennas Propagation Soc. Int. Symp. (2008).

[18] J.H. Hwang, J.B. Sung, C.H. Hyoung, et al., Analysis of signal interference in human body communication using human body as transmission medium, IEEE Antennas Propagation Soc. Int. Symp. (2006) 495−498.

[19] J.H. Hwang, T.W. Kang, S.W. Kang, Receptive properties of the human body of emitted electromagnetic waves for energy harvesting, IEEE Antennas Propagation Soc. Int. Symp. (2012).

[20] J.H. Hwang, C.H. Hyoung, K.H. Park, et al., Energy harvesting from ambient electromagnetic wave using human body as antenna, Electron. Lett. 49 (2) (2013) 149−151.

[21] J.H. Hwang, Channel model for body area network (BAN), IEEE (2010), P802.15-08-0780-10-006.

[22] K. Fujii, D. Ishide, M. Takaashi, et al., A study on the frequency characteristic of a transmission channel using human body for the wearable devices, Int. Symp. Antenna Propagation (2005) 359−362.

[23] J. Wang, D. Ishide, M. Takaashi, et al., Characterization and performance of high-frequency pulse transmission for human body area communications, IEICE Trans. Commun. E90-B (6) (2007) 1344–1350.

[24] K. Fujii, M. Takahashi, K. Ito, et al., Study on the transmission mechanism for wearable for wearable device using the human body as a transmission channel, IEICE Trans. Commun. E88-B (6) (2005) 2401–2410.

[25] K. Fujii, M. Takahashi, K. Ito, Electric field distributions of wearable devices using the human body as a transmission channel, IEEE Trans. Antenna Propagation 55 (7) (2007) 2080–2087.

[26] J.H. Hwang, T.W. Kang, S.W. Kang, Measurement Results of Human Body's Signal Loss with Multiple Subjects for Human Body Communication, IEEE Antennas Propagation Soc. Int. Symp. (2011) 1666–1669.

[27] A. Fazzi, S. Ouzounov, J.V.D. Homberg., A 2.75 mW wideband correlation-based transceiver for body-coupled communication, IEEE Int. Solid State Circuits Conf.-Dig. Tech. Pap. (2009) 204–205, 205a.

[28] C.H. Hyoung, J.B. Sung, J.H. Hwang, et al., A novel system for intrabody communication: touch-and-play. Circuits and Systems, IEEE Int. Symp. Circuits Syst. (2006) 1343–1346.

[29] Hyoung, C. H. (2012). Analog Front-End for Human Body Communications and Feasibility Study for Medical Services. Ph. D. Dissertation, Korea Advanced Institude of Science and Technology, Korea.

[30] J.H. Hwang, J.B. Sung, S.E. Kim, et al., Effect of load impedance on the signal loss of human body communication, IEEE Int. Symp. Antenna Propagation (2007) 3217–3220.

[31] M. Shinagawa, M. Fukumoto, K. Ochiai, H. Kyuragi, A near-field-sensing transceiver for intrabody communication based on the electrooptic effect, IEEE Trans. Instrum. Meas. 53 (6) (2004) 1533–1538.

[32] S.J. Song, N.J. Cho, S.Y. Kim, H.J. Yoo, A 4.8-mW 10-Mb/s wideband signaling receiver analog front-end for human body communication, 32nd Eur. Solid Stat Circuits Conf. (2006) 488–491.

[33] Intromedic web-site: <www.intromedic.com/eng/sub_products_2.html>, (Last Accessed: 27.06.14).

[34] Renesas web-site: <www.renesas.com/edge_ol/features/08/index.jsp>, (Last Accessed: 27.06.14).

[35] NTT web-site: <https://www.ntt-review.jp/archive/ntttechnical.php?contents=ntr201003sf1.html>, (Last Accessed: 27.06.14).

[36] Sony web-site: <www.sonycsl.co.jp/IL/projects/wearable_key/>, (Last Accessed: 27.06.14).

[37] IEC web-site: <www.iec.ch/dyn/www/f?p = 103:23:0::::FSP_ORG_ID,FSP_LANG_ID:1251,25 Hyong>, (Last Accessed: 27.06.14).

Channel Models for On-Body Communications

H.G. Sandalidis[1] and I. Maglogiannis[2]

[1]Department of Computer Science and Biomedical Informatics, University of Thessaly, Lamia, Greece, [2]Department of Digital Systems, University of Piraeus, Piraeus, Greece

1. INTRODUCTION

The adoption of wearable systems in modern healthcare telemonitoring systems has been considered a medical challenge towards the highest level of quality of life. The current state-of-the-art technologies in wearable computing, wireless telemedical platforms, and wireless sensors allow easy and unobtrusive electronic measurement of several vital signals and health conditions regardless of the time and the place the patient needs condition monitoring. These measurements can be either stored locally on a monitoring wearable device for later transmission or directly transmitted, e.g., over the public phone network, to a medical center. Such architecture is depicted in Figure 1, where the on-body network resides on the body of the monitored person.

Wave propagation in the human body area is considered a complicated process. In wireless body area networks (WBANs), propagation phenomena include, among others, the frequency-dependent electromagnetic transmission, the strong absorption and scattering from human tissues, and high losses for non-line-of-site links, as well as the appearance of frequent shadowing effects [1]. The need to find reliable and simultaneously simplistic models in the majority of frequency bands where WBANs operate is the driving force that has led to the appearance of many studies on the topic in recent years [2].

Channel models may be classified into two main categories: analytical and empirical. Analytical models require a detailed description of the propagation environment and attempt to achieve a precise modeling of the power attenuation at a specific position. Empirical models, on the other hand, attempt to foresee the propagation characteristics

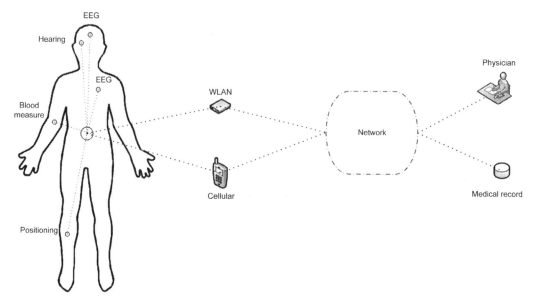

FIGURE 1 A typical architecture of a patient monitoring system using wearable sensors. On-body wearable sensors wirelessly communicate with wearable (mobile phone) or stationary (WLAN) gateway devices to relay sensor data to remote locations.

based on a given set of measurements. Thus, they are more suitable for complex environments and applications such as body area networking [3].

Transmission in WBANs is significantly affected by human body tissues and is frequency dependent. Signals are attenuated and delayed to a large extent and reach the receiver as a sum of several components traveling in multiple paths. The simplest formula used to empirically capture the key mechanisms of signal attenuation in the body area is the Friis transmission formula, i.e., [4],

$$PL(d)_{dB} = 10n\log(d) + C, \tag{1}$$

where $PL(d)_{dB}$ is the path loss in decibels at transmission distance d, n is the path loss exponent, and C is a constant.

Shadowing phenomena are likely to appear due to the human body's movements and the variations in the surroundings of human body parts as well. Moreover, multi-path reflections make the channel response look like a series of pulses. Propagation paths can also experience fading due to energy absorption, reflection, and diffraction mechanisms, as well as due to multi-path effects of the surrounding environments [3,5].

All the above phenomena have a different degree of influence on WBAN behavior according to the frequency band of operation. Usually, WBANs are designed to operate in the bands of 400 MHz, 600 MHz, 900 MHz, 2.4 GHz, and 3.1 to 10.6 GHz. Due to the different channel characteristics for each of the above frequency ranges, proper channel modeling is required in each case.

This chapter aims to summarize the fundamental channel models introduced in the literature concerning wearable WBANs. The basic models were launched by the IEEE 802.15.6 subgroup, which was created to develop standards for BANs [6]. Most of them provide adequate performance and were generated after a number of proper measurements in indoor environments. The group examined different scenarios for channel modeling depending on the location of communication nodes, i.e., implant, on-body, and external.

The chapter particularly focuses on the fundamental channel models for on-body communications launched by either the IEEE or other independent researchers. Basic experimental setups are referred to in detail. Both line-of-sight (LOS) and non-line-of-sight (NLOS) channels, depending on the way the electromagnetic waves propagate between communication nodes, are investigated. In view of the above, this chapter can be a valuable tool for researchers and designers who are working in this area and who desire to build their own WBAN.

2. IEEE 802.15.6 TG6 STANDARD MODELS

The IEEE 802.15.6 standard has been under development since 2007. Particularly, the Task Group TG6 has aimed to develop BAN standards optimized for low-power devices for medical and non-medical devices that can be placed inside or on the surface of the human body [6].

Communications in the standard are broadly defined, including narrowband and ultra-wideband (UWB) communications. Narrowband is appropriate for electronic healthcare applications since bio-signals at these frequencies are less attenuated from the human body. However, they have small bandwidth, and symbol interference is likely to be caused due to multi-path effects. Ultra wideband (UWB) in IEEE 802.15.6 requires large bandwidth, about 499 MHz. Table 1 shows the carrier frequency bands and channel bandwidths proposed for WBAN by the IEEE group [7].

TABLE 1 Frequency Bands and Bandwidths

Narrowband Communications		Ultra-Wideband Communications	
Frequency Range	**Bandwidth**	**Frequency Range**	**Bandwidth**
402–405 MHz	300 KHz	3.2–4.7 GHz	499 MHz
420–450 MHz	300 KHz	6.2–10.2 GHz	499 MHz
863–870 MHz	400 KHz		
902–928 MHz	500 KHz		
950–956 MHz	400 KHz		
2360–2400 MHz	1 MHz		
2400–2483.5 MHz	1 MHz		

A significant number of documents were submitted to the channel modeling subgroup or presented and discussed at the IEEE 802.15.6 meetings [8]. The most representative propagation models for on-body communications are presented in the following subsections.

2.1 Aoyagi et al. Model

One of the fundamental channel models of the IEEE 802.15.6 subgroup was provided by Aoyagi et al. It is based on the simple Friis formula, where an additional term was added in order to describe shadowing [9]. This additional term N is a Gaussian distributed function with a mean value 0 dB and standard deviation σ_N dB, which can be determined based on statistical fitting [4]. Particularly, the path loss observed at a receiving node located at distance d from a transmitter is expressed as

$$PL(d)_{dB} = \alpha \log(d) + b + N, \qquad (2)$$

where α and b are coefficients of linear fitting.

Based on that model, the authors provided a set of measurements considering several frequency bands suitable for WBAN operation. More details on the measurement setup, derivation, and data analysis are given in [9]. Experiments took place in different environments, i.e., in a hospital room and an anechoic chamber. The anechoic chamber provides ideal transmission conditions since it eliminates any possible reflections from the surroundings.

During the experiment process, several antennas were located on the human body, including the left hand, left upper arm, left ear, head, shoulder, chest, right rib, left waist, thigh, and ankle. The received measurements consisted of a proper set of empirical results that were properly fitted to the linear model of (2) using the least square method. The measurements took possible human body turnarounds into account, i.e., LOS and NLOS situations were considered in the experiment as well.

The propagation distance between a transmitter and receiver was scaled and varied from 100 to 1,000 mm. Different frequency bands were used and the results obtained are depicted in Table 2. Roughly speaking, five submodels were distinguished, namely A, B, C, D, and E, which refer to the five frequency bands where experiments took place, i.e., 400–450 MHz, 608–614 MHz, 950–956 MHz, 2.4–2.5 GHz, and 3.1–10.6 GHz, accordingly. The parameters, A, B, and σ_N for each sub-model differ from each other and are clearly outlined for the two environments (hospital room and anechoic chamber). It can be shown that the path loss effects caused by the presence of the human body are greater in an anechoic chamber, in comparison to the hospital room, due to the reflections from the walls. This can be verified in [9, Figure 3] where analytical path loss results for all models are illustrated using proper diagrams.

2.2 Dolmans and Fort Model

A more complicated model was suggested by Dolmans and Fort [10]. The measurements took place in an office environment for 915 MHz and 2.45 GHz.

TABLE 2 Model Parameters

Parameters	Hospital Room	Anechoic Chamber
MODEL A: 400–450 MHz		
A	3	22.6
B	34.6	−7.85
σ_N	4.63	5.60
MODEL B: 608–614 MHz		
A	16.7	17.2
B	−0.45	1.61
σ_N	5.99	6.96
MODEL C: 950–956 MHz		
A	15.5	28.8
B	5.38	−23.5
σ_N	5.35	11.7
MODEL D: 2.4–2.5 GHz		
A	6.6	29.3
B	36.1	−16.8
σ_N	3.80	6.89
MODEL E: 3.1–10.6 GHz		
A	19.2	34.1
B	3.38	−31.4
σ_N	4.40	4.85

The channel parameters were obtained using measurements from receiving nodes placed in front of and on the back of a human body. The transmitter was worn at approximately shoulder height. Several path-loss models were tested to fit the measurement data. The combined exponential-linear saturation model was found to give the best fit with the measurement data. That happened since the path loss follows an exponential decay around the perimeter of the body and flattens out as distance increases, due to energy received from multi-path reflections of the indoor environment [8].

The path loss according to that model is expressed as

$$PL(d)_{dB} = -10\log(P_0^{-m_0 d} + P_1) + \sigma_P n_P, \tag{3}$$

where P_0 is the average loss close to the antenna, m_0 is the average decay rate in dB/cm for the surface wave moving around the perimeter of the body, P_1 is the mean attenuation of components radiated away from the body and reflected back to the receiver, σ_P is the log-normal variance in dB around the average representing the variations measured at different body and room locations, and n_P is a Gaussian variable of zero mean and unit variance.

The authors also provided a model of the flat small-scale fading observed at the measured data. They examined several distributions such as Rayleigh, lognormal, Nakagami-m, and Rician using maximum-likelihood parameter estimates and found that the Rician distribution reflects the most adequate model. The Rician distribution is characterized by a parameter, K, defined as the ratio of the specular component to the random multi-path component powers [10]. This parameter decreases when the receiver is moved away. The K-factor is expressed in dB as

$$K_{dB} = K_0 - m_K P_{dB} + \sigma_K n_K, \tag{4}$$

where K_0 is the fit with measurement data for the K-factor for low path loss, m_k is the slope of the linear correlation between path loss and K-factor, P_{dB} is the path loss in dB, σ_k is the log-normal variance of the measured data between path loss and K-factor, and n_k is the Gaussian variable with zero mean and unit variance.

Regarding narrowband systems, where frequency selective fading appears, the authors estimated the delay spread from the cumulative density function for antenna separations of 15 and 45 cm. The delay spread is modeled with a normal distribution. Table 3 summarizes the parameter values for the two frequency bands where the model is valid.

2.3 Miniutti et al. Model

This model deals with measurements of on-body narrowband wireless channels at the frequencies of 820 MHz and 2.36 GHz [11]. The experiments took place in an office environment where several antennas were located on the human body. The aim was to estimate the path loss assuming three different human actions, i.e., standing, walking, and running. As the authors declare, the innovation of their study compared to other ones lies in the examination of continuous movement of the human body on the wireless channel. Table 4 lists the body locations and the separations in cm of transmitters and receivers used.

The multi-path transmission around the body and the surrounding environment also causes a fading appearance. A significant amount of fading also occurs due to the movement of the human body. The distribution of the normalized received power was described using a probability density function that obtains the best match for all scenarios. The Gamma distribution was proven the best fit to average fade duration. Moreover, the same distribution fitted to a dB scale was the best fit to fade magnitude [8].

An analytical description of the experimental process and the instrumentation equipment is provided in [11]. The path loss at a given time t is obtained as

$$PL(d)_{dB} = P_{tX} - P_{rX} + G_{amplifiers} - L_{cable}, \tag{5}$$

where P_{tX} is the transmitted power, P_{rX} the RMS received power, $G_{amplifiers}$ the amplifier gain, and L_{cable} the cable loss. Table 5 shows the average path loss for each action and antenna placement for the two frequency bands. In general, the path loss is greater at 2.36 GHz than at 820 MHz.

TABLE 3   Model Parameters

Parameters	915 MHz	2.45 GHz
Path Loss		
P_0 [dB]	−1.9	−25.8
m_0 [dB/cm]	2.1	2.0
P_1 [dB]	−59.4	−71.3
σ_P [dB]	3.2	3.6
Flat Small-scale Fading		
K_0 [dB]	40.1	30.6
m_K [dB]	0.61	0.43
σ_K [dB]	2.4	3.4
Frequency Selective Small-Scale Fading		
Parameters of the Mean Calue of the Delay Spread		
Distance [cm]	t_{rms} *[ns]*	t_{rms} *[ns]*
15	3	6
45	9	16
Parameters of the 90% Cumulative Value of the Delay Spread		
Distance [cm]	t_{rms} *[ns]*	t_{rms} *[ns]*
15	5	11
45	15	22

TABLE 4   Distance Between Transmitters and Receivers

Receiver Location	Transmitter Location					
	Chest	**Right Wrist**	**Left Wrist**	**Right Ankle**	**Left Ankle**	**Back**
Right hip	38	11	30	90	93	45
Chest		36		115		25 (50)

2.4 Astrin Model

Astrin performed measurements of a body channel taken at 13.56 MHz in order to be used for an IEEE standard [12]. The body channel at that frequency has a link loss similar to free space and can be used for body area network applications. However, due to the small available bandwidth, only low data rates of a few kbps can be transmitted.

TABLE 5   Average Path Loss [dB]

	820 MHz								
Action	**Receiver at Right Hip, Transmitter at:**						**Receiver at Chest, Transmitter at:**		
	Chest	**Right Wrist**	**Left Wrist**	**Right Ankle**	**Left Ankle**	**Back**	**Back**	**Right Wrist**	**Right Ankle**
Standing	57.4	50.2	59.8	54.3	68.7	61.8	66.3	54.5	54.3
Walking	52.9	38.4	63.6	48.1	55.5	57.1	63.8	51.3	56.9
Running	44.1	37.2	60.2	48.9	54.2	62.3	66.3	49.4	54.1
	2.36 GHz								
Action	**Receiver at Right Hip, Transmitter at:**						**Receiver at Chest, Transmitter at:**		
	Chest	**Right Wrist**	**Left Wrist**	**Right Ankle**	**Left Ankle**	**Back**	**Back**	**Right Wrist**	**Right Ankle**
Standing	65.3	44.5	74.7	60.9	70.7	75.3	73.0	70.5	66.3
Walking	59.1	47.3	59.8	53.9	58.5	67.4	72.0	64.9	62.4
Running	55.9	36.3	52.5	55.0	59.0	68.5	71.7	57.4	63.3

TABLE 6   Measurement Results

Description	Signal Amplitude Reduction	dB Loss in Relation to Air
Through the hand	3.3%	−0.15
Through the wrist	2.8%	−0.12
Torso, front to back	3.4%	−0.15
Through the thigh	1.9%	−0.08
Through the ankle	2.8%	−0.12
Left ear to right ear	2.0%	−0.09
Left ear to right ear, wearing metal glasses	1.5%	−0.07

TABLE 7   Signal Values vs. Distance

Distance (in)	3	4	5	6	7	8	9	10	11	12
Received signal pk-pk (mv)	6.93	4.74	2.93	1.86	1.25	0.837	0.633	0.471	0.384	0.305

Therefore, the BANs operating at those frequencies can be used to signal an emergency condition or to transmit reliably a "wake up" signal to a sleeping BAN node [12].

The measurements were taken at specific locations on a live human body and are presented in Table 6. The signals received at various distances are shown in Table 7.

TABLE 8 PDP Model Parameters

α_1	γ_0	$-4.6\,\mathrm{dB}$
	Γ	59.7
	σ_S	5.02 dB
t_1	$1/\lambda$	1.85 ns
L	$\bar{L}$	38.1

2.5 Aoyagi et al., Power Delay Profile Model

A power delay profile (PDP) model for the range 3.1 to 10.6 GHz was proposed by Aoyagi et al. [9]. The power delay profile of a channel represents the average power of the received signal in terms of the delay with respect to the first arrival path in multi-path transmission. Since signals in WBANs are usually transmitted following multiple paths, the channel response seems like a series of pulses. This mainly appears at these specific bands where highly frequency-selective channels are observed [9].

The PDP model is characterized by the following equations:

$$h(t) = \sum_{l=0}^{L-1} \alpha_l e^{j\phi_l} \delta(t - t_l)$$

$$10\log|\alpha_l|^2 = \begin{cases} 0 & l = 0 \\ \gamma_0 + 10\log\left(e^{-\frac{t_l}{\Gamma}}\right) + S & l \neq 0 \end{cases},$$

$$p(t_l|t_{l-1}) = \lambda e^{-\lambda(t_l - t_{l-1})}$$

$$p(L) = \frac{\bar{L}^L e^{\bar{L}}}{L!}$$

(6)

where φ_l is the phase for the l-th path and follows a uniform distribution over $(0, 2\pi)$, α_l is the path amplitude for the l-th path, t_l is the path arrival time, L is the number of the arrival paths, $\delta(t)$ is the Dirac function, Γ is an exponential decay with a Rician factor γ_0, S is a normal distribution with zero mean and standard deviation of σ_s, λ is the path arrival rate, and $\bar{L}$ is the average number of the L. The corresponding parameters are given in Table 8.

2.6 Dolmans and Fort Wideband Model

Dolmans and Fort proposed an ultra wideband (UWB) model in the range 3 to 10 GHz that satisfies the requirements for IEEE standards [10]. Wideband models are more complex than narrowband ones since there are more reflections in the channel. In the UWB range the propagation though the body is negligible and signal transmission is achieved through diffraction around the body and reflections from the surrounding environment.

TABLE 9 Path Loss Parameters for Different Antenna-body Separations

	Around the Torso			Along the Torso	
	0 mm	5 mm	10 mm	0 mm	5 mm
n	5.8	5.9	6.0	3.1	3.1
d_0 (m)	0.1	0.1	0.1	0.1	0.1
P_0dB	56.1	48.4	45.8A	56.5	44.6

TABLE 10 Body Posture Results at 3.1−5.1 GHz

	Action	Left Ear	Right Ear	Left Wrist	Right Wrist	Right Waist	Left Ankle	Right Ankle
Anechoic Chamber	Standing	62.2	61.6	64.9	67.5	64.2	72.8	69.1
	Sitting	71.9	65.7	55.6	69.9	65.9	76.3	73.9
Office	Standing	73.8	70.4	61.4	70.9	74.3	76.4	68.3
	Sitting	62.3	72.1	65.6	76.3	74.7	79.8	75.7

TABLE 11 Body Posture Results at 7.25−8.5 GHz

	Action	Left Ear	Right Ear	Left Wrist	Right Wrist	Right Waist	Left Ankle	Right Ankle
Anechoic Chamber	Standing	81.7	83.9	69.3	63.9	76.1	76.8	77.1
	Sitting	84.4	85.2	73.6	75.5	82.38	79.8	95.2
Office	Standing	75.5	75.6	80.6	66.4	71.9	74.4	75.8
	Sitting	67.1	84.5	67.0	67.1	67.5	70.6	83.4

The authors performed experiments in an anechoic chamber. A set of antennas was placed on a human body. The measurements were performed in six planes separated by 7 cm along the vertical axis of the torso. The reader is referred to [10] for more details.

The empirical power decay model was used and found to fit the numerical results obtained in an adequate manner, i.e.,

$$PL(d)_{dB} = P_0[dB] + 10n\log\frac{d}{d_0}. \tag{7}$$

Table 9 shows the corresponding parameters. It is clearly seen that a much lower exponent is measured when the propagation is along the front ($n \approx 3$) rather than around the torso ($n \approx 6$).

TABLE 12   Body Movement Results at 3.1—5.1 GHz

		Anechoic Chamber		Office	
		Mean	**Stdev**	**Mean**	**Stdev**
Left wrist	Forward direction	65.7	4.3	71.7	5.8
	Side direction	70.9	6.0	76.6	7.7
Right wrist	Forward direction	72.9	4.4	73.6	2.3
	Side direction	71.8	5.9	75.1	3.1
Left ankle	Forward direction	74.6	2.5	76.4	0.1
	Side direction	67.9	6.8	70.4	8.4
Right ankle	Forward direction	70.8	2.5	69.9	2.3
	Side direction	72.8	5.3	69.6	1.9

TABLE 13   Body Movement Results at 7.25—8.5 GHz

		Anechoic Chamber		Office	
		Mean	**Stdev**	**Mean**	**Stdev**
Left wrist	Forward direction	76.9	6.6	74.1	3.4
	Side direction	77.4	10.1	78.1	4.7
Right wrist	Forward direction	79.7	7.1	73.0	7.1
	Side direction	79.5	8.1	75.9	6.0
Left ankle	Forward direction	82.4	7.9	75.9	2.2
	Side direction	77.4	0.8	75.7	1.9
Right ankle	Forward direction	85.4	11.7	77.5	2.5
	Side direction	82.0	6.9	78.1	3.3

2.7 Kang et al. Model

Kang et al. proposed a UWB body surface model at the frequency bands of 3.1 to 5.1 GHz, and 7.25 to 8.5 GHz, respectively [13]. At first the body posture effects were taken into account. The transmitter antenna was located in the left waist and receiver antennas were placed in various positions. Measurements were carried out in an anechoic chamber and office environments. The results are shown in Tables 10 and 11. All values are in dB scale.

Then the effect of body movement was examined, i.e., an arm and a leg were moved in a forward or side direction during measurements. Tables 12 and 13 summarize the experimental results.

TABLE 14   Distances Between the Transmitter and Receivers

Position	Distance d [mm]
Right wrist	440~525
Right upper arm	360
Left ear	710
Head	650
Shoulder	310
Chest	230
Right rib	183
Left waist	140
Thigh	340
Ankle	815~940

2.8 Kim et al. Model

Kim et al. proposed a dynamic statistical channel model for the IEEE 802.15.6 [14]. The measurements were performed in a radio anechoic chamber using a real-time channel sounding system at 4.5 GHz and bandwidth of 120 MHz. Measurements focused on capturing the fading effect that appears due to the movement of the human body. The transmitter was fixed on or around the navel and the measurements were conducted one by one at 10 receiver positions placed at the specific locations shown in Table 14.

The authors tried to fit the measurement results by using some well-known probability density functions such as normal, log-normal, and Weibull distributions, and found the best match of them. In general, it was observed that the normal distribution provides a best fit of the still postures, the log-normal distribution shows a good match in cases of still postures and small movements, whereas the Weibull distribution can represent much better large movement behaviors. Table 15 summarizes the findings.

Tables 16, 17, and 18 present the various parameters for the three distributions. The values of the log-normal distribution in Table 16 are in decibels. The values in parentheses are negative log-likelihood values by which the best fit distribution were obtained (less means better fit).

3. INDEPENDENT STUDIES

Since the subject of body area networking is an appealing research topic, a large number of independent studies have appeared in the technical literature in recent years. Accurate propagation models for body surface communication have been derived by research groups around the world, and these models can be applied to either general or specific-purpose WBAN architectures. It is obvious that a full presentation of these works cannot be properly done here, therefore, only some indicative studies are discussed in the following subsections.

TABLE 15 Fitting Distributions

Position	Still	Walking	Up-Down
Right wrist	Normal	Weibull	Weibull
Right upper arm	Log-normal	Weibull	Weibull
Head	Weibull	Log-normal	Log-normal
Right ear	Normal	Log-normal	Weibull
Shoulder	Log-normal	Weibull	Weibull
Chest	Log-normal	Log-normal	Weibull
Right rib	Log-normal	Log-normal	Weibull
Left waist	Normal	Log-normal	Weibull
Right thigh	Log-normal	Log-normal	Weibull
Right ankle	Log-normal	Weibull	Weibull

TABLE 16 Fitting Results for Normal Distribution $f(x|\mu,\sigma) = \frac{1}{\sigma\sqrt{2\pi}}e^{-\frac{(x-\mu)^2}{\sigma^2}}$

Position	Still μ/s [dB] (− LogL)	Walking μ/s [dB] (− LogL)	Up−down μ/s [dB] (−LogL)
Right wrist	1.0000/0.1279 (− 6303)	−3.0875/4.4063 (28972)	8.9461/5.1576 (30220)
Right upper arm	1.0000/0.1500 (− 4735)	3.2443/1.7012 (19332)	7.3374/6.1415 (32061)
Head	1.0000/0.0666 (− 12706)	0.4477/0.2445 (4535)	0.9058/1.2629 (16207)
Right ear	1.0000/0.1349 (− 5826)	0.7303/0.3014 (101)	0.8205/0.8232 (12148)
Shoulder	1.0000/0.0335 (− 19782)	2.6849/1.4627 (2153)	1.5437/1.3265 (17519)
Chest	1.0000/0.3481 (3600)	3.6360/2.4688 (17849)	7.1082/9.2594 (36609)
Right rib	1.0000/0.1654 (− 3812)	0.8572/0.2744 (1256)	4.7043/2.9597 (24784)
Left waist	1.0000/0.0399 (− 17978)	0.7264/0.1609 (− 4018)	0.5458/0.3105 (2456)
Right thigh	1.0000/0.0964 (− 9254)	0.6500/0.4388 (5831)	1.1357/0.6949 (10272)
Right ankle	1.0000/0.0787 (− 11025)	1.6070/0.9968 (14248)	1.2489/1.2220 (16122)

3.1 CWC Oulu University Model

The Centre of Wireless Communications (CWC) at Oulu University of Finland performed extensive measurements at the Oulu University Hospital, focusing on the UWB band [15]. The aim of the work was to generate realistic WBAN channel models to be used in designing WBAN applications for hospital use. The results obtained were compared with the ones provided by the IEEE group.

TABLE 17   Fitting Results for lognormal Distribution $f(x|\mu,\sigma) = \frac{10/\ln10}{x\sigma\sqrt{2\pi}}e^{-\frac{(10\log x-\mu)^2}{\sigma^2}}$

Position	Still μ/s [dB] (− LogL)	Walking μ/s [dB] (− LogL)	Up−down μ/s [dB] (− LogL)
Right wrist	−0.0652/0.9531 (−1114.1)	−3.9908/11.7727 (14948)	8.4456/3.4836 (31047)
Right upper arm	−0.0491/0.6544 (−4800.9)	4.4491/2.4972 (18735)	6.3009/5.6692 (31093)
Head	−0.0100/0.2969 (−12476)	−3.5392/3.2052 (2983.8)	−3.6932/5.3385 (7601)
Right ear	−0.0400/0.5921 (−5815.6)	−4.1465/2.4654 (−1005.5)	−3.3130/5.2458 (8384.1)
Shoulder	−0.0024/0.1453 (−19806)	−1.8011/2.0549 (2507.8)	0.2165/4.1023 (14537)
Chest	−0.2359/1.4019 (2315)	3.6360/2.4688 (16779)	5.0788/6.4077 (29907)
Right rib	−0.0470/0.6043 (−5657.3)	−0.8883/1.3994 (818.79)	5.7033/3.2101 (24051)
Left waist	−0.0035/0.1742 (−17942)	−1.4965/0.9770 (−4109.5)	−3.5588/3.3466 (3337.8)
Right thigh	−0.0191/ 0.4017 (−9717)	−2.6924/2.6453 (2970.7)	−0.5995/3.7712 (11098)
Right ankle	−0.0131/0.3352 (−11251)	0.9424/3.7030 (14861)	−1.1303/4.7167 (12357)

TABLE 18   Fitting Results for Weibull Distribution $f(x|a,b) = \frac{b}{\alpha}\left(\frac{x}{\alpha}\right)^{b-1}e^{-\left(\frac{x}{\alpha}\right)^b}$

Position	Still a/b (− LogL)	Walking a/b (− LogL)	Up−down a/b (− LogL)
Right wrist	1.0478/7.7411 (−5770.3)	1.4690/0.4510 (14629)	10.0006/1.7319(29786)
Right upper arm	1.0655/7.3618 (−4529.3)	3.6759/2.0325 (18625)	7.5368/1.0787 (29631)
Head	1.0297/17.9473 (−13029)	0.6360/1.5519 (3061.7)	0.8015/0.8228 (8458.3)
Right ear	1.0594/7.7870 (−5394.6)	0.5071/1.9516 (−840.12)	0.8224/1.0055 (7959.7)
Shoulder	1.0165/29.8877 (−18940)	0.8237/2.6536 (1913.3)	1.6534/1.2203 (14452)
Chest	1.1203/2.9913 (3487.2)	3.0419/1.9575 (16909)	6.3808/0.8247 (29393)
Right rib	1.0742/4.9107 (−1979.2)	0.9539/3.2009 (1357.2)	5.2709/1.6440 (23654)
Left waist	1.0190/28.9055 (−17946)	0.7915/4.9572 (−3903.2)	0.6117/1.7907 (1823.7)
Right thigh	1.0464/9.6410 (−7951.2)	0.7325/1.6297 (3867.9)	1.2626/1.6251 (9539.3)
Right ankle	1.0379/11.8572 (−9813.1)	1.7880/1.6227 (13323)	1.2867/1.0759 (12124)

The measurements took place at an anechoic chamber, classroom, and different hospital rooms. Three hospital cases were examined, i.e., a surgery room, a conventional ward room, and a corridor. The last two environments took into account the specific propagation characteristics that are valid, whereas the anechoic chamber was used to validate the measurement system performance. The frequency range lies between 3.1 GHz and 10 GHz, and 100 consecutive frequency responses were taken using a proper vector network analyzer [15]. Table 19 presents the parameters used during the experimental process.

TABLE 19 Experimental Parameters

Parameter	Value
Frequency band	UWB
Bandwidth	6.9 GHz
IF Bandwidth of VNA	3 KHz
Number of points over the band	1601
Max detectable delay	231 ns
Sweep time	800 ms
Average noise floor	−120 dBm
Transmit power	0 dBm
Tx and Rx cable loss	7.96 dB

Two situations where a patient was either lying down or standing were considered. For more information about the experimental process and the measurements setup scenarios the reader is referred to [16] and the references therein. Roughly speaking, two cases were examined, A1 and A2, referred to as on-on-links and on-off-links. In A1 cases, the antenna positions were changed to cover more links.

The channel model was deduced from the measurement data yielding to the double-cluster model. The fast decaying first cluster corresponds to the effect of a human body and the second cluster occurs from the surrounding reflections [17]. The path amplitude decay is given as

$$
10\log|\alpha_l| = \begin{cases} 0, & l=0 \\ \gamma_{01} + 10\log\left(e^{-\frac{\tau_l}{\Gamma_1}}\right), & 1 \le l \le l_1, \\ \sum_{m=1}^{M}\left(\gamma_{02m} + 10\log\left(e^{-\frac{\tau_l}{\Gamma_{2m}}}\right)\right) & l_2 \le l \le L-1, \end{cases} \tag{8}
$$

where γ_{01}, γ_{02m} are the Rician factors and Γ_1, Γ_{2m} the exponential decaying factors for the two clusters. M is the number of sub-clusters within the second cluster and l_1 and l_2 are the number of multi-path components in the first and second cluster, respectively. The amplitude variations are modeled with a log-normal distribution having zero-mean and standard deviation, σ. The time difference between the consecutive arriving paths follows the exponential distribution, i.e.,

$$
p(t_l|t_{l-1}) = \begin{cases} \lambda_1 e^{-\lambda_1(t_l-t_{l-1})}, & 1 \le l \le l_1 \\ \lambda_2 e^{-\lambda_2(t_l-t_{l-1})}, & l_2 \le l \le L-1 \end{cases}, \tag{9}
$$

where t is the arrival time following a Poisson distribution and λ is the rate of path arrival in a cluster. Finally, the number of arrival paths L also follows a Poisson distribution as

$$p(L) = \frac{\mu_L^L e^{\mu_L}}{L!}, \tag{10}$$

where μ_L is the average of L. Table 20 shows values for these parameters found using the least square method.

3.2 Wang et al. Model

Wang et al. derived a channel model in the UWB band for on-body communications. The model takes into account the statistical variations of several body postures and movements [18]. It was derived based on a realistic human body model, and numerical electromagnetic field analysis techniques were used. Specifically, a dipole transmitter was placed on the left chest and five receivers were located on the right chest, the left and right waists, and the two ears. Nine postures were used for standing, ten for walking and running, and six for sitting. To characterize the UWB model the authors applied a proper modification of the Saleh-Valenzuela model [19]. A detailed analysis can be found in [2,18].

At first, the path loss for the five links was estimated based on a simple path loss formula; the results are shown in Table 21.

Then, the average power delay profile was shown to decay exponentially whereas the log-normal distribution was found to provide the best fit to the power distribution in the

TABLE 20   Parameter Values

	Conventional ward				Corridor	
	Standing		Laying down		Standing	
	A1	A2	A1	A2	A1	A2
$\gamma 01$ (dB)	−61	−74	−64	−65	−47	−27
$\gamma 02$ (dB)	−91, −82, 19, −87, −6, −99	−83	−85	−84	−82	−82
$\Gamma 1$	1.11	6.67	3.12	4.14	0.77	1.47
$\Gamma 2$	30.30, 31.25, 2.44, 29.41, 4.55, 108.7	31.25	32.26	29.41	24.39	24.39
σ_{X1} (dB)	2.45	4.41	6.31	4.86	3.75	1.96
σ_{X2} (dB)	2.07, 2.21, 1.62, 1.44, 1.2, 0.91	2.8	3.5	2.79	4.04	2.46
$1/\lambda_1$	3.717	8	4.764	6.024	6.024	6.024
$1/\lambda_2$	6.125	5.43	6.369	8	1.667	1.667
μL	324	323	324	323	324	324

TABLE 21 Path Loss vs. Distance

Receiver	Distance (m)	Path Loss (dB)
Right ear	0.31	61.7
Left ear	0.26	58.8
Right chest	0.16	50.8
Right waist	0.56	71.4
Left waist	0.53	70.5

multi-paths. The authors found that the inter-path delay that is related to the temporal delay between two successive paths follows the inverse Gaussian distribution, i.e.,

$$f(x, \mu, \lambda) = \left(\frac{\lambda}{2\pi x^3}\right)^{\frac{1}{2}} e^{-\frac{\lambda(x-\mu)^2}{2\mu^2 x}}. \tag{11}$$

The parameters of the model are depicted in Table 22, where γ is the time constant for power decay, σ is the standard deviation of power distribution, τ_0 is a constant representing the mean arrival time of first path, τ_k-τ_{k-1} represents the inter-path delay following the inverse Gaussian distribution, and Ω_0 is the mean power gain of the first path.

A discrete time impulse response function applied to the five transmission links with the corresponding parameter values is then implemented based on the modified Saleh-Valenzuela model:

$$h(t) = \sum_{k=0}^{K} a_k \delta(t - t_k), \tag{12}$$

where α_k is the multi-path power gain and t_K is the delay of the kth multi-path component corresponding to the arrival time of the first path. The model implementation is thoroughly discussed in [18].

3.3 Reusens et al. Model

Reusens et al. investigated the propagation channel between two half-wavelength dipoles at 2.45 GHz, placed near a human body [20]. Propagation measurements for different parts of a body were performed on real humans in a multi-path environment. Path loss was also numerically investigated with simulations. The channel parameters were extracted from the measurement and simulation data.

For the measurements along the arm, the transmitter was located on the wrist and the receiver at various positions towards the shoulder. The measurements for the back and the torso were performed by placing the transmitter at different positions at the shoulder height and the moving receiver below the transmitter. For the leg the transmitter was located at the ankle and the receiver was moved toward the knee. Moreover, an average

TABLE 22   Model Parameters

Parameters	Right Ear	Left Ear	Right Chest	Left Waist	Right Waist
γ (ns)	0.38	0.26	0.21	0.30	0.47
σ (dB)	7.5	12.56	15.6	8.46	7.87
τ_0 (ns)	1.46	0.92	0.68	1.89	2.01
τ_k-τ_{k-1} (ns) μ_τ	0.30	0.56	0.37	0.38	0.33
τ_k-τ_{k-1} (ns) λ_τ	1.08	0.45	1.43	0.75	0.85
Ω_0 (dB)	−60.7	−62.1	−53.3	−71.5	−69.9

TABLE 23   Measurements Values

Parameter	Arm	Leg	Back	Torso	Whole Body
d_0 (cm)	10	10	10	10	10
$P_{0,dB}$ (dB)	32.2	32.5	36.8	41.2	35.2
N	3.35	3.45	2.18	3.23	3.11
σ (dB)	4.1	5.3	5.6	6.1	6.1

path loss model for the whole human body was obtained through fitting of all measurement data. Eq. (7) was used to estimate the path loss between the transmitting and the receiving antennas as a function of the distance. Table 23 presents the parameter values of the fitted path-loss models.

Simulations with a realistic human body phantom were also performed and were in agreement with the measurements. The cumulative distribution functions of the deviation of measured path loss and models were found to follow the lognormal distribution.

3.4 Queen's University of Belfast Models

The wireless communications research group at the Queen's University of Belfast is quite active on channel modeling for on-body communications. Over the years, the group has performed extensive research in order to examine the propagation aspects of wearable communications systems for a range of environments [21]. Much of the research focus for WBAN applications has been centered on the unlicensed industrial, scientific, and medical (ISM) bands at 868 MHz and 2.45 GHz.

In [22], stationary and mobile user scenarios for on-body communications at 868 MHz were investigated. Twelve on-body propagation paths, located on the upper torso and limbs, were considered. Measurements were taken in an anechoic chamber, open office area, and hallway environments. The authors focused on the fading effect appearance.

They identified that two-thirds of the paths were Nakagami distributed while the remaining were Rician. The majority of the Nakagami paths occurred when the user was stationary. The performance of three diversity combination schemes for two-branch spatial on-body diversity systems was investigated in [23].

In [24], fading characteristics or a WBAN operating in an outdoor environment at 2.45 GHz were presented. Both stationary and moving users were examined. When the user was stationary, a small amount of fading was observed for stationary users. When the user was moving, small-scale fading significantly increased, characterized by the Nakagami distribution.

Several other studies have also been performed. Interested readers are directed to the group website [25] for more information.

3.5 University of Birmingham and Queen Mary University Models

Experiments on on-body communications were also performed by research teams of the universities of Birmingham and Queen Mary, University of London [3]. One such experiment is discussed in [26] where the propagation path loss of an on-body channel was measured at 2.45 GHz. First, the antennas were attached to the body at a number of positions and orientations, and measurements were taken inside an anechoic chamber. Next, measurements inside of a laboratory were performed that indicated significant differences in path loss. Finally, additional measurements were also taken while walking on the campus of the University of Birmingham. Movements caused variations in the distance between the antennas, which made the channel variability quite severe.

In [27] the path gain and its variations were examined in some realistic scenarios. Short-term and long-term fading statistics were investigated. It was shown that the short-term fading was best described by a Rician distribution with the K-value log-normally distributed and the long-term fading by a gamma distribution.

4. CONCLUSIONS

Channel modeling is a necessary and important task to evaluate the performance of a WBAN network. Some of the most significant studies on the topic that have appeared in the literature were presented in this chapter. The majority of them were submitted following the recommendations of the IEEE 802.16.6 subgroup or presented at IEEE BAN meetings. Most of the IEEE as well as other models were derived from a set of experiments using wearable sensors in several body positions and at a specific frequency band. Efforts have been made to make the models as realistic as possible, despite the adversities that often appear due to the complex nature of the human body as a signal transmission medium.

The models presented could help designers predict the performance of a network they wish to design. Some models such as the Aoyagi et al. are quite simple and generic; one has to determine mainly the frequency of operation and the environment in which he wishes to build the network. More detailed information is given by the other models, but the designer should pay attention to the particular environment characteristics, parameters, and network topologies. Of course, several other requirements on the physical and MAC layer briefly presented in [28] should be satisfied; e.g., operation in a power-constrained environment, appropriate data rates, coexistence with other wireless network devices, etc.

The extraction of an adequate channel model is the first step in the performance evaluation of a WBAN network. The clearest view of physical layer performance is obtained by examining suitable metrics such as the bit error rate, the outage probability of point-to-point links following the path losses suggested by the above channel models. In view of the above, such a detailed analysis is more than necessary.

However, the general drawback of the propagation WBAN models discussed in this chapter is that they refer to specific node topologies, different environments, e.g., anechoic chamber, etc., and they were tested to be applicable to specific frequency bands. Therefore, a derivation of a generalized model seems to be a challenging research direction. That model should be obtained by fitting properly a large number of measurements collected by different human body locations. The model deduction process should meet the general guidelines and know-how techniques that are already used in the area of RF communications, e.g., the model should be as simple as possible and contain the minimum necessary parameters such as transmission distance, operating frequency, factors describing the environment, etc. In such a case, it would be quite interesting to regulate rules about the path-loss behavior in terms of frequency increase in discrete steps.

Acknowledgements

The present study has been co-financed by the European Union (European Social Fund − ESF) and Greek national funds through the Operational Program "Education and Lifelong Learning" of the National Strategic Reference Framework (NSRF) - Research Funding Program: Thales\Interdisciplinary Research in Affective Computing for Biological Activity Recognition in Assistive Environments. Investing in knowledge society through the European Social Fund.

References

[1] D.J. Cook, W. Song, Ambient intelligence and wearable computing: sensors on the body, in the home and beyond, J. Ambient Intell. Smart Environ. 1 (2009) 83−86.

[2] J. Wang, Q. Wang, Body Area Communications: Channel Modeling, Communication Systems, and EMC, John Wiley, 2013.

[3] P.S. Hall, Y. Hao, Antennas and Propagation for Body-Centric Wireless Communications, Artech House, 2006.

[4] A. Goldsmith, Wireless Communications, Cambridge University Press, 2005.

[5] S. Ullah, et al., A comprehensive survey of wireless body area networks: on PHY, MAC, and network layer solutions, J. Med. Syst. 36 (2012) 1065−1094.

[6] 802.15.6-2012 - IEEE Standard for Local and metropolitan area networks - Part 15.6: Wireless Body Area Networks, 2012.

[7] D.B. Smith, D. Miniutti, T.A. Lamahewa, L.W. Hanlen, Propagation Models for Body Area Networks: A Survey and New Outlook, <http://www.nicta.com.au/pub?doc = 5775>.

[8] K.Y. Yazdandoost, K. Sayrafian-Pour, Channel model for body area network. IEEE P802.15-08-0780-09-0006, 2009.

[9] T. Aoyagi, et al., Channel model for wearable and implantable WBANs. IEEE 802.15-08-0416-04-0006, 2008.

[10] G. Dolmans, A. Fort, Channel models WBAN-Holst centre/IMEC-NL. IEEE 802.15-08-0418-01-0006, 2008.

[11] D. Miniutti, et al., Narrowband channel characterization for body area network. IEEE 802.15-08-0421-00-0006, 2008.

[12] A. Astrin, Measurements of body channel at 13.5 MHz. IEEE 802.15-08-0590-00-0006, 2008.

[13] N.-G. Kang, C. Cho, S.-H. Park, E.T. Won, Channel model for WBANs. IEEE 802.15-08-0781-00-0006, 2008.

[14] M. Kim, et al., Statistical property of dynamic BAN channel gain at 4.5 GHz. IEEE 802-.15-08-0489-01-0006, 2008.

[15] M. Hämäläinen, A. Taparugssanagorn, J. Iinatti, On the WBAN radio channel modelling for medical applications. Proceedings of the 5th European Conference on Antennas and Propagation (EUCAP). Rome Italy, 2011, pp. 2967–2971.

[16] A. Taparugssanagorn, A. Rabbachin, M. Hämäläinen, J. Saloranta, J. Iinatti, A review of channel modelling for wireless body area networks in wireless medical communications. Proc. 11th Int. Symp. on Wireless Personal Multimedia Communications (WPMC) 2008, Lapland Finland, 2008.

[17] H. Viittala, M. Hämäläinen, J. Iinatti, A. Taparugssanagorn, Different experimental WBAN channel models and IEEE802.15.6 models: Comparison and effects. 2nd International Symposium on Applied Sciences in Biomedical and Communication Technologies, (ISABEL 2009). Bratislava, Slovakia, 2009.

[18] Q. Wang, T. Tayamachi, I. Kimura, J. Wang, An on-body channel model for UWB body area communications for various postures, IEEE Trans. Antennas Propagation 57 (2009) 991–998.

[19] A.F. Molisch, et al., A comprehensive standardized model for ultrawideband propagation channels, IEEE Trans. Antennas Propagation 54 (2006) 3151–3166.

[20] Reusens, et al., Characterization of on-body communication channel and energy efficient topology design for wireless body area networks, IEEE Trans. Inf. Technol. Biomed. 13 (2009) 933–945.

[21] S.L. Cotton, W.G. Scanlon, Wireless body area networks - technology, implementation and applications, in: M. Yuce, J. Khan (Eds.), Wireless Body Area Networks, CRC Press Taylor & Francis Group, 2012, pp. 323–348.

[22] S.L. Cotton, W.G. Scanlon, A statistical analysis of indoor multipath fading for a narrowband wireless body area network. IEEE 17th International Symposium on Personal, Indoor and Mobile Radio Communications (PIMRC). Helsinki Finland, 2006.

[23] S.L. Cotton, W.G. Scanlon, Characterization and modeling of on-body spatial diversity within indoor environments at 868 MHz, IEEE Trans. Wireless Commun. 8 (2009) 176–185.

[24] S.L. Cotton, W.G. Scanlon, Characterization of the on-body channel in an outdoor environment at 2.45 GHz. 2009. EuCAP 2009. 3rd European Conference on Antennas and Propagation. Berlin Germany, 2009, pp. 722–725.

[25] <http://www.ee.qub.ac.uk/wireless/index.html>, (Last Accessed: 27.06.14).

[26] P.S. Hall, et al., Antennas and propagation for on-body communication systems, IEEE Antennas Propagation Mag. 49 (2007) 41–58.

[27] Y.I. Nechayev, Z.H. Hu, P.S. Hall, Short-term and long-term fading of on-body transmission channels at 2.45 GHz, 2009. LAPC 2009. Loughborough Antennas & Propagation Conference. Loughborough UK, 2009, pp. 657–660.

[28] J.Y. Khan, M.R. Yuce, G. Bulger, B. Harding, Wireless Body Area Network (WBAN) design techniques and performance evaluation, J. Med. Syst. 36 (2012) 1441–1457.

Trust Establishment in Wireless Body Area Networks

Shucheng Yu[1], Ming Li[2], and Lu Shi[1]

[1]University of Arkansas at Little Rock, Little Rock, Arkansas, USA,
[2]Utah State University, Logan, Utah, USA

1. INTRODUCTION

In recent years, WBAN technology has been increasingly applied in the healthcare domain. A number of medical devices, wearable or implantable, are integrated into WBANs to monitor patients' health status, treat patients with automatic therapies, and so on and so forth. Such WBAN devices include pacemakers, implantable cardioverter defibrillators (ICDs), implantable drug pumps, ECG/EMG/EEG monitoring devices, etc. Today there are over 3 million pacemakers and over 1.7 million ICDs in use according to recent research [1]. Other applications of WBAN extend to sports, military, or security. With the popularity of WBAN applications, however, great privacy and security risks about WBAN devices exist, which may hinder the wide adoption of WBAN technology in real life scenarios. These risks include the possibility of exposing sensitive personal information, wireless hijack attacks in which attackers manipulate physiological data or commands transmitted between legitimate devices, the vulnerability to physical device compromise attacks or replacement of WBAN devices, etc. In order to address these risks, fundamental security mechanisms shall be in place such that each WBAN device has the confidence that it is communicating with legitimate peers, and every physiological measurement or command is sent to authentic WBAN device(s) without being modified or overheard by unintended parties. To this end, it is important to design secure and practical security mechanisms that allow WBAN devices to authenticate each other (i.e., verify that each device is valid and trustworthy) and establish secret key(s) (i.e., generate shared cryptographic key(s)) for protecting the subsequent communications. Major challenges exist with establishing secure communications in WBAN mainly due to the following factors: 1) resource-constrained or heterogeneous WBAN; 2) users lacking the expertise to conduct complex operations for security bootstrapping; and 3) compatibility with billions of commercial-off-the-shelf WBAN devices that have been on the market already. This chapter reviews state-of-the-art techniques of WBAN security.

2. WBAN DEVICE AUTHENTICATION TECHNIQUES

Among various security measures, authentication is the fundamental step towards the initial trust establishment (e.g., key generation) and subsequent secure communications in WBANs. With effective authentication, attackers will fail to pretend to be valid sensor nodes and join the WBAN for private information-involved communication, thereby avoiding either wrong reports or false commands by attackers, which may put the patient's safety at risk. Unfortunately, WBAN devices are not designed with enough security considerations in current practices. The situation might be more severe for healthcare applications involving patient information in which the lack of security solutions has been shown to result in the possibility of fatal consequences. In [2], researchers utilized wireless access to steal personal information from a common cardiac defibrillator, and demonstrated that fatal heart rhythms can be induced. In [3,4], the dosage levels of an insulin pump were remotely adjusted by hackers with knowledge of the pump's serial number, which could kill the patient wearing the pump. Being aware of the data privacy or security risks, the FDA has begun to urge manufacturers to tighten security measures on wireless medical devices [5]. Therefore, an effective node-authentication mechanism is the key to WBAN's security and user safety.

Although great efforts have been made on authentication in wireless networks, the same issue remains a challenge in WBAN due to its unique features and stringent application-level requirements. Conventionally, authentication is achieved based on pre-distributed secret keys among sensor nodes in a network. Such key distribution in wireless sensor networks (WSNs) is extensively described in the literature [6−11]. However, if this method is directly applied to a WBAN, end-users are required to trust the entire chain of the distribution process, in which untrustworthy users may be involved to hamper the trust establishment or even launch attacks. In addition, general users of WBAN devices are expected to have little knowledge of the WBAN technology, implying that high usability, i.e., ideally "plug-and-play," is desired. To be specific, the authentication process should be simplified, automatic, and transparent to users. Thus, it is highly desirable that in WBANs node trustworthiness is established and evaluated without assuming any prior security context among nodes.

Since WBAN devices are ubiquitous, they are likely to be physically compromised. Consequently, pre-shared secret materials in the devices, e.g., keys, may be disclosed to attackers. This would allow attackers to disguise themselves as legitimate sensor nodes in the network and further render traditional cryptographic authentication mechanisms ineffective. From this aspect, node authentication mechanisms in WBAN should have minimal reliance on cryptography.

Finally, resources, such as hardware, energy, and user interfaces, are extremely limited for low-end WBAN sensor devices. This imposes additional requirements on authentication mechanisms in terms of communication and computation costs. Moreover, most of the existing non-cryptographic authentication mechanisms require advanced hardware such as multiple antennas [12], or significant modifications to the system software. It is important to note that hardware requirements should be minimized in the WBAN, not only because of extra cost but also due to compatibility with legacy systems.

The following sections review cryptographic and non-cryptographic authentication mechanisms in WBANs, respectively.

2.1 Cryptographic Authentication Mechanisms in WBAN

In this section, cryptographic authentication mechanisms in WBAN are investigated according to the type of cryptography. Existing crypto-based authentication schemes can be classified into the following categories: symmetric key-based (SKC-based) authentication and public key-based (PKC-based) authentication. While the high computational cost of asymmetric cryptography makes symmetric encryption the more viable option, key-distribution in symmetric encryption is challenging.

2.1.1 Symmetric Key-Based Authentication

As an efficient choice for distributed access control in WBANs, symmetric key-based authentication relies on the trust establishment by the prior security context [13–17]. By predistributing key materials, a pairwise key can be easily generated between the user and any authorized entity. And then authentication can be implemented by using the authentication key.

However, SKC-based authentication schemes suffer from several disadvantages [18]. First, since compromised nodes are harmful to the system, it is desirable to detect and revoke compromised nodes in a timely fashion. For most schemes, either high computational overhead or complex key management in the revocation of nodes is involved; therefore it is difficult to achieve fine-grained access control. Second, such schemes are subject to user collusion in which attackers collude to exchange and derive keys and other sensitive information, or exhibit agreed-upon behavior in the authentication process. Finally, if a WBAN device is compromised physically, the data and prior security context stored in it will possibly be exposed to unauthenticated users.

It is also noteworthy that there exist alternatives to SKC-based authentication that do not assume preshared secrets and additional hardware devices, while enjoying higher usability [19,20]. For example, the group device pairing in [19] allows a group of WBAN devices to establish a common group key based on symmetric key cryptography. Each device authenticates itself to the whole group as a valid member by visually human-aided verification. However, the group device pairing technique assumes the existence of an additional out-of-band (OOB) secure channel to facilitate human-aided verification, which may not be intuitive to use.

To avoid the above issues, [21] proposed lightweight source and data authentication schemes in a routing framework for WBANs by utilizing hash-chain techniques. Since source authentication requires decryption operations, it is much more costly than data authentication only with equality checks involved. Therefore, source authentication is disabled if the neighbor set is not changing based on the prediction results. However, according to [21], source authentication only achieves 70% accuracy in terms of filtering false requests.

2.1.2 *Public Key-Based Authentication*

Among public key-based authentication schemes, attribute-based encryption (ABE) and identity-based encryption (IBE) are the most common techniques.

ABE achieves flexible one-to-many encryption based on attributes. The decryption of a ciphertext is possible only if the set of attributes of the user key matches the attributes of the ciphertext [22]. In this way, only a group of users satisfying a certain access protocol can read the ciphertext. A crucial security feature of ABE is collusion-resistance, so any user key cannot be derived by collusion. The features of ABE make it a good candidate for authentication in WBANs.

Ciphertext policy ABE was proposed in [23] to provide role-based access control on encrypted data in WBANs. The secret is split among secret key components belonging to different attributes owned by a user to provide collusion resistance. By employing ciphertext policy ABE, [24] proposes the primitive functions to implement a secret-sharing scheme, which provides message authenticity. In [25,26], ABE was exploited to secure the communications between a WBAN and its external users. While [26] achieved secure communication between the data controller and an external user by fuzzy ABE, [25] utilized the basic ABE to self-protect electronic medical records (EMRs) on mobile devices and offline communications.

Besides ABE, IBE has also been actively studied and widely applied in cryptography research in which the user's public key carries unique information about the user's identity in the form of an ASCII string. One common feature of the IBE-based schemes is that no prior trust/key predistribution is required between individual users. Conventional IBE primitives are computationally demanding and cannot be efficiently implemented on sensor devices in WBAN. To solve this problem, [27,28] proposed a lightweight IBE-based access control scheme to balance security and privacy in WBANs, which was built upon elliptic curve cryptography (ECC). The limitation of [27,28] is that the master secret key becomes vulnerable to compromise if more than a limited number of secret keys are released. The Boneh–Franklin IBE algorithm was applied in [29] to achieve a faster authenticated key establishment and encryption scheme for WBANs with less energy consumption and optimal memory requirement.

However, it is important to note that a trusted certificate authority, called the private key generator is involved with private key generation, and it is capable of decrypting messages without authorization. Therefore, the private key generator must be highly trusted. If the private key generator is compromised, all the messages protected by the public-private key pairs are also compromised, which makes the generator a target to attackers.

2.2 Non-Cryptographic Authentication Mechanisms in WBAN

From another perspective, non-cryptographic methods provide an alternative way of authentication without key predistribution and non-intuitive user participation. In addition, most non-cryptographic schemes have simpler protocols with less complicated computation. In general, current non-cryptographic authentication mechanisms are mainly divided into the following categories based on type of techniques: biometric-based authentication, channel-based authentication, proximity-based authentication, and a combination of authentication schemes.

2.2.1 Biometric-Based Authentication

Since security context in traditional authentication methods can be easily stolen, shared, or forged, biometric systems are explored to provide more reliable authentication with unique traits physically linked to the user. Common traits include anatomical traits such as fingerprint, iris, electrocardiogram (EEG), etc., and behavioral traits such as signature, gait, etc. To assist authentication, physiological values of a biometric trait are measured using appropriate sensors, and features are extracted from the measurements to create a template stored in the system database. When a user device is to be authenticated, its current corresponding physiological values are compared with the template to verify the identity [12,30]. Physiological values of the same biometric trait can also be collected and compared at sender and receiver sides located on different body parts. To achieve better security, instead of a static template, [31] extracted the time-variant features from the ECG signals, which then were used for message authentication. Common accelerometer data extracted from body motion was used for authentication in [32], but specialized sensing hardware was required for every sensor.

Although these methods do not rely on preshared secrets, biometric-based authentication is vulnerable in two aspects. First, it is hard for body sensors in different positions to measure the same physiological signal with the same accuracy. Second, a spoof attack may occur using a counterfeit biometric trait that is not obtained from a live person, e.g., a gummy finger, photograph, or mask of a face, or even a dismembered finger from a legitimate user.

2.2.2 Channel-Based Authentication

Recently, there has been an increasing interest in exploring received signal strength (RSS) measurements for authentication [33–35]. Generally, RSS values tend to vary over time due to mobility and channel environments. Channel-based solutions leverage such RSS variations to achieve authentication in a different way. Zeng et al. [36] proposed to compare the lists of temporal RSS variation to handle identity-based attacks, where an intruder trying to impersonate another user B that is communicating with A can be detected by A. To protect communication between two devices, the secure device-pairing scheme proposed in [35] utilizes differential RSS to perform proximity detection for limiting malicious senders by distance, but the proposed method required additional hardware — at least two receiver antennas. In addition, the use of RSS in authentication can be compromised by adversaries with array antennas that can launch a beam-forming attack to spoof location. In other words, by sending different signal strengths in the directions of different access points, an attacker attempts to appear similar to another node instead of its own actual location. Furthermore, a link's RSS can be eavesdropped.

In addition to utilizing RSS, other channel-based authentication schemes build a signature for each device's wireless channel. For example, the temporal link signature in [37] uses channel impulse response (CIR) information to uniquely identify the link between a transmitter and a receiver. Then, if an attacker at a different location pretends to be the transmitter, the change in the physical channel will be detected due to link distinction. It is also difficult to infer the link signature measurements from interactions between nodes. But this method requires a learning phase and advanced hardware platforms such as GNU radio.

2.2.3 Proximity-Based Authentication

Several proximity-based authentication schemes are based on co-location detection. Amigo in [38] extended the Diffie—Hellman key exchange with authentication of co-located devices. Specifically, after a shared secret is derived by Diffie—Hellman key exchange, both devices monitor the radio environment for a short period of time and exchange a signature including its RSS of that environment with the other device. Then at each device, similarity detection between the received signature and its own signature is performed independently to determine proximity. Ensemble technique [34] provides proximity-based authentication by monitoring the transmissions and analyzing RSS. Similarly, to securely pair wireless devices in proximity with one another, Mathur et al. [39] proposed a co-location based pairing scheme by exploiting shared time-varying environmental signals that were used to generate a common cryptographic key for the devices authenticating each other's physical proximity for further confidential communication. However, the main drawback of the methods in [34,38,39] is that the devices are required to be within a half wavelength distance of each other, which is restrictive for sensor devices deployed in a WBAN.

Other methods exploit secure ranging techniques, such as distance bounding [40], to determine a device's proximity. For example, in [41], based on ultrasonic distance bounding, an implanted medical device can limit access to its resources only to devices that are in its close proximity. But the common concern with the RF distance bounding technique is that specialized/advanced hardware must be involved; otherwise high accuracy cannot be achieved. Although [42] proposed the first design of RF distance bounding that can be realized fully using a wireless channel, it requires multi-radio capabilities and additional hardware, which may not be available on general WBAN devices, especially legacy devices.

2.2.4 Other Authentication Schemes

To distinguish legitimate WBAN devices on/near body from imposters, [43,44] propose a channel-based and proximity-based authentication scheme — BANA — based on the observation that an off-body attacker has obviously distinct RSS variation behavior with an on-body sensor. The advantage of BANA lies in the fact that it is simple, lightweight, and does not require any additional hardware, but still promises effectiveness, efficiency, and applicability in real-life scenarios. It is important to note that body movements are required in BANA to obtain unique channel characteristics for authentication.

2.3 Summary of Authentication Methods

Due to unique characteristics of WBANs, effective and efficient authentication techniques are required to guarantee security and privacy in WBAN applications.

Table 1 provides a summary of current authentication schemes in WBANs. Each cryptographic or non-cryptographic authentication scheme has its pros and cons, which also imply potential for improvement. Cryptographic authentication schemes may be best suited to the nature of self-deployable WBANs, but they generally suffer from extensive computation and rely on pre-shared secrets that may be disclosed to unauthorized users if

TABLE 1   Summary of current authentication methods in WBAN

	Comparison Criteria	Cryptographic Authentication based on:		Non-Cryptographic Authentication based on:		
		SKC	PKC	Biometric	Channel	Proximity
Security	Preshared Secrets	Yes	Partial	No	No	No
	Vulnerability	Key management, collusion, physically compromise	PKG must be highly trusted	Forging biometric trait	Beam forming attack, location spoof attack	
Usability	Human Interaction	Partial	No	Yes	Partial	Partial
Cost	Additional Hardware	No	No	Require specialized sensing hardware	May require	May require
	Other Requirements	May require OOB channel	N/A	Consistent physiological signal	Channel reciprocity	Within half wavelength distance

physically compromised. With the development of biometric techniques and comprehensive study of channel characteristics, authentication schemes based on non-cryptographic methods are substantially explored. Complex, but distinguishable, human body characteristics provide an ideal way for authentication, but they also raise user privacy concerns. Wireless channels are also complicated and difficult to predict, and greatly affected by environment interference, so the performance of channel based-authentication schemes is subject to channel conditions. In proximity based-authentication schemes, WBAN devices must be in close proximity to be authenticated, but location spoof by attackers could undermine security. While prior secret is no longer necessary in non-cryptographic solutions, there may exist extra requirements to implement them in real-life scenarios, such as advanced hardware, human involvement, etc.

Different levels of security should be considered in authentication schemes. For example, in healthcare applications, life-threatening requests shall be distinguished from others and given highest security priority. Appropriate privacy-protection measures are needed along with authentication.

3. SECRET KEY ESTABLISHMENT IN WBAN

In order to protect the physiological data and the command messages transmitted in WBAN, a secret key shall be established between WBAN devices. One straightforward approach for key establishment is to preload shared secret keys in the devices prior to WBAN deployment. In practice, however, this approach may fail due to the following factors: first, devices in the same WBAN can be made by different manufactures and preloading the same secret keys to these devices at the time of manufacturing is impractical; second, it is impractical for general WBAN users who may lack the necessary expertise to load secret keys to WBAN devices; third, WBAN devices may lack the necessary interface (such as USB) for loading the keys, especially for miniature devices; last but not least, preloading the same secret keys will put the entire WBAN under threat once one of the WBAN devices is compromised. Considering practical WBAN applications, a body of research has been focusing on establishing secret keys for WBAN on the fly based on various resources such as biometric signal of the patient, patient's body motion, wireless channel characteristics, etc.

3.1 Secret Key Establishment Based on Biometrics or Motion

Many schemes have been proposed to measure and compare physiological information collected by the sensors, such as electrocardiogram (ECG), photoplethysmogram (PPG), iris, and fingerprint, to assist authentication and key establishment without a priori distribution of keying material. For key generation combined with authentication, schemes in [45–48] establish physiological data-based keys between devices for verification. For example, [45] proposes to use a secure environmental value as the source of random secret key information. Specific examples of secure environmental value include physiological data such as inter-pulse-interval (IPI) and heart rate variability. Secret keys can be extracted from ECG data measured by on-body ECG sensors [48–50] or a combination of PPG and ECG signals [46]. Physiological variables such as blood glucose, blood pressure, temperature, etc., were considered in [47] to obfuscate an arbitrary secret key for securing data. Due to their uniqueness, randomness, and time-sensitiveness, physiological data can act as the dependable source for both device authentication and secret key extraction. The major drawback of biometric-based techniques, however, is that the biometrics derived from physiological features are usually accompanied with high degrees of noise and variability inherently present in the signals. It is also difficult to guarantee consistent physiological signal measurements with the same accuracy for sensors located in different positions on the human body. Moreover, not all the physiological parameters have the same level of entropy for key generation. According to [51], for example, heart rate is not a good choice because its level of entropy is not satisfactory.

Schemes in [32,52,53] exploited the movement patterns when shaking devices together for authentication, and generated shared secret keys based on the measured acceleration data in the shaking process. Similar to biometric-based methods, these schemes require specialized sensing hardware and human participation.

3.2 Secret Key Establishment Based on Wireless Channel Characteristics

Use of a wireless channel for authentication and/or key generation has been of great interest recently [54,55]. One of the key research topics in this area is improvement of the key generation rate. Lai et al. [56] exploited the random channels associated with relay nodes in the wireless network as additional random sources for the key generation between two nodes with one-hop relay nodes.

3.3 Authenticated Secret Key Establishment in WBAN

3.3.1 *Challenges*

While device authentication and secret key establishment are important for WBAN security, simultaneously achieving both of them is required in practical applications, i.e., WBAN devices need to establish shared secret keys with authenticated peers for secure communications. Many existing methods achieve this utilizing non-wireless channels and under constrained scenarios. For example, physiological data can be used for both device authentication and secret key extraction [45–48]. WBAN devices can be shaken together and take advantage of the same motion pattern measured by the devices for both authentication and secret key extraction [32,53]. However, these solutions are not able to support general commercial-off-the-shelf WBAN devices due to the additional hardware that is needed for measuring the biometric or motion data. An authenticated secret key extraction solution compatible with commercial off-the-shelf devices shall make minimal assumptions on device hardware and only utilize widely available resources such as wireless channel measurements. However, it is challenging to use wireless channels alone for simultaneously realizing authentication and key generation. In particular, the dilemma exists since authentication usually requires proximity, while fast key generation requires channel fading that proximity cannot provide.

3.3.2 *ASK-BAN: Authenticated Secret Key Establishment Utilizing Channel Characteristics for Wireless BAN*

To simultaneously achieve device authentication and fast secret key extraction, ASK-BAN takes advantage of the following characteristics for wireless channels among on-body devices:

1. On-Body channels exhibit obviously different variations (Figure 1, in which the control unit (CU) of the WBAN is deployed to the front of the body and in clear line-of-sight location to sensor nodes S1, S2, and S3).
2. Channels between LOS on-body devices tend to be much more stable than those in NLOS locations. For example, sensor S3 has more stable RSSs for its channel to sensor S4 than other nodes since S3 and S4 are both on the back of the subject and in close LOS locations to each other.

Based on these channel characteristics, ASK-BAN proposes a multi-hop authenticated secret key extraction solution between the CU and on-body sensors with the help of trusted sensors as relay nodes. For multi-hop authentication, ASK-BAN observes that trust relationship is transitive: if RSS variations between A-B and that between B-C are both stable, i.e., A trusts B and B trusts C, then A can trust C with high confidence, and A-B-C is a "trust path" between A and C. Therefore, to authenticate a node ASK-BAN looks for a multi-hop "trust path" between the CU and that node. For secret key extraction, the main challenge is to achieve a high key generation rate during the authentication process. To this end, between each on-body sensor and CU, ASK-BAN exploits possible multi-hop paths that exhibit relatively large RSS variations. Therefore, ASK-BAN adopts five steps for authenticated secret key extraction:

Step 1: Pairwise Key Generation and Initial Authentication. By letting each node broadcast a known message, each pair of nodes measure RSS for packets received from

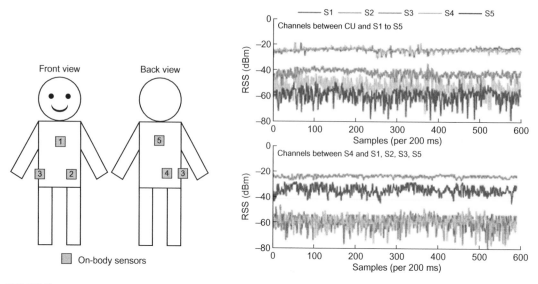

FIGURE 1   WBAN sensor deployment (left) and measured RSS values.

each other and obtain a shared secret key, the length of which is used as the estimation to the secrecy capacity of the channel between the pair. Meanwhile, each node uses the measured RSS values to authenticate all the other nodes with BANA [43,44].

Step 2: Broadcasting Authenticated Secrecy Capacity. During this step each node broadcasts the information regarding every other node i: a) trust value (Yes or No) for node i; b) the secrecy capacity for the channel between node i and itself.

Step 3: Discovering Multi-hop Trusted Paths with Maximum Multi-hop Secrecy Capacity. On receiving the broadcast message from all other nodes in step 2, each node obtains a weighted graph, with each edge indicating the trust relationship between the connected pair and the weight, meaning the secrecy capacity for the channel. By running the max-flow algorithm, each node is able to find multi-hop trusted paths between itself and CU that has the maximum secrecy capacity.

Step 4: Broadcasting Aggregated Secret Key. For each max-flow path, each node, if it is on the path, broadcasts the XORed value of secret keys (obtained in step 1) shared with its previous-hop and next-hop, respectively. After having collected all the XORed values for the max-flow paths, each node and CU are able to derive a shared secret key that has the maximum length.

3.4 Wireless In-Band Trust Establishment in WBAN

The ASK-BAN channel characteristic-based authentication approach assumes all the devices on/near the body are trusted. However, this may not hold in all cases. For example, a careful attacker may sneakily place a malicious sensor near the patient's body or attach another sensor onto the patient's body. In such scenarios, the malicious device can join the BAN easily. To deal with this, in addition to proving device proximity, we need to achieve a higher security goal — demonstrative identification. That is, the user/patient should be able to verify the devices forming the BAN are exactly those she designated to be. Here we assume the designated devices are all benign, because otherwise there is no way to protect the established secret keys. Moreover, the user can count the number of devices correctly.

3.4.1 Related Work

It is well known that the simple Diffie–Hellman key exchange over the wireless channel suffers from the man-in-the-middle attack, as the unprotected wireless signal is subjected to malicious modifications (such as bit flipping and message overshadowing [57]). Thus, in the past decade, various researchers have proposed secure channel-based approaches to work around this problem, which is usually called "secure device pairing" [58]. Secure device pairing relies on the security (authentication) properties of some auxiliary out-of-band (OOB) channel in one way or another. For example, well-known OOB channels include USB connections [59], infrared [60], visual [61], audio, faraday cage [62], etc. However, all these schemes require non-trivial human support, and the devices to be paired should possess common additional hardware such as USB ports, screens, keypads, LEDs, accelerometers, etc. This assumption is often strong and impractical, because all these schemes are often obtrusive to use and not scalable, and are against the global trend for device miniaturization. Moreover, it is commonly believed that human implemented OOB channels can only tolerate up to 10 devices [61]. The human-implemented OOB channel and requirement for advanced hardware have been major obstacles against the practical adoption of those protocols.

There is a growth of interest in using merely wireless in-band communication to achieve authentication and protect message integrity. The Integrity code (I-code) was proposed by Capkun et al. [57], and tamper-evident pairing was proposed by Gollakota et al. [63]. The I-code primitive protects the integrity of every message sent over the insecure wireless channel. It assumes the infeasibility of signal cancellation, and exploits unidirectional error detection codes to provide message tamper-evidence. The I-code method can be applied to key establishment, satellite signal authentication, etc. On the other hand, tamper-evident pairing is an in-band device pairing protocol for 802.11 devices, which uses a tamper-evident announcement that protects the message integrity by embedding cryptographic authentication information (e.g., a hash) into the physical signals, such that any tampering with it will be caught by the receiver.

Though the concept of the above is appealing, there are two limitations. First, security of I-code and tamper-evident pairing is based on the infeasibility of energy cancellation. But these methods only achieve a weak security guarantee, since recently Popper et al. [64] proposed a stronger yet practical correlated signal cancellation attack using a pair of directional antennas. Second, it is difficult to apply these methods to securely initialize multiple constrained devices such as medical sensors due to the scalability issue. I-code and tamper-evident-announcement are both one-to-one message authentication primitives suitable for pairwise communication. If implemented on a sensor platform with 250 kbps transmission rate, an I-coded message requires 0.5 s to transmit 50 bits on a ZigBee sensor platform, given a slot length of 5 ms [57]. In tamper-evident-announcement each synchronization packet must be at least 19 ms long [63]. In addition, the number of "ON_OFF" slots is large (roughly equals a hash length). This yields a total of more than 750 ms for each tamper-evident-announcement. Thus, direct usage or simple extension of I-code or tamper-evident pairing is not scalable to a large group of constrained devices, whereas the delay is critical in many real-world BAN applications.

3.4.2 "Chorus": Authenticated Message Comparison over Wireless Channel

We aim at making ad hoc trust initialization work strictly in-band and scalable to a group of devices, by introducing a novel physical-layer primitive called "Chorus," which achieves authenticated message comparison over the insecure wireless channel, and use it to construct secure group authenticated key agreement protocols. The Chorus is partially inspired from I-code and tamper-evident pairing in that it exploits the infeasibility of signal cancellation and unidirectional error detection codes. However, it also combines an idea similar to I-code with the concept of empirical OOB channels used in message authentication protocols to achieve key authentication and confirmation. In most of the group message authentication protocols, the role of OOB channels is to achieve secure comparison: an authentication string (AS) is typically derived by each device from the protocol transcript (messages to be authenticated); when all nodes' ASs are equal to each other all devices should accept the received messages, and whenever any nodes' ASs are not equal all devices should reject the received messages.

The key idea of Chorus is to let N devices compare the equality of their fixed-length strings by simultaneously emitting specially encoded signals, such that any differences among the strings will be detected by all the devices. Chorus only outputs 1 bit of information (accept-all strings are equal, or reject-some strings are different). Due to the unidirectional property of the wireless channel (attacker can only flip a "0" to "1" but not vice versa), changing the comparison result from reject to accept is impossible except negligible probability. This makes Chorus an ideal replacement for traditional OOB channels. The detailed steps are as follows:

1. Chorus starts with a synchronization packet sent by one node (called coordinator), which contains random content and is longer than a usual packet. All other nodes detect the existence of this packet via threshold energy detection.
2. After a short period when the sync packet ends, the coordinator broadcasts a short CTS_TO_SELF packet of length T_{cts}, which reserves the channel for the time period until Chorus concludes, by suppressing unwanted interference from other co-existing devices.
3. Comparison phase: Each node i encodes its bit string s_i (of length l) using Manchester coding to obtain a $2l$ bit string ($0 \rightarrow 01$ and $1 \rightarrow 10$), and map each encoded bit (1/0) into an ON/OFF slot, respectively (of the same duration T_s). During each time slot $1 \leq j \leq 2l$, if it is an ON slot for a node, a short packet with random content is transmitted, simultaneously with everyone else ("chorus"); but if j is an OFF slot for a node, it remains silent and listens to the channel. If $\forall 1 \leq j \leq 2l$, a node i does not detect energy in any of its own OFF slots, it accepts the received messages, otherwise it rejects the received messages.

A sample timing diagram of a Chorus run is depicted in Figure 2, where node N's string is " 1110⋯ " which differs from others' strings ("1100⋯") by one bit. The encoded strings are " 10101001⋯ " and " 10100101⋯ ," respectively. This can be detected by all nodes (including N itself), because N will detect the aggregated signal of all other nodes during its 6th (OFF) slot, while all other nodes detect energy during their 5th (OFF) slot.

Different from I-code, in Chorus when each node sends its own signal, it cannot receive others' signals (we do not assume full-duplex transceivers). It seems that half of the information is lost. Thus the question is whether non-spoofing property can still be achieved. We can show that it is indeed the case as long as an adversary can only flip a "0" to a "1" bit but not vice versa (i.e., if signal cancellation is infeasible) [65]. In addition, when we consider an attacker that can only inject a signal generated by itself, it can be shown that the realization of basic Chorus is secure against such type of attacks [65].

We also need to defend against correlated signal cancellation attacks where the attacker does not generate its own signal. A practical attack was proposed by Popper et al. [64]. It is based on signal relaying, i.e., the attacker is located at a distance away from both the sender and receiver, and utilizes a pair of directional antennas to relay the sender's signal to the receiver. If the attacker creates a phase delay for the carrier signal on the relay channel that is a multiple of π and with the same signal amplitude, the received signal strength can be completely attenuated. This attack doesn't depend on the packet content and modulation, while it mainly works under stable and predictable channel environments (e.g., static indoor scenarios).

In [65], we proposed to make novel use of uncoordinated frequency hopping (UFH) [66] to protect Chorus from the above type of attack. The basic idea is to make the probability of cancellation arbitrarily small by randomly hopping over multiple frequencies (minislots) within each slot. This is due to the observation that the key factor for an attacker to succeed is to create a phase difference of $\Delta\phi = (2k-1)\pi, k = 1, 2, ...$, which can be defeated by changing the frequency. We assume the processing delay at the attacker is negligible. For each node, in each OFF slot, as long as it detects energy during at least one of the minislots, this node will reject the received messages. Otherwise, if it does not detect energy in any OFF slot, it outputs accept.

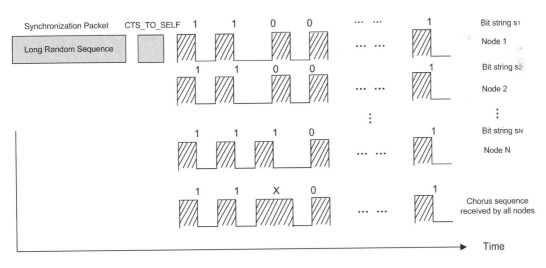

FIGURE 2   An example execution of basic Chorus using Manchester coding.

Based on Chorus, secure in-band trust initialization protocols were designed [65], where all the messages to be authenticated are exchanged using the normal high-bandwidth wireless transmission, with only one run of Chorus in the end of the protocols. Such protocols can achieve greater scalability than previous solutions and are suitable for constrained devices. Specifically, in the setting of BANs, as long as the user has a correct device count for the designated devices, she can input it into the controller and start the trust initialization protocol. Since the use of Chorus makes it resistant to man-in-the-middle attacks, even if an attacker passes BANA's proximity verification it cannot successfully impersonate a legitimate device (any attempts to block or tamper with the messages from a legitimate device can be detected).

4. SUMMARY

In this chapter, we focused on the important problem of initial trust establishment in WBANs. It is the prerequisite for security and privacy protection in WBANs, without which patients' safety can be at risk. It is also challenging due to the simultaneous requirements of high security, efficiency, and usability. We surveyed state-of-the-art solutions to trust establishment and analyzed their strengths and weaknesses in the context of WBANs. We then identified the need for establishing trust based on only wireless channels, without using any secure out-of-band channels. This is necessary to free the user from active participation and achieve "plug-and-play," while eliminating the need for additional hardware interfaces. We presented two of our initial solutions in this direction, namely ASK-BAN and Chorus. The former authenticates BAN devices and generates secret keys among them based on their co-location/proximity with regard to the human body, exploiting channel characteristics. The latter authenticates an arbitrary group of devices designated by the human user through creating an authenticated string comparison based on recent advances in anti-signal-cancellation in the wireless channel. Both solutions involve little or no user effort, and are lightweight. Their security is analyzed via analysis and experiments.

Regarding practical implementation of security mechanisms in WBANs, the selection of proper methodologies depends on various factors. There is no single method that suits all scenarios. One needs to jointly consider the application level functionalities, system security requirements, hardware/software/power/physical constraints, usability requirements, and the tradeoffs among them. For example, traditional encryption may not be a feasible choice in some legacy medical devices such as implanted ones, as software updates require the device be taken out. The channel-based trust establishment approaches proposed in this chapter may be more suited for resource-constrained wearable sensor devices that do not have common sensing capabilities for biometrics, while enjoying a similar level of usability.

In the future, work can focus on relaxing the assumption of trusted devices on the body as the sensor devices can be compromised. For example, compromise detection mechanisms and trust establishment protocols that are resilient to compromised devices.

References

[1] Implantable Medical Devices: Hacking Humans, <https://www.blackhat.com/us-13/briefings.html>, (Last Accessed: 01.07.14).

[2] Scientists work to keep hackers out of implanted medical devices, <http://www.cnn.com/2010/TECH/04/16/medical.device.security/index.html>, (Last Accessed: 01.07.14).

[3] Medical Device Hacking Prompts Concern, <http://www.cyberprivacynews.com/2011/08/medical-device-hacking-prompts-concern/>, (Last Accessed: 01.07.14).

[4] Black Hat: Insulin pumps can be hacked, <http://www.scmagazine.com/black-hat-insulin-pumps-can-be-hacked/printarticle/209106/>, (Last Accessed: 01.07.14).

[5] How hackers can kill you, <http://situationroom.blogs.cnn.com/2013/06/15/how-hackers-can-kill-you/>, (Last Accessed: 01.07.14).

[6] L. Eschenauer, V.D. Gligor, A key-management scheme for distributed sensor networks, Proc. 9th ACM conference on Computer and communications security, ACM, New York, NY, USA, 2002, pp. 41–47.

[7] H. Chan, A. Perrig, D. Song, Random key predistribution schemes for sensor networks," in Security and Privacy, 2003, Proc. 2003 Symp (2003) 197–213.

[8] W. Du, J. Deng, Y.S. Han, P.K. Varshney, J. Katz, A. Khalili, A pairwise key predistribution scheme for wireless sensor networks, ACM Trans. Inf. Syst. Secur. 8 (2) (2005) 228–258.

[9] D. Liu, P. Ning, Establishing pairwise keys in distributed sensor networks, Proc. 10th ACM conference on Computer and Communications Security, ACM, New York, NY, USA, 2003, pp. 52–61.

[10] D. Liu, P. Ning, W. Du, Group-based key predistribution for wireless sensor networks, ACM Trans. Sen. Netw. 4 (2) (2008) 11:1–11:30.

[11] A. Perrig, R. Szewczyk, J.D. Tygar, V. Wen, D.E. Culler, Spins: security protocols for sensor networks, Wirel. Netw. 8 (5) (2002) 521–534.

[12] A.K. Jain, K. Nandakumar, Biometric Authentication: System Security and User Privacy, Computer 45 (11) (2012) 87–92.

[13] K. Malasri, L. Wang, Addressing security in medical sensor networks, Proc. 1st ACM SIGMOBILE International Workshop on Systems and Networking Support for Healthcare and Assisted Living Environments, ACM, New York, NY, USA, 2007, pp. 7–12.

[14] S.A. Devi, R.V. Babu, B.S. Rao, A new approach for evolution of end to end in wireless sensor network, Int. J. Comput. Sci. Eng. 3 (6) (2011) 2531–2543.

[15] M. Mana, M. Feham, B.A. Bensaber, A light weight protocol to provide location privacy in wireless body area networks, Int. J. Netw. Secur. Appl. 3 (2) (2011) 1–11.

[16] O. Delgado-Mohatar, A. Fuster-Sabater, J.M. Sierra, A light- weight authentication scheme for wireless sensor networks, Ad Hoc Netw. 9 (5) (2011) 727–735.

[17] T. Zia, A. Zomaya, A lightweight security framework for wireless sensor networks, J. Wireless Mobile Netw. Ubiquitous Comput. Dependable Appl. 2 (2011) 53–73.

[18] M. Li, W. Lou, K. Ren, Data security and privacy in wireless body area networks, Wireless Commun. IEEE 17 (1) (2010) 51–58.

[19] M. Li, S. Yu, W. Lou, K. Ren, Group device pairing based secure sensor association and key management for body area networks, INFOCOM 2010 Proc. IEEE (2010) 1–9.

[20] M. Li, S. Yu, J.D. Guttman, W. Lou, K. Ren, Secure ad-hoc trust initialization and key management in wireless body area networks, ACM Trans. Sens. Netw. (TOSN) (2012).

[21] X. Liang, X. Li, Q. Shen, R. Lu, X. Lin, X. Shen, et al., Exploiting prediction to enable secure and reliable routing in wireless body area networks, INFOCOM 2012 Proc. IEEE (2012) 388–396.

[22] <http://en.wikipedia.org/wiki/Attribute-based_encryption>, (Last Accessed: 01.07.14).

[23] J. Bethencourt, A. Sahai, B. Waters, Ciphertext-Policy Attribute-Based Encryption, Proc. IEEE Symp. Secur. Priv. (2007).

[24] C. Hu, F. Zhang, X. Cheng, X. Liao, D. Chen, Securing communications between external users and wireless body area networks, Proceedings of the 2nd ACM Workshop on Hot Topics on Wireless Network Security and Privacy (HotWiSec '13), ACM, New York, NY, USA, 2013, pp. 31–36.

[25] J. Akinyele, M. Pagano, M. Green, C. Lehmann, Z. Peterson, A. Rubin, Securing electronic medical records using attribute-based encryption on mobile devices, pages 75–86. Proceedings of the 1st ACM Workshop on Security and Privacy in Smartphones and Mobile Devices, ACM, 2011

[26] C. Hu, N. Zhang, H. Li, X. Cheng, X. Liao, Body area network security: A fuzzy attribute-based signcryption scheme. to appear in IEEE Journal on Selected Areas in Communications (JSAC), Spec. Issue Emerg. Technol. Commun. (2012).

[27] C.C. Tan, H. Wang, S. Zhong, Q. Li, Body sensor network security: an identity-based cryptography approach, Proceedings of the First ACM Conference on Wireless Network Security (WiSec '08), ACM, New York, NY, USA, 2008, pp. 148–153.

[28] C.C. Tan, H. Wang, S. Zhong, Q. Li, IBE-Lite: A Lightweight Identity-Based Cryptography for Body Sensor Networks, Info. Technol. Biomed. IEEE Trans. 13 (6) (2009) 926–932.

[29] C. Rong, H. Cheng, Authenticated health monitoring scheme for wireless body sensor networks, Proceedings of the 7th International Conference on Body Area Networks (BodyNets '12), ICST (Institute for Computer Sciences, Social-Informatics and Telecommunications Engineering), ICST, Brussels, Belgium, Belgium, 2012, pp. 31–35.

[30] X. Hei, X. Du, Biometric-based two-level secure access control for implantable medical devices during emergencies, 30th IEEE Int. Conf. Comput. Commun. (INFOCOM 2011) (2011) 346–350, Shanghai, P. R. China.

[31] Z. Zhang, H. Wang, A.V. Vasilakos, H. Fang, ECG-Cryptography and Authentication in Body Area Networks, Info. Technol. Biomed. IEEE Trans. 16 (6) (2012) 1070–1078.

[32] R. Mayrhofer, H. Gellersen, Shake Well Before Use: Intuitive and Secure Pairing of Mobile Devices, Mobile Comput. IEEE Trans. 8 (6) (2009) 792–806.

[33] A. Varshavsky, A. Scannell, A. LaMarca, E. De Lara, Amigo: proximity-based authentication of mobile devices, Proc. 9th Inter- National Conference on Ubiquitous Computing, Springer-Verlag, Berlin, Heidelberg, 2007, pp. 253–270.

[34] A. Kalamandeen, A. Scannell, E. de Lara, A. Sheth, A. LaMarca, Ensemble: cooperative proximity-based authentication, Applications, and Services. Proc. 8th International Conference on Mobile Systems, ACM, New York, NY, USA, 2010, pp. 331–344.

[35] L. Cai, K. Zeng, H. Chen, P. Mohapatra, Good neighbor: Ad hoc pairing of nearby wireless devices by multiple antennas, Netw. Distributed Syst. Secur. Symp. (2011).

[36] K. Zeng, K. Govindan, P. Mohapatra, Non-cryptographic authentication and identification in wireless networks [security and privacy in emerging wireless networks], IEEE Wireless Commun. 17 (2010) 56–62.

[37] N. Patwari, S.K. Kasera, Robust location distinction using temporal link signatures, Proc. 13th Annual ACM International Conference on Mobile Computing and Networking, ACM, New York, NY, USA, 2007, pp. 111–122.

[38] A. Varshavsky, A. Scannell, A. LaMarca, E.D.e. Lara, Amigo: proximity-based authentication of mobile devices, Proc. 9th Inter-National Conference on Ubiquitous Computing, Springer-Verlag, Berlin, Heidelberg, 2007, pp. 253–270.

[39] S. Mathur, R. Miller, A. Varshavsky, W. Trappe, N. Mandayam, Proximate: proximity-based secure pairing using ambient wireless signals, applications, and services. Proc. 9th International Conference on Mobile Systems, ACM, New York, NY, USA, 2011, pp. 211–224.

[40] S. Brands, D. Chaum, Distance-bounding protocols, Workshop on the Theory and Application of Cryptographic Techniques on Advances In Cryptology, Springer-Verlag New York, Inc., Secaucus, NJ, USA, 1994, pp. 344–359.

[41] K.B. Rasmussen, C. Castelluccia, T.S. Heydt-Benjamin, S. Capkun, Proximity-based access control for implantable medical devices, Proc. 16th ACM Conference on Computer and Communications Security, ACM, New York, NY, USA, 2009, pp. 410–419.

[42] K.B. Rasmussen, S.C. apkun, Realization of rf distance bounding, Proc. 19th USENIX Conference on Security, USENIX Association, Berkeley, CA, USA, 2010, pp. 25–25.

[43] L. Shi, M. Li, S. Yu, J. Yuan, BANA: body area network authentication exploiting channel characteristics, Proceedings of the fifth ACM conference on Security and Privacy in Wireless and Mobile Networks (WISEC '12), ACM, New York, NY, USA, 2012, pp. 27–38.

[44] L. Shi, M. Li, S. Yu, J. Yuan, BANA: Body Area Network Authentication Exploiting Channel Characteristics, Selected Areas, Commun. IEEE J. 31 (9) (2013) 1803–1816.

[45] K. Singh, V. Muthukkumarasamy, Authenticated key establishment protocols for a home health care system. In Intelligent Sensors, Sensor Networks and Information, 2007, ISSNIP 2007. 3rd Int. Conf. (2007) 353–358.

[46] K. Venkatasubramanian, A. Banerjee, S. Gupta, Pska:Usable and secure key agreement scheme for body area networks, Info. Technol. Biomed. IEEE Trans. 14 (1) (2010) 60−68.

[47] K.K. Venkatasubramanian, S.K.S. Gupta, Physiological value-based efficient usable security solutions for body sensor networks, ACM Trans. Sen. Netw. 6 (4) (2010) 31:1−31:36.

[48] F. Xu, Z. Qin, C. Tan, B. Wang, Q. Li., Imdguard: Securing implantable medical devices with the external wearable guardian. In INFOCOM, 2011 Proc. IEEE (2011) 1862−1870.

[49] C. Poon, Y. Zhang, S. Bao, A novel biometrics method to secure wireless body area sensor networks for tele-medicine and m-health, IEEE Commun. Mag. 44 (4) (2006) 73−81.

[50] S. Bao, C. Poon, Y. Zhang, L. Shen, Using the timing information of heartbeats as an entity identiffer to secure body sensor network, IEEE Trans. Inf. Technol. Biomed. 12 (6) (2008) 772−779.

[51] S. Cherukuri, K. Venkatasubramanian, S. Gupta., Biosec: a biometric based approach for securing communication in wireless networks of biosensors implanted in the human body, Parallel Process. Workshops 2003 Proc. 2003 Int. Conf. (2003) 432−439.

[52] D. Bichler, G. Stromberg, M. Huemer, M. Low, Key generation based on acceleration data of shaking processes, Proceedings of the 9th International Conference on Ubiquitous Computing, UbiComp'07, Springer-Verlag, Berlin, Heidelberg, 2007.

[53] R. Mayrohofer, H. Gellersen, Shake well before use: authentication based on accelerometer data, Proceedings of the 5th International Conference on Pervasive Computing, Pervasive'07, Springer-Verlag, Berlin, Heidelberg, 2007.

[54] S. Mathur, W. Trappe, N. Mandayam, C. Ye, A. Reznik., Radio-telepathy: extracting a secret key from an unauthenticated wireless channel, pages 128−139. Proceedings of the 14th ACM International Conference on Mobile Computing and Networking, MobiCom '08, ACM, New York, NY, USA, 2008

[55] S. Jana, S. Premnath, M. Clark, S. Kasera, N. Patwari, S. Krishnamurthy, On the effectiveness of secret key extraction from wireless signal strength in real environments, pages 321−332. Proceedings of the 15th Annual International Conference on Mobile Computing and Networking, Mobicom'09, ACM, 2009

[56] L. Lai, Y. Liang, and W. Du. Phy-based cooperative key generation in wireless networks. In Communication, Control, and Computing (Allerton), 2011 49th Annual Allerton Conference on.

[57] S. Capkun, M. Cagalj, R. Rengaswamy, I. Tsigkogiannis, J.-P. Hubaux, M. Srivastava, Integrity codes: Message integrity protection and authentication over insecure channels, IEEE Trans. Dependable Secure Comput. 5 (4) (2008) 208−223.

[58] M. Li, W. Lou, K. Ren, in: H. Tilborg, S. Jajodia (Eds.), "Secure Device Pairing," in Encyclopedia of Cryptography and Security, Second ed., Springer, 2011.

[59] F. Stajano, R.J. Anderson., The resurrecting duckling: Security issues for ad-hoc wireless networks, IWSP '00 (2000) 172−194.

[60] D. Balfanz, D.K. Smetters, P. Stewart, H.C. Wong, Talking to strangers: authentication in ad-hoc wireless networks, NDSS '02 (2002).

[61] C.-H. O. Chen, C.-W. Chen, C. Kuo, Y.-H. Lai, J.M. McCune, A. Studer, et al., Gangs: gather, authenticate 'n group securely, MobiCom '08 (2008) 92−103.

[62] C. Kuo, M. Luk, R. Negi, A. Perrig., Message-in-a-bottle: user-friendly and secure key deployment for sensor nodes, SenSys '07 (2007) 233−246.

[63] S. Gollakota, N. Ahmed, N. Zeldovich, D. Katabi, Secure in-band wireless pairing, pages 16−16. USENIX, SEC'11, USENIX Association, Berkeley, CA, USA, 2011

[64] C.P. opper, N.O. Tippenhauer, B. Danev, S. Capkun, Investigation of signal and message manipulations on the wireless channel, ESORICS'11 (2011) 40−59.

[65] Y. Hou, M. Li, J.D. Guttman, Chorus: Scalable In-band Trust Initialization for Multiple Constrained Devices over the Insecure Wireless Channel, The sixth ACM Conf Secur Priv Wireless Mobile Netw. (ACM WiSec 2013) (2013) 17−19, Budapest, Hungary.

[66] M. Strasser, S. Capkun, C. Popper, M. Cagalj, Jamming-resistant key establishment using uncoordinated frequency hopping, IEEE S & P (2008) 64−78, IEEE.

6.4

Wireless Body Area Networks

Paolo Barsocchi and Francesco Potortì

ISTI-CNR, Pisa, Italy

1. INTRODUCTION

Localization of devices and people has been recognized as one of the main building blocks of context aware systems [1–4], which have one of their main application fields in ambient assisted living (AAL). Indeed, the user position can be used for detecting user activities, activating devices, opening doors, etc. While in outdoor scenarios GNSS (global navigation satellite systems) constitute a reliable and easily available technology, in indoor scenarios GNSS is largely unavailable.

The indoor environment lacks a system that possesses the excellent performance parameters of outdoor GNSS in terms of global coverage, high accuracy, short latency, high availability, high integrity, and low user costs. Like indoor settings, certain outdoor environments are not well covered by GNSS due to insufficient views to the open sky.

Therefore, the indoor localization problem is solved by means of ad hoc solutions, among which one of the most promising is based on wearable technologies. Improvements in indoor positioning performance have the potential to create opportunities for businesses. However, system performances differ greatly, because the environments have a number of substantial dissimilarities. Indoor environments are particularly challenging for localization systems for several reasons:

- Severe multi-path from signal reflection from walls and furniture
- Non-line-of-sight (NLoS) conditions
- High attenuation and signal scattering due to high density of obstacles
- Fast temporal changes due to the presence of people and opening of doors
- High demand for precision and accuracy

493

On the other hand, indoor settings facilitate positioning and navigation in many ways:

- Small coverage areas
- Small weather influences such as small temperature gradients
- Fixed geometric constraints from planar surfaces and orthogonality of walls
- Infrastructure such as electricity, Internet access, walls suitable for target mounting
- Low dynamics due to slow walking and driving speeds

Indoor localization systems can be classified based on the signal types and/or technologies (infrared, ultrasound, ultra-wideband, RFID, packet radio), signal metrics (AOA — angle of arrival, TOA — time of arrival, TDOA — time difference of arrival, and RSS — received signal strength) and the metric processing methods (range-based and range-free algorithms) [5]. Each solution has advantages and shortcomings, which, in most cases, can be summarized in a trade-off between several metrics (such as accuracy, installation complexity, etc.).

In the following we will discuss which metrics are important for evaluating indoor localization systems, the technologies, and the main localization algorithms (namely, range-free and range-based algorithms).

2. EVALUATION METRICS

In order to evaluate the localization systems a set of criteria weighted according to their relevance and importance for a given applications must be defined. The combination of different metrics makes it possible to evaluate the overall performance of a localization system.

EvAAL (Evaluating AAL Systems through Competitive Benchmarking) [6] is an ongoing effort toward establishing benchmarks and evaluation metrics for comparing ambient-assisted living solutions. Each year one or more international competitions are organized, with the long-term goal of evaluating complete AAL systems. In the last three years an international competition on indoor localization has been organized and five metrics identified.

Two of them are objectively measurable (*hard*) quantities, namely *accuracy* and *availability*. Accuracy is the classical measurement of the goodness of a localization system, based on samples of the distance between the point where the system locates the user and the point where the user really is (error distance). The error series should be reduced to a scalar score, and the literature is rich in methods to reach this result. Of the 195 papers of the first edition of the *Indoor Positioning and Indoor Navigation* (IPIN, 2010) conference, 115 works describe real or simulated systems that are amenable to being evaluated by measuring some kind of metrics. The metrics taken into account in these works are visual path comparison, usually as a graph that shows the real and the estimated path (32% of cases), mean error (31%), cumulative distribution function CDF (20%), a quantile value (11%), and finally error variance (5%).

Availability is a measure of how well the system performs at providing regularly spaced measurements: this is especially significant for experimental or prototypal systems. In other words, the availability measures the capacity of the systems to produce fresh data continuously. As such, it is simply computed as the ratio between the number of received samples and the number of expected samples (one every half a second for the EvAAL competition).

Besides hard quantities, some soft ones need to be considered, namely installation complexity, user acceptance, and interoperability with AAL systems.

Installation complexity is a measure of effort required to install the AAL localization system in a home. It is measured as a function of the person minutes of work needed to complete the installation.

User acceptance expresses how much the localization system is invasive in the user's daily life and thereby the impact perceived by the user. This parameter is estimated with a simple questionnaire (available on the EvAAL website) that considers aspects of usability like the presence and invasiveness of the tags, the visibility of the installation within the environment, and the complexity of maintenance procedures.

Interoperability measures how much the system is easy to integrate with other systems. This parameter depends on the scenario in which the localization system is evaluated, since localization can be exploited by other applications to offer advanced services. In EvAAL the interoperability is also measured with a questionnaire that takes into account aspects like the availability of APIs and documentation, the licensing scheme, the presence of testing tools, and the portability among different operating systems.

All these metrics must be taken into account to discuss the performance of an indoor localization system. The rationale behind the choice of these metrics is multi-faceted. First of all, the accuracy is the most important metric to assess the performance of a localization and tracking system. In general indoor applications do not need high precision, and giving accuracy too much importance diminishes the significance of the evaluation performance with respect to real-life systems. The availability is also an important metric because an unresponsive system can be as difficult to manage as an inaccurate one. Therefore, the weight to give to the hard metrics together should not be more than the weight of metrics related to interaction with the main stakeholders for a given system: system integrators, installers, and final users.

3. TECHNOLOGIES

Many different technologies are used for localization purposes. Most of them require the user to be localized to wear or carry some sort of device.

If the on-body devices need to communicate or otherwise respond to external (not on-body) devices we speak of on-body *active* systems. The external devices are typically part of some infrastructure that is deployed in the environment and which is able to track users. Being infrastructure-based, these systems can be arbitrarily powerful: they can interact with other systems in a variety of ways, they can provide user interfaces (e.g., a graphical web interface), and they can carry on sophisticated computations on data received from on-body devices.

If the on-body devices gather information from the environment and interpret it without the need to actively communicate anything or respond to some external device, we speak of *on-body passive* systems. The information gathered by the on-body system can be generated by some external devices for the purpose of allowing localization, or else can be opportunistically gathered from the environment and the body movements.

Last, we speak about *device-free* systems where no on-body device is necessary to achieve localization. In this case, as in the case of active systems, an external infrastructure senses the position and movements of the user.

Each of the above categories can use several different technologies, which are treated in the following sections. Currently, no single low-cost method is accurate or general enough to provide satisfactory performance. Practical systems need to *fuse* information created by more subsystems in order to obtain satisfactory performance at low cost. This topic is treated in the following section.

3.1 Wearable Active Transducers

Active transducers include those that respond to an external stimulus and those that autonomously generate some sort of beacon signal that is read and interpreted by external devices, which are generally part of an external (not on-body) infrastructure.

Three technologies are treated here: ultrasound capsules, RFID tags, and UWB (ultra-wide band). Of these, RFID are the least obtrusive, since the tags are very small and easily embeddable in clothes. Ultrasound suffers from the size of capsules, the power needed by the associated hardware, and the fact that the signal is blocked by the body itself. UWB systems are still immature and most require bulky equipment, but newer systems promise high precision with small, low-cost devices.

3.1.1 *Ultrasound Systems*

A number of different methods are available to exploit ultrasound systems. The most common one is to measure the time of flight (TOA, time of arrival) of pulses exchanged by fixed devices installed in the environment (on walls, ceilings, or furniture) and one or more devices carried by the user. The device carried by the user may be an emitter or a receiver. Time synchronization between emitter and receivers is achieved by the use of an ancillary radio channel. Choosing to put the receiver on the user increases the required computational power on the on-body device but addresses privacy concerns. A system using this method is iLoc [7].

Other methods do not require any radio connection, and exploit the difference between received times (TDOA).

3.1.2 *RFID Tags*

RFID is a technology that uses relatively bulky, high-powered *readers* and many *tags* that are very small and inexpensive. Tags are available in tiny cases that allow embedding in clothes or shoes. When the reader sends a radio signal, the tags send a signed radio response, uniquely identifying each tag. Tags can be *active* (battery operated) or *passive* (powerless); in the latter case they work by accumulating a tiny fraction of the energy emitted by the reader through a resonant circuit and use it to retransmit back their ID.

RFID systems with on-body tags are generally used as proximity sensors. A very rough estimate of the distance can be obtained by measuring the RSS (received signal strength) from the RFID tag, which is generally not enough to compute the user position. RFID systems with on-body tags are useful as elements of localization systems that fuse information coming from several subsystems.

More accurate positioning based on the various RSS methods discussed further in the chapter has been investigated. Many RFID tags can be installed in the environment, with the user carrying a reader, but as of today readers are too bulky to be considered on-body devices.

3.1.3 Ultra-Wide Band Systems

Devices are expected to appear on the market that follow the ToA positioning method standardized in IEEE 802.15.4a. Such devices would allow accurate trilateration techniques to be applied as long as the user carries a compliant device on-body.

3.2 Packet Radio Systems (Active and Passive)

Packet radio methods are the most flexible methods for active and passive localization. They are treated in greater detail in the next section. Technologies include proprietary systems and those based on the low-power Bluetooth (IEEE 802.15.1) and the ZigBee MAC (IEEE 802.15.4) standards.

Packet radio technologies are built for communications, not localization. Transmitters and receivers are not custom-built, but are standard communications nodes. As such, only methods based on received signal strength and in some cases time of flight are possible. Despite these shortcomings, methods based on packet radio are very popular in research circles and have recently achieved good levels of generality and accuracy. The reason why these devices are attractive is that they are mass-produced, featuring low prices, small size, and long battery life. Moreover, all these characteristics have been constantly improving in the last ten years, and trends suggest that they will keep improving for the foreseeable future. This also means that it is realistic to think that the presence of small devices communicating via packet radios is going to be ubiquitous in indoor and outdoor environments. As a consequence, it is possible to think of forthcoming feature-rich smart environments where localization is a service provided by already existing devices, without the need for any specific hardware installation.

Packet-radio systems generally use methods based on RSS, because all receivers provide this measurement on a per-packet basis, as a means to evaluate the link quality. While it is in principle possible to use time of flight as the basis for localization systems based on packet radios, the current devices do not offer generally available and accurate methods to estimate it.

Bluetooth and IEEE 802.15.4 are standards for low-to-medium bandwidth communication over short distances. They operate in the 2.4 GHz band, which is unlicensed worldwide, and both are designed to be resistant to interference from other devices working in the same unlicensed radio band. While wifi is also in principle usable for the same purpose, it is not commonly adopted for wearable systems, because the power requirements of wifi are currently much higher than that of competing standards. However, the widespread use of smartphones may change this trend; in fact, a wireless network-rich environment such as on office, where many wifi access points are available, can be exploited by passive smartphone-based systems.

Active wearable systems based on packet radio work by exchanging packets with external devices that operate RSS measurements on the exchanged packets and use those measurements as the base data in various RSS-based methods are discussed below. The great majority of proposed systems fall into this category.

On the other hand, passive packet-radio systems only receive packets. In principle, they can work anonymously, by simply sniffing network traffic from the environment without exchanging any data at all with external devices. In order for this to work, the on-body device needs a priori knowledge of the position of the transmitting devices that are installed in the environment.

3.3 Wearable Passive Systems

Passive systems are based on a variety of technologies. One is packet radio, which was discussed in the previous section. One more passive technology that can be described as on-body is the one used in high-level gesture recognition systems that consist of light reflectors on different parts of the body.

In this section we speak about *inertial systems* (accelerometers and gyroscopes), *compasses*, and atmospheric *pressure sensors*. These are low-power sensors that can all be found on high-end smartphones available today, and can be expected to be found on all smartphones in the near future and in many portable and wearable or embedded devices. The low-price availability of these sensors is recent, and is a consequence of the advances in MEMS (micro electro-mechanical systems) manufacturing based on silicon.

Inertial sensors are able to measure 3-D linear acceleration and angular velocity, thus in principle allowing the realization of a complete 3-D *dead reckoning* localization system, i.e., a system that integrates the velocity to compute position change from a starting point. However, the accuracy of these cheap sensors, especially their output drift characteristics, requires more effort to obtain reliable positioning than simply integrating their output. On the other hand, recent advances in algorithms for PDR (*pedestrian dead reckoning*) provide an accuracy and a precision that was unthinkable only a few years ago, and permit PDR to be the main building block for practically usable systems, at least in indoor environments.

Electronic compasses are a natural complement to inertial systems. While linear accelerometers provide an absolute reference for the vertical (Earth's gravity) direction, compasses provide an absolute horizontal reference based on the Earth's magnetic field. However, the magnetic reference is not as reliable, because the magnetic field in indoor environments is strongly perturbed by many metallic materials used for construction and furniture, thus making it difficult to obtain a reliable measurement of the North direction. Some systems use the perturbations as a stable fingerprinting method [8].

Atmospheric pressure sensors are another recent addition to the array of cheaply available MEMS on consumer devices. Their sensitivity is sufficient to reliably detect fast altitude changes, such as those caused by the user moving to a different floor in a building.

3.4 Device-Free Localization

Device-free systems do not require any device to be worn or carried by the user, so one cannot speak of wearable sensors. We only provide a quick overview of technologies for device-free localization.

Simplest of all are the systems based on traditional PIR (presence infra-red) sensors. These can signal the presence of moving persons in their field of view.

UWB (ultra-wide band) experimental localization systems that are based on the principle of radar have been demonstrated that use one emitter and two receiver antennas (or the other way around) that collect the signal reflected by the user's body. These systems are purposely built, and thus expensive [9].

Video cameras offer the possibility of very precise localization and user detection by distinguishing among users. Special care should be used in these systems where privacy concerns are of interest.

Multiple microphones deployed in the environment can be used to identify the source of noise. They can be used to detect the position of a person who speaks aloud.

Laser-based systems coupled with a videocamera have found their way to the mass market with affordable prices. The most notable device is the Kinect and similar devices, which include software for ranging and posture identification.

Radio tomographic imaging (RTI) is a technique proposed by Wilson and Patwari that is able to exploit arrays of inexpensive packet-radio sensors to precisely identify the position of users in an indoor environment [10].

3.5 Hybrid Techniques

The above technologies are not necessarily used alone: the responses from each are put together using *data fusion* techniques. In fact, any working system that is not only for laboratory research is forced to use more than one method to obtain acceptable performance.

3.5.1 Data Fusion

Three main mathematical techniques exist for data fusion techniques: Kalman filters, Bayesian inference, and particle filters; each is, in fact, a wide class of methods that must be adapted to specific needs.

Kalman filters are recursive algorithms using continuous variables that are widely used in navigation, especially in their nonlinear version called the *extended Kalman filter*. For highly nonlinear problems, such as those commonly found in indoor localization, *unscented Kalman filters* are typically used instead.

Bayesian inference is a broad class of statistical techniques that use the Bayes rule to update the estimate of a position each time some new information (*evidence*) is acquired, in a way that depends on a measure of reliability of that information.

Particle filters are methods that have gained a lot of interest for data fusion of information for personal localization, especially in indoor environments. This technique was born in robotics circles; it is based on a Bayesian update rule applied to a discrete localization grid where a *particle sworm* made of tens or even hundreds of points is located. Time is discrete: at each cycle each particle moves independently of the others, with random velocity and direction. It is then assigned a *weight* dependent on its probability of belonging to some position distribution. Low-weight particles are removed and replaced with new particles created in proximity to the surviving ones. Particle filters are conceptually simple, and easy to implement and to modify with new constraints and information. However, they are computation heavy, so they are not appropriate for implementation on small devices. Recently, implementation of particle filters for PDR (pedestrian dead reckoning) has been demonstrated on smartphones (see, e.g., [11]), with the help of external references like wifi access point positioning or manual intervention.

3.5.2 SLAM (Simultaneous Location and Mapping)

Like particle filtering, discussed above, this technique comes from robotics. It works by examining the environment and making a picture of it; then, while moving in the environment, by tracing the movement by means of external reference and inertial sensors and simultaneously drawing a map of the environment. Adjustments to past and present estimated position are done every time a location is revisited, thus closing a loop of movements.

This technique is very computation-intensive and thus not currently usable on a small portable device, but it can be used by an external machine that receives data provided by wearable sensors. A method was very recently proposed for SLAM on an indoor environment, which only requires a sensor to be put on the shoe of the user [12]. The authors hope that improvements in the algorithm will make it possible to use the sensors inside a smartphone carried by the user.

4. WEARABLE RADIOS

Wearable-based solutions estimate the (unknown) location of the mobile sensors (hereafter also called mobiles) with respect to a set of fixed sensors (called anchors), whose position is known. The position estimation of a mobile can be achieved by using two different approaches, either range-based or range-free. The former consists of protocols that use absolute point-to-point distance estimates for calculating the location. The latter makes no assumption about the availability or validity of such information. The effectiveness of these two localization approaches depends on the accuracy required by the applications that use the location information. Acknowledging that the range-free solutions have a coarse accuracy [13], these techniques are unsuitable in applications where location accuracy is one of the main requirements. On the other hand, range-based localization exploits measurements of physical quantities related to beacon packets exchanged between the mobile and the anchors. Radio signal measurements are typically the received signal strength (RSS), the angle of arrival (AOA), the time of arrival (TOA), or the time difference of arrival (TDOA). Although AOA or TDOA can guarantee a high localization precision, they require specific and complex hardware. This is a major drawback in particular in AAL applications, which are deeply involved with users' monitoring and thus may suffer from complex and invasive hardware.

There are at least two possible scenarios where RSS techniques are suited better than other radio signal measurement techniques. In the first, futuristic scenario, wireless sensors are ubiquitous in the environment and on the user's body for health monitoring. In this case, no additional equipment is required to localize the user. In the second scenario, various sensors (i.e., gyroscopes, accelerometers, compass, and pressure sensor) are placed on the user's body to recognize the movements. In the common case of sensors using wireless communications, exploiting the RSS measurements can give an additional source of information (such as the position of the user) at no additional equipment cost. For these reasons in the following sections of this chapter we will refer only to RSS-based localization techniques.

4.1 Range-Free

In this section we present the most used and referenced range-free localization schemes that use radio connectivity to infer proximity to a set of anchor nodes. These schemes make no assumptions about the availability or validity of point-to-point distance estimation.

4.1.1 Centroid

The centroid scheme was proposed in [14]. This localization scheme assumes that a set of anchor nodes (A_i, $1 \leq i \leq n$), with overlapping regions of coverage, exist in the deployment area of the WSN. The main idea is to treat the anchor nodes, located at (X_i, Y_i), as physical points of equal mass and to find the center of gravity (centroid) of all these masses:

$$(X_G, Y_G) = \left(\frac{\sum_{i=1}^n X_i}{n}, \frac{\sum_{i=1}^n Y_i}{n} \right) \tag{1}$$

An example of how the centroid scheme works is shown in Figure 1, where a sensor node N_k is within communication range to four anchor nodes, $A_1...A_4$. The node N_k localizes itself to the centroid of the quadrilateral A_1 A_2 A_3 A_4 (for the case of a quadrilateral, the centroid is at the point of intersection of the bimedians — the lines connecting the middle points of opposite sides).

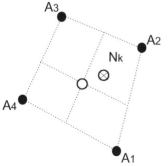

FIGURE 1 Centroid localization-node N_k localized as the centroid of the A_1 A_2 A_3 A_4 quadrilateral.

4.1.2 APIT

APIT [15] is an area-based range-free localization scheme. It assumes that a small number of nodes, called anchors, are equipped with high-powered transmitters and know their location, obtained via GPS or some other mechanism. Using beacons from these anchors, APIT employs a novel area-based approach to perform location estimation by marking the environment with triangular regions between anchor nodes as shown in Figure 2. A node's presence inside or outside of these triangular regions allows a node to narrow down the area in which it can potentially reside. By using different combinations of anchors, the size of the estimated area in which a node resides can be reduced to provide a good location estimate.

The method used to narrow down the possible area in which a target node resides is called the point-in-triangulation test (PIT). For three given anchors, A, B, and C, the PIT chooses whether a point M with an unknown position is inside triangle △ABC or not. APIT repeats the PIT with different anchor combinations until all combinations are exhausted or the required accuracy is achieved. At this point, APIT calculates the centroid of the intersection of all of the triangles in which a node resides to find its estimated position. In [16] the authors implemented the APIT system on an outdoor experimental testbed showing that at least 80% of nodes lie within a one-hop region of their estimated areas. Both simulation and experimental results have verified that APIT is a promising technique for range-free localization in large outdoor sensor networks.

4.1.3 SeRLoc

SeRLoc [17] is another area-based range-free localization method. SeRLoc assumes two types of nodes: normal nodes and locators (i.e., anchors). Normal nodes are equipped with

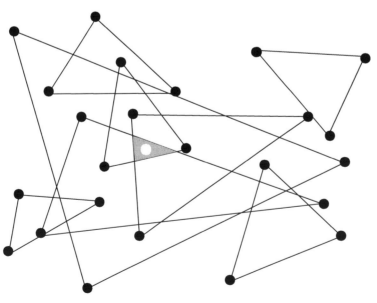

FIGURE 2 Area-based APIT algorithm overview.

omnidirectional antennas, while locators are equipped with directional sectored antennas (locations of locators are known a priori). In SeRLoc, a sensor estimates its location based on the information transmitted by the locators. Figure 3 shows the main idea, with node N_k within radio range to locators A_1, A_2, and A_3.

SeRLoc localizes the sensor nodes in four steps. First, a locator transmits directional beacons within a sector. Each beacon contains the locator's position and the angles of the sector boundary lines. A normal node collects the beacons from all locators it hears. Second, it identifies an approximate search area within which it is located based on the coordinates of the locators heard. Third, it computes the overlapping sector region using a majority vote scheme. Finally, SeRLoc estimates a node location as the centroid of the overlapping region.

We note that SeRLoc is unique in its secure design. It can deal with various kinds of attacks including wormhole and Sybil attacks. We do not describe its security features here except to note that the authors prove in [17] that their approach is more secure, robust, and accurate in the presence of attacks, compared with other state-of-the-art solutions that largely ignore this issue.

4.2 Range-Based

In this section we make reference to the first scenario described in section 4, where the environment is rich in sensors deployed for reasons other than localization. The idea is to exploit the RSS measurements of these sensors.

The main range-based indoor localization approaches that exploit the RSS are based on fingerprint and on signal propagation models. In both cases a mobile sensor is localized by means of a set of anchors that exchange beacon packets with the mobile in order to collect sequences of RSS values. In particular, at a given instant of time, the system computes a tuple of RSS values (one RSS value obtained from each anchor), which is used to

FIGURE 3 SerLoc Localization.

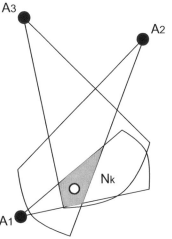

estimate the position of the mobile sensor at that time. The fingerprint schemes, also referred to as *pattern matching*, require a preliminary system calibration procedure (an off-line phase [18–22]). This phase is executed after the deployment of the anchors, and it consists of performing a set of RSS measurements at a set of points in the environment. These points correspond to possible locations of the mobile sensors that should be localized. For each point a tuple of RSS values is produced, which is stored in a database. During the localization procedure (the online phase), every time a new RSS tuple associated to a mobile is produced, the localization system compares it with those stored in the database to find the most likely position of the mobile.

4.2.1 Fingerprinting

Location fingerprinting differs from other localization principles. Instead of computing the distances between the user and the anchors and triangulating the user's location, the location of the user is determined by comparing the obtained RSS values to a radio map. The radio map is constructed in an offline phase and it contains the measured RSS patterns at certain locations. This way the characteristics of the signal propagation in indoor environments are captured and the modeling of the complex signal propagation is avoided. The point of fingerprinting is that it does not require knowledge either of the transmitters' location or of the characteristics of the environment. Only the measurements, which imply the characteristics of the environment, i.e., the RSS values, are needed. However, the offline phase can be computationally intensive, and the radio maps have to be stored in memory.

4.2.1.1 RADIO MAP

The construction of the radio map begins by dividing the area of interest into cells with the help of a floor plan. RSSI values of the radio signals transmitted by APs are collected in calibration points inside the cells for a certain period of time and stored in the radio map. The ith element in the radio map has the form $M_i = (B_i, a_{ij}, \theta_i), i = 1, \ldots \ldots \ldots, M$, where B_i is the ith cell, whose center p_i is the ith calibration point. Vector a_{ij} holds the RSSI values measured from the anchor j. The parameter θ_i contains any other information needed in the location estimation phase. This can be, for example, the orientation $\theta_i = d_i \in$ {north, south, east, west} of the mobile, such as in the RADAR system [23]. We denote the ith fingerprint by R_i and the set of all fingerprints by $R = \{R_1, \ldots, R_M\}$, so the ith element of the radio map is $M_i = (B_i, R_i)$.

The radio map can be modified or preprocessed before applying it in the location estimation phase. The motivation can be the reduction of the memory requirements of the radio map or the reduction of the computational cost of location estimation. In addition, the fingerprint can also include information about the distribution, either a histogram for each transmitter or a more simplified parameter such as mean or variance. Once the database of fingerprints exists, a device calculates position by recording a fingerprint and "matching" to the database. This usually consists of measuring a "distance" between the measured RSS fingerprint and each fingerprint in the database.

4.2.1.2 LOCATION ESTIMATION

Given the radio map, the objective of the location estimation phase is to infer the state (location) of the mobile device from the received measurements vector y, which includes RSSI samples y_j from several anchors.

4.2.1.2.1 DETERMINISTIC APPROACHES

In the deterministic approaches, the state u is assumed to be a non-random vector [23]. The main objective is to compute the estimate $\hat{u}$ of the state at every time step. Usually the estimate is a linear combination of the calibration points p_i, i.e.,

$$\hat{u} = \sum_{i=1}^{M} \frac{w_i}{\sum_{j=1}^{M} w_j} p_i \tag{2}$$

where all weights w_i are non-negative. One possible weight w_i is the inverse of the norm of the RSSI innovation [24], i.e.,

$$w_i = \frac{1}{||\acute{y} - \acute{a}_i||} \tag{3}$$

where $\acute{y}$ is the measurement vector, $\acute{a}_i$ is the vector of the means of the RSSI values of each AP at the ith calibration point, and the norm $|| \cdot ||$ is arbitrary. Examples of possible norms are given in Table 1. The Euclidean norm (2-norm) is widely used, but the Manhattan norm (1-norm) is also common [23–25].

The estimator, that keeps the K biggest weights and sets the others to zero, is called the weighted K-nearest neighbor method (WKNN) [24]. The WKNN with equal weights is called the K-nearest neighbor method (KNN) [23]. The simplest method, where K = 1, is called the nearest neighbor method (NN) [27]. In general, the KNN and the WKNN can perform better than the NN method, particularly with parameter values K = 3 and K = 4 [24]. However, if the density of the radio map is high, the NN method can perform as well as the more complicated methods.

4.2.1.2.2 PROBABILISTIC APPROACHES

In the probabilistic (or statistical) approaches the state u is assumed to be a random vector [28]. The idea in the probabilistic framework is to compute the conditional probability distribution function $p_{u \vee y}(u|y) \triangleq p(u|y)$ (posterior) of the state u given measurements y. The posterior contains all the necessary information to compute an arbitrary estimation of the state. Using the Bayes rule we get

$$p(u|y) = \frac{p(y|u)p(u)}{p(y)} \tag{4}$$

where $p(y|u)$ is the likelihood, $p(u)$ is the prior and $p(y)$ is a normalizing constant. A conventional choice for the prior [28] is the uniform distribution, i.e.,

$$p(u) = \frac{\sum_{i=1}^{M} \psi_{B_i}(u)}{\sum_{j=1}^{M} |B_j|} \tag{5}$$

TABLE 1   Possible Norms

Name	$\|u\| (u \in R^n)$
p-norm [23]	$\|u\|_p = \left(\sum\limits_{i=1}^{n} \|u_i\|^p \right)^{\frac{1}{p}}$
modified p-norm [25]	$\|u\|_{mp} = \left(\sum\limits_{i=1}^{n} \frac{1}{w_i} \|u_i\|^p \right)^{\frac{1}{p}}$
infinity-norm	$\|u\|_\infty = max_i(\|u_i\|)$
Mahalanobis-norm (S is the covariance matrix) [26]	$\|u\|_M = \sqrt{u^T S^{-1} u}$

where $|B_i|$ is the area of B_i and

$$\psi_{B_i}(u) = \begin{cases} 1, u \in B_i \\ 0, u \notin B_i \end{cases} \tag{6}$$

Assuming that the measurements collected at the calibration point represent the distribution of the RSSI in the whole cell (i.e., the likelihood is constant inside each cell B_i) the likelihood is

$$p(y|u) = \sum_{i=1}^{M} p(y|i)\psi_{B_i}(u) \tag{7}$$

where $p(y|i) = p_{v_i}(y - a_i)$ and $v_i = y - a_i$. We assume that the components of the random vector v_i are independent. Thus,

$$p(y|i) = \prod_{j=1}^{n} p_{v_{ij}}(y_j - a_{ij}) \tag{8}$$

where $y \in R^n$.

There are several approaches for computing the likelihood p(y|i). Examples of these methods are given in Table 2 [31].

Substituting prior and likelihood equations into the posterior equation we get

$$p(u|y) = \sum_{i=1}^{M} \beta_i \frac{\psi_{B_i}(u)}{|B_j|} \tag{9}$$

From this posterior we can compute an estimate of the state u. One possible estimate is the maximum a posteriori (MAP) estimate. In this case, it is the same as the maximum-likelihood (ML) estimate, because the prior is uniform. The posterior is piecewise constant, and thus the MAP estimator is ambiguous. If the posterior has the maximum value only in one cell, say B_i, it is reasonable to use the center of the cell as the MAP estimate, i.e.,

$$\hat{u}_{MAP} = p_i \tag{10}$$

TABLE 2 Likelihood Functions

Name of the Method	$p(y|i) = \prod_{j=1}^{n} p_{v_{ij}}(y_j - a_{ij})$
Gaussian [29]	$p_{v_{ij}}(x) = \frac{1}{\sqrt{2\pi\hat{\sigma}_{ij}^2}} exp\left(\frac{-x^2}{\hat{\sigma}_{ij}^2}\right)$
Log-normal [30]	$p_{v_{ij}}(x) = \frac{1}{x\sqrt{2\pi\hat{\sigma}_{ij}^2}} exp\left(\frac{-\log(x) - \mu^2}{\hat{\sigma}_{ij}^2}\right)$
Inverse function	$p_{v_{ij}}(x) = \begin{cases} \dfrac{2 - |x|}{2\log(t) + 3}, |x| \leq 1 \\ \dfrac{1}{(2\log(t) + 3)|x|}, 1 < |x| \leq t \\ 0, t < |x| \end{cases}$
Exponential	$p_{v_{ij}}(x) = \frac{1}{2}e^{-|x|}$

Another commonly used estimate is the mean of the posterior, i.e.,

$$\hat{u}_{MEAN} = \sum_{i=1}^{M} \beta_i p_i \qquad (11)$$

Clearly, with fingerprinting there is a trade-off between the number of points in the grid (the larger this number, the more accurate the localization is) and the overhead due to the offline phase. In practice, the main drawback of this method is the high number of extensive and accurate measurements required during the offline phase to create the database. In fact, the creation of the database is not automatic: it is a human-based, time-consuming procedure and this is a practical barrier to wide adoption, because it makes the method unsuitable for rapid or ad hoc deployment. For these reasons, methods alternative to fingerprinting make use of signal propagation models, which are analytical models that relate RSS measures with distances [14,23,26,32].

4.2.2 Propagation Model

RSS is dependent on the environment, because the radio frequency signal suffers from reflection, diffraction, and multi-path effects that make the signal strength rather noisy. Consequently, localization systems based on signal propagation models need to calibrate parameters of the propagation model. The calibration procedure works in two phases: the training phase and the estimation phase. In the training phase, a set of RSS values are measured at a set of points in the area of interest; in the estimation phase, this information is used to estimate the propagation model parameters. The relationship between RSS and the reciprocal distance between anchors and mobile is then established by means of the calibrated propagation model.

Once a number of distance estimates between the mobile and the anchors are available, the most used algorithm to infer the mobile position is based on multi-lateration. As for the fingerprinting techniques, the localization based on the signal propagation model is expensive in terms of calibration time, since the calibration is not automatic. Moreover, the described calibration technique does not solve the time-dependent problem, since wireless channel variations affect the propagation model parameters, and this fact can significantly impact on the quality of the localization system. However, in recent works this problem has been solved by an automatic calibration procedure (called virtual calibration) of the signal propagation model that is only based on the RSSIs measured among the anchors and that can be executed periodically and automatically (i.e., without human intervention) [33].

4.2.2.1 INDOOR SIGNAL PROPAGATION MODEL

Signal propagation is affected by many factors such as antenna height, antenna gain, antenna radiation pattern, transmitter-receiver distance, reflection, multi-path transmission, non-line-of-sight, obstructions, vegetation scattering and diffraction, RSS measurement uncertainty, etc. [34]. All these factors have to be considered when the distance estimation is calculated from the signal strength measurement. Several propagation models have been proposed [35]. They are characterized by different levels of complexity, depending on how many physical phenomena are considered. Basically, simpler models can be used when the environment characteristics are close to ideal conditions. Unfortunately, in most cases it is necessary to consider specific features of signal propagation in radio channels. In the literature most researchers model the indoor path loss with the one-slope model [36], which assumes a linear dependence between the path loss (dB) and the logarithm of the distance d between the transmitter and the receiver:

$$L(d)\big|_{dB} = l_0 + 10\alpha \, \log_{10} d \tag{12}$$

where l_0 is the path loss at a reference distance of 1 m and α is the power decay index (also called path loss exponent). A generalization of the one-slope model is the two-slope model suggested in [37] to approximate the two-ray propagation model. Usually, in an urban environment, the two-slope model is characterized by a break point that separates the various properties of propagation in near and far regions relative to the transmitter: the path loss exponent changes when the distance d is greater than the break point. In particular, authors in [37] describe the existence of a transition region where the break point b is such that

$$\frac{\pi h_t h_r}{\lambda} < b < \frac{4\pi h_t h_r}{\lambda} \tag{13}$$

where h_t is the transmitter antenna height, h_r the receiver antenna height, and λ is the wavelength of the radio signal. However, in a typical sensor network scenario, the break point distance is on the order of hundreds of meters; therefore, in practice, the one-slope and the two-slope models are equivalent in indoor scenarios where the rooms are only a few square meters in size. Although the one-slope model is simple to use, it does not adequately account for the propagation characteristics in indoor environments. Indeed, one further generalization of the one-slope model consists of adding an attenuation term due to losses introduced by walls and floors penetrated by the direct path:

$$L(d)\big|_{dB} = l_0 + 10\alpha \, \log_{10} d + WAF\big|_{dB} + FAF\big|_{dB} \tag{14}$$

where FAF$|_{db}$ is the floor attenuation factor, and WAF$|_{db}$ is the wall attenuation factor expressed as

$$WAF|_{dB} = \sum_{i=1}^{N} K_i l_i \qquad (15)$$

where k_i is the number of penetrated walls of type i, and l_i is the attenuation due to the wall of type i. Without loss in generality, assuming that sensors are all located on the same floor, the attenuation term due to the propagation among different floors can be neglected. A similar model was proposed in [38], where a multi-wall component is introduced, which includes the number of normal and fireproof doors and their status (open/closed) met by the direct paths. The received power RSS is obtained as the difference between the transmitted power P^t and $L(d)$, i.e.,

$$RSS = P^t - L(d)\Big|_{dB} = P^t - l_0 - 10\alpha \log_{10}d - \sum_{i=1}^{N} K_i l_i \qquad (16)$$

Letting $r_0 = P^t - l_0$ be the RSSI at the reference distance of 1 m,

$$RSS = r_0 - 10\alpha \log_{10}d - \sum_{i=1}^{N} K_i l_i \qquad (17)$$

This equation is used during the calibration procedure to estimate the propagation model parameters (r_0, α, l_i). In [33] the authors show that they are able to estimate the parameters of the chosen propagation model by exploiting the communications among anchors, without performing a preliminary measurement campaign.

4.2.2.2 MULTI-LATERATION ALGORITHM

In the last two decades, many localization techniques have been developed for wireless sensor network applications [39,40]. Among them, multi-lateration is seen to be one of the most popular localization techniques. Multi-lateration localization techniques form the basis of many other more sophisticated localization algorithms such as the iterative and the collaborative multi-lateration technique [39,41]. Multi-lateration is a simple localization technique that is based on distance measurements from multiple anchor nodes to the sensor node to be localized.

The sensor node location estimated by the multi-lateration localization technique is the one that minimizes the sum of squared distances between a hypothesized sensor location to all the anchor locations. Multi-lateration is a nonlinear optimization problem, although suboptimal linear optimization approaches can be applied. One of the major limitations of the classical multi-lateration localization approach is that it assumes that the anchor locations are error free and the only errors are in the distance measurements. This assumption is invalid in practice because the locations of anchor nodes are always inaccurate. There are, in general, two sources of errors for anchor location errors. The first source of error is due to the measurement errors of the anchor nodes. The second source of anchor location error arises when the locations of anchor nodes are only estimates. Some localization approaches such as AHLoS (ad-hoc localization system) [39] use an iterative process. For example, in AHLoS, some sensor nodes in the network first estimate their own positions based on broadcasted beacon positions. They then become anchors and broadcast their estimated positions to other nearby nodes for localization. The iterative process can lead to error propagation, resulting in large sensor location estimation errors as networks grow. The classical multi-lateration localization algorithms do not take into account and ignore the anchor location errors even though they exist. Recently, a number of approaches have been proposed to deal with the problem of anchor localization uncertainty for multi-lateration. In [42], the authors formulated the problem of multi-lateration in the presence of anchor position errors using the constrained least square technique (CTLS). The problem is then solved by using the Newton iterative algorithm. In [43], semi-definite programming (SDP) algorithms were proposed for sensor localization in the presence of anchor position uncertainties, and the corresponding Cramér-Rao lower bound (CRLB) was derived. In [44,45], a distributed localization algorithm was proposed based on second-order cone programming relaxation. In [46], an expectation-maximization (EM) estimator for localization of sensor nodes with erroneous anchor positions was proposed, which iteratively refines the anchor positions and estimates the sensor locations. A common disadvantage of the above mentioned approaches is that they all involve solving a nonlinear optimization problem.

Figure 4 shows the geometric relationship between the anchors and the node to be localized, where M anchors are used. In the figure, $\{x, y\}$ are the unknown coordinates of the sensor node, $\{u_m, v_m\}$ are the known coordinates of the m-th anchor and d_m is the estimated distance between the sensor node and the m-th anchor. Both the anchor locations and the distance measurements are assumed to contain errors. The multi-lateration localization approach is to estimate $\{x, y\}$ given $\{u_m, v_m, d_m; m = 1,2,...,M\}$. Denote

$$g_m(\underline{x}) = \sqrt{(x-u_m)^2 + (y-v_m)^2} \qquad (18)$$

In general, multi-lateration minimizes the following sum of squared errors between the measured distances and hypothetical ones based on the unknown sensor node location [47]:

$$\min_x \sum_m (g_m(\underline{x}) - d_m)^2 \qquad (19)$$

Another technique [48] computes the intersection points among the three circles centered in a_1, a_2, and a_3; up to six intersection points can occur. Then, it estimates the position of the mobile $\{x, y\}$ as the center of mass between these intersection points, i.e.,

$$(x,y) = \left(\frac{\sum_{i=1}^n X_i}{n}, \frac{\sum_{i=1}^n Y_i}{n} \right) \qquad (20)$$

In [29] context information provided by smart devices deployed in the environment is exploited by the localization algorithm in order to improve the performance in terms

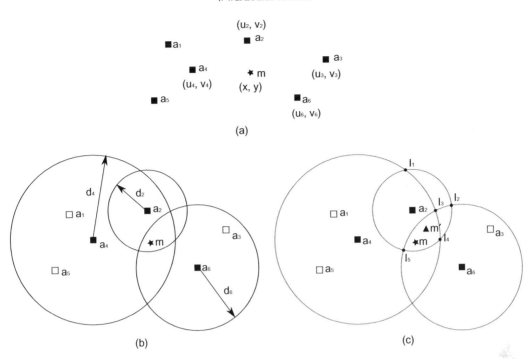

FIGURE 4   Anchor nodes and node to be localized: (a) the anchors are deployed in the environment; (b) the estimated distance among each anchor and the mobile has been evaluated; (c) the position of the mobile m' is estimated.

of localization error. In [49] the authors propose a decision tree algorithm to select the candidate intersection points that better provide the location of the mobile.

4.2.2.3 MAXIMUM-LIKELIHOOD DISTANCE ESTIMATOR AND CRAMÉR-RAO BOUNDS

The Cramér-Rao bound can be used to compute the theoretical attainable performance of localization systems that exploit RSS measurements. In this section, we derive the expression for the maximum-likelihood estimators (MLEs) and the Cramér-Rao bound (CRB) for the estimation of the distance between the mobile and the anchors. The properties of the bound on localization error may help in designing efficient localization algorithms; moreover, it provides suggestions for a positioning system design by revealing error trends associated with the system deployment. We derive the expressions for both the case of a single RSS value and for the more interesting cases of multiple RSS values used to estimate the distance.

4.2.2.3.1 RSS INFORMATION: SINGLE SAMPLE CASE The mobile device periodically receives the beacon signal sent by a given anchor and registers its RSS value r, which can be modeled as a Gaussian random variable. Thus, we have

$$r \ N(R, o^2) \tag{21}$$

and, from the relationship between RSS distance already explained in section 4.2.2.1, i.e., $RSS = r_0 - 10\alpha \log_{10}d - \sum_{i=1}^{N} K_i l_i$, the probability density function of r conditioned on the distance d between the mobile and the anchor is

$$f_d(r) \propto \frac{1}{d}\exp\left[\frac{(r-R)^2}{-2o^2}\right] \tag{22}$$

Then, the maximum-likelihood estimator of d is

$$\hat{d} = \arg\max_d f_d(r) \tag{23}$$

In order to evaluate such maximum, we differentiate the log-likelihood function, obtaining the following function:

$$\frac{\partial}{\partial d}lnf_d(r) = A\sum_{i=1}^{N}\frac{(r-R)}{o^2 d}, \text{ being } A = 10\alpha log_{10}e \tag{24}$$

therefore

$$\frac{\partial}{\partial d}lnf_d(r) = 0 \Rightarrow r - R = 0 \tag{25}$$

and thus

$$\hat{d} = 10^{(r_0 - r - L_w)/10\alpha} \tag{26}$$

The Fisher information I measures the amount of information that a random variable carries about an unknown parameter. Here the random variable is r, and d is the unknown parameter. The inverse of the Fisher information, known as the Cramér-Rao bound, is the minimum variance that can be achieved when estimating d by using any unbiased estimator. By evaluating the Fisher information for RSS measurements (details to compute the Fisher information are presented in [36]) we can identify the minimum theoretical error for the distance estimation. The Fisher information associated with RSS measurements is

$$I_d = E_d\left[\left(\frac{\partial}{\partial d}lnf_d(r)\right)^2\right] = \frac{A^2}{o^2 d^2}E_d\left[(r-R)^2\right] = \left(\frac{A}{\sigma d}\right)^2 \tag{27}$$

hence, the CRB entails

$$\sqrt{Var(\hat{d})} \geq \frac{\sigma d}{A} \tag{28}$$

4.2.2.3.2 RSS INFORMATION: MULTIPLE SAMPLES CASE In order to improve localization accuracy, multiple RSS samples are normally used to estimate the distance between anchors and mobile; as already stated, we assume that the RSS values r are i.i.d. random variables. We assume that the mobile device did not move significantly in the measurement time span during which the RSS samples are collected. Being $(r_1, \ldots, r_N)$ independent random variables, the joint PDF conditioned on the distance d is

$$f_d(r_1, \ldots, r_N) \propto \prod_{i=1}^{N} \frac{1}{d} \exp\left[\frac{(r_i - R)^2}{-2o^2}\right] \tag{29}$$

Following the same steps as in the previous case, we obtain

$$\frac{\partial}{\partial d} \ln f_d(r_1, \ldots, r_N) = A \sum_{i=1}^{N} \frac{(r_i - R)}{o^2 d} \tag{30}$$

therefore

$$\frac{\partial}{\partial d} \ln f_d(r_1, \ldots, r_N) = 0 \Rightarrow \sum_{i=1}^{N} r_i - R = 0 \Rightarrow \frac{\sum_{i=1}^{N} r_i}{N} = R \tag{31}$$

and thus the MLE of the distance is

$$\hat{d} = 10^{(r_0 - r_i - L_w)/10\alpha} \tag{32}$$

The Fischer information is

$$I_d = E_d\left[\left(\frac{\partial}{\partial d} \ln f_d(r_1, \ldots, r_N)\right)^2\right] = \frac{A^2}{o^2 d^2} E_d\left[\sum_{i=1}^{N} (r_i - R)^2\right] = \left(\frac{A}{od}\right)^2 No^2 = \left(\frac{A}{od}\right)^2 N \tag{33}$$

hence, the CRB is

$$\sqrt{\mathrm{Var}(\hat{d})} \geq \frac{od}{A\sqrt{N}} \tag{34}$$

5. CONCLUSIONS

The world of personal localization using wearable devices or no devices at all is rapidly evolving, yet no single solution has yet emerged as the killer application. Working commercial or prototyping solutions that are general enough to be of practical usage in indoor environments are based on RSS measurements of packet radio signals, ultrasound, and MEMS (micro electro-mechanical systems) such as those found in smartphones. A fusion of inputs from more than one technology is commonly used, but the most powerful fusion systems cannot run on the wearable devices: rather, they run on external servers that are part of the local infrastructure.

Exceptions to this rule are starting to appear: PDR (pedestrian dead reckoning) systems have reached a level of maturity that allows them to run on a smartphone at a level of accuracy sufficient for navigating a building even without assistance from the environment, but with the option to fuse information obtained from the local infrastructure.

Any recommendation pointing to a specific technology would be obsolete the moment it is written, as low-price commercial offerings are starting to come out and academic research is fast progressing. As an example, let's examine the criteria to consider for evaluating a smartphone-based PDR system. Since these systems are resource intensive, it is important to verify the requirements imposed by the application on the smartphone, in terms of CPU power, sensor availability, and most importantly, battery consumption. Systems that do not require a preliminary training phase should be preferred; preference can turn into a requirement if the application is going to be used occasionally, for example, for visitors. It should be possible to have the choice of downloading a complete map of the environment, for offline use, or getting partial maps on demand, in order to speed up starting times and minimize memory and network usage. Some systems are able to track the position even without a map: this can be useful to get back on one's step, or to superimpose the recorded path once the user obtains a map of the environment, or even to create a map of the unknown environment. The more sources of information the system manages to fuse together, the better: for example, an application could be able to run on a low-range smartphone without a compass or a pressure sensor by exploiting a map of wifi access points in the environment, at the price of lower accuracy and higher battery consumption.

In the future we foresee several methods being used on different devices, and different sources of information being exploited depending on the computing capabilities of the portable or wearable devices. More specifically, we see two trends. The first, consequent to the diffusion of ever more powerful smartphones, will push toward the use of PDR methods fused with compass and atmospheric pressure information for smartphones. The second, because of ubiquity of devices communicating through packet radio in smart environments, will push toward RSS-based methods to be implemented in small devices collaborating with local infrastructure. The two trends will merge and coexist in flexible ways, depending on the capabilities of the wearable devices and on the services offered by the local infrastructure.

References

[1] R. Want, A. Hopper, V. Falcao, J. Gibbons, The active badge location system, ACM Trans. Info. Syst. 10 (1) (1992) 91–102.
[2] G.D. Abowd, C.G. Atkeson, J. Hong, S. Long, R. Kooper, M. Pinkerton, Cyberguide: a mobile context-aware tour guide, Wireless Netw. 3 (5) (1997) 421–433.
[3] Y. Sumi, T. Etani, S. Fels, N. Simonet, K. Kobayashi, K. Mase, C-map: Building a context-aware mobile assistant for exhibition tours, Community Computing and Support Systems, Lecture Notes in Computer Science, Springer- Verlag, London, UK, 1998, pp 137–154.
[4] K. Cheverst, N. Davies, K. Mitchell, A. Friday, C. Efstratiou, Developing a context-aware electronic tourist guide: some issues and experiences, Proceedings of the SIGCHI conference on Human factors in computing systems (CHI '00), ACM Press, New York, NY, USA, 2000, pp 17–24.
[5] A.R. Jiménez Ruiz, F. Seco Granja, J.C. Prieto Honorato, J.I. Guevara Rosas., Accurate pedestrian indoor navigation by tightly coupling foot-mounted imu and RFID measurements, IEEE T. Instrum. Meas. 61 (1) (2012) 178–189.
[6] Paolo B, Stefano C, Francesco F, Francesco P, Evaluating AAL solutions through competitive benchmarking: the localization competition, IEEE Pervasive Comput Mag 12 (No. 4) (2013) 72–79.

[7] S. Knauth, L. Kaufmann, C. Jost, R. Kistler, A. Klapproth, The iLoc Ultrasound Indoor Localization System at the EvAAL 2011 Competition. Evaluating AAL Systems Through Competitive Benchmarking, Indoor Localization Track. Commun. Comput. Info. Sci. 309 (2012) 52—64.

[8] Robertson P, Frassl M, Angermann M, Doniec M, Julian, B.J, Garcia Puyol M. Simultaneous Localization and Mapping for pedestrians using distortions of the local magnetic field intensity in large indoor environments, Indoor Positioning and Indoor Navigation (IPIN), 2013 International Conference on, vol., no., pp.1,10, 28-31 2013.

[9] Pietrzyk M.M, von der Grun T. Experimental validation of a TOA UWB ranging platform with the energy detection receiver, Indoor Positioning and Indoor Navigation (IPIN), 2010 International Conference on , vol., no., pp.1,8, 15-17, 2010.

[10] J. Wilson, N. Patwari., Radio Tomographic Imaging with Wireless Networks, IEEE Trans. Mobile Comput. 9 (5) (2009) 621—632.

[11] < http://movea.com >, (Last Accessed: 03.07.14).

[12] M. Angermann, P. Robertson, FootSLAM: Pedestrian Simultaneous Localization and Mapping Without Exteroceptive Sensors Hitchhiking on Human Perception and Cognition, Proc. IEEE 100 (2012) 1840—1848.

[13] G. Giorgetti, S.K.S. Gupta, G. Manes, Localization using signal strength: to range or not to range? MELT '08, Proceedings of the First ACM International Workshop on Mobile Entity Localization and Tracking in GPS-Less Environments, ACM, New York, NY, USA, 2008, pp. 91—96.

[14] N. Bulusu, J. Heidemann, D. Estrin, GPS-less low cost outdoor localization for very small devices, IEEE Pers. Commun. Mag. 7 (5) (2000) 28—34.

[15] T. He, C. Huang, B. Blum, J.A. Stankovic, T. Abdelzaher, Range-Free localization schemes in large scale sensor networks, ACM Int. Conf. Mobile Comput. Netw. (2003) (Mobicom).

[16] V R. Chandrasekhar, W K.G. Seah, Z. A. Eu and A P. Venkatesh. (2010). Range-free Area Localization Scheme for Wireless Sensor Networks. Wireless Sensor Networks: Application - Centric Design, Chapter 17, 14, ISBN 978-953-307-321-7.

[17] Lazos, L., and Poovendran, R. SeRLoc: Secure range-independent localization for wireless sensor networks. In ACM Workshop on Wireless Security (WiSe) (2004).]

[18] T. Christ, P. Godwin, R. Lavigne, A prison guard duress alarm location system, Int. Carnahan Conf. Secur. Technol. (1993) 106—116.

[19] K. Lorincz, M. Welsh, Motetrack: a robust, decentralized approach to RF-based location tracking, Pers. Ubiquitous Comput. 11 (6) (2006) 489—503.

[20] King T, Kopf S, Haenselmann T, Lubberger C, Effelsberg W. (2006). Compass: a probabilistic indoor positioning system based on 802.11 and digital compasses. Proceedings of the First ACM International Workshop on Wireless Network Testbeds, Experimental evaluation and Characterization (WiNTECH), Los Angeles, CA, USA.

[21] M. Youssef, A. Agrawala, The WLAN location determination system. 0MobiSys '05, Proceedings of the 3rd International Conference on Mobile Systems, Applications, and Services, ACM, New York, NY, USA, 2005, , pp. 205—218.

[22] A. Papapostolou, H. Chaouchi, Wife: wireless indoor positioning based on fingerprint evaluation, 8th Int. IFIP-TC 6 Netw. Conf. (2009) 234—247.

[23] P. Bahl, V.N. Padmanabhan., Radar: An in-building RF-based user location and tracking system, in INFOCOM 2000, Nineteenth Annu. Joint Conf. IEEE Comput. Commun. Soc. 2 (10) (2000) 775—784.

[24] B. Li, J. Salter, A. G. Dempster, and C. Rizos, (2006). "Indoor positioning techniques based on wireless LAN," School of Surveying and Spatial Information Systems, UNSW, Sydney, Australia, Tech. Rep.

[25] P. Prasithsangaree, P. Krishnamurthy, and P. Chrysanthis. (2002). "On indoor position location with wireless LANs," Telecommunications Program, University of Pittsburgh PA 15260, Tech. Rep.

[26] W.M. Yeung, J. Zhou, J.K. Ng, Enhanced fingerprint-based location estimation system in wireless LAN environment, in: M.K. Denko, C.-S. Shih, K.-C. Li, et al. (Eds.), Emerging Directions in Embedded and Ubiquitous Computing, Springer, 2007, pp. 273—284.

[27] S. Saha, K. Chauhuri, D. Sanghi, and P. Bhagwat, (2003). "Location determination of a mobile device using IEEE 802.11b access point signals," Department of Computer Science and Engineering, Tech. Rep.

[28] T. Roos, P. Myllymaki, H. Tirri, P. Misikangas, J. Sievanen., A probabilistic approach to WLAN user location estimation, Int. J. Wireless Info. Netw. 9 (3) (2002) 155—163.

[29] A. Haeberlen, E. Flannery, A.M. Ladd, A. Rudys, D.S. Wallach, L.E. Kavraki., Practical robust localization over large-scale 802.11 wireless networks, MobiCom'04 (2004).

[30] K. Kaemarungsi, Distribution of WLAN received signal strength indication for indoor location determination, National Electronics and Computer Technology Center, Thailand, 2006, Tech. Rep.

[31] R.O. Duda, P.E. Hart, D.G. Stork., Pattern Classification, John Wiley, Sons Inc., 2001.

[32] X. An, J. Wang, R.V. Prasad, I.G.M.M. Niemegeers, Opt: online person tracking system for context-awareness in wireless personal network, Proceedings of the 2nd International Workshop on Multi-Hop Ad Hoc Networks: From Theory to Reality (REALMAN '06), ACM, New York, NY, USA, 2006, , pp. 47−54.

[33] P. Barsocchi, S. Lenzi, S. Chessa, F. Furfari, Automatic virtual Calibration of Range-Based Indoor Localization Systems, Wireless Commun. Mobile Comput. 12 (17) (2012) 1546−1557.

[34] T. Stoyanova, F. Kerasiotis, A. Prayati, G. Papadopoulos, A Practical RF Propagation Model for Wireless Network Sensors, Third Int. Conf. Sens. Technol. Appl. SENSORCOMM '09 (2009) 194−199.

[35] T. Rappaport, Wireless Communications: Principles and Practice, second Edition, Prentice Hall, 2001.

[36] N. Patwari, I.A.O. Hero, M. Perkins, N. Correal, R. O'Dea, Relative location estimation in wireless sensor networks, IEEE Trans. Signal Process. 51 (8) (2003) 2137−2148.

[37] E. Green, M. Hata, Microcellular propagation measurements in an urban environment, IEEE Int. Symp. Pers. Indoor Mobile Radio Commun. (1991) 324−328.

[38] A. Borrelli, C. Monti, M. Vari, F. Mazzenga, Channel models for IEEE 802.11b indoor system design, IEEE Int. Conf. Commun. 6 (2004) 3701−3705.

[39] A. Savvides, C.C. Han, M.B. Srivastava., Dynamic fine-grained localization in ad hoc networks of sensors, Proc. 7th Annu. ACM/IEEE Int. Conf. Mobile Comput. Netw. (MobiCom'01) (2001) 166−179, Rome, Italy.

[40] G. Mao, B. Fidan, B.D.O. Anderson., Wireless sensor network localization techniques, Comput. Netw. 51 (issue 10) (2007) 2529−2553.

[41] A. Savvides, H. Park, and M. B. Srivastava. (2002). The bits and flops of the N-hop multilateration primitive for node localization problems. Proc. the First ACM International Workshop on Wireless Sensor Networks and Applications, pp. 112-121, Atlanta, Georgia, USA.

[42] J. Wan, N. Yu, R. Feng, Y. Wu, C. Su., Localization refinement for wireless sensor networks, Comput. Commun. 32 (Issues 13-14) (2009) 1515−1524.

[43] K.W.K. Lui, W.-K. Ma, H.C. So, F.K.W. Chan, Semi-definite programming algorithms for sensor network node localization with uncertainties in anchor positions and/or propagation speed, IEEE Trans. Signal Process. 57 (2) (2009) 752−763.

[44] S. Srirangarajan, A.H. Tewfik, Z.Q. Luo, Distributed sensor network localization with inaccurate anchor positions and noisy distance information, Proc. 2007 IEEE Int. Conf. Acoust. Speech Signal Process. (ICASSP'07) 3 (2007) 521−524, Honolulu, HI, USA.

[45] S. Srirangarajan, A.H. Twefic, Z.Q. Luo, Distributed sensor network localization using SOCP relaxation,", IEEE Trans. Wireless Commun. 7 (12) (2008) 4886−4894.

[46] M. Leng, Y.-C. Wu, Localization of wireless sensor nodes with erroneous anchors via EM algorithm,", Proc. 2010 IEEE Global Telecom-Munications Conf. (GLOBECOM'10) (2010) 1−5, Miami, Fl, USA.

[47] Y. Zhou, J. Li, L. Lamont, Multilateration Localization in the Presence of Anchor Location Uncertainties, Globecom 2012-Ad Hoc Sens. Netw. Symp. (2012).

[48] P. Barsocchi, S. Chessa, F. Potortì, F. Furfari, E. Ferro., Context Driven Enhancement of RSS-based Localization Systems, IEEE Symp. Comput. Commun. (ISCC) (2011) 463−468.

[49] Khalid K. Almuzaini and T. Aaron Gulliver. (2010). Range-based Localization in Wireless Networks using Decision Trees. IEEE Globecom 2010 Workshop on Heterogeneous, Multi-hop Wireless and Mobile Networks.

Fundamentals of Wearable Sensors for the Monitoring of Physical and Physiological Changes in Daily Life

Masaaki Makikawa[1], Naruhiro Shiozawa[1], and Shima Okada[2]

[1]Ritsumeikan University, Kyoto, Japan, [2]Kinki University, Osaka, Japan

1. INTRODUCTION

This chapter provides an overview of wearable sensors for the monitoring of physical and physiological changes in daily life, including their fundamentals and applications.

Here arises one question: Should all physical or physiological sensors be wearable? Table 1 shows the classification of biological signal monitoring by its measurement frequency. As shown in this table, many types of biosignals need not be measured continuously for 24 hours. Almost no user may want to know his/her body weight change at each step while walking. In addition, the reasons for monitoring are not limited to "health check to find any signs of abnormality," but vary with application. For example, "checking the level of fitness," "acquiring changes in disease condition in daily life," and "as a part of the clinical diagnosis and/or treatment" are all parts of clinical monitoring. In the workplace "monitoring of health condition to prevent accidents can also be an important aspect of ambulatory monitoring. Besides health-related monitoring activities, "quantitative evaluation of comfort/discomfort for new industrial products" and "human-machine interface for the visually or auditory handicapped" can be easily imagined as applications of wearable sensors. Every day wearable sensor technology becomes more important in our daily lives.

Figure 1 shows the current situation of a wearable digital biological monitoring device and its near future. As shown in Figure 1(b) the goal of this type of mobile monitoring system is to realize a "virtual doctor" system that watches our health 24 hours and sends emergency information to a real, responsible doctor. Figure 1(a) shows a current research or commercially available smartphone-based health-monitoring device. Smartphones are widely used, and it is

TABLE 1   Classification of Biological Signal Monitoring.

Monitoring Frequency	Applications	Monitoring Purposes
Once a day — Once a year	Physical examination, Periodic medical check	Health check to find any sign of the abnormality
A few minutes every day	Blood pressure manometer	Health check to find any sign of the abnormality
A few hours every day	Step counter, Fitness monitoring	Check of the attainment level of health goal
	Monitoring of cardiovascular system, Sleep meter	Health check to find any sign of the abnormality
	Evaluation of clothing and shoe	Quantitative evaluation of comfort/discomfort
Always during being active	Prediction of fall	Attack monitoring for accident prevention
	Detection of heart abnormality	Health check to find any sign of the abnormality
	Detection of drowsy driving	
	Detection of gas leak	
Full time monitoring for a certain period	Holter monitor	Provision of daily health data for clinical diagnosis
	Quantitative evaluation of drug effect	Monitoring of disease condition for continuous treatment
Almost Always	Brain-machine interface	New human-machine interface

possible to take advantage of their capabilities for mobile health monitoring by adding biological sensors. Here, the sensor is interfaced to the smartphone through a popular digital interface such as USB. Hence, digitization of biological sensor data is important.

2. WEARABLE SENSORS FOR PHYSIOLOGICAL SIGNAL MEASUREMENT

This section describes frequently used biosignal sensors for health monitoring in daily life. In addition, motion sensors and physical activity sensors are introduced in other chapters of the book.

2.1 Fundamentals of the Three Types of Electrodes and ECG Monitoring by Non-Contact Electrodes

Electrodes are essential in the measurement of bioelectric phenomena, such as heart activity and muscle activity. Table 2 shows the classification of electrodes. There are three types of

(a)

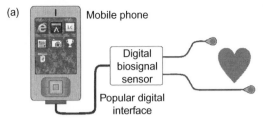

Mobile phone

Digital biosignal sensor

Popular digital interface

FIGURE 1   Mobile digital biosignal monitor: (a) its current situation and (b) its future.

(b)

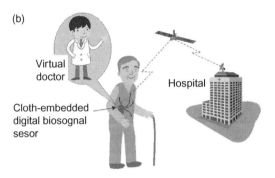

Virtual doctor

Hospital

Cloth-embedded digital biosognal sesor

TABLE 2   Three Types of Electrodes

Type	Use	Defect
Skin electrode	This is most popular electrode and is used to measure ECG, EMG, and EEG. Electro-conductive paste is needed to keep constant adhesion between electrode and skin.	Skin irritation caused by paste. Drying of paste during long time measurement.
Capacitivey-coupled electrode	This is reaching a practical level in recent years. ECG can be measured through clothing	This cannot be used in case of air existing between electrode and skin
Dry electrode	This has intermediate properties between capacitively-coupled electrode and the skin electrode. Paste is not required and skin irritation is negligible.	Adhesion to the skin should be assured.
		Electrodes on the skin induce discomfort of the subject.

electrodes: the most popular "skin electrodes," "capacitively coupled electrodes," which have recently become practical, and "dry electrodes," which have intermediate properties between the skin electrodes and capacitively coupled types. Description of "needle electrodes" is omitted here, because this type is invasive and imposes unnecessary pain to the user.

Figure 2 shows an equivalent electrical circuit of skin, capacitively coupled, and dry electrodes. As shown in this figure, the skin electrode can be roughly compared to a

(a) Skin electrode

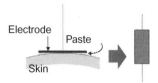

(b) Capacitively-coupled electrode

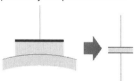

(c) Dry electrode

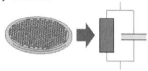

FIGURE 2   Three types of electrode and their electrical equivalent circuits. a) skin electrode, b) capacitively-coupled electrode, c) dry electrode.

resistor connecting skin with the signal cable, although the actual circuit is more complicated. For a capacitively coupled electrode, a capacitor is the coupling component, and for a dry electrode a parallel resistor and a capacitor comprise the coupling circuit. As shown in Figure 2(c), there are many small projections on the surface of the dry electrode to have better adherence because an electro-conductive paste is not used for this electrode. This dry electrode was developed to monitor a biosignal for a long time without any skin redness or skin irritation caused by the paste or drying of the paste itself.

An illustration of the operating principle of the capacitively coupled electrode is shown in Figure 3. As a capacitor in electromagnetism, the principle of this electrode is that charges are generated on the metal surface in response to the electrical fluctuations occurring at a distance. In practice, a capacitor can be assumed between metal plates and electrical body signal source such as the heart.

Figure 3(b) shows the equivalent circuit of a system consisting of body signal source V_{heart}, input impedance of a preamplifier, (resistance: R_{in}, capacitance: C_{in}), capacitance between a capacitively coupled electrode and signal source C_E, and capacitance C_G between body and ground. Then the voltage generated at the input of preamplifier V_{in} is calculated as

$$\frac{V_{in}}{V_{heart}} = \frac{1}{\frac{1}{j\omega C_E R_{in}} + \frac{C_{in}}{C_E} + \frac{1}{j\omega C_G R_{in}} + \frac{C_{in}}{C_G} + 1} \qquad (1)$$

Here, if the input resistance of preamplifier is large enough, and capacitance is small enough, i.e., $C_{in} \langle\langle C_E, C_G, R_{in} \gg 1$, then the voltage V_{in} becomes almost equal to the voltage of signal source V_{heart} as shown Eq. (2), and V_{heart} can be taken out as V_{in}.

$$\frac{V_{in}}{V_{heart}} \cong 1 \qquad (2)$$

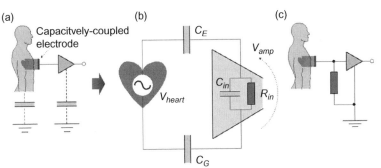

FIGURE 3 Measurement principle of the capacitively coupled electrode.

C_E : Capacitance between biosignal source and electrode
C_G : Capacitance between body, earth and circuit
C_{in} : Input capacitance of preamplifier
R_{in} : Input resistance of preamplifier
V_{heart} : Voltage of biosignal source
V_{in} : Voltage received by preamplifier

Input impedance and capacitance of the op-amp, which we used in our laboratory, are 10 TΩ and 1 pF. Such input impedance is orders of magnitude greater than that of a representative commonly available op-amp such as 741 (2 MΩ and 1.4 pF, respectively). Technologies have remarkably progressed and now allow for indirect measurement of bioelectric signals.

The limitation of the capacitively coupled electrode is that the static electricity generated by the friction between clothes and electrodes accumulate on the surface of the electrode, and the amplifier's output voltage is often saturated. To avoid this saturation, a resistor is inserted between the electrode and ground to release this static charge, as shown in Figure 3(c). This resistor breaks the relation shown in Eq. (2). Therefore, the selection of the resistor is rather complex. In addition, the capacitor formed between skin and electrode should be at least 30 pF based on our experiences or the biosignal will disappear in the environmental noise.

Figure 4 shows examples of ECG measured by capacitively coupled electrodes (red) and ECG measured simultaneously by skin electrodes (green); Figure 4(a) and (b) show the measurement with bipolar leads using a differential amplifier, and all electrodes are the capacitively coupled type. No ground electrode was used in the case of Figure 4(b). Figure 4(c) shows the measurement with a single electrode and a single-ended amplifier. While the signal in Figure 4(c) is noisy, the peaks of R-wave are still detectable. As shown in this figure, heart activity can be acquired by one metal plate set near the human body. From our experiences, a 1 mm layer of air between the skin and the electrode makes the measurement impossible, because the dielectric constant of air is slightly greater than that of vacuum or 1. However, measurement is possible when some dielectric material with a higher dielectric constant (such as cloth or plastic film) exists between the skin and the electrode. For example, we could measure ECG if a bundle of copy paper of 5.0 mm thickness is placed between the skin and the electrode.

Figure 5(a) shows a configuration of a system to measure ECG of a driver while he is driving. An electro-conductive cloth is placed on the surface of the driver seat as an

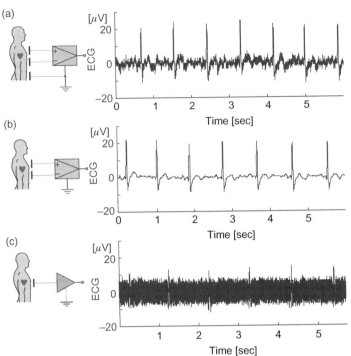

FIGURE 4   Examples of ECG measured by capacitively coupled electrodes.

indirect electrode, i.e., capacitively coupled electrode, and a metal band is attached around the steering as a ground electrode. Thus, in this system ECG is measured and amplified by a single-ended type amplifier while the subject is wearing normal clothing and driving a car. The seat surface is the place where the best contact between the skin and the electrode is expected while driving, and steering is the only place where direct contact between the body and electrode is expected. Moreover, the environment inside the car is noise-less electrically as in a shielded room, and a single-ended amplifier can be used. Figure 5(b) shows an example of ECG measured by this system. The middle plot is the original data and magnified plots of two periods are shown above and below. The top plot is the plot when the car was on a bumpy road, and the bottom plot is when the driver turned the steering wheel. The colored plot shows the processed data using digital band pass filter of 10 to 60 Hz. As shown in this plot ECG of a driver can be measured except at the moment when the driver changes his hand to grip the steering wheel. At this moment, noise contaminates the ECG data because the ground is unstable.

2.2 Biosignal Sensors for Cardiovascular System Monitoring

Sensors for measuring the change in physiological condition in daily life will be described in this section. A wide range of biosignals should be measured in "physiology," but not so many kinds of biosignals are necessary to be monitored in daily life. Here,

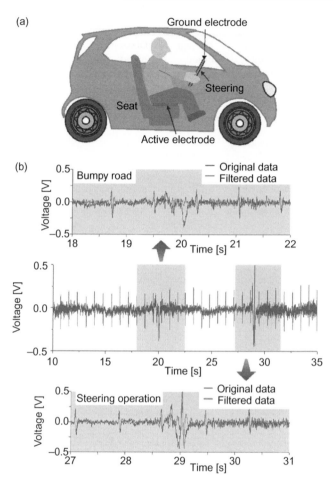

FIGURE 5   Non-contact and unconscious ECG monitoring system for car driver.

sensors for the measurement of blood pressure, pulse volume, blood flow, SPO2 (saturation of peripheral oxygen), respiration, and body temperature are described.

2.2.1 Measurement of Blood Pressure

Blood pressure, i.e., the pressure in the artery, pulsates periodically in response to the heartbeat. Hypertension is the cause of cardiovascular diseases, and is considered one of the three major underlying causes for stroke, together with diabetes and hypercholesterolemia. Therefore, the importance of measuring blood pressure has been recognized for a long time, and it is considered one of the four clinical vital signs. Blood pressure measurement is divided into two methods: the indirect method to measure the blood pressure without invading the body and the direct method to measure by inserting a catheter into an artery. The direct method is usually not suitable for routine blood pressure monitoring.

Table 3 shows the representative indirect blood pressure measurement methods. The principle of these methods is the same: the arterial pressure is measured by increasing and

TABLE 3   Various Indirect Blood Pressure Measuring Methods

Metrology	Designer	Systolic Blood Pressure	Average Blood Pressure	Diastolic Blood Pressure	Sphygmotono-gram
Cuff-oscillometric method: Detects blood vessel pulsation vibration that is propagated to the cuff	Marey, 1876	Estimable	Estimable	Non-applicable	Non-applicable
Palpatory method: Detects blood vessel pulsation of the distal side of cuff	Riva-Rocci, 1896	Measurable	Non-applicable	Non-applicable	Non-applicable
Flash method: Detects the color change of the skin	Gaertner, 1899	Measurable	Non-applicable	Non-applicable	Non-applicable
Auscultatory method: Detects a change in tone Korotkoff sound that occurs within the artery. Mercury sphygmomanometer.	Korotkoff, 1905	Measurable	Non-applicable	Measurable	Non-applicable
Oscillometric method: Detects vibration of the cuff instead of Korotkoff sound. Home blood pressure monitor	Mauck, et al. 1980	Measurable	Non-applicable	Measurable	Non-applicable
Ultrasound Kinetoarteriography: Measured by ultrasound changes in blood flow velocity	Ware & Laenger, 1967	Measurable	Non-applicable	Estimable	Non-applicable
Volume-Oscillometric method: Detection of volume pulse wave just below the pressurization portion	Yamakoshi, et al., 1982	Measurable	Measurable	Estimable	Non-applicable
Arterial Tonometry: Utilization of the pressure of equilibrium through the artery wall	Mackay, 1962	Estimable	Estimable	Estimable	Measurable
Volume-compensation method: Controlled to maintain a no load condition is always the artery wall	Yamakoshi, et al., 1980	Measurable	Measurable	Measurable	Measurable

decreasing the pressure of an inflatable cuff (Riva-Rocci cuff) wrapped around the arm. The first trial to measure blood pressure, which is omitted in the table, was conducted by E.J. Marey in 1876. He measured the systolic pressure by detecting the heartbeat transmitted through the cuff.

Figure 6 shows the measurement principle of the "volume compensation method." As shown in the figure, a cuff is also used in this method, but the cuff pressure generating circuit and a control circuit for precisely controlling the cuff pressure are different from the widely used auscultatory method. Blood volume in the finger is detected by the optical sensor and LED and the blood volume is kept constant by varying the cuff pressure, which should match the variation of blood pressure.

Figure 7 introduces a blood pressure measurement method using a pulse wave and ECG. The relationship between flow rate and its velocity is used as the operating principle

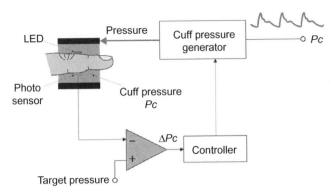

FIGURE 6 Measurement principle of continuous blood pressure changes due to volume-compensation method.

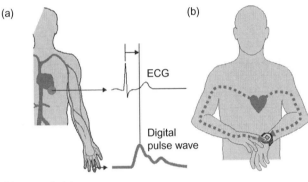

FIGURE 7 Estimation of blood pressure by simultaneous measurement of the pulse wave and ECG.

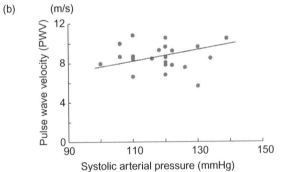

of this method, which is well known in fluid dynamics. For example, water flow rate in a water pipe is proportional to water pressure. Therefore, blood pressure can be estimated from the velocity of blood flow. Figure 7 shows a method for estimating blood pressure by simultaneous measurement of the pulse wave and ECG. As shown in Figure 7(a), blood-flow velocity can be known from the peak time of the R wave of ECG indicating the timing of blood flow from the heart ventricle, and the peak time of the pulse wave arrival time of the blood flow to the periphery. Figure 6(b) shows the relationship between the pulse wave propagation velocity and systolic blood pressure. As shown in this figure, the propagation velocity of the pulse wave increases according to the blood pressure. Figure 7(c) is a blood pressure measuring watch a Japanese company has been making commercially available. This watch is equipped with a pulse-wave sensor consisting of an optical sensor and light source, and also equipped with one electrode on the face and another on the back. ECG can be measured when the user puts his/her hand on the surface electrode.

However, there is a serious problem with this method. If the artery is modeled as a rigid pipe, it is possible to determine the absolute value of the systolic blood pressure. However, an artery is actually an elastic body, and its elastic modulus changes according to arterial sclerosis, which can significantly affect the blood flow velocity. This method can be used if there is little change in the artery characteristics due to aging and appropriate calibration is done for each user. This method is not very practical for daily monitoring, but it can be used for chronic monitoring when one is only interested in the changes in blood pressure and not its actual value.

2.2.2 Plethysmogram and Oximeter

The pulse wave is composed of a pressure wave generated by the contraction of heart ventricle and its reflection from the periphery, which is different from the blood flow. Blood-flow velocity in the aorta is about 1 m/sec, and the velocity of the pulse wave is about 10 m/sec, i.e., 10 times faster than blood flow itself. Figure 8 shows the configuration of a plethysmograph. As shown in this figure, the actual pulse wave is measured as the volume change of blood at the capillaries using near-infrared light. This figure demonstrates plethysmography at the fingertip, in which infrared light emitted by an LED is transmitted through the fingertip and received by a phototransistor on the opposite side. Because hemoglobin in the red blood cell absorbs infrared light well, a signal synchronized with the heartbeat can be obtained as output of the phototransistor (Figure 8). Here, P_1 is the pressure wave generated by the heartbeat and P_2 is its reflection wave from the periphery. Since the reflection wave P_2 becomes large with respect to the pressure wave P_1 as the arteries become hardened, the pulse wave is often used as an indicator of arteriosclerosis.

The measurement path commonly used for plethysmogram on fingertip is shown in Figure 8. Earlobe, forehead, and wrist are also used as measurement sites in adults, and toes, palms, and ankles are of interest for newborns.

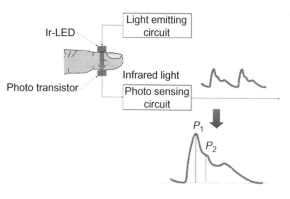

FIGURE 8 Measurement of photoelectric plethysmography.

(a)

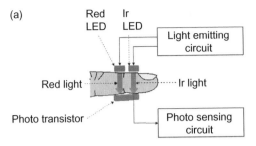

FIGURE 9   (a) Pulse oximetry and (b) absorption coefficient of oxygenated hemoglobin and hemoglobin for each light wavelength.

(b)

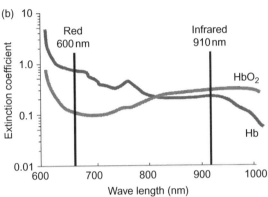

Plethysmography utilizes the fact that blood absorbs infrared light. How about light of a different wave length? Figure 9(b) shows the extinction coefficient of deoxyhemoglobin and oxyhemoglobin as a function of wavelength. Hemoglobin is a protein having an iron atom in the center, which carries oxygen and makes hemoglobin red. As shown in Figure 9(b), the extinction coefficients of deoxyhemoglobin and oxyhemoglobin are different. Pulse oxymetry was developed using this phenomenon to measure the oxygen saturation in the blood. In this system two sources of light of different wavelengths, i.e., infrared light of 910 nm and red light of 660 nm, are used. The oxygen saturation in the blood can be estimated from the difference in absorption of these two wavelengths. By looking at the ratio of the amplitudes of pulses at these two wavelengths, one can determine the relative oxygen saturation of the hemoglobin in the blood in the capillaries being illuminated. Pulse oxmetry is used for not only patient management in hospitals during surgery, or ICU, but also for home oxygen therapy and mountain climbing.

2.2.3 Respirometry

Respiration rate is clinically important as it is also one of the four vital signs. Breathing measurement or respirometry in daily life is classified by two different categories of measurement. One is the measurement in daily activities such as during working, and the other is the measurement of breathing during sleep. The respiration sensor is different according to the purpose of the measurement and where it is being done.

A respiration measurement band shown in Figure 10(a) and respiration pickup sensor in Figure 10(b) installed in the nostril are used for breathing pattern monitoring. The

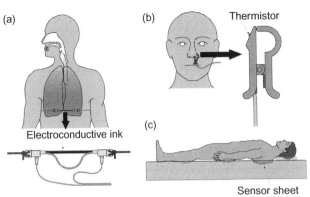

(a)

(b)   Thermistor

(c)   Sensor sheet

Electroconductive ink

FIGURE 10   Various respiration sensors.

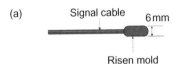

(a)   Signal cable   6 mm

Risen mold

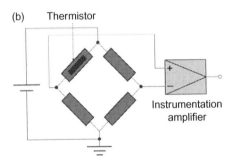

(b)   Thermistor

Instrumentation amplifier

FIGURE 11   Thermistor sensor and its signal conditioning circuit.

respiration measurement band is composed of an elastic tube filled with conductive ink and a string, and it is wrapped around the rib cage to record changes in the chest circumference due to breathing. A thermistor is set at the tip of the nasal respiration sensor. The thermistor senses the change in the temperature due to air intake and exhalation. Both sensors measure the actual respiration waveform from which the respiration rate and apnea can be determined. Therefore, these sensors are popular in daily monitoring. A biometric sensor sheet shown in Figure 10(c) was developed to monitor the respiration during sleep. Respiration monitoring during sleep should be unrestrained and unconscious and the sensor should be soft, flexible, and can be installed in a pillow or under the sheets. Details of this sheet sensor will be described in the next section.

2.2.4 Measurement of Body Temperature

Temperature is the third of the four vital signs in clinical medicine. Temperature sensors are classified into contact type, e.g., thermistor and thermocouple, and non-contact type, which senses infrared light emitted from the body surface based on the principle of blackbody radiation. Figure 11 shows a thermistor most popular in body-temperature measurement. The thermistor utilizes the property that resistance of semiconductor or metal oxide changes with temperature. As shown in Figure 11(a), the thermistor is small, resin-molded, and inexpensive, and hence thermistors are commonly used to measure core body temperature by rectal placement during routine activities of human subjects. A bridge circuit shown in Figure 11(b) is commonly used for temperature measurement by a thermistor.

The core body temperature in humans is different from body surface temperature. Human beings are homeothermal animals, and the core temperature is controlled to be nearly constant. On the other hand the skin surface temperature varies greatly depending on the environment. Therefore, core temperature is required to be measured in daily monitoring. Rectal temperature is representative of the core temperature, but this measurement is not comfortable for long-term observation. The eardrum temperature, gullet temperature, sublingual temperature, and axillary temperature are often selected as related to core temperature. However, considering temperature monitoring in daily life, the gullet, sublingual, and axillary temperature measurements are invasive and sensitive to body motion. Hence, an earphone-type eardrum temperature monitoring device is now available commercially.

Figure 12 shows another representative temperature sensor, i.e., a thermocouple and a thermopile, which is an application of several thermocouples. As shown in Figure 12(a), when both ends of two different kinds of metal wire are connected, thermally a small voltage is generated by the Seebeck effect when a temperature difference exists between two connected ends of wires. This is the measurement principle of the thermocouple. However, this sensor is difficult to use in daily monitoring. Figure 12(b) shows the structure of a thermopile, in which about 100 thermocouples are sequentially set to attain high sensitivity to small increases of temperature at the hot junction. This sensor is small, lightweight, low power, and is suitable for daily monitoring. Figure 12(c) shows a wearable body temperature monitoring device that senses the tympanum temperature by a thermopile set inside.

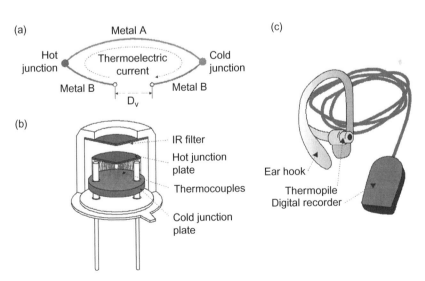

FIGURE 12   Thermocouple and thermopile.

2.3 Biosignal Sensors for Sleep Monitoring

Human beings spend one-third of each day sleeping, and sleep is essential for people to maintain healthy physical functions. Daily long-term monitoring of physical condition during sleep and sleep quality is important for health management. This section introduces the ballistic cardiogram, non-contact ECG, sheet type respiratory and body motion sensors, and video camera monitoring.

2.3.1 Ballistic Cardiogram

A number of methods have been proposed for the non-restrictive measurement of the resting heart rate in a lying position, along with detection of resting heart rate and respiration by changes in electrical capacitance. In recent years, many new sensors have been developed to offer comparatively easy operation and installation. Methods developed to date include: non-restrictive measurement of respiratory and heart rates from the internal pressure changes in an air mat affected by body motion due to respiration or heart activity, and non-restrictive measurement of respiratory and heart rate using sheets or pillows with a wire-shaped respiratory pickup sensor attached to an air mat. These methods are known as "ballistocardiography". Ballistocardiography is a method of measuring delicate body motions caused by heart activity and blood movement in blood vessels. In general, the motion transmitted to the back of a subject in a supine position is measured by ballistocardiography as shown in Figure 13. The basic problem with ballistocardiography is that it is influenced by body weight and motion, movement of organs inside the abdomen, and the distortion of fat and musculature.

2.3.2 Non-Restrictive Heart-Rate Measurement

Long-term monitoring of heart rates is important for health management. Heart rate is the fourth clinical vital sign. An analysis of heart-rate variability can facilitate the early discovery of a variety of illnesses and health conditions. For example, high blood pressure, diabetes and obesity are accompanied by a decrease in heart-rate variability, while psychological stress, lack of exercise and other factors are also known to affect heart-rate variability. Research has also been conducted in relation to the prognosis of heart diseases, particularly following myocardial infarction. In many cases, researchers analyzed heart-rate variability by using the resting heart rate because the resting heart-rate variability can be an index of sympathetic or parasympathetic dominance. To analyze heart rate for such

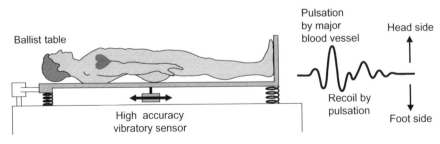

FIGURE 13   The basis of the ballistocardiogram (BCG).

(a)

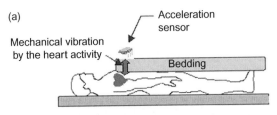

FIGURE 14   An outline of the experimental motion heartbeat detection system in a down quilt.

(b)

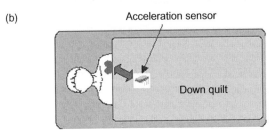

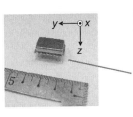

Data item	Value
Range of detection	$\pm 1\,ms^{-2}$
Sensitivity of detection	$1.9\,V/ms^{-2}$
Response frequency	$0.8\sim1000\,Hz$
Resonance frequency	$10\sim15\,kHz$
Size	W20.5*D12.5*H5 mm
Electrical requirements	0.5 mA
Weight	4 gr

FIGURE 15   Characteristics of acceleration sensor and the specifications of the acceleration.

prognosis of lifestyle-related diseases, we should measure the long-term resting heart rate. In this section, we show a non-restrictive measurement method using only an acceleration sensor set inside a down quilt. This method was easy for use at home.

We introduce the system to achieve the method of non-restrictive measurement of resting heart rate in a lying position. Figure 14 shows an outline of the experimental system. A small acceleration sensor is installed in the middle layer of a down quilt. When the subject lies down and is covered with the down quilt, the minute mechanical vibrations resulting from the subject's heart activity are transmitted to the down quilt, and the acceleration sensor picks up this vibration. The acceleration sensor is sewn into the down quilt directly above the left side of the anterior chest, which is the location of the heart. This system employs a piezoelectric acceleration sensor and was set as its sensitivity axis (z-axis) oriented to the Earth's gravity. As shown in Figure 14, the sensitive axis of the acceleration sensor installed in the down quilt is always perpendicular to the surface of the quilt, so the sensitive axis of the acceleration sensor still faces the heart when the subject is in a lateral position. The sensitivity of measurements using this experimental system depends more on the distance from the heart to the sensor than on the subject's position.

An image of the acceleration sensor and its features are shown on the left half of Figure 15. The right-hand side of Figure 15 shows the specifications of the piezoelectric acceleration sensor (MA3-01Aa, Micro Stone Co., Ltd.). This table indicates that the sensor

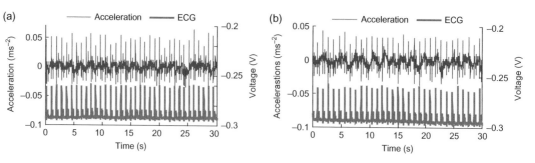

FIGURE 16   R-R interval in 5 minutes in (a) supine position and (b) lateral position.

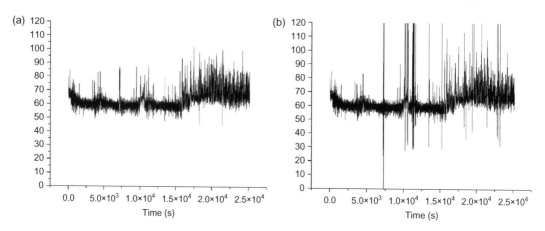

FIGURE 17   Results of R-R interval through the night (a) by ECG measurement and (b) acceleration sensor.

has a built-in amplifier and a comparatively high-sensitivity output of 1.9 V per 1 ms^{-2} of acceleration. This sensor also features high linearity and a response frequency over 0.8 Hz, so it prevents the heart activity vibrations from overlapping with the vibration caused from respiration. Furthermore, the sensor is small and light, so even when installed in the down quilt, it did not disturb the subject, and a low-consumption current of 0.5 mA allows for long-term measurements under battery power. The original resonant frequency is between 10 and 15 KHz, which is sufficiently high in comparison to the frequency of mechanical vibrations generated by the heart's activity. Heart activity data obtained with this acceleration sensor was taken at 200 Hz by a 12-bit A/D conversion board (PCI-3153, Interface Corp.).

Figure 16 shows an example of a detected heart-rate signal in a supine position by using the system and ECG for 30 seconds. As shown in this graph, high-frequency noise overlaps the heart-rate signal by the experimental system, but the peak is obtained at the same position as the ECG-R wave.

Using this method makes it possible to achieve non-restrictive heart-rate measurement throughout the night. Figure 17 shows the calculated R-R interval during sleep.

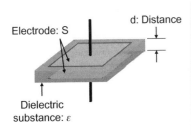

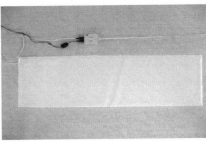

FIGURE 18 An outline of a capacitive mat-type sensor.

2.3.3 *Sheet Type Respiratory and Body Motion Sensor*

From the standpoint of ensuring the safety of patients in hospitals, a system to monitor the condition of sleep in patients is becoming necessary. Sleep-monitoring systems have been developed in recent years that are useful for managing physiological conditions by measuring the respiratory movement and heart rate of sleeping patients using highly sensitive pressure sensors placed on the bed. These technologies are extremely useful as monitoring systems because they do not restrict the patient during sleep.

In this section, a sheet-type sensor for measuring heart rate, respiratory movement, and body movement based on electrostatic capacitance is introduced. The outline of this sheet-type sensor is shown in Figure 18. A capacitor is formed with a conducting electrode on both sides of an insulating sheet. Displacement of the minute pressure, which arises from a human's heartbeats, respiratory movement, and body movement, is changed into the capacitance change between the electrodes. Here, ε is dielectric constant, area of polar plates is S, distance between polar plates is set to d, and capacity value is set to C in Figure 18, as shown in Eq. (3). If the area of an electrode is constant, the capacity value C will change with the distance d between polar plates:

$$C = \frac{\varepsilon \cdot S}{d} \tag{3}$$

Body movement and respiratory movement are measurable as a comparatively big displacement. However, the displacement by a heartbeat is very minute, and it is difficult to detect heartbeat signal. In order to detect the capacitance change by the minute displacement originating from a heartbeat with sufficient accuracy, the LC oscillation-type detector circuit was used. This circuit changes minute vibrations into the frequency domain. Figure 19 shows an example of a detected heart-rate signal and respiratory movement.

2.3.4 *Video Camera*

When wearing some sensors, even adults feel constrained and tied-up, and for children it is a cause of significant stress. Therefore, it was necessary to develop new measurement and assessment systems to enable unconscious monitoring of daily sleep. As one such approach, we developed a method to monitor sleep using a video camera and image analysis. This method monitors sleep stage with only body movement obtained from video

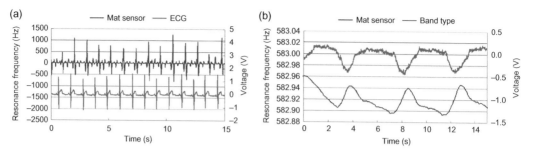

FIGURE 19   Measurements of (a) heart rate and (b) respiratory movement.

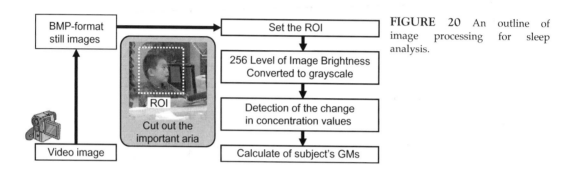

FIGURE 20 An outline of image processing for sleep analysis.

images recorded. With this method, it is possible to conduct completely non-contact monitoring and estimate all-night sleep stages such as awake, slow-wave sleep (SWS), and rapid eye movement (REM) sleep. This method focused on the strong relationship between sleep stage and body movement. The body movements during sleep were divided into the following three states by Fukumoto et al.: twitch movements (TMs), localized movements (LMs), and gross movements (GMs). Among these three movements, GMs indicate the body movements that occur for a longer time period, such as turning over in bed, which usually continues for over 2 sec. GMs are most directly related to the sleep-wake cycle.

Figure 20 shows the outline of the video processing. The video output was converted to BMP-format still images and analyzed. As mentioned above, GMs continue for over 2 sec. Therefore, 1 fps was considered to be an appropriate value for the frame rate. The region of interest (ROI) was specified for high-speed processing. The ROI is used to distinguish the necessary region from the overall image. ROI processing enables high-speed processing by eliminating excess noise and decreasing the image size. After ROI processing, the image was converted to gray scale, and the changes in the gray values between frames were detected using image processing.

GMs are thought to be inhibited during SWS. Therefore, we predicted that SWS could be classified by extracting the continuous time of non-movement. Regarding the GM data, a "section" of body movement is defined as the interval from the period when GMs are inhibited, then activated, then inhibited again. In each section, the time at which activated

GMs are inhibited is t_0, the time at which GMs are activated from the inhibited state is t_{MTin}, and the time at which the inhibition of GMs is restarted is t_{MTout}. As such, the continuous time of non-movement, at time t, in each section is expressed as

$$MT_{rest} = \begin{cases} t_{MTin} - t_0, |t_0 \le t \le t_{MTin} \\ 0, |t_{MTin} \le t \le t_{MTout} \end{cases} \tag{4}$$

The wake stage could be extracted by considering parameters related to the amount of GMs when they were not controlled. Based on the notion that sleep stage is correlated with GMs immediately before the stage, the simple moving average of a 60 sec interval could be calculated. The reference section for the moving average was set as 60 sec immediately before the section because we intended for at least one epoch to be included.

The linear discriminant function method was utilized to estimate the sleep stage from body movements during sleep using video image processing. Linear discriminant analysis can be used to distinguish no fewer than two groups of data clusters. In the current analysis, the independent variables were the continuous time of non-movement and the moving average amount of body movement. The response variables were the three clusters wake, light and REM, and SWS.

Eq. (5) and Eq. (6) define the linear discriminant function. Here, x_1 and x_2 represent the continuous time of non-movement and the moving average amount of body movement, respectively. We are able to distinguish the three groups by solving for the coefficients a_1, a_2, a_3, a_4, and the constants $const_1$, $const_2$. Moreover, Eq. (7) and Eq. (8) define the linear discriminant function. Eq. (6) distinguishes the discriminant functions of wake and categories other than wake, while Eq. (8) is the discriminant function that distinguishes SWS and categories other than SWS. In these equations, z_1 and z_2 are the linear discriminant scores, and the division of categories is determined by the positive and negative numbers of z_1, z_2:

$$0 = a_1 x_1 + a_2 x_2 + const_1 \tag{5}$$

$$0 = a_3 x_1 + a_4 x_2 + const_2 \tag{6}$$

$$z_1 = a_1 x_1 + a_2 x_2 + const_1 \tag{7}$$

$$z_2 = a_3 x_1 + a_4 x_2 + const_2 \tag{8}$$

Below we explain the calculation method for a_1, a_2, and $const_1$ in Eq. (5). As demonstrated in Eq. (9) and Eq. (10), we calculate the total sum of squares S_T of the difference between the discriminant score z_1 and the average z_{all} of all discriminant groups, as well as the total sum of squares S_{wake} of the difference between the discriminant score z_1 and the average z_{wake} of the discriminant group wake. Parameter n_{all} indicates the number of the linear discriminant scores of all discriminant groups, while n_{wake} indicates the number of linear discriminant scores of the discriminant group wake. Moreover, S_T indicates all variations within the data, while S_{wake} indicates the variations among the groups of the discriminant group wake:

$$S_T = \sum_{i=0}^{n_{all}} (z_i - \bar{z}_{all}) \tag{9}$$

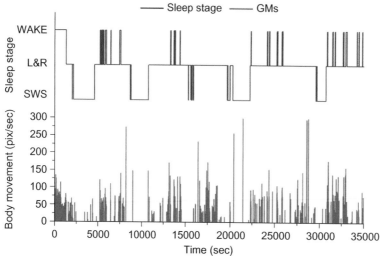

FIGURE 21 Sleep stages (wake, light and REM, and SWS) determined by polysomnogram and the gross body movement determined by difference image processing.

$$S_{wake} = \sum_{j=0}^{n_{wake}} (z_j - \bar{z}_{wake}) \tag{10}$$

The value $F(a_1, a_2)$ is the variation rate, which is determined by dividing S_{wake} (the variation among the group of the discriminant group wake) by the total variation S_T. This is demonstrated in Eq. (11). To divide the discriminant group wake and other groups optimally, it is necessary to calculate a_1, a_2, which will determine the maximum value of the variation rate $F(a_1, a_2)$. Therefore, as indicated in Eq. (12) we intend for a_1, a_2 to give us the value 0 when $F(a_1, a_2)$ has been partially differentiated by a_1, a_2, respectively. The calculation methods for a_3, a_4, and $const_2$ are the same:

$$F(a_1, a_2) = S_{wake}/S_T \tag{11}$$

$$\begin{cases} \partial F(a_1, a_2)/\partial a_1 \\ \partial F(a_1, a_2)/\partial a_2 \end{cases} \tag{12}$$

Figure 21 shows an example of an estimated sleep stage determined by a polysomnogram and linear discriminant.

2.4 Wearable Sensors for Physical Activity Measurement

Physical activity in daily life is the most preferable monitoring aspect because it is the nature of human beings to move and our daily activities are closely related to our health. There are many kinds of motion sensors to measure physical activities, such as

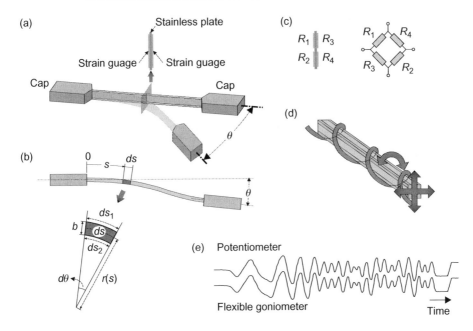

FIGURE 22 Flexible goniometer.

accelerometers, gyrosensors, goniometers, GPS sensors, etc. However, these sensors are discussed in other chapters of this book. Therefore, three new topics are introduced in this chapter. The first topic is "flexible goniomer," the second is "joint motion measurement using two accelerometers set both near sides around the joint," and the last is "flexible force sensors during walking."

2.4.1 *Flexible Goniometer for Articular Motion*

Morimoto developed a flexible goniometer as an articular motion sensor that does not make use of a rotary encoder to measure human articular motion shown in Figure 22(a). Figure 22(b) illustrates the configuration and the operating principles of a single-axis flexible goniometer. As shown in this figure, the goniometer has a strain gauge attached to both sides of the flexible plate between the end blocks. It generates a value that depends on the angle θ between the two end blocks. The operating principles of the flexible goniometer are illustrated in Figure 22(c). When the angle between the end blocks at both ends is θ, and when the locus at a distance s is considered to be ds, the curvature radius will be $r(s)$ and the central angle will be $d\theta$. During extension and contraction, ds_1, ds_2, and ds for the top, bottom, and center of this arc are given by

$$\begin{cases} ds_1 = (r(s) + b/2) \cdot d\theta \\ ds = r(s) \cdot d\theta \\ ds_2 = (r(s) - b/2) \cdot d\theta \end{cases} \tag{13}$$

The extensions ε_1, ε_2 for the top and the bottom are given by

$$\begin{cases} \varepsilon_1 = \dfrac{ds_1 - ds}{ds} = \dfrac{b}{2r(s)} \\[2ex] \varepsilon_2 = \dfrac{ds_2 - ds}{ds} = -\dfrac{b}{2r(s)} \end{cases} \tag{14}$$

Accordingly, the extension and the contraction $\Delta\ell_1$, $\Delta\ell_2$ for the measured part, in its entirety, are given by

$$\begin{cases} \Delta\ell_1 = \int_0^\ell \varepsilon_1 ds = \int_0^\theta \dfrac{b}{2r(s)} r(s) d\theta = \dfrac{b}{2}\theta \\[2ex] \Delta\ell_2 = \int_0^\ell \varepsilon_2 ds = \int_0^\theta -\dfrac{b}{2r(s)} r(s) d\theta = -\dfrac{b}{2}\theta \end{cases} \tag{15}$$

The extension $\Delta\ell$ for the measured part, in its entirety, is found to be proportional to the angle θ, formed by the fixed ends at both the ends, for both the top and bottom sides. However, in this form, the extension and the contraction of the fixed ends, i.e., the extension and the contraction of the measured part in its entirety, can be mistaken to be a change in the angle θ. Moreover, as longitudinal twists in the measured part can also be a cause for miscalculation of the angle, an accurate measurement of the angles can be made by setting up a bridge circuit using four strain gauges, as shown in Figure 22(d). The use of the flexible goniometer can be extended to measure the bi-axial rotational angles by attaching strain gauges to the top and bottom sides, as well as the left and right sides of the flexible square rod, as shown in Figure 22(e).

2.4.2 Joint Motion Measurement Using Two Accelerometers Set Near Both Sides of the Joint

The last topic is a unique joint motion monitoring method using accelerometers, which we developed previously. Figure 23(a) is the schematic representation of the principle and the equipment to measure the angle change of a one-axis joint, such as the elbow. In this method, two accelerometers are set on both sides of the joint and as close as possible ("both-near-sides") to its joint axis.

In this case, it is assumed that the acceleration $\mathbf{a_1}$ loaded on the accelerometer 1 is largely similar to acceleration $\mathbf{a_2}$ loaded on the accelerometer 2 under the below

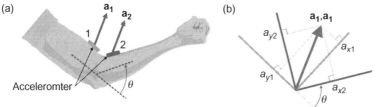

FIGURE 23 Principle and equipment to measure one axis joint motion.

conditions; 1) the rotation radius around the joint is small, and the centrifugal force and the turning force loaded on sensor 1 can be regarded as zero, and 2) when this system including two limbs and its joint rotates on other center, these two sensors receive the same magnitude of centrifugal force and turning force.

In Figure 23, the x-axis and y-axis components of acceleration $\mathbf{a_1}$ are defined as (a_{x1}, a_{y1}), those of acceleration $\mathbf{a_2}$ as (a_{x2}, a_{y2}), and the joint angle as θ. Using the rotation matrix of θ, the relation between (a_{x1}, a_{y1}) and (a_{x2}, a_{y2}) can be described by

$$\begin{pmatrix} a_{x1} \\ a_{y1} \end{pmatrix} = \begin{pmatrix} \cos\theta & -\sin\theta \\ \sin\theta & \cos\theta \end{pmatrix} \begin{pmatrix} a_{x2} \\ a_{y2} \end{pmatrix} \tag{16}$$

Thus, the joint angle θ can easily be solved as follows:

$$\tan\theta = \frac{a_{x2} \cdot a_{y1} - a_{x1} \cdot a_{y2}}{a_{x1} \cdot a_{x2} - a_{y1} \cdot a_{y2}} \tag{17}$$

Figure 24 shows the method to measure three-axis joint motion, such as a shoulder joint. In the same manner as the monitoring of one-axis joint, two accelerometers 1 and 2 are set at "both-near-sides" around the joint. The accelerometer set on the body is adjusted so that the z-axis is in the vertical direction, the y-axis in the longitudinal direction and x-axis in the horizontal direction. The accelerometer set on the upper limb is adjusted so that the x-, y- and z-axis correspond to that of the accelerometer on the body.

The components on the x- , y- and z-axis of the acceleration measured by the accelerometer mounted on the shoulder are defined as (a_{x1}, a_{y1}, a_{z1}) and that on the upper limb are defined as (a_{x2}, a_{y2}, a_{z2}). In this case, the acceleration loaded on the accelerometer mounted on the shoulder is much the same as that in the upper limb in the same manner as the monitoring of a one-axis joint. Using the rotation matrix R_{xyz}, the relation between the components (a_{x1}, a_{y1}, a_{z1}) and (a_{x2}, a_{y2}, a_{z2}) is described by

$$\begin{pmatrix} a_{x1} \\ a_{y1} \\ a_{z1} \end{pmatrix} = R_{zyx} \begin{pmatrix} a_{x2} \\ a_{y2} \\ a_{z2} \end{pmatrix} \tag{18}$$

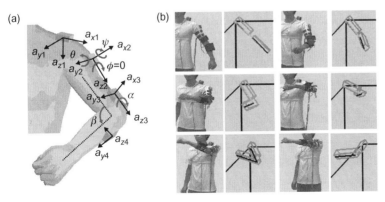

FIGURE 24 (a) Principle and equipment to measure three axis joint motion and (b) example of upper-limb motion monitoring.

Here the rotation matrix R_{xyz} is given by

$$
\begin{aligned}
R_{xyz} &= \begin{pmatrix} 1 & 0 & 0 \\ 0 & \cos\psi & -\sin\psi \\ 0 & \sin\psi & \cos\psi \end{pmatrix} \begin{pmatrix} \cos\theta & 0 & -\sin\theta \\ 0 & 1 & 0 \\ \sin\theta & 0 & \cos\theta \end{pmatrix} \begin{pmatrix} \cos\varphi & -\sin\varphi & 0 \\ \sin\varphi & \cos\varphi & 0 \\ 0 & 0 & 1 \end{pmatrix} \\[2mm]
&= \begin{pmatrix} \cos\varphi\,\text{cod}\,\theta & \cos\theta\sin\varphi & \sin\theta \\ -\sin\psi\sin\theta\cos\varphi - \sin\varphi\cos\psi & -\sin\theta\sin\psi\sin\varphi + \cos\psi\cos\varphi & \sin\psi\cos\theta \\ -\sin\theta\cos\psi\cos\varphi + \sin\psi\sin\varphi & -\sin\varphi\sin\theta\cos\psi - \cos\varphi\sin\psi & \cos\theta\,\text{cod}\,\psi \end{pmatrix}
\end{aligned}
\tag{19}
$$

However, angle ψ, φ, θ can't be calculated from this equation, because this equation is indeterminate. To get the solution of equation (18), one remarkable feature of the shoulder joint should be considered, i.e. the skin of the arm, where the accelerometer 2 is set, shows almost no motion when the subject rotates his arm around the z-axis. Therefore, the rotation angle φ can be regarded as zero, and the rotation matrix R_{xyz} can be modified as shown in equation (20), and the flexion/extension angle ψ and abduction/adduction angle θ can easily be obtained from this equation (18).

$$
R_{xyz} = \begin{pmatrix} \cos\theta & 0 & \sin\theta \\ -\sin\psi\sin\theta & \cos\psi & \sin\psi\cos\theta \\ -\sin\theta\cos\psi & -\sin\psi & \cos\theta\cos\psi \end{pmatrix}
\tag{20}
$$

On the other hand, the rotation angle φ can be decided as the twist angle α of the upper arm, which is calculated from outputs of accelerometer 2 and 3. Of course the flexion/extension angle β of the elbow can be easily calculated as mentioned before.

Figure 25(b) shows an example of upper limb motion monitoring. In this case a series of limb motions are measured, calculated, and reproduced in real time on a personal computer display as a stick figure picture.

2.4.3 Force Monitoring During Walking

In kinematics, which explains the human movement and motion from a mechanical standpoint, the external forces acting on the subjects need to be found. While walking, the external force of the ground acts on a person through their feet. This force is called the ground reaction force. This ground reaction force must be measured for the purpose of the kinematic gait analysis. However, a typical ground reaction force gauge is set up on the floor and cannot measure the ground reaction force in the course of daily life. To measure the load on the soles of the feet while measuring the daily behavior, sensors need

(a)   (b)

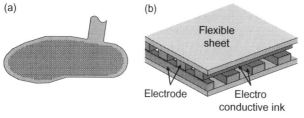

FIGURE 25   Pressure distribution under foot sensor.

to be fitted to measure the external forces on footwear such as shoes or slippers. To that end, we have developed a foot sole pressure distribution sensor, as illustrated in Figure 25. For these sensors, the electrode bands are printed on the top and the bottom of a flexible sheet, so that they intersect with each other. A conductive ink is applied on the surface of all the electrodes shown in Figure 25(a). When pressure is applied, the ink is compressed and the resistance value for the conductive ink in that area drops. This is measured as the variation in resistance between grid ends and can be obtained for each grid point as pressure values.

3. SUMMARY

It is seen that there can be a wide variety of sensors that can pick up clinical vital signs as well as other physiological and physical signals from the human body. In most examples, these sensors were wearable or non-obstructably located in the environment of the human subject. At the beginning of this chapter, we asked if such sensing systems needed to be wearable. There is no unique answer to such a question. It all depends on how the information from the sensor, or more correctly the group of sensors, is to be used. How much will the instrumentation affect the behavior and physiology of the subject? In answering such a question, we need to remember the fundamental principle of physiological measurement: make sure that the instrumentation doesn't affect the system being measured or at least only minimally affects it.

Wearing Sensors Inside and Outside of the Human Body for the Early Detection of Diseases

Carmen C.Y. Poon[1,2], *Yali Zheng*[2], *Ningqi Luo*[2], *Xiaorong Ding*[2], *and Yuan Ting Zhang*[2,3]

[1]Department of Surgery, The Chinese University of Hong Kong, Hong Kong SAR, China, [2]Department of Electronic Engineering, The Chinese University of Hong Kong, Hong Kong SAR, China, [3]Key Laboratory for Health Informatics of Chinese Academy of Science (HICAS), Shenzhen, China

1. INTRODUCTION

Advancing health informatics has been identified by the U.S. National Academy of Engineering as one of the fourteen grand engineering challenges of the twenty-first century, with wearable sensors and devices being highlighted as one of the key technologies in this field [1]. Wearable sensors and devices can be used to monitor chronic diseases continuously for early detection of symptoms as well as to capture acute and transient features that are difficult to be picked up during infrequent and ad hoc hospital visits.

The most classical example of wearable sensors for disease diagnosis is the Holter monitor or ambulatory electrocardiogram (ECG) monitoring device, which captures a subject's electrical activities of the heart continuously over a 24-hour or even longer periods [2]. These systems have huge clinical applications since heart abnormalities are often paroxysmal and asymptomatic, and therefore, short-time and sporadic recording of ECG is often inadequate for detecting symptoms, while longer term ECG monitoring over several hours or even several days will provide better and accurate diagnosis. Since the Holter monitor was developed over 50 years ago, wearable sensors are advancing at a tremendously fast speed, demonstrating their potentials in various medical fields. In this chapter, recent advancements in wearable sensors for cardiovascular, neurological, and gastrointestinal diseases will be reviewed and discussed. In particular, the last section of this chapter outlines a promising future direction of wearable sensors, which by using recent flexible technologies, sensors can be worn non-invasively inside the human body for sensing information and dispensing drugs in the gastrointestinal (GI) tracts.

2. CARDIOVASCULAR DISEASES

Cardiovascular diseases (CVDs) are the leading cause of death worldwide. About 17.3 million people died from CVDs in 2008, representing 30% of global deaths. Over 80% of deaths resulting from CVDs occurred in developing countries. It is anticipated that the number of people who die from CVDs will increase to 23.3 million by 2030, as reported by the World Health Organization [3]. The social and economic burden to society resulting from CVDs is huge worldwide.

CVDs are a collection of disorders of the cardiac system and blood vessels, including coronary heart disease, atrial fibrillation, heart failure, and peripheral arterial diseases. CVDs are mainly attributable to conventional risk factors, such as unhealthy diet, physical inactivity, tobacco use, and harmful use of alcohol [4]. These factors may present as high blood pressure (BP), raised blood glucose, and elevated blood lipids in the individuals. In particular, high BP, which is also known as hypertension, has been identified as a major dominant risk factor of CVDs, affecting 1 billion people globally [5]. One of the difficulties in controlling high BP is it usually develops without obvious symptoms or warning signs, hence people are usually oblivious and unaware of their high BP until severe symptoms appear. Hypertension is therefore known as the silent killer. In addition to absolute BP, BP variability within 24 hours has been reported to be an independent predictor of incidence of cardiac events [6]. Hence, long-term and continuous monitoring of BP can bring new insights into the cause of sudden cardiac deaths.

Prevention of CVDs partly relies on whether these abovementioned risk factors can be identified and controlled at an early stage. Wearable sensors, if designed to be worn unobtrusively, will encourage individuals to use them more frequently and will therefore increase the chance of detecting symptoms at an earlier stage. They also have the ability to disclose dynamic changes in physiological status and sense and monitor health status of individuals remotely.

2.1 Monitoring Risk Factors of Cardiovascular Diseases

In this subsection, monitoring of BP will be used as an example for discussion. It is now widely accepted that 24-hour ambulatory monitoring gives a better prediction of risk than office measurements and is useful for diagnosing white-coat hypertension, a phenomenon in which elevated BP is observed during clinical visits, but not in other settings [7]. Furthermore, continuous BP measurements have the advantage of detecting BP variability. Although the volume-clamp method has been used for beat-to-beat non-invasive BP monitoring, its cuff-based design prevents it from being used over extensive periods and is considered undesirable. Alternatively, a cuff-less BP measurement approach based on the measurement of pulse transit time or pulse arrival time has been proposed as a method for measuring BP without a cuff [8,9]. The method utilizes the fact that the transmission speed of the pulse along the arteries is related to pressure-dependent mechanical properties of the artery wall and therefore can be used as an estimate of BP. As shown in Figure 1, the sensors required for this method can be designed into a wearable garment or armband for the measurement of blood pressure [10,11].

In addition to using wearable sensors to monitor the cardiovascular parameters, there are also wearable sensors developed for fitness control and smoking supervision to alert the user to change his lifestyle as obesity and smoking are two major risk factors of CVDs. The mobile personal trainer (MOPET) system [12] has been investigated to supervise a physical fitness activity with motivation and health advice for the user. Lopez et al. [13] have recently developed an automatic wearable cigarette tracker. The tracker can detect smoking events through monitoring the cigarette-to-mouth hand gestures and recognize characteristic patterns of respiration during smoke inhalations. The system is flexible and non-invasive with applicability in free-living conditions over extended periods of time.

To summarize, wearable systems and related wireless technologies can offer the optimal solution for detecting the key risk factors of CVDs because they can provide continuous service almost anytime at anyplace with convenience, comfort and low cost, and summon medical assistance when required.

2.2 Diagnosis of Cardiovascular Diseases

Given the transient nature of symptoms of CVDs, wearable sensors and systems are also useful for improving the diagnosis of CVDs. Atrial fibrillation (AF) is the most common type of cardiac arrhythmia, caused by rapid, disorganized electrical impulses of the heart. AF can lead to stroke and heart failure. The incidence of AF is higher in elderly adults, and is often a complication after cardiac surgery without accompanying symptoms [14]. AF can be diagnosed from ECG, usually with the absence of P waves, rapid heart rate and irregular

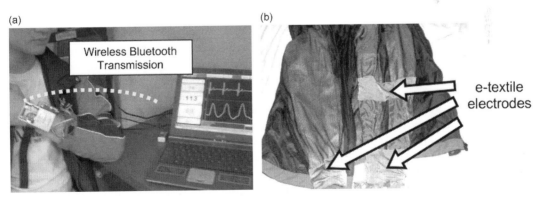

FIGURE 1   Wearable garments for pulse-transit-time-based method for cuff-less blood-pressure monitoring [10,11].

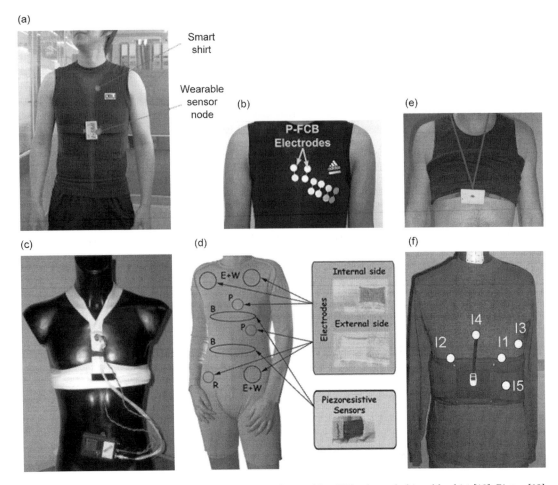

FIGURE 2 Wearable garments for ECG monitoring. Smart shirts [11], planar-fashionable shirt [18], Biotex [19], Lobin [21], Wealthy [22], and Protex [23].

heart rhythms. The incidence, time, symptoms, and risk factors of AF can be monitored and analyzed by using a wearable cardiac event recorder [15]. Equipped with an automatic warning expert system, it can activate the emergency medical alarm system, and thus allow pre-emptive actions to be taken with an aim of preventing sudden deaths.

In addition to the diagnosis of AF, ECG can also provide important information for diagnosing and assessing other major risk factors of chronic cardiac diseases. Wearable systems for collecting ECG have been developed from e-textile materials, e.g., in the form of smart shirts [11]. The most important components of wearable monitoring systems are the wearable sensors, which are attached to the individual by integration to the wearable garment [16]. The fabric-based active electrodes have been developed to embed into clothing for ECG monitoring [17], e.g., based on a compact planar-fashionable circuit board [18]. Figure 2 shows a number of wearable ECG systems that have been developed, e.g.,

Biotex [19], Smart Vest [20], Lobin [21], Wealthy [22], and Protex [23]. Many of these systems have integrated sensor design to allow them to measure not only ECG but also other physiological signals such as photoplethysmogram (PPG) and respiration.

Coronary heart disease (CHD), also known as coronary artery atherosclerosis, is caused by the blockages of blood vessels of the heart resulting in chest pain due to a lack of blood supply to the heart, i.e., angina. With uncontrolled blockage, parts of the heart may die, resulting in myocardial infarction (MI). Although many people with CHD have symptoms such as chest pain or shortness of breath, CHD can also be asymptomatic and can lead to sudden deaths as a result of rupture of atherosclerotic plaque. Major risk factors of CHD include age, high BP, smoking, and diabetes. Identification of individuals subject to high risks of CHD helps to create management plans for controlling the risk factors for early intervention. It has been reported that the baseline ECG abnormalities [24] and pulse pressure [25] can be used for screening risks of CHD.

Peripheral arteries disease (PAD) is a slowly progressive disease mainly due to atherosclerosis and further affects vessels such as the aorta and arteries of the lower extremities. PAD can lead to CAD, stroke, MI, and death from other vascular causes. Hypertension, diabetes, and smoking are the key risk factors associated with PAD [26]. Ankle-brachial index (ABI), the ratio of the BP in the lower legs to the BP in the arms, is a widely used index to assess asymptomatic PAD. PAD is determined with ABI < 0.9 in either of the lower limbs. Early determination of PAD is necessary for timely intervention to improve prognosis. As technology for sensing PPG becomes mature, it is possible to build wearable systems for diagnosis of PAD [27], which is a promising alternative to the more costly and less convenient measurement method of ABI by continuous-wave Doppler.

In addition to measuring ABI, wearable systems comprised of PPG sensors can also provide a patient's heart rate, oxygen saturation, and heart rate variability. PPG detects blood-volume changes in the micro-vascular bed of tissue. It has the advantages of being miniature and lightweight. Various PPG wearable systems have been developed based on different configurations, e.g., in the form of a ring, ear-wearable gadget, eyeglass, and garment. One of the pioneer developments in wearable devices is the PPG sensor ring developed by Asada et al. [28]. Several studies focused on PPG monitoring systems as ear-worn devices, such as the wearable in-ear measuring system (IN-MONIT) with micro-optic reflective sensor [29], a motion-tolerant magnetic earring sensor with an adaptive noise cancellation method [30], and an earphone system connected to a mobile phone [31] have been done. Wearable systems in the form of garments with sensors embedded in a hat, glove, and sock have also been developed [32]. Objects such as eyeglasses [33] are also a good platform for PPG monitoring with discomfort reduced by clips to the finger or ear. Figure 3 shows examples of some of the state-of-the-art wearable sensors for PPG measurement.

2.3 Summary and Future Development

Despite advancements in medicine and technologies, CVDs remain one of the major leading causes of death worldwide. CVDs such as AF and MI are transient in nature and therefore the use of unobtrusive wearable systems in an out-of-hospital setting to collect information for diagnosis has obvious advantages over the classic bulky alternatives that

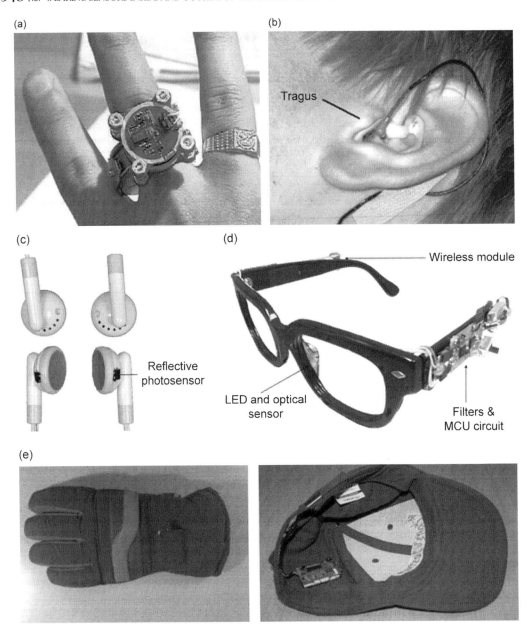

FIGURE 3 Typical wearable systems with PPG sensors for monitoring cardiac activity, ring PPG sensor [28], in-ear PPG sensor [29], earphone PPG [31], eyeglasses [33], and garments [32] with integrated wireless PPG sensors.

can be used only in the hospital. In the future, it is anticipated that these sensors will be seamlessly fused with our daily lives to enable prompt and pervasive management of cardiovascular diseases as well as many other abnormal health threatening conditions.

3. NEUROLOGICAL DISEASES

Neurological diseases such as Parkinson's disease and stroke are amongst the major causes of disability. Neurological rehabilitation provides therapeutic exercise to help patients to restore motor functions. Due to the shortage of hospital-centered rehabilitation resources, home-based rehabilitation therapies have gained interest. In this respect, wearable devices can play a role by providing remote monitoring in home settings of the mobility and physical functioning of patients with neurological disorders. It can significantly reduce healthcare cost and is useful in early symptom detection, which makes prompt and effective intervention possible. Since the present clinical tools for the assessment of these patients' physical performance are either based on self-report or observation, wearable sensors can also provide an objective alternative to assessing the condition of these patients by monitoring their daily activities. Most importantly, incorporated with actuators, wearable devices (or wearable robots) can be used as rehabilitative and assistive devices for disabled people for the treatment of motor disorders. Furthermore, wearable devices can be used for the prediction of sudden unexpected events related to chronic neurological diseases such as seizures. By continuous monitoring with wearable devices, some transient and covert features or symptoms related to these acute events can be captured for the understanding of pathology and the prediction of incidents.

3.1 Motor Activity Monitoring and Intervention for Neurological Rehabilitation

Micro-electromechanical inertial sensors such as accelerometers (ACCs) and gyroscopes are widely adopted for the monitoring of motor activities. ACC sensors measure changes in velocity and displacement while gyroscopes measure changes in orientation such as rotational displacement, velocity, and acceleration. These sensors have been extensively used for the continuous and automatic monitoring of movement disorders and functional activities [34]. Patel et al. first developed an integrated platform (SHIMMER) with a wearable system and algorithm to estimate the severity of three different Parkinson's symptoms (i.e., tremor, bradykinesia, and dyskinesia) [35], as shown in Figure 4(a). Each sensor node in the SHIMMER platform consists of a triaxial ACC sensor, a microprocessor, a radio transmitter, and a MicroSD card slot. Three feature types, including root-mean-square value, the data range value, and two frequency-domain features, were extracted to estimate the clinical score from the ACC signals recorded from patients during ten different motor tasks. The platform showed promising estimation results in terms of clinical score compared to that derived from visual inspection of video recordings. Recently, the same research team developed a Web-based system (MercuryLive) based on SHIMMER for the monitoring of patients with Parkinson's disease in home settings [36]. It contains three tiers, including central server, patient host, and clinician host. The clinician tier can access the sensor data to estimate the clinical score remotely. Rigas adopted a set of wearable sensors to detect and assess tremor in a ubiquitous environment [37]. Two sets of features extracted from tremor activity (3−12 Hz component in the measured ACC signals) were incorporated into a hidden Markov model for tremor severity recognition, and the

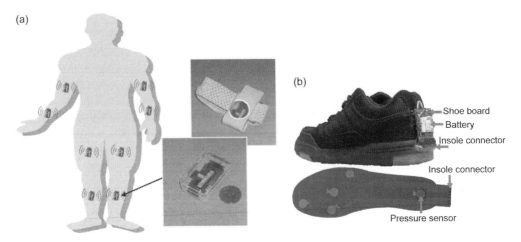

FIGURE 4 (a) SHIMMER platform developed by Harvard Medical School consists of sensor nodes located on upper and lower limbs [35], and (b) a wearable shoe for the rehabilitation of stroke patients developed by Clarkson University [38].

results showed high accuracy in tremor quantification and high specificity in distinguishing tremor activity from other motor symptoms. A wearable shoe-based device shown in Figure 4(b) has also been designed for the rehabilitation of stroke patients [38].

Other novel-sensing methods have also been proposed to measure motor activities. A wearable system using a capacitive sensing method for locomotion monitoring and classification has been developed in [39]. The wearable strain-sensing shirt in Figure 5 with conductive elastomers distributed over arm, forearm, and shoulders was designed for motion analysis in the context of neurological rehabilitation [40]. A set of representative rehabilitation exercises (i.e., gleno-humeral flexion of shoulder on a sagittal plane, lateral abduction, and external rotation) was designed to evaluate the performance of the system in terms of posture recognition and classification.

In terms of the classification of various motor activities, ACC sensors have inherent limitations in differentiating between an active versus a passive performance of a movement. EMG, which is monotonically related to muscle torque, has the advantage to distinguish between active and passive movements. Therefore, the combination of EMG and ACC information is expected to significantly improve the classification of performance. Roy et al. investigated the feasibility of a hybrid-surface EMG (sEMG) and ACC wearable sensor system for automatic classification of daily activities in patients with stroke [41]. The results showed that the hybrid configuration can achieve a mean sensitivity and specificity of 95% and 99.7% for the identification tasks and misclassification error of less than 10% for non-identification tasks. A comparison of the classification performance of the different configurations is shown in Figure 6. The results of this study proved that the inclusion of both EMG and ACC can achieve higher classification accuracy.

In addition to motor activity monitoring, wearable systems can also be designed with real-time feedback to control human-machine interaction to assist daily activities for patients suffering from motor disorders. Tremor is considered the most common motor

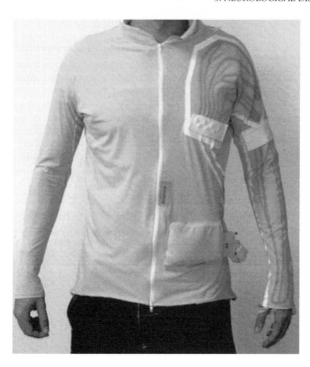

FIGURE 5 Garment prototype with printed strain sensors shown in gray for posture recognition in the context of neurological rehabilitation [40].

disorder caused by neurological diseases, and a variety of wearable systems for tremor assessment and suppression have been previously developed [42,43]. A representative wearable robotic exoskeleton for orthotic tremor suppression (WOTAS) [44] was designed to measure and suppress tremor by using two gyroscopes placed distally and proximally to the elbow joint. To depress tremor without affecting concomitant voluntary movement, one critical issue is tremor characterization, i.e., extracting instantaneous tremor parameters from the raw motion data to generate control command to drive a neuroprothesis or the human muscle itself in real time. Tremor was modeled as a sinusoidal signal of frequency ω_0 plus M harmonics to calculate the estimation error ε_k as follows:

$$\varepsilon_k = s_k - \sum_{r=1}^{M}[w_{r_k}\sin(r\omega_{0_k}k) + w_{r+M_k}\cos(r\omega_{0_k}k)] \tag{1}$$

The tremor frequency and amplitude can then be estimated by recursive method, i.e.,

$$w_{0_{k+1}} = w_{0_k} + 2\mu_0\varepsilon_k\sum_{r=1}^{M}r(w_{r_k}x_{M+r_k} - w_{M+r_k}x_{r_k})$$

where

$$x_{r_k} = \begin{cases} \sin\left(r\sum_{t=1}^{k}w_{0_t}\right), & \text{for } 1 \leq r \leq M \\ \cos\left((r-M)\sum_{t=1}^{k}w_{0_t}\right), & \text{for } M+1 \leq r \leq 2M \end{cases} \tag{2}$$

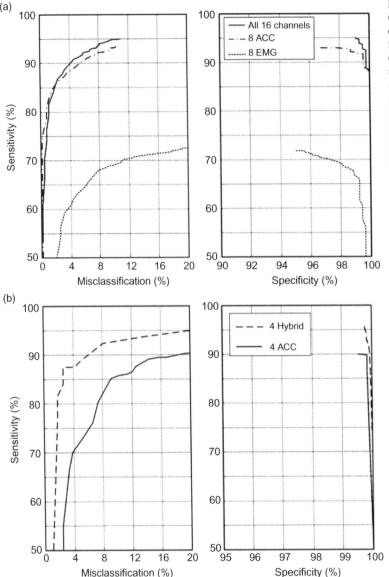

FIGURE 6
(a) Classification performance of identification and non-identification tasks for data collected from eight sEMG sensors (dotted line), eight ACC sensors (dash/dot line), and the combination of the sixteen sensors (solid line). (b) Classification performance for data from four hybrid sensors and four ACC sensors [41].

Separation of the voluntary and tremorous movement is needed before the characterization. Active and passive tremor reduction strategies, i.e., impedance control and notch filtering, were implemented and validated on the system and can achieve 40% of tremor power reduction for all users [44]. Similar wearable biofeedback systems have also been developed, such as an EMG-controlled exoskeletal orthosis wearable system for exercise training after stroke [45], wearable intelligent systems with real-time vibro-tactile feedback for the training of motor functions [46], and for posture correction [47] in rehabilitative and protective applications.

3.2 Seizure Activity Monitoring for Epilepsy Patients

Mechanisms leading to sudden and unexpected death following a seizure in patients with chronic epilepsy are not fully understood in medicine [48]. Cardiac arrhythmia, respiratory dysfunction, dysregulation of systemic or cerebral circulation, and seizure-induced hormonal and metabolic changes can be potential pathomechanisms of the disease. The occurrences of seizures are unpredictable and random. For inpatients, the forewarning of seizure activity is often monitored by EEG video hospital systems, which is a bulky system that greatly restricts the activities of the patient. Wearable devices with ACC sensors are another promising alternative for seizure-activity monitoring with predefined algorithms for real-time analysis of the ACC signals. Sensitivity and false alarm rate are the most important indices of wearable systems for these applications. A user-friendly designed wrist-type device for seizure detection was developed by Danish Care Technology ApS, which contains a three-axis ACC sensor, a microprocessor, and battery, as shown in Figure 7 [49]. A number of clinical studies have been conducted to validate the feasibility and accuracy of this sort of wearable device. In a prospective multi-center study including 73 patients of generalized tonic-clonic seizures (GTCS), this wrist-type device showed a mean sensitivity of 91% and a false alarm rate of 0.2/day. Another clinical study [50] showed that a bracelet alarm device with a three-axis for epilepsy monitoring could identify tonic, clonic, and tonic-clonic seizures with similar sensitivity (20 of 22 seizures) and low false alarm rate (8 false alarms during 1,692 hours of monitoring).

Patients with GTSC have higher risk for injuries and sudden deaths that cannot be detected by the abovementioned devices [49]. Poh et al. recently developed a wrist-worn device with skin conductance electrodes for long-term electrodermal activity (EDA) recordings that reflected sympathetic nerve activities [51]. The performance of this device was compared to an FDA-approved device during the baseline, task, and recovery conditions and found high correlations in all states. The performance of the device with two different electrode materials, i.e., Ag/AgCl and conductive fabric electrodes, was also compared and the results showed that the fabric electrodes performed as well as Ag/AgCl under baseline state, but were less promising during stressor task. This wearable device opens up opportunities to monitor autonomic nerve activities unobtrusively during daily activities, which would be very meaningful for the long-term monitoring of psychological and neurological conditions. The author conducted a clinical study to explore the clinical value of the EDA recordings in the prediction of sudden death in epilepsy [52].

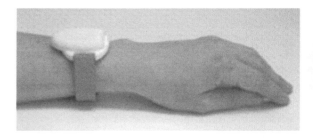

FIGURE 7   The wrist-worn accelerometry device and the control unit for seizure detection [49].

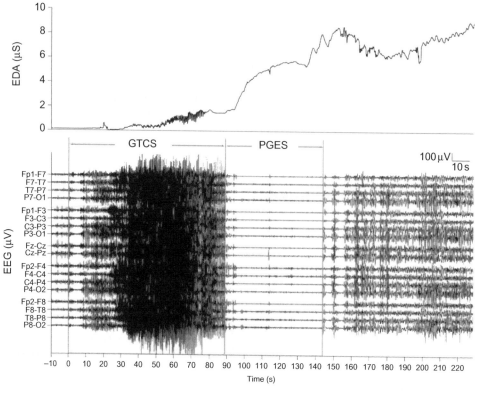

FIGURE 8 An example of EDA and EEG recordings during a tonic-clonic seizure. The increase of EDA amplitude during the event is associated with the duration of postictal EEG suppression [52].

As shown in Figure 8, high correlation (r = 0.81, p = 0.003) was found between the increase in EDA amplitude and the duration of EEG suppression, which has been reported in a previous study being significantly prolonged in earlier GTCS of patients who died later of epilepsy-related sudden death. The findings of this study indicate that sudden death in epilepsy may be correlated with postictal autonomic dysfunction, and autonomic parameters such as EDA amplitude might be able to provide valuable information regarding the severity of the seizure for prompt decision and action.

3.3 Summary and Future Development

This section presents an overview of recent developments in the field of wearable sensors and systems related to the monitoring and intervention of neurological diseases such as stroke, Parkinson's disease, and epilepsy. At present, applications mainly focus on the remote monitoring and assessment of motor functions as well as real-time intervention during the rehabilitation process. An emerging application in this area of wearable devices is to capture critical moments and extract information related to the occurrences of acute events for predicting future events. For future developments, these systems should be designed to be applied comfortably on a daily basis. A leading EU-funded project in this area was launched aiming to develop unobtrusive systems for monitoring of daily activities and training of motor functions for stroke survivors using smart-textile technology implemented on shoes, trousers, shirts, and gloves [53]. With recent advances in flexible electronics, skin-attachable devices such as the epidermal electronics, which can measure EMG and other electrophysiological signals on the skin [54], and the electronic artificial skin (e-skin) using organic materials, also provide promising solutions for unobtrusive implementations of these systems [55].

4. GASTROINTESTINAL DISEASES

Although the gastrointestinal (GI) tract is an organ situated inside the human body, it is exposed to the external environment as much as the human skin. Conventionally, sensing inside the GI tract is accomplished by implantable devices, e.g., esophageal pH monitoring, which is the gold standard for diagnosing gastroesophageal reflux disease (GERD) [56]. GERD is a chronic symptom of mucosal damage caused by stomach acid coming up from the stomach into the esophagus. It is increasingly recognized worldwide for its linkage to the development of esophageal carcinoma. The majority of GERD cases do not have endoscopic abnormalities and a complete evaluation with biopsies, esophageal motility, and 24-hour ambulatory pH monitoring are frequently needed before therapy is initiated. An ambulatory wireless pH monitoring system called Bravo™ (Given Imaging Ltd.) can be clipped to the lining of the esophagus for continuously measuring pH over a 48-hour period in an ambulatory setting such that acid reflux variables, including total reflux time, number of reflux episodes, and total percentage time of $pH < 4$, can be estimated and collected as the symptom-association probability scores for the diagnosis of GERD [57]. The sensor eventually falls off the esophageal lining after several days and is passed in the stool.

In order to anchor the sensor to the esophagus, the sensing capsule is designed with a pin. Recent advances in flexible optoelectronics, however, have opened up a new direction for designing wearable sensors to be worn non-invasively on irregular body surfaces or the GI wall for the early and prompt detection of diseases. In this section, designing flexible wearable sensors will be briefly discussed.

4.1 On-Body Wearable Sensor Design Based on Flexible Electronics

Wearable sensors can be fabricated on flexible substrate with a reasonable degree of stretchability such that they can conform to the complex and intricate body surface [55]. At present, studies in this area are focused on the design of on-body sensors, e.g., a high-pressure sensitivity sensor based on flexible polymer transistor for measuring blood pressure [58,59] and multi-sensor epidermal electronics system (EES) based on flexible silicone substrate [54,60]. The integrated EES is designed with temperature sensor and electrodes for measuring ECG/EMG with an aim to adhering to the human skin for monitoring

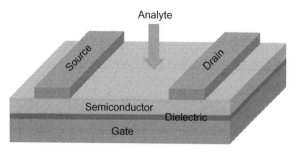

FIGURE 9 Illustration of a field-effect transistor as a chemical and biological sensor.

health conditions in our daily lives. An ultra-thin sensing film for mobile electronics application, healthcare, and biomedical systems has also been proposed [61].

4.2 In-Body Chemical and Biological Flexible Sensors and Systems

Chemical and biological sensors are widely used in many different medical fields, including the sensing of pH inside the GI tract [62]. Chemical sensors measure concentration of a specific component of a chemical reaction while a biological sensor is used to detect analytes, such as proteins, DNA, and antibodies based on physiochemical measurements. Figure 9 shows a typical sensor based on a field-effect transistor (FET).

The channel current of the sensor shown in Figure 9 can be described by the following equation [63]:

$$I_{DS} = \begin{cases} \dfrac{W}{L}\mu C_i(V_G - V_t)V_{DS} & V_{DS} \leq V_G - V_t \\ \dfrac{W}{2L}\mu C_i(V_G - V_t)^2 & V_{DS} > V_G - V_t \end{cases} \tag{3}$$

where I_{DS}, V_G, and V_t are channel current, gate voltage, and threshold voltage, respectively, W and L are the width and length of the channel, respectively, μ is the carrier mobility, and C_i is the capacitance of the gate insulator. When there is a change in chemical concentration or a component, drain and source current I_{DS} will change, as the chemical diffuses into the semiconductor layer and changes the carrier mobility (μ) and threshold voltage (V_t). Information about the chemical reaction of a system can therefore be reflected by the chemical concentration or component. The conductivity of an organic semiconductor is sensitive to ion-doping. Therefore, some organic field-effect transistors have been used for ion sensing. For example, poly(3-hexylthiophene-2,5-diyl) can be used as an active layer of the field-effect transistor to detect the concentration of K^+ and H^+ ions. This type of sensor can be useful for measuring the acidity in the GI tract, where gastric acid, a digestive fluid formed in the stomach, is composed of hydrochloric acid (HCl) (around 0.5%, or 5,000 parts per million) as high as 0.1 M, potassium chloride (KCl), and sodium chloride (NaCl).

The sensor and supporting components together form an electronics system, as illustrated in Figure 10. The sensor is the key component of the electronics system, defining the function of it, while the interconnector connects the sensor to other sensors or contact

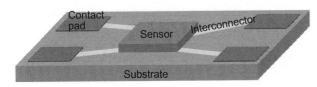

FIGURE 10 Components of an electronic sensory unit.

pads, and the substrate supports the sensor and the device as a whole. Compared to integrated circuits, the sensor is analogous to the die, the substrate is analogous to the frame, and the interconnector is the wire bond.

4.2.1 Technical Challenges for Fabricating Flexible Sensory Systems

Two major challenges arise for fabricating flexible electronics systems: 1) to maintain high performances while keeping the system flexible, and 2) to transfer the system and adhere it to the body surface seamlessly and firmly.

4.3 Design of High-Performance Flexible Sensory Systems

Since bending or changes in the shape of the system will affect the performance of the sensor, the sensor should be designed as a rigid component. When the sensor is kept small enough, the irregular body surface will appear as a smooth plane to the sensor, even though the body surface such as the human skin will have wrinkles, creases, and pits with amplitudes and feature sizes of 15 to 100 μm and 40 to 1,000 μm, respectively.

In order to keep the entire system as flexible as possible, the substrate and interconnector should be made of flexible materials with a reasonable degree of stretchability. Polydimethylsiloxane (PMDS) is a suitable choice for use as flexible substrate. PDMS (A: B = 30:1) has low Young's modulus of 145 kPa at the thickness of 0.6 mm, which is similar to human epidermis modulus, 140 to 600 kPa [54]. Low modulus assures the PDMS film has high stretchability. Meanwhile, PDMS has good biocompatibility, which will ensure the flexible system can be used for *in vivo* applications.

The challenge for designing the interconnector is to ensure stretchability while maintaining conductivity. In particular, the stretching of solid material lengthens chemical bonds and results in changing the distance between atoms and decreasing conductivity. Although crystalline structures of metals make them a good electricity conductor, they are hard to mold since the internal bonds are unbendable. Therefore, good conductor materials do not stretch well and stretchable materials are not good conductors. Recent studies reported the use of cross-stacked super-aligned carbon nanotube films [64] and the polymer-embedded carbon nanotube ribbons [65] as the conductor materials. Another study reported the development of a high conductivity and stretchability polymer by doping Au nanoparticles in polyurethane and suggested that the high conductivity of the polymer under stretch results from the dynamic self-organization of the nanoparticles under stress [66]. A simpler way has also been proposed by using conventional gold conductor and changing the straight line to filamentary serpentine shape [60].

(a)
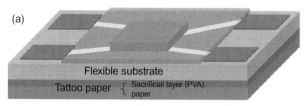

(b)
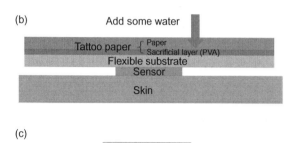

(c)

FIGURE 11 Illustration of the transfer of the device to body surface.

4.4 Transfer and Adherence of the Flexible System to Irregular Body Surface Seamlessly and Firmly

Ensuring good contacts to the body surface is another challenge in designing flexible electronics systems, such that they can adhere to the skin seamlessly and firmly to reduce the effect of motion artifacts.

Since the substrate is flexible, if it is thin enough, it can be designed with an adhesive tape to attach the system to the body surface by van der Waals interactions. Nevertheless, the thin substrate will easily collapse if it is not supported by a strong material. Tattoo papers have been therefore proposed to be used as a support structure [67]. The tattoo paper is covered by a thin sacrificial layer of PVA, which is water soluble. The transferring process is illustrated in Figure 11. The device is fabricated on the flexible substrate, which is supported by tattoo paper and transferred onto the skin. When water is rushed on the surface, the water-soluble sacrificial layer of PVA will detach from the tattoo paper and adhere to the body surface.

Though this device-transferring technique is especially suitable for epidermal electronics, it can also be used for sensing inside the human body. For example, fabrication of a flexible electrode array on thin polyimide substrate for mapping brain activity *in vivo* has been reported [68].

Fabrication of flexible and wearable sensors to seamlessly and non-invasively attach to the GI tracts for detection of bleeding and pH is still at the research stage. There are several major technical challenges that need to be overcome. First, in order to ensure good contact with the irregular GI tract and to maintain robust measurement during GI peristalsis, fabrication of flexible and stretchable sensors of different types is needed, e.g., flexible and stretchable light-emitting diodes and optical sensors to detect bleeding. Second, materials used to support the flexible electronics specialized for adhesion to the mucosa layer of the GI tract need to be further studied. At present, there are several types of muco-adhesive materials commercially available on the market. Previous studies compared the performance of some of them, such as Carbopol 971 P, polycarbophil, Carrageenan type λ, and Sodium carboxymethylcellulose [69]. Carbopol 971 P, a high molecular weight cross-linked polymer of acrylic acid (Noveon Inc.), is one of the outstanding materials. It can attach to the mucus via physical bonds [70], and its powdered polymers have a long history of safe use in cosmetic and pharmaceutical products [71]. Third, transfer of the sensors to the GI wall will require novel and skillful endoscopic techniques as this is clearly much more difficult than using them as EES. Overall, development of sensors worn on the inside of the body is a multi-disciplinary problem with high research potential.

5. CONCLUSION

Health informatics, particularly wearable sensors and systems, have emerged as a promising field of research. Conventionally, wearable and implantable sensory systems were thought to be designed for sensing information on the body surface and inside the human body, respectively. Recent advances in flexible optoelectronics have, however, opened up a new phase where sensors can be worn non-invasively inside the human body, particularly along the GI tract, for sensing information as well as dispensing drugs. This advancement will refine the scope of wearable systems and enhance the potential applications for early detection of cardiovascular, neurological, and gastrointestinal diseases.

Acknowledgment

This work was supported by the Hong Kong Innovation and Technology Commission (ITS/159/11 and ITS/197/12), CUHK Direct Grant (No. 4054058), and in part by the 973 Project Fund (2010CB732606) and the Guangdong LCHT Innovation Research Team Fund in China.

References

[1] U.S. National Academcy of Engineering. (2013). Advance Health Informatics. Available: <http://www.engineeringchallenges.org/cms/8996/8938.aspx>. (Last Accessed: 06.07.14).
[2] M. Turakhia, et al., Clinical experience and diagnostic yield from a national registry of 14-day ambulatory ECG patch monitoring, J. Am. Coll. Cardiol. 59 (2012), E646-E646.
[3] WHO. (2013). Cardiovascular diseases (CVDs). Available: <http://www.who.int/mediacentre/factsheets/fs317/en/>. (Last Accessed: 06.07.14).
[4] S. Yusuf, et al., Global burden of cardiovascular diseases Part II: Variations in cardiovascular disease by specific ethnic groups and geographic regions and prevention strategies, Circulation 104 (2001) 2855–2864.
[5] WHO. A global brief on Hypertension [Online].
[6] G. Mancia, Short-and long-term blood pressure variability present and future, Hypertension 60 (2012) 512–517.
[7] T.G. Pickering, et al., Recommendations for blood pressure measurement in humans and experimental animals-Part 1: Blood pressure measurement in humans-A statement for professionals from the Subcommittee

of Professional and Public Education of the American Heart Association Council on High Blood Pressure Research, Hypertension 45 (2005) 142−161.

[8] M.H. Pollak, P.A. Obrist, Aortic-radial pulse transit time and ECG Q-wave to radial pulse wave interval as indices of beat-by-beat blood pressure change, Psychophysiology 20 (1983) 21−28.

[9] C.C.Y. Poon, Y.T. Zhang, Cuff-less and noninvasive measurements of arterial blood pressure by pulse transit time, 27th Annu. Int. Conf. IEEE Eng. Med. Biol. Soc. (2005) 5877−5880.

[10] C.H. Chan, et al., A hybrid body sensor network for continuous and long-term measurement of arterial blood pressure, Proc. 4th IEEE-EMBS Int. Summer School Sym. Med. Devices. Biosens. (2007) 121−123.

[11] Y.D. Lee, W.Y. Chung, Wireless sensor network based wearable smart shirt for ubiquitous health and activity monitoring, Sens. Actuators B Chem. 140 (2009) 390−395.

[12] F. Buttussi, L. Chittaro, MOPET: A context-aware and user-adaptive wearable system for fitness training, Artif. Intell. Med. 42 (2008) 153−163.

[13] P. Lopez Meyer, et al., Monitoring of cigarette smoking using wearable sensors and Support Vector Machines 2013.

[14] M. Funk, et al., Incidence, timing, symptoms, and risk factors for atrial fibrillation after cardiac surgery, Am. J. Crit. Care vol. 12 (2003) 424−433.

[15] C.-T. Lin, et al., An intelligent telecardiology system using a wearable and wireless ECG to detect atrial fibrillation, Info. Technol. Biomed. IEEE Trans. 14 (2010) 726−733.

[16] M.M. Baig, et al., A comprehensive survey of wearable and wireless ECG monitoring systems for older adults, Med. Biol. Eng. Comput. (2013) 1−11.

[17] C.R. Merritt, et al., Fabric-based active electrode design and fabrication for health monitoring clothing, Info. Technol. Biomed. IEEE Trans. 13 (2009) 274−280.

[18] J. Yoo, et al., A wearable ECG acquisition system with compact planar-fashionable circuit board-based shirt, Info. Technol. Biomed. IEEE Trans. 13 (2009) 897−902.

[19] S. Coyle, et al., BIOTEX—biosensing textiles for personalised healthcare management, Info. Technol. Biomed. IEEE Trans. 14 (2010) 364−370.

[20] P. Pandian, et al., Smart Vest: Wearable multi-parameter remote physiological monitoring system, Med. Eng. Phys. 30 (2008) 466−477.

[21] G. López, et al., LOBIN: E-textile and wireless-sensor-network-based platform for healthcare monitoring in future hospital environments, Info. Technol. Biomed. IEEE Trans. 14 (2010) 1446−1458.

[22] R. Paradiso, et al., A wearable health care system based on knitted integrated sensors, Info. Technol. Biomed. IEEE Trans. 9 (2005) 337−344.

[23] D. Curone, et al., Smart garments for emergency operators: the ProeTEX project, Info. Technol. Biomed. IEEE Trans. 14 (2010) 694−701.

[24] V.A. Moyer, Screening for Coronary Heart Disease With Electrocardiography: US Preventive Services Task Force Recommendation Statement, Ann. Intern. Med. 157 (2012) 512−518.

[25] S.S. Franklin, et al., Is pulse pressure useful in predicting risk for coronary heart disease? The Framingham Heart Study, Circulation 100 (1999) 354−360.

[26] R.L. Muir, Peripheral arterial disease: Pathophysiology, risk factors, diagnosis, treatment, and prevention, J. Vasc. Nurs. 27 (2009) 26−30.

[27] C.P. Oates, et al., Beyond the Ankle-Brachial Pressure Index for the Diagnosis of Peripheral Arterial Disease—Time for A New Look at Photoplethysmography, Angiology 64 (2013) 492−493.

[28] H.H. Asada, et al., Mobile monitoring with wearable photoplethysmographic biosensors, Eng. Med. Biol. Mag. IEEE 22 (2003) 28−40.

[29] S. Vogel, et al., In-ear vital signs monitoring using a novel microoptic reflective sensor, Info. Technol. Biomed. IEEE Trans. 13 (2009) 882−889.

[30] M.-Z. Poh, et al., Motion-tolerant magnetic earring sensor and wireless earpiece for wearable photoplethysmography, Info. Technol. Biomed. IEEE Trans. 14 (2010) 786−794.

[31] M. Poh, et al., Cardiovascular monitoring using earphones and a mobile device 2011.

[32] J. Spigulis, et al., Wearable wireless photoplethysmography sensors, Photonics Europe (2008), 69912O−69912O-7.

[33] Y. Zheng, et al., "A clip-free eyeglasses-based wearable monitoring device for measuring photoplethysmograhic signals," in Engineering in Medicine and Biology Society (EMBC), 2012 Annual International Conference of the IEEE, 2012, pp. 5022−5025.

[34] G. Grimaldi, M. Manto, Neurological tremor: Sensors, signal processing and emerging applications, Sensors 10 (2010) 1399—1422.

[35] S. Patel, et al., Monitoring Motor Fluctuations in Patients With Parkinson's Disease Using Wearable Sensors, IEEE Trans. Info. Technol. Biomed. 13 (2009) 864—873.

[36] B.R. Chen, et al., A Web-Based System for Home Monitoring of Patients With Parkinson's Disease Using Wearable Sensors, IEEE Trans. Biomed. Eng. 58 (2011) 831—836.

[37] G. Rigas, et al., Assessment of tremor activity in the Parkinson's disease using a set of wearable sensors, IEEE Trans. Info. Technol. Biomed. 16 (2012) 478—487.

[38] S. R. Edgar, et al., "Wearable shoe-based device for rehabilitation of stroke patients," in Engineering in Medicine and Biology Society (EMBC), 2010 Annual International Conference of the IEEE, 2010, pp. 3772—3775.

[39] B. Chen, et al., Locomotion mode classification using a wearable capacitive sensing system, Neural Syst. Rehabil. Eng. IEEE Trans. 21 (2013) 744—755.

[40] T. Giorgino, et al., Sensor evaluation for wearable strain gauges in neurological rehabilitation, Neural Syst. Rehabil. Eng. IEEE Trans. 17 (2009) 409—415.

[41] S.H. Roy, et al., A combined sEMG and accelerometer system for monitoring functional activity in stroke, IEEE Trans. Neural Syst. Rehabil. Eng. 17 (2009) 585—594.

[42] J. Kotovsky, M.J. Rosen, A wearable tremor-suppression orthosis, J. Rehabil. Res. Dev. 35 (1998) 373—387.

[43] M. Manto, et al., Evaluation of a wearable orthosis and an associated algorithm for tremor suppression, Physiol. Meas. 28 (2007) 415—425.

[44] E. Rocon, et al., Design and validation of a rehabilitation robotic exoskeleton for tremor assessment and suppression, IEEE Trans. Neural Syst. Rehabil. Eng. 15 (2007) 367—378.

[45] J. Stein, et al., Electromyography-controlled exoskeletal upper-limb-powered orthosis for exercise training after stroke, Am. J. Phys. Med. Rehabil. 86 (2007) 255—261.

[46] J. Lieberman, C. Breazeal, TIKL: Development of a Wearable Vibrotactile Feedback Suit for Improved Human Motor Learning, Robot. IEEE Trans. 23 (2007) 919—926.

[47] A.A. Gopalai, S. Senanayake, A wearable real-time intelligent posture corrective system using vibrotactile feedback, IEEE-Asme Trans. Mechatronics 16 (2011) 827—834.

[48] O. Devinsky, CURRENT CONCEPTS sudden, unexpected death in epilepsy, N. Engl. J. Med. 365 (2011) 1801—1811.

[49] S. Beniczky, et al., Detection of generalized tonicclonic seizures by a wireless wrist accelerometer: A prospective, multicenter study, Epilepsia 54 (2013) e58—e61.

[50] U. Kramer, et al., A novel portable seizure detection alarm system: preliminary results, J. Clin. Neurophysiol. 28 (2011) 36—38.

[51] P. Ming-Zher, et al., A wearable sensor for unobtrusive, long-term assessment of electrodermal activity, IEEE Trans. Biomed. Eng. 57 (2010) 1243—1252.

[52] M.-Z. Poh, et al., Autonomic changes with seizures correlate with postictal EEG suppression, Neurology (2012).

[53] (2011). Training and monitoring of daily-life physical INTERACTION with the environment after stroke (FP7-287351, EU funded project). Available: <http://www.smartex.it/index.php/en/research/projects/european-projects/22-projects/82-interaction>. (Last Accessed: 06.07.14).

[54] D.-H. Kim, et al., Epidermal electronics, Science 333 (2011) 838—843.

[55] T. Someya, Stretchable Electronics, WILEY-VCH, 2012.

[56] K.R. DeVault, et al., Updated guidelines for the diagnosis and treatment of gastroesophageal reflux disease, Am. J. Gastroenterol. 94 (1999) 1434—1442.

[57] B. Chander, et al., 24 Versus 48-hour bravo pH monitoring, J. Clin. Gastroenterol. 46 (2012) 197—200.

[58] S.C. Mannsfeld, et al., Highly sensitive flexible pressure sensors with microstructured rubber dielectric layers, Nat. Mater. 9 (2010) 859—864.

[59] G. Schwartz, et al., Flexible polymer transistors with high pressure sensitivity for application in electronic skin and health monitoring, Nat. commun. 4 (2013) 1859.

[60] W.H. Yeo, et al., Multifunctional epidermal electronics printed directly onto the skin, Adv. Mater. (2013).

[61] M. Kaltenbrunner, et al., An ultra-lightweight design for imperceptible plastic electronics, Nature 499 (2013) 458—463.

[62] P. Lin, F. Yan, Organic thin-film transistors for chemical and biological sensing, Adv. mater. 24 (2012) 34–51.

[63] G. Horowitz, Organic field-effect transistors, Adv. Mater. 10 (1998) 365–377.

[64] K. Liu, et al., Cross-stacked superaligned carbon nanotube films for transparent and stretchable conductors, Adv. Funct. Mater. 21 (2011) 2721–2728.

[65] Y. Zhang, et al., Polymer-embedded carbon nanotube ribbons for stretchable conductors, Adv. Mater. 22 (2010) 3027–3031.

[66] Y. Kim, et al., Stretchable nanoparticle conductors with self-organized conductive pathways, Nature 500 (2013) 59–63.

[67] J.R. Windmiller, J. Wang, Wearable electrochemical sensors and biosensors: a review, Electroanalysis 25 (2013) 29–46.

[68] J. Viventi, et al., Flexible, foldable, actively multiplexed, high-density electrode array for mapping brain activity in vivo, Nat. Neurosci. 14 (2011) 1599–1605.

[69] D. Dodou, P. Breedveld, P.A. Wieringa, Stick, unstick, restick sticky films in the colon, Minim. Invasive Ther. Allied Technol. 15 (2006) 286–295.

[70] L. Corporation, "Toxicology of the Carbopol Polymers as a Class," October 17, 2013.

[71] D.-H. Kim, N. Lu, Y. Huang, J.A. Rogers, Materials for stretchable electronics in bioinspired and biointegrated devices, MRS Bull 37 (2012) 226–235.

Wearable and Non-Invasive Assistive Technologies

Maysam Ghovanloo[1] and Xueliang Huo[2]

[1]GT-Bionics Lab, School of Electrical and Computer Engineering, Georgia Institute of Technology, Atlanta, Georgia, USA, [2]Interactive Entertainment Business, Microsoft, Redmond, Washington, USA

1. ASSISTIVE DEVICES FOR INDIVIDUALS WITH SEVERE PARALYSIS

The number of people with paralysis is increasing among all age groups. A study initiated by the Christopher and Dana Reeve Foundation showed that one in fifty people in the United States is living with paralysis. Figure 1 shows major causes of paralysis from spinal cord injuries (SCI) to neuromuscular disorders. Sixteen percent of these individuals (about one million) have stated that they are unable to move and cannot live without continuous help [1]. Moreover, the National Institutes of Health in the United States report that 11,000 cases of severe SCIs from automotive accidents, acts of violence, and falls add every year to this population. Sadly, 55% of individuals with spinal cord injuries are between 16 and 30 years old, and they will need special care for the rest of their lives [2].

Assistive technology (AT) is "any item, piece of equipment or product system whether acquired commercially off the shelf, modified or customized that is used to increase, maintain or improve functional capabilities of individuals with disabilities" [3]. ATs are either worn by users (i.e., wearable head tracker) or mounted on their wheelchairs (i.e., sip-n-puff) or bed for easy access. They can enable individuals with different types of disabilities, particularly severe paralysis, to communicate their intentions to other devices in their environments, particularly computers. This will ease the individuals' need for receiving continuous help, thus reducing the burden on their family members, releasing their dedicated caregivers, and reducing their healthcare and assisted-living costs. It may also help them to be employed and experience active, independent, and productive lives.

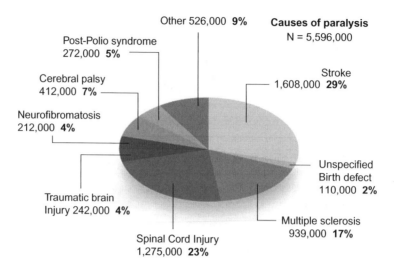

FIGURE 1   Causes of paralysis in the U.S. [1].

Among ATs, those providing alternative control for computer access and wheeled mobility are considered the most important for today's active lifestyle since they can improve the users' quality of life (QoL) by easing two major limitations: effective communication and independent mobility [4]. It is generally accepted that once individuals with disability are "enabled" to move around and effectively access computers and/or smartphones, they can virtually do most if not all of the things that able-bodied individuals with educational, administrative, or scholarly careers do on a daily basis [5,6]. This has resulted in a considerable amount of ongoing research towards developing new ATs that can potentially take advantage of any remaining abilities of these individuals, such as head motion, eye movements, muscle contractions, and even brain signals, to provide this population with alternative means to interact with computers and electronic gadgets. However, up until now, very few ATs have made a successful transition outside of the research laboratories into the consumer market to be widely used by severely disabled individuals, and many of those that are commercialized still have numerous shortcomings and difficulties.

1.1 Sip-n-Puff

Sip-n-puff is a simple, low-cost, switch-based AT, which allows its user to control a powered wheelchair (PWC) [7] or manipulate a mouse cursor [8] by blowing and sucking through a straw (Figure 2). However, it is slow, cumbersome for complicated commands, and offers very limited flexibility, degrees of freedom (DoF), and adaptability to user abilities. It only has a limited number of direct choices (four commands: soft sip, hard sip, soft puff, hard puff), which should be entered one-at-a-time in series. Another major limitation of sip-n-puff is the lack of proportional control, as opposed to a joystick, which can provide a much easier and smoother control over different movements, such as acceleration

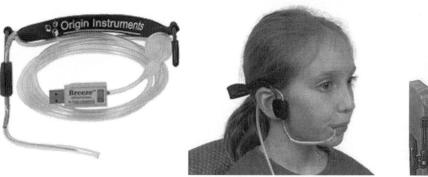

FIGURE 2   Sip-n-puff devices require users to suck and blow through a straw that is connected to a pneumatic switch or pressure sensor to control powered wheelchairs [7,8].

and deceleration of a PWC. Sip-n-puff needs regular cleaning and maintenance due to being exposed to saliva and food residue. It requires diaphragm control and may not benefit those who continuously use ventilators, such as those with SCI at level C2.

1.2 Head Controllers

Another group of ATs, known as head-pointers, are designed to replace arm and hand functions with head movements. Many of these devices are dedicated to emulating a computer mouse as a mean for users to control a cursor movement on the computer screen [9–15]. Some of them can also be used for wheelchair operation by embedding switch sensors in the head rest [16] or detecting head motion using inertial sensors [17,18]. Figure 3 shows a variety of such devices that are based on different tracking mechanisms.

One limitation of head-controlled assistive devices is that only individuals in whom head movement is not inhibited may benefit from them [9]. Many of those with quadriplegia and locked-in syndrome do not have good head movements, or if they do, their muscles are weak. Therefore, they may not benefit from any of these devices. Another limitation of head controllers is that the user's head should always be in the range within the reach of the device sensors, otherwise the AT would not be accessible to the user. For example, the head mounted sensors, reflectors, and laser beams, or even the user's facial features, are not accessible when he/she is not sitting in front of the computer/webcam or when lying in bed. Also, the use of head-controlled assistive devices for a long period of time can be quite fatiguing since they exhaust the user's neck muscles, which may already be weak as a result of the disability. They are also susceptible to inertial forces applied to the head when the wheelchair is in motion.

1.3 Eye-Tracking Systems

Another category of assistive devices operate by tracking eye movements and eye gaze, or more precisely, by detecting corneal reflections and tracking pupil position [19–22].

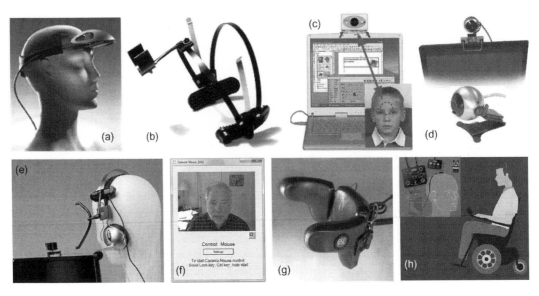

FIGURE 3   Different types of head movement-based assistive devices: (a) Boost Tracer based on head acceleration measured by gyroscopes [10], (b) Headmaster based on the intensity of ultrasonic sounds received by three
head-mounted microphones [11], (c) Headmouse based on infrared reflection received from a head-dot [12],
(d) Tracker Pro based on infrared reflection similar to Headmouse [13], (e) TrackIR based on infrared emission
received from an active unit mounted on the headset [14], (f) Camera Mouse based on tracking the movements
of a user-defined facial feature within a webcam field of view [15], (g) Head array wheelchair controller based
on proximity sensors [16], and (h) Magitek wheelchair controller based on head movements measured by
accelerometers [17].

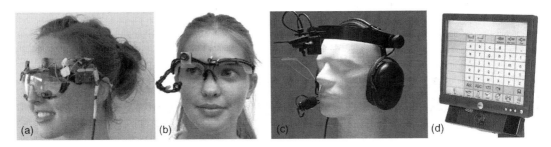

FIGURE 4   Different eye tracking devices: (a) Wearable EOG-based eye tracker operates by interpreting bio-
electrical signals recorded using surface electrodes attached to the skin around user's eyes [26], (b) Lightweight
wearable eye tracking headgear tracks the eye movement using a micro-lens camera [20], (c) iView X HED wearable video-based eye tracker from SensoMotoric Instruments (SMI) works by tracking the pupil position based on
corneal reflection [21], and (d) EyeTech TM3 computer-mounted eye tracker based on corneal reflection similar
to iView [22].

In these devices, a camera placed in front of the users, or a miniature optical sensor worn
by the users, captures the light, usually artificially generated infrared light, reflected from
the cornea, lens, or retinal blood vessel (Figure 4). The information is further analyzed to
extract the eye movement from the changes in reflection, and translated to move a cursor

on the computer screen. Electro-oculograms (EOG) have also been utilized for detecting the eye movements to generate control commands for both computer access and wheelchair control [23–27].

Since eyes have evolved as sensory parts of our body, a drawback of the eye-tracking systems is that they affect the users' normal vision by requiring extra eye movements that sometimes interfere with the users' visual tasks. Despite some recent progress, the "Midas touch" problem, which refers to initiation of unintended commands, is an issue when the user just looks at some point and the system considers that as a command [27]. The EOG-based method, shown in Figure 4(a), requires facial surface electrodes and a bulky signal-processing unit attached to the goggles that are unsightly and give the user a strange look. This might make the user feel uncomfortable in public. In general, camera-based eye trackers (Figure 4(b)–(d)) are sensitive to the ambient light condition, and generally not suitable or safe for the outdoor environment and wheelchair control. The computer-mounted eye tracking method always requires a camera or display in front of the users for detection or visual feedback, respectively. This is similar to the limitations of the head-controlled devices, requiring the user's head to remain within a certain range.

1.4 Electromyography-Based Controllers

Electromyogram (EMG) is referred to the electrical signals that are generated by muscle fibers during muscle contractions [28]. EMG-based control systems monitor EMG signals from a targeted group of muscles, typically facial [28], neck [29], or shoulder muscles [30], which are associated with the movements or twitches that the user is still able to perform. Customized signal-processing algorithms can recognize the EMG patterns associated with each movement and produce a set of discrete control commands that can be used to move the mouse cursor and perform selection for computer interaction [28,29] or replace joystick function to manipulate a wheelchair [30,31].

EMG-based systems are relatively error-prone and need complex muscular interactions [31]. These systems require highly specialized hardware and sophisticated signal-processing algorithms, therefore resulting in low portability [32]. Proper positioning of EMG electrodes for good contact with the skin, which sometimes need special gels or adhesives, and taking them off, are time consuming and cumbersome. Moreover, facial electrode attachment suffers the same cosmetic problem as an EOG-based eye tracker.

1.5 Voice Controllers

There are environmental controllers that utilize voice commands as input. Speech recognition software, such as Siri from Apple, Dragon Naturally Speaking [33] from Nuance, and Talking Desktop [34], are effective in particular aspects of computer access, such as text entry or individual commands (copy, paste, delete, etc.). However, they are not efficient in cursor navigation and are sensitive to accents, dialects, and environment noise. There are non-speech, sound-based voice controllers, such as the Vocal Joystick [35], developed for cursor control by mapping different sounds to specific cursor movement directions while associating the energy (loudness) of the sound to the velocity of cursor movement. In these devices, language specificity and accent sensitivity has been eliminated. However, users might feel uncomfortable and awkward when making such sounds in public places, particularly when people are supposed to be quiet, e.g., in a church or library. There are researchers working on developing voice-based controllers for PWC manipulation [36,37]. These devices can provide reasonable bandwidth and have relatively short response time. However, they are not safe enough to operate the wheelchair independently and therefore have to rely on an additional autonomous navigation system to avoid collisions. A common problem associated with almost all voice-based controllers is that they can work properly in indoor and quiet environments, but become inefficient and even completely useless in noisy outdoor environments. Voice-based systems can also raise privacy concerns when trying to type an email or text message in public.

1.6 Brain-Computer Interfaces

A group of assistive devices, known as brain-computer interfaces (BCIs), directly tap into the source of all volitional control, the brain. Such BCIs can potentially provide broad coverage among users with various disabilities. However, depending on how close the electrodes are placed with respect to the brain, there is always a compromise between invasiveness and information bandwidth. Non-invasive BCIs sense surface electrical signals generated from forehead muscles and electro-encephalogram (EEG) activities (Figure 5(a) −(c)) [38−44]. Limited bandwidth and susceptibility to noise and motion artifacts have prevented these devices from being used for important tasks such as navigating PWCs in outdoor environments, which need short reaction times and high reliability. EEG-based BCIs need extensive training and concentration before adequate control is obtained or retained. They also have considerable setup time to prepare the scalp sites for good electrode contact, properly place and later remove the electrodes or electrode cap, and finally clean the skin. There is the risk of harmful skin breakdown or allergic reaction if electrodes and gels remain on the scalp for extended periods of time. Several groups are developing dry electrodes [45]. However, the quality of recording is not as good as the standard electrodes. Invasive BCIs (Figure 5(d)−(e)), on the other hand, require planar (electrocorticogram − ECoG) or intracortical electrodes to record brain signals, often from the motor cortex area, while many patients may not be quite ready to undergo a complicated brain surgery for the sake of regaining access to their environment [46−50]. Also, long-term and reliable recording beyond two to three years is yet to be demonstrated.

1.7 Tongue-Operated Devices

Figure 6 shows a group of ATs that operate based on the users' voluntary oral movements. Tongue-Touch-Keypad (TTK) (a) is a switch-based device with nine keys that require tongue pressure [51]. Tongue-Mouse (b) is like a touch pad that is operated by tongue [52]. Tongue-Point (c) and Integra-Mouse (d) are modified joysticks to be used with tongue and mouth [53,54]. They need tongue pressure and may cause fatigue and irritation over long-term use. Think-A-Move (e) measures pressure changes in the ear

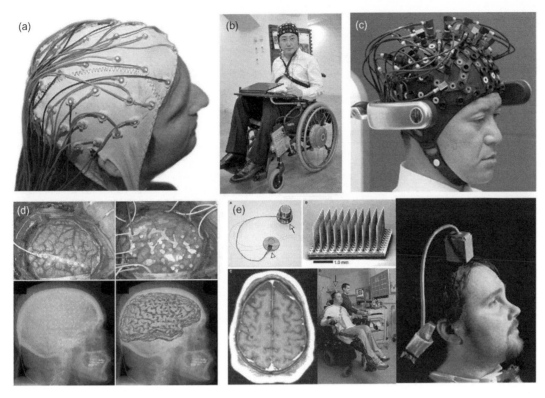

FIGURE 5 Some of the existing brain computer interfaces (BCIs): (a) Noninvasive surface EEG-based BCI, (b) BSI-Toyota EEG-based BCI for wheelchair control [42], (c) Honda BCI system combining EEG with NIR [43], (d) Invasive BCIs utilizing electrocorticogram (ECoG) signal [48], and (e) BrainGate invasive BCI based on the neural signals detected by intracranial microelectrodes [46].

canal as a result of tongue movements [55]. It offers only 1-D control with limited DoF. Figure 6(f) shows an optical tongue gesture detector, which does not need any attachments to the tongue [56]. A potential problem with this device is the high probability of unintended commands during speech or ingestion. Yet there is also an inductive intraoral controller, which is similar to TTK with 18 inductive switches that are activated with a metallic activation unit on the tongue [57]. However, these technologies either need bulky objects inserted into users' mouths, preventing them from eating or talking while using the devices, or require tongue and lip contact and pressure, which may cause fatigue and irritation over long-term use.

In summary, the existing ATs either provide their users with very slow and limited control over their environment or they are highly invasive and in early stages of development. Therefore, there is clearly an urgent need for developing a new wearable assistive device that can meet the following requirements to better assist its end users:

- Take full advantage of a user's existing capability
- Either non-invasive or minimally invasive

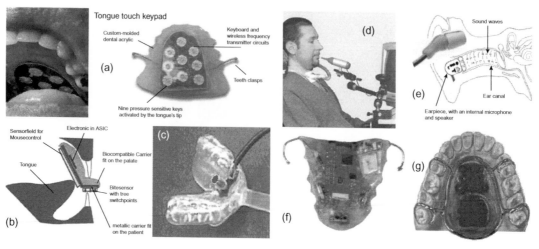

FIGURE 6   Tongue-operated ATs: (a) Tongue Touch Keypad (TTK) [51], (b) Tongue-Mouse [52], (c) Tongue-Point [53], (d) Integra-Mouse [54] (e) Think-A-Move [55], (f) Optical tongue gesture detector [56], and (g) Inductive tongue-computer interface [57].

- Powerful and can be used to interface with computer, wheelchair, different electronic devices, and environment
- User friendly, putting no or little physical and mental burden on end users
- Wearable and robust and can be used under different conditions
- Cosmetically acceptable
- Cost effective

2. WHY USE THE TONGUE FOR WEARABLE TECHNOLOGY?

The motor homunculus in Figure 7 shows that tongue and mouth occupy a significant amount of sensory and motor cortex in the human brain that rivals that of the fingers and the hands. Hence, they are inherently capable of sophisticated motor control and manipulation tasks with many degrees of freedom, which is evident from its role in speech and ingestion [58]. The tongue is connected to the brain via the hypoglossal nerve, which generally escapes severe damage in SCI and most neuromuscular diseases. As a result, even patients with high level SCIs still maintain intact tongue control capabilities. The tongue can move rapidly and accurately within the oral cavity, which indicates its high capacity for wideband indirect communication with the brain. Its motion is intuitive, and unlike EEG-based BCIs does not require thinking or concentration. The tongue muscle has a low rate of perceived exertion and does not fatigue easily. Therefore, a tongue-based device can be used continuously for several hours as long as it allows the tongue to freely move within the oral space. The motoneurons controlling tongue muscles receive a wealth of vestibular input, allowing the tongue body position to be reflexively adjusted with changes in the body position. Therefore, tongue-operated devices can be easily used anywhere, and in any position, such as sitting on a wheelchair or lying in bed. Another advantage of using

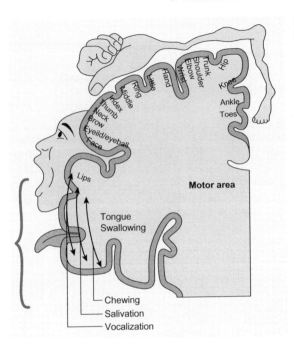

FIGURE 7 Tongue and mouth in the motor homunculus [58].

the tongue is that the tongue location inside the mouth can afford its users considerable privacy, which is especially important for people with disabilities, who do not want to be considered different from their able-bodied counterparts. Finally, unlike some BCIs that use neural signals from the motor cortex, picked up by implanted electrode arrays on the brain surface, non-invasive access to the tongue motion is readily available without penetrating the skull.

3. WIRELESS TRACKING OF TONGUE MOTION

A tongue drive system (TDS) is a minimally invasive, unobtrusive, tongue-operated, wireless and wearable assistive technology that can enable people with severe paralysis to control their environment, such as access computers or driving wheelchairs, using nothing but their volitional tongue movements. The system infers the users' intentions based on the voluntary positions, gestures, and movements of their tongues and translates them into certain user-defined commands that are simultaneously available to the TDS users in real time. These commands can be used to control the movements of a cursor on the PC screen, thereby substituting for a mouse or touchpad, or to substitute for the joystick function in a powered wheelchair (PWC) [59].

Conceptually, a TDS consists of an array of magnetic sensors, either mounted on a dental retainer inside the mouth, similar to an orthodontic brace (intraoral TDS or iTDS), as shown in Figure 8(a), or on a headset outside the mouth, similar to a head-worn microphone (external TDS, eTDS), as shown in Figure 8(b), plus a small permanent magnetic

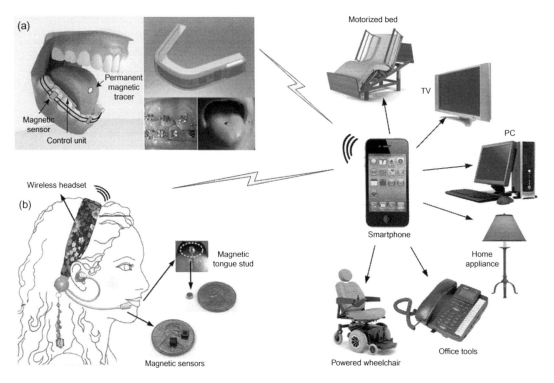

FIGURE 8 Block diagram of the Tongue Drive System: (a) Intraoral TDS (iTDS) with magnetic sensors and control unit located on a dental retainer, and (b) External TDS (eTDS) with all the electronics mounted on a headset.

tracer. The magnetic tracer, which is the size of a grain of rice, can be temporarily attached to the tongue using tissue adhesives. However, for long-term use the user should receive a tongue piercing and wear a magnetic tongue stud with the magnetic tracer embedded in it. Alternatively, the tracer can be coated with biocompatible materials, such as titanium or gold, and implanted under the tongue mucosa. The magnetic field generated by the tracer varies inside and around the mouth with the tongue movements. These variations can be detected by the magnetic sensors and wirelessly transmitted to a smartphone or a PC, which can be worn by the user or attached to his/her PWC. A sensor signal-processing (SSP) algorithm running on the PC/smartphone classifies the sensor signals and converts them into user-defined control commands, which are then wirelessly communicated to the target devices in the user's environment [59].

Alternatively, the magnetic tracer can be considered a magnetic dipole because its size is much smaller than the distance between sensors. The position and orientation of the magnetic tracer inside the oral cavity can be accurately tracked using measured magnetic field strength at known sensor locations and dipole equation. Various iterative optimization algorithms, such as Swamp Particle, Powell, DIRECT, and Nelder-Mead, have been implemented to solve such high-order nonlinear problems [60]. The trajectory of the magnetic tracer, which represents the movement of the tongue, can be used to define tongue gesture commands to provide users with a theoretically unlimited number of commands. It can also be utilized for more advanced proportional control.

A key advantage of the TDS is that a few magnetic sensors and an inherently wireless small magnetic tracer can capture a large number of tongue movements, each of which can represent a specific command. A set of dedicated tongue movements can be tailored to each individual user based on his/her mouth anatomy, preferences, lifestyle, and remaining abilities, and mapped onto a set of customized functions for environmental access. Therefore, a TDS can benefit a wide range of potential users with different types of disabilities because of its flexible and adaptive operating mechanism. By tracking tongue movements in real-time, a TDS also has the potential to provide its users with proportional control, which is easier, smoother, and more natural than the switch-based control for complex tasks such as maneuvering a PWC in confined spaces. Using TDS does not require users' tongues to touch or push against anything. This can significantly reduce tongue fatigue, which is an important factor that affects AT acceptability, and therefore results in greater user satisfaction and technology adoption rate. The TDS headset can be equipped with additional transducers, such as a microphone or motion sensors, and combined with commercial voice recognition software and a customized graphical user interface (GUI) to create a "single" integrated, multi-modal, multi-functional system, which can be effectively used in a variety of environments for multiple purposes [61].

4. WEARABLE TONGUE DRIVE SYSTEM

Figure 9 shows the latest wearable eTDS prototype [59], which includes: 1) A magnetic tracer, 2) a wireless headset built on headgear to mechanically support an array of four 3-axial magnetic sensors and a control unit that combines and packetizes the acquired magnetic field measurement raw data before wireless transmission, 3) a wireless receiver that receives the data packets from the headset and delivers them to the PC or

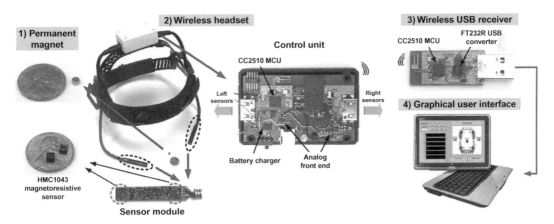

FIGURE 9 Major components of the external Tongue Drive System (eTDS).

smartphone, and 4) a GUI running on the PC/smartphone, which includes high through-put data streaming drivers and the SSP algorithm for filtering and classifying the magnetic sensor signals.

4.1 Permanent Magnetic Tracer

The TDS magnetic tracer was made of an alloy, $Nd_2Fe_{14}B$, known as a rare-earth perma-nent magnet, which has one of the highest residual magnetic strength values, resulting in a small-sized tracer without sacrificing the signal-to-noise ratio (SNR). Currently, we use a disc-shaped magnetic tracer ($\Phi 5\,mm \times 1.6\,mm$) with $Br = 14{,}500$ Gauss from K&J Magnetics.

4.2 Wireless Headset

The wireless headset has been equipped with a pair of goosenecks, each of which mechanically supports two 3-axial anisotropic magneto-resistive (AMR) HMC1043 sensors (Honeywell) near the subjects' cheeks, symmetrical to the sagittal plane to detect the mag-netic field variation due to tongue motion. It also has a wireless control unit to packetize and wirelessly transmit the data samples and a pair of rechargeable batteries. We used commercially available headgear, shown in Figure 9, for most human subject trials. However, we have also developed a custom-designed headset using 3-D printing technology [62].

As shown in the eTDS headset block diagram in Figure 10, each HMC1043 consists of three orthogonal AMR Wheatstone bridges, which resistances change in the presence of a magnetic field in parallel with its sensing direction, and result in differential output voltages. These outputs are multi-plexed before being amplified by a low noise

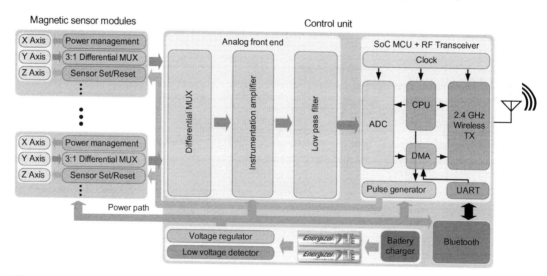

FIGURE 10　The block diagram of the eTDS wireless headset.

instrumentation amplifier, INA331 (Texas Instruments), with a gain of 200 V/V. A micro-controller unit (MCU) with a built-in 2.4 GHz RF transceiver (CC2510, TI) samples each sensor output at 50 Hz, using its on-chip 12-bit analog-to-digital converter (ADC), while turning on only one sensor at a time to save power. Each sensor is duty cycled at 2%, which results in a total on-time of only 8%. To avoid sensitivity and linearity degradations in the presence of strong fields (>20 Gauss) when the magnetic tracer is very close to the sensor (<1 cm), the MCU generates a short pulse (2 µs) to reset the sensor right before its output is sampled.

If users hold their tongues close to the left-back module (<1 cm) for >3 s, the TDS status switches between operational and standby modes. When the system is in the operational mode, all four sensor outputs are sampled at 50 Hz, and the results are packed into the data frame for RF transmission. In the standby mode, the MCU only samples the left-back side module at 1 Hz and turns off the RF transceiver to save power.

A simple but effective wireless handshaking has been implemented between the headset and the wireless receiver to establish a dedicated wireless connection between the two devices without interference from nearby eTDS headsets. When the eTDS headset is turned on, it enters an initialization mode by default and broadcasts a handshaking request packet containing a specific header and its unique network ID using a basic frequency channel (2.45 GHz) at 1 s time intervals for one minute. If the headset receives a response packet back from a nearby USB receiver within the initialization period, it will update its frequency channel, standby threshold, and other operating parameters that are included in the response packet. Then it sends an acknowledgement packet back to the transceiver to complete the handshaking. The headset then switches to normal operating mode using the received parameters. Otherwise, the headset will enter the standby mode by blinking a red LED to indicate that the initialization has failed and the power cycle should be repeated.

The power management circuitry includes a pair of AAA Ni-Mn batteries, a voltage regulator, a low-voltage detector, and a battery charger. The system consumes roughly 6.5 mA at 2.5 V supply, and can run for more than 120 hours following a full charge.

4.3 Wireless USB Receiver

Figure 9 shows a prototype wireless USB receiver dongle, which has the same type of MCU as the eTDS headset (CC2510). The receiver has two operating modes: handshaking and normal. In the handshaking mode, the receiver first listens to any incoming handshaking request packets from the eTDS headsets within range (~ 10 m). If it receives a handshaking request packet with a valid network ID, it will scan through all available frequency channels and choose the least crowded one as the communication channel for that specific headset. The receiver then switches to transmit mode and sends a handshaking response packet to the headset, before switching back to receiver mode and waiting for the confirmation of the acknowledgment packet. If an acknowledgment is received within 5 s, the receiver will update its frequency channel to the same frequency as the eTDS headset and enters the normal operating mode of receiving data packets. Otherwise, it will notify the user via PC/smartphone that the handshaking has failed. In the normal mode, the CC2510

MCU wirelessly receives the RF packets from the headset through the 2.4 GHz wireless link, extracts the sensor output from the packets, and then delivers them to the PC/smartphone through USB.

4.4 Graphical User Interface

The current GUI for computer access has been developed in the LabVIEW environment for testing and demonstration purposes. Generally, there is no need to present the eTDS users with a specific GUI. Because as long as the SSP engine is running in the background, the eTDS can directly substitute the mouse and keyboard functions in the Windows operating system and provide users with access to all the applications or software on the PC.

In the PWC GUI, a universal wheelchair control protocol has been implemented based on two state vectors: one for linear movements and one for rotations. The speed and direction of the wheelchair movements or rotations are proportional to the absolute values and polarities of these two state vectors, respectively. Five commands are defined in the PWC GUI to modify the analog state vectors, resulting in the wheelchair moving forward (FD) or backward (BD), turning right (TR) or left (TL) and stopping/neutral (N). Each command increments/decrements its associated state vector by a certain amount until a predefined maximum/minimum level is reached. The neutral command (N), which is issued automatically when the tongue returns back to its resting position, always returns the state vectors back to zero. Therefore, by simply returning their tongues to their resting positions, the users can bring the wheelchair to a standstill.

Based on the above rules, we implemented two wheelchair control strategies: discrete and continuous. In the discrete control strategy, the state vectors are mutually exclusive, i.e., only one state vector can be non-zero at any time. If a new command changes the current state, e.g., from FD to TR, the old state vector (linear) has to be gradually reduced/increased to zero before the new vector (rotation) can be changed. Hence, the user is not allowed to change the moving direction of the wheelchair before stopping. This was a safety feature particularly for novice users at the cost of reducing the wheelchair agility. In the continuous control strategy, on the other hand, the state vectors are no longer mutually exclusive and the users are allowed to steer the wheelchair to the left or right as it is moving forward or backward. Thus, the wheelchair movements are continuous and much smoother, making it possible to follow a curve, for example.

5. SENSOR SIGNAL-PROCESSING ALGORITHM

The sensor signal-processing (SSP) algorithm, which has been implemented in C, has three main components: external magnetic interference (EMI) attenuation, feature extraction (FE), and command classification.

5.1 External Magnetic Interference Attenuation

The EMI attenuation is a pre-processing function to enhance the signal-to-noise ratio (SNR) by minimizing interference from the ambient magnetic field, such as the Earth's magnetic field (EMF) and focus on the field generated by the magnetic tracer on the tongue. A stereo differential noise cancellation technique was implemented and proved to be inherently robust against EMI. In this method, the outputs of each 3-axial sensor module are mathematically transformed to orient the sensor module in parallel to the module on the opposite side of the sagittal plane. EMI sources are often far from the sensors and result in common-mode signals in each sensor module and the virtual replica of the opposite side module. On the other hand, signals resulting from the movements of the magnetic tracer that is located in between the two sensor modules are differential in nature unless the tracer moves symmetrically with respect to both sensor modules along the sagittal plane. Therefore, if the transformed outputs of each sensor pair are subtracted, the EMI common-mode signal is significantly attenuated, while the differential-mode is amplified. As a result, the SNR is greatly improved [63].

5.2 Feature Extraction

The FE algorithm, which is based on principal component analysis (PCA), is used to reduce the dimensions of the incoming sensor data and accelerate computations. During the feature identification (also known as training) session, users associate a preferred tongue gesture or position to each TDS command and repeat that command 10 times in 3 s intervals by moving their tongue from its resting position to the desired position after receiving a visual cue from the GUI. Meanwhile, the sensor outputs are recorded in each repetition and labeled with the executed command. Once training is completed, the FE extracts the most significant features of the sensor waveforms for each specific command offline in order to reduce the dimensions of the incoming data. The labeled samples are then used to form a cluster for each command in the virtual PCA feature space. During normal TDS operation, the same FE algorithm runs over the incoming raw sensor data and reflects them onto the PCA feature space by calculating the principal component vector for each sample in real time. These vectors contain the most significant features that help in discriminating different clusters (TDS commands).

5.3 Command Classification

The k-nearest neighbors (KNN) classifier is used within the PCA feature space to calculate the proximity of the incoming data points to the clusters formed during the training session. The KNN algorithm inflates a virtual sphere from the position of incoming data point in the PCA space until it contains k-nearest classified training points. Then it associates the new data point to the command that has the majority of the training points inside that sphere. The current eTDS prototype supports six individual tongue commands for mouse control, including four directional commands (UP, DOWN, LEFT, and RIGHT) and two selection commands (LEFT-SELECT and RIGHT-SELECT), which are simultaneously available to the user, plus a neutral command defined at the tongue-resting position. The same commands can be used for wheelchair navigation when the associated application is activated on the smartphone (accelerate, decelerate, turn-left, and turn-right). Selection commands in this are used to control the wheelchair power-seating functions.

To improve the classification accuracy, we have also employed a two-stage classification algorithm that can distinguish between six different control commands inside the oral cavity with near absolute accuracy [64]. The first phase mainly focuses on performing a hundred percent accurate discrimination between left (left, up, left-select) versus right side (right, down, right-select) commands. This is done by calculating the Euclidean distance of an upcoming point to the left and right command-positions, which are averaged from training trials. These distances are normalized to compensate for any asymmetry in users' left vs. right-side commands and then compared to produce a left/right decision. Based on the outcome of the first stage, the second stage of classification is applied on either the left or right side of the oral cavity to detect and discriminate between left, up, left-select, and neutral commands on the left side, or right, down, right-select, and neutral commands on the right side. The second stage of classification begins with de-noising the raw data using the noise cancellation technique mentioned above. The filtered data is fed to a group of linear and nonlinear classifiers consisting of linear, diagonal linear, quadratic, diagonal quadratic, Mahalanobis minimum distance, and KNN classifiers. At the end, the outputs of all classifiers are combined following a majority voting schema to provide a final result.

6. DUAL-MODE TONGUE DRIVE SYSTEM

6.1 The Advantages of Multi-Modal System

It has been understood that an interface that is designed around only one input modality may not be fast and flexible enough to meet the diverse needs of the end users in today's hectic and demanding lifestyles [65]. Most existing assistive devices, including our TDS, operate well for a narrow set of specific tasks, under a specific set of environmental conditions, for users with a specific set of remaining abilities. Due to the wide variety of tasks in daily life, various types and levels of disabilities, a multitude of environmental conditions, and a diversity of user goals and preferences, ATs that work perfectly well for one set of tasks, users, and environments, might show poor performance in other tasks or environments by the same users, or even completely lose their functionality when used for other applications by other users. In addition to the environmental and operating conditions, the performance of the single-mode ATs can be further degraded by the users' condition, such as fatigue, spasms, weakness, accent, etc.

A multi-modal device that expands the user access beyond one input channel, on the other hand, can potentially improve the speed of access by increasing the information transfer bandwidth between users and computers [66]. A clear proof of this fact is the use of both mouse/touchpad and keyboard by the majority of able-bodied users on their desktop or laptop machines. In addition, multi-modal devices increase the number of alternatives available to users to accomplish a certain task, giving users the ability to switch among different input modalities, based on their convenience, familiarity, and environmental conditions [67]. Multi-modal devices can also provide their users with more options to cope with fatigue. This is an important factor that improves the acceptability of ATs and can result in greater user satisfaction and technology adoption.

6.2 The Concept of Dual-Mode Tongue Drive System

The dTDS, the block diagram for which is shown in Figure 11, operates based on the information collected from two independent input channels: free voluntary tongue motion and speech. The two input channels are processed independently, while being simultaneously accessible to the users.

The primary dTDS modality involves tracking tongue motion in 3-D oral space using a small magnetic tracer attached to the tongue via adhesives, piercing, or implantation, and an array of magnetic sensors, similar to the original TDS. The secondary dTDS input modality is based on the user's speech, captured using a microphone, conditioned, digitized, and wirelessly transmitted to the smartphone/PC along with the magnetic sensor data. Both TDS and speech recognition (SR) modalities are simultaneously accessible to the dTDS users, particularly for mouse navigation and typing, respectively, and they have the flexibility to choose their desired input mode for any specific task without external assistance. The tongue-based primary modality is always active and regarded as the default input modality. The tongue commands, however, can be used to enable/disable the speech-based secondary modality via the dTDS graphical user interface (GUI) to reduce the system power consumption and extend battery lifetime.

By taking advantage of the strength of both TDS and SR technologies, dTDS can provide people with severe disabilities with a more efficacious, flexible, and reliable computer access tool that can be used in a wider variety of personal and environmental conditions. Specifically, the dTDS can help its users by 1) increasing the speed of access by using each modality for its optimal target tasks and functions; 2) allowing users to select either technology depending on the personal and environmental conditions, such as weakness,

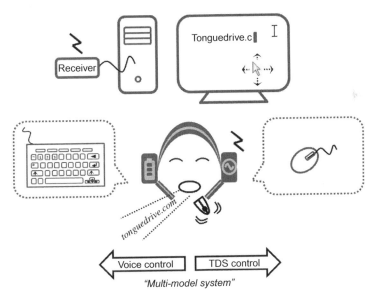

FIGURE 11   The block diagram of dual-mode Tongue Drive System (dTDS).

fatigue, acoustic noise and privacy [66]; 3) and by providing users with a higher level of independence by eliminating the need for switching from one AT to another, which often requires receiving assistance from a caregiver.

6.2.1 Wearable dTDS Prototype

The dTDS prototype, built on a customized wireless headset, is an enhanced version of the original TDS with the necessary hardware for a two-way wireless audio link to acquire and transmit users' vocal commands, while providing them with auditory feedback through an earphone. Figure 12 shows the main components of the dTDS prototype, with two major improvements from the original TDS: 1) A custom-designed wireless headset, fabricated through 3-D rapid prototyping, which mechanically supports an array of four 3-axial magnetic sensors and a microphone plus their interfacing circuitry to measure magnetic field and acoustic signals (a control unit combines and packetizes the acquired raw data before wireless transmission), and 2) a wireless transceiver acting like a bi-directional wireless gateway to exchange audio/data packets between the headset and the PC or smartphone.

6.2.1.1 WIRELESS HEADSET

A customized wireless headset was designed to combine aesthetics with user comfort, mechanical strength, and stable positioning of the sensors. The headset was also designed to offer flexibility and adjustability to adapt to the user's head anatomy, while enabling proper positioning of the magnetic sensors and the microphone near users' cheeks [62].

The headset has a pair of adjustable sensor poles, each of which holds a pair of 3-axial magneto-impedance (MI) sensors (AMI306, Aichi Steel) near the subjects' cheeks, similar to TDS, to measure the magnetic field strength. A low-power MCU (CC2510) with a built-in 2.4 GHz RF transceiver communicates with each sensor through the I2C interface to acquire samples at 50 Hz, while turning on only one sensor at a time to save power.

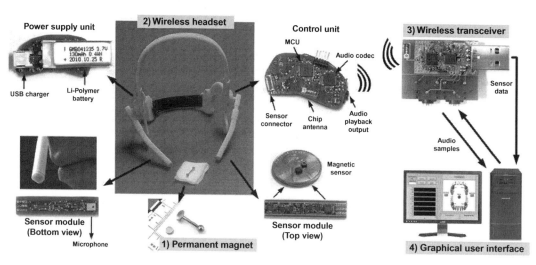

FIGURE 12   Major components of the dual-mode Tongue Drive System (dTDS).

When all four sensors are sampled, the results are packed into one magnetic data frame to be ready for RF transmission.

The acoustic signal acquisition is managed by an audio codec (TLV320-AIC3204, TI) and delivered through the built-in inter-IC sound (I2S) interface of the CC2510 MCU [61]. A miniaturized SiSonic MEMS microphone (Knowles) was placed near the tip of the right sensor board, as shown in Figure 6, to capture the acoustic signal. Digitized audio samples are compressed to an 8-bit format using the CC2510 built-in μ-Law compression hardware to save the RF bandwidth. Once a complete audio data frame consisting of 54 samples has been acquired in 6.75 ms, the MCU assembles an RF packet containing one audio and one magnetic data frame and transmits it wirelessly [61].

After sending each RF packet, the MCU expects to receive a back telemetry packet including one data frame and one optional audio frame, which depends on whether the uplink audio channel from the transceiver to the headset has been activated or not. The data frame contains control commands from the PC/smartphone to switch on/off the speech modality. The audio frame in the back telemetry packet contains digitized sound signals from the PC/smartphone. The MCU extracts the audio samples from the back telemetry packet and sends them to the playback DAC of the audio codec through the I2S interface to generate audible analog audio signals if the user attaches an earphone to the headset audio jack. The CC2510 MCU can handle an RF data rate of 500 kbps, which is sufficient for bi-directional data and audio transmission.

Power-management circuitry includes a miniaturized 130 mAh lithium-polymer battery, a voltage regulator, a low voltage detector and a battery charger. dTDS consumes either 6 or 35 mA from a 3 V supply depending on whether the bi-directional audio channel is off or on. This would allow the system to be used continuously for 20 or 4 hours in the uni-modal TDS or dTDS modes, respectively.

6.2.1.2 WIRELESS USB TRANSCEIVER

Figure 12 shows a prototype of the transceiver equipped with a USB port and two audio jacks to interface the magnetic sensor data and acoustic signals with the PC, respectively. The transceiver has two operating modes: handshaking and normal. The handshaking mode is similar to the TDS USB receiver, explained in section 4.3. The normal mode is slightly different. In this mode, the transceiver works like a bi-directional wireless gateway to exchange data and audio samples between the dTDS headset and the PC/smartphone. The magnetic data within the headset packets are extracted and delivered to the PC/ smartphone through the USB port. The audio data, however, is streamed into a playback audio codec via its I2S interface and converted to an analog audio signal, which is then delivered to the microphone input of the PC/smartphone through a 3.5 mm audio jack (see Figure 12). The transceiver can also receive analog audio output from the PC/smartphone headphone jack and digitize it using the same audio codec and I2S interface. These audio samples are compressed using the CC2510 built-in μ-law compression hardware and packaged in an audio frame. The transceiver also receives data packets from the computer, which contain the dTDS operating parameters used to program the dTDS headset on the fly. The data packet is combined with the audio frame to form a back telemetry RF packet, which is then wirelessly sent back to the headset.

Table 1 summarizes some of the key features of the dTDS prototype.

TABLE 1   Dual-mode Tongue Drive System Hardware Specifications

Specification	Value
MAGNETIC TRACER	
Material	$Nd_2Fe_{14}B$ rare-earth magnet
Size (diameter and thickness)	$\varnothing$ 3 mm $\times$ 1.6 mm
Residual magnetic strength	14500 Gauss
MAGNETIC SENSORS	
Type	Aichi Steel AMI306 MI sensor
Dimensions	$2.0 \times 2.0 \times 1.0$ mm^3
Sensitivity / range	600 LSB/Gauss/ $\pm$ 300 μT
MICROPHONE	
Type	SiSonic SPM0408HE5H
Dimensions	$4.7 \times 3.8 \times 1.1$ mm^3
Sensitivity / SNR	-22 dB / 59 dB
CONTROL UNIT	
Microcontroller	TI–CC2510 SoC
Wireless frequency / data rate	2.4 GHz/500 kbps
Sampling rate	50 sample/s/sensor
Number of sensors /duty cycle	4/8%
Audio codec / interface	TLV320AIC3204/I2 S
Audio sampling rate / resolution / compression	8 ksps/16 bits / μ-Law
Operating voltage / total current	3 V/35 mA (audio on)
	3 V/6 mA (audio off)
Dimensions	36×16 mm^2
HEADSET	
Rapid prototyping material	Objet VeroGray resin
Total weight	90 g (including battery)

7. CLINICAL ASSESSMENT

The performance of the eTDS prototype was evaluated by thirteen patients with high-level SCI (C2-C5) [68]. Trials consisted of two independent sessions: computer access (CA) and PWC navigation session (PWCN). This section only reports the experimental procedure and important results from both sessions [68].

7.1 Subjects

Thirteen human subjects (four females and nine males) aged 18 to 64 years old with SCI (C2~C5) were recruited from the Shepherd Center inpatient (11) and outpatient (2) populations. Informed consent was obtained from all subjects. All trials were carried out in the SCI unit of the Shepherd Center with approvals from the Georgia Institute of Technology and the Shepherd Center IRBs.

7.2 Magnet Attachment

A new permanent magnet was sanitized using 70% isopropyl rubbing alcohol, dried and attached to a 20 cm thread of dental floss using superglue. The upper surface of the magnet was softened by adding a layer of medical-grade silicone rubber (Nusil Technology) to prevent possible harm to the subjects' teeth and gums. The subjects' tongue surface was dried for better adherence, and the bottom of the magnet was attached to the subjects' tongues, about 1 cm from the tip, using a dental adhesive. The other end of the dental floss was tied to the eTDS headset during the trials to prevent the tracer from being accidentally swallowed or aspired if it was detached from the subject's tongue.

7.3 Command Definition

To facilitate command classification, subjects were advised to choose their tongue positions for different commands as diversely as possible. They were also asked to refrain from defining the TDS commands in the midline of the mouth (over the sagittal plane) because those positions are shared with the tongue's natural movements during speech, breathing, and coughing. The recommended tongue positions were as follows: touching the roots of the lower-left teeth with the tip of the tongue for "left," lower-right teeth for "right," upper-left teeth for "up," upper-right teeth for "down," left cheek for "left-click," and right cheek for "double-click."

7.4 Training Session

During this session, a customized GUI prompted subjects to execute each command by turning on its associated indicator on the screen in 3 s intervals. Subjects were asked to issue the command by moving their tongue from its resting position to the corresponding command position when the command light was on, and returning it back to its resting position when the light went off. This procedure was repeated 10 times for the entire set of 6 commands plus the tongue resting position, resulting in a total of 70 data points.

7.5 Response Time Measurement

This experiment was designed to provide a quantitative measure of the TDS performance by measuring how quickly and accurately a command can be issued from the time it is intended by the user. This period, which is referred to as the TDS response time, T, along with the correct command selection probability within T, were used to calculate the information transfer rate (ITR) for the TDS, which is a widely accepted measure for evaluating and comparing the performance of different BCIs. The ITR indicates the amount of information that is communicated between a user and a computer within a certain time period. There are various definitions for the ITR. The one we used has been detailed in [39].

A dedicated GUI was developed for this experiment to randomly select 1 out of 6 commands and turn its indicator on. Subjects were asked to issue the command within T, on an audiovisual cue [68]. The GUI also provided subjects with an additional real-time visual feedback by changing the size of a bar associated with each command, indicating how close the tongue was to the position of that specific command. During the test, t was changed from 2 s to 1.5, 1.2, and 1.0 s, and 40 commands were issued at each time interval. The mean probability of correct choices (PCC) for each T was recorded.

7.5.1 Powered Wheelchair Navigation

In the PWCN session, subjects were transferred to a Q6000 powered wheelchair with a 12″ laptop placed on a wheelchair tray in front of them, as shown in Figure 13(a). They were asked to define four commands (FD, BD, TR, TL) to control the wheelchair state vectors in addition to the tongue resting position (N) for stopping. Then they were required to navigate the wheelchair, using TDS, through an obstacle course, as fast as

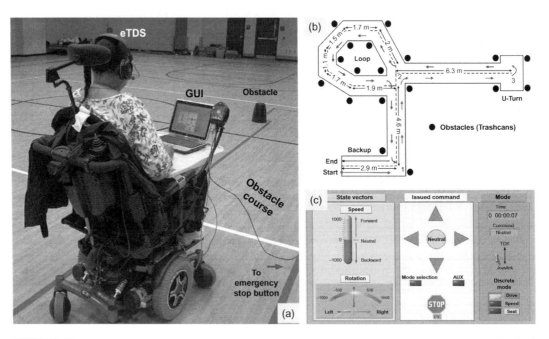

FIGURE 13   (a) A subject with SCI at level C4, wearing the eTDS prototype and navigating a powered wheelchair through an obstacle course. (b) Plan of the powered wheelchair navigation track in the obstacle course showing dimensions, location of obstacles, and approximate powered wheelchair trajectory. (c) The GUI provides users with visual feedback on the commands that have been selected.

possible, while avoiding obstacles or running off the track. Three slightly different courses were utilized. However, they were all close to the layout shown in Figure 13(b). The average track length was 38.9 ± 3.9 m with 10.9 ± 1.0 turns.

During the experiment, the laptop lid was initially opened to provide the subjects with visual feedback (shown in Figure 13(c)). However, later it was closed to help them see the track more easily. Subjects were required to repeat each experiment at least twice for discrete and continuous control strategies, with and without visual feedback. The navigation time, number of collisions, and number of issued commands were recorded for each trial.

7.5.2 Results

All subjects, including three who had very limited computer experience, managed to complete all the required CA and PWCN tasks without difficulty. Figure 14 shows subjects' performance in completing the response time measurement task. On average, a reasonable PCC of 82% was achieved with $T = 1$ s, yielding ITR = 95 bits/min. Table 2 compares the response time, number of commands, and calculated ITR of a few tongue computer interfaces (TCIs) and BCIs that are reported in the literature. It can be seen that the TDS offers a much better ITR compared to BCIs and TCIs due to its rapid response time.

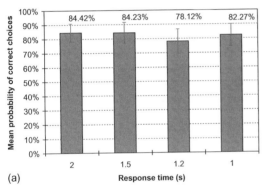 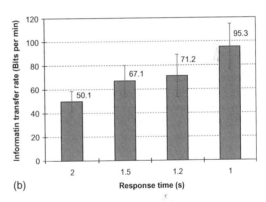

FIGURE 14   Response time measurement results: (a) Mean probability of correct choice (PCC) vs. response time, and (b) The eTDS information transfer rate (ITR) vs. response time.

TABLE 2   Comparison Between the Tongue Drive System and Other BCIs/TCIs*

Reference	Type	Number of Commands	Response Time (s)	ITR (Bits/Min)
[39]	EEG-BCI	$2 - 4$	$3 - 4$	25
[69]	TTK-TCI*	9	3.5	40
[57]	TCI*	5	2.4	58
TDS	TCI*	6	1.0	95

TCI: Tongue Computer Interface

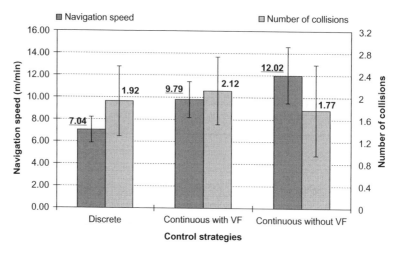

FIGURE 15   Average navigation speed and number of collisions for discrete and continuous control strategies, with and without visual feedback (VF).

Figure 15 shows the results of the PWCN session, including the average navigation speed and number of collisions along with their 95% confidence intervals using different control strategies. In general, the continuous control strategy was much more efficient than the discrete control. Subjects consistently performed better without visual feedback by navigating faster with fewer collisions. These results demonstrate that subjects could easily remember and correctly issue the TDS tongue commands without requiring a computer screen in front of them, which may distract their attention or block their sight. Improved performance without visual feedback can also be attributed to the learning effect because it always followed the trials with visual feedback.

8. FUTURE DIRECTIONS

The work performed to date has created a solid theoretical and technical basis for the development of the TDS in the context of wearable assistive technology. However, a considerable amount of work remains to be done before TDS can be accepted, used, and appreciated by its end users on a daily basis.

8.1 Intraoral TDS

An intraoral version of the TDS (iTDS) is under development [70]. In this new generation of TDS, the size of electronic components will radically shrink to the level that they can be hermetically sealed and embedded in a dental retainer, to be worn comfortably inside the mouth. The iTDS dental retainers can be customized to the users' oral anatomy by orthodontists to firmly clasp to their teeth and reduce the range of displacements.

The iTDS can significantly improve the reliability, performance, safety, and acceptability of this assistive technology by resolving the mechanical stability problem while being completely inconspicuous, hidden inside the mouth.

8.2 Multi-Modal TDS

The performance and end-user coverage of the current TDS will be further enhanced by adding other input modalities, such as head control using commercial motion sensors. In the current dTDS, the commands from different modalities, e.g., tongue and voice commands, are used to operate their dedicated devices or complete dedicated tasks individually. In addition, these commands can be fused together to enrich the control of one device at a time and achieve a higher control accuracy and bandwidth in demanding tasks such as being able to activate numerous controls on a gaming console as well as various shortcuts.

8.3 Data Compression and Sensor Fusion

The RF transceiver is the most power hungry part of the TDS. In order to reduce the power consumption and extend the battery lifetime of the TDS, the active time of the RF transceiver must be reduced. It can be achieved by compressing sensor outputs (magnetometer, accelerometer, gyroscope, ambient light sensor, microphone, and camera) using software or a hardware codec to reduce RF packet size and therefore save wireless transmission power. Alternatively, the data can even be processed locally to generate control commands by a low-power digital signal processor (DSP) incorporated in the control unit. In this case, the RF transceiver only needs to transmit control commands instead of raw sensor output, resulting in a significant reduction of wireless transmission bandwidth and power. This requires a highly efficient sensor fusion and processing algorithm to be implemented in a local DSP.

8.4 SSP Algorithm Improvement

The current SSP algorithm will be optimized to improve the command classification accuracy and to solve the "junk commands" problem. These are random unintended commands that are sometimes issued as the user moves his/her tongue from one command to another. New SSP algorithms, such as those based on support vector machines (SVMs), should be explored and evaluated to increase the number of tongue commands, possibly from six (coarse mode) to twelve (fine mode). Proportional control capability should definitely be explored and added to the current TDS to provide its end users with much easier, smoother, and more natural control over computer mouse cursor or powered wheelchairs.

8.5 Environmental Control

The TDS can also be used as an input device for electronic aids to daily life (EADLs) or environmental control units (EDUs) to interact and manipulate electronic appliances such as a television, radio, CD player, lights, fan, etc., in a smart home environments. The commercially available EDAL devices receive their control commands from a central controller, i.e., a computer, a touch-screen terminal, or simply an array of switches, and then communicate with the remote devices through RF, infrared, ultrasonic, or power lines using the widely accepted X10 protocol. In the TDS, a PC or a smartphone that runs the SSP algorithm can communicate with the EADL devices through USB or wireless link after converting TDS commands into a format that can be recognized by these commercial devices. In this way, a new set of functions for environmental control can be added to the TDS with minor modifications.

References

[1] Christopher and Dana Reeve Foundation. One degree of separation: Paralysis and spinal cord injury in the United States, Available from: <http://www.christopherreeve.org/site/c.ddJFKRNoFiG/b.5091685/k.58BD/One_Degree_of_Separation.htm>. (Last Accessed: 10.07.14).

[2] National Institute of Neurological Disorders and Stroke (NINDS), NIH. Spinal cord injury: Hope through research. Available from: <http://www.ninds.nih.gov/disorders/sci/detail_sci.htm>. (Last Accessed: 10.07.14).

[3] The US technology-related assistance for individuals with disabilities act of 1988, Section 3.1. Public Law 100-407. (Aug. 1988, renewed in 1998 in the Clinton Assistive Technology Act.) Available from: <http://section508.gov/index.cfm?fuseAction = AssistAct>. (Last Accessed: 10.07.14).

[4] Carlson, D. and Ehrlich, N. (2005). Assistive technology and information technology use and need by persons with disabilities in the United States. Report of U.S. Department of Education, National Institute on Disability and Rehabilitation, Washington, D.C.

[5] A.M. Cook, J.M. Polgar, Cook and Hussey's Assistive Technologies: Principles and Practice, third ed., St. Louis: Mosby, 2007.

[6] M.J. Scherer, Living in the State of Stuck: How Assistive Technology Impacts the Lives of People with Disabilities, forth ed., Brookline Book, MA, 2005.

[7] Therafin Corp. Sip-N-Puff. Available from: <http://www.therafin.com/sipnpuff.htm>. (Last Accessed: 10.07.14).

[8] Origin Instruments Corp. Sip and Puff Switch. Available from: <http://www.orin.com/access/sip_puff/>. (Last Accessed: 10.07.14).

[9] C. Pereira, R. Neto, A. Reynaldo, M. Luzo, R. Oliveira, Development and evaluation of a head-controlled human-computer interface with mouse-like functions for physically disabled users, Clin. Sci. 64 (2009) 975–981.

[10] Boost Technology. Boost Tracer. Available from: <http://www.boosttechnology.com/>. (Last Accessed: 10.07.14).

[11] D. Anson, G. Lawler, A. Kissinger, M. Timko, J. Tuminski, B. Drew, The efficacy of three head pointing devices for a mouse emulation task, Assist. Tech. 14 (2002) 140–150.

[12] Origin Instruments Corp. Headmouse Extreme. Available from: <http://www.orin.com/access/headmouse>. (Last Accessed: 10.07.14).

[13] Madentec Limited. Tracker Pro Wireless Head Tracking. Available from: <http://www.ablenetinc.com/Assistive-Technology/Computer-Access/TrackerPro>. (Last Accessed: 10.07.14).

[14] Natural Point. TrackIR. Available from: <http://www.naturalpoint.com/trackir>. (Last Accessed: 10.07.14).

[15] Camera Mouse. CameraMouse. Available from: <http://www.cameramouse.org>. (Last Accessed: 10.07.14).

[16] Adaptive Switch Labs Inc. ASL Head Array. Available from: <http://www.asl-inc.com/products/product_detail.php?prod=103>. (Last Accessed: 10.07.14).

[17] Magitek.com., LLC. Magitek Human Interface Drive Controls. Available from: <http://www.magitek.com>. (Last Accessed: 10.07.14).

[18] D.A. Craig, H.T. Nguyen, Wireless real-time head movement system using a personal digital assistant (PDA) for control of a power wheelchair, Proc. IEEE Eng. Med. Biol. Conf. (2005) 772–775.

[19] Y.L. Chen, F.T. Tang, W.H. Chang, M.K. Wong, Y.Y. Shih, T.S. Kuo, The new design of an infrared-controlled human–computer interface for the disabled, IEEE Trans. Rehab. Eng. 7 (1999) 474–481.

[20] J.S. Babcock, J.B. Pelz, Building a lightweight eye tracking headgear, Proc. 2004 Symp. Eye Track. Res. Appl. (2004) 109–114.

[21] SensoMotoric Instruments. IVIEW X™ HED. Available from: <http://www.smivision.com/en/gaze-and-eye-tracking-systems/products/iview-x-hed.html>. (Last Accessed: 10.07.14).

[22] Eyetech Digital System, Mesa, AZ. Available from: <http://www.eyetechds.com/assistivetech/products/qg3.htm>.

[23] R. Barea, L. Boquete, M. Mazo, E. Lopez, System for assisted mobility using eye movements based on electro-oculography, IEEE Trans. Rehab. Eng. 10 (2002) 209–218.

[24] C. Law, M. Leung, Y. Xu, S. Tso, A cap as interface for wheelchair control, IEEE/RSJ Intl. Conf. Intell. Robots Syst. 2 (2002) 1439–1444.

[25] Y. Chen, W.S. Newman, A human-robot interface based on electrooculography, Proc. Int. Conf. Robot. Automation 1 (2004) 243–248.

[26] A. Bulling, D. Roggen, G. Troster, Wearable EOG goggles: Seamless sensing and context-awareness in every-day environments, J. Ambient Intell. Smart Environ. (JAISE) 1 (2009) 157–171.

[27] R. Jacob, The use of eye movements in human-computer interaction techniques: what you look at is what you get, ACM Trans. Info. Syst. (TOIS) 9 (1991) 152–169.

[28] C. Chin, A. Barreto, G. Gremades, M. Adjouadi, Integrated electromyogram and eye-gaze tracking cursor control system for computer users with motor disabilities, J. Rehabil. Res. Dev. 45 (2008) 161–174.

[29] G. Chang, W. Kang, J. Luh, C. Cheng, J. Lai, J. Chen, et al., Real-time implementation of electromyogram pattern recognition as a control command of man-machine interface, Med. Eng. Phys. 18 (1996) 529–537.

[30] I. Moon, M. Lee, J. Chu, M. Mun, Wearable emg-based hci for electric-powered wheelchair users with motor disabilities, Proc. Intl. IEEE Conf. Robot. Automation (2005) 2649–2654.

[31] T. Felzer, R. Nordman, Alternative wheelchair control, Proc. Intl. IEEE-BAIS Symp. Res. on Assistive Tech. (2007) 67–74.

[32] J. Music, M. Cecic, M. Bonkovic, Testing inertial sensor performance as hands-free human-computer interface, WSEAS Trans. Comput. 8 (2009) 715–724.

[33] Nuance. Dragon voice recognition software. Available from: <http://www.nuance.com>, cited in Oct. 2013. (Last Accessed: 10.07.14).

[34] Talking Desktop Software. Talking Desktop Voice Recognition Software. Available from: <http://www.talkingdesktop.com>, cited in Oct. 2013.

[35] S. Harada, J.A. Landay, J. Malkin, X. Li, J.A. Bilmes, The vocal joystick: evaluation of voice-based cursor control techniques, Proc. ACM Conf. Comput. Accessibility – CHI (2006) 197–204.

[36] R. Simpson, S. Levine, Voice control of a powered wheelchair, IEEE Trans. Rehab. Eng. 10 (2002) 122–125.

[37] Pacnik, G., Benkic, K., and Brecko, B., Voice operated intelligent wheelchair – VOIC. Proc. ISIE 3, 1221–1226.

[38] J.J. Vidal, Toward direct brain–computer communication, Annu. Rev. Biophys. Bioeng. 2 (1973) 157–180.

[39] J.R. Wolpaw, N. Birbaumer, D.J. McFarland, G. Pfurtscheller, T.M. Vaughan, Brain-computer interfaces for communication and control, Clin. Neurophysiol. 113 (2002) 767–791.

[40] M.M. Moore, Real-world applications for brain-computer interface technology, IEEE Trans. Rehabil. Eng. 11 (2003) 162–165.

[41] D.J. McFarland, D.J. Krusienski, W.A. Sarnacki, J.R. Wolpaw, Emulation of computer mouse control with a noninvasive brain–computer interface, J. Neural Eng. 5 (2008) 101–110.

[42] K. Choi, A. Cichocki, Control of a wheelchair by motor imagery in real time, Lect. Notes Comput. Sci. 5326 (2008) 330–337.

[43] R. Bogue, Brain-computer interfaces: control by thought, Ind. Robot Int. J. 37 (2010) 126–132.

[44] S. Coyle, T. Ward, C. Markham, G. McDarby, On the suitability of near-infrared (NIR) systems for next-generation brain–computer interfaces, Physiol. Meas. 25 (2004) 815–822.

[45] Y.M. Chi, Y.-T. Wang, Y. Wang, C. Maier, T.-P. Jung, G. Cauwenberghs, Dry and noncontact EEG sensors for mobile brain–computer interfaces, IEEE Trans. Neural Sys. Rehab. Eng. 20 (2012) 228–235.

[46] L.R. Hochberg, et al., Neuronal ensemble control of prosthetic devices by a human with tetraplegia, Nature 442 (2006) 164–171.

[47] J.P. Donoghue, Bridging the brain to the world: a perspective on neural interface systems, Neuron 60 (2008) 511–521.

[48] G. Schalk, et al., Two-dimensional movement control using electrocorticographic signals in humans, J. Neural Eng. 5 (2008) 74–83.

[49] M. Velliste, S. Perel, M.C. Spalding, A.S. Whitford, A.B. Schwartz, Cortical control of a prosthetic arm for self-feeding, Nature 453 (2008) 1098−1101.

[50] P.R. Kennedy, D. Andreasen, P. Ehirim, B. King, T. Kirby, H. Mao, et al., Using human extra-cortical local field potentials to control a switch, J. Neural Eng. 1 (2004) 72−77.

[51] TongueTouch Keypad™ (TTK), New Abilities Inc., [Online]. Available from: <http://www.newabilities.com/>. (Last Accessed: 10.07.14).

[52] W. Nutt, C. Arlanch, S. Nigg, G. Staufert, Tongue-mouse for quadriplegics, J. Micromech. Microeng. 8 (1998) 155−157.

[53] C. Salem, S. Zhai, An isometric tongue pointing device, Proc. CHI 97 (1997) 22−27.

[54] USB Integra Mouse, Tash Inc., [Online]. Available from: <http://www.tashinc.com/catalog/ca_usb_integra_mouse.html>.

[55] R. Vaidyanathan, B. Chung, L. Gupta, H. Kook, S. Kota, J.D. West, A tongue-movement communication and control concept for hands-free human-machine interfaces, IEEE Trans. Systems, Man. Cybern. A Syst. Hum. 37 (2007) 533−546.

[56] S. Saponas, D. Kelly, B.A. Parviz, D.S. Tan, Optically sensing tongue gestures for computer input, Proc. ACM Symp. User Interface Softw. Technol. (2009) 177−180.

[57] L.N.S.A Struijk, An inductive tongue computer interface for control of computers and assistive devices, IEEE Trans. Biomed. Eng. 53 (2006) 2594−2597.

[58] E.R. Kandel, J.H. Schwartz, T.M. Jessell, Principles of Neural Science, forth ed., McGraw-Hill, New York, 2000.

[59] X. Huo, J. Wang, M. Ghovanloo, A magneto-inductive sensor based wireless tongue-computer interface, IEEE Trans. Neural Syst. Rehabil. Eng. 16 (2008) 497−504.

[60] J. Wang, X. Huo, M. Ghovanloo, A Modified Particle Swarm Optimization Method for Real-Time Magnetic Tracking of Tongue Motion, Proc. IEEE 30th Eng. Med. Biol. Conf. (2008).

[61] X. Huo, H. Park, J. Kim, M. Ghovanloo, A dual-mode human computer interface combining speech and tongue motion for people with severe disabilities, IEEE Trans. Neural Syst. Rehabil. Eng. (2013).

[62] H. Park, J. Kim, X. Huo, I.O. Wang, M. Ghovanloo, New ergonomic headset for tongue-drive system with wireless smartphone interface, Proc. 33rd IEEE Eng. Med. Biol. Conf. (2011) 7344−7347.

[63] X. Huo, J. Wang, M. Ghovanloo, A wireless tongue-computer interface using stereo differential magnetic field measurement, Proc. 29th IEEE Eng. Med. Biol. Conf. (2007) 5723−5726.

[64] E.B. Sadeghian, X. Huo, M. Ghonvaloo, Command detection and classification in tongue drive assistive technology, Proc. IEEE 33rd Eng. Med. Biol. Conf. (2011) 5465−5468.

[65] S. Keates, P. Robinson, The use of gestures in multimodal input, Proc. 3rd Intl. ACM Conf. Assist. Tech. (1998) 35−42.

[66] F. Shein, N. Brownlow, J. Treviranus, P. Pames, Climbing out of the rut: The future of interface technology, Proc. of the Visions Conf.: Augmentative and Alternative Comm. in the Next Decade, University of Delaware/Alfred I. duPont Institute, Wilmington, DE, 1990.

[67] M. Baljko, The contrastive evaluation of unimodal and multimodal interfaces for voice output communication aids, Proc. 7th Intl. Conf. Multimodal Interfaces (2005) 301−308.

[68] X. Huo, M. Ghovanloo, Evaluation of a wireless wearable tongue−computer interface by individuals with high-level spinal cord injuries, J. Neural Eng. 7 (2010) 497−504.

[69] C. Lau, S. O'Leary, Comparison of computer interface devices for persons with severe physical disabilities, Am. J. Occup. Ther. 47 (1993) 1022−1030.

[70] H. Park, et al., A wireless magnetoresistive sensing system for an intraoral tongue-computer in terface, IEEE Trans. Biomed. Circ. Syst. 6 (2012) 571−585.

[71] X. Huo, J. Wang, M. Ghovanloo, Using unconstrained tongue motion as an alternative control mechanism for wheeled mobility, IEEE Trans. Biomed. Eng. 56 (2009) 1719−1726.

[72] C. Cheng, X. Huo, M. Ghovanloo, Towards a magnetic localization system for 3-D tracking of tongue movements in speech-language therapy, Proc. IEEE 31st Eng. Med. Biol. Conf. (2009) 563−566.

[73] X. Huo, J. Wang, M. Ghovanloo, A magnetic wireless tongue-computer interface, Proc. Intl. IEEE EMBS Conf. Neural Eng. (2007) 322−326.

[74] A. Smith, J. Dunaway, P. Demasco, D. Peichl, Multimodal input for computer access and alternative communication, Proc. 2nd Annu. ACM Conf. Assist. Tech. (1996) 80−85.

7.4

Detection and Characterization of Food Intake by Wearable Sensors

Juan M. Fontana and Edward Sazonov

The University of Alabama, Tuscaloosa, Alabama, USA

1. INTRODUCTION

Food intake is the primary source of energy and nutrients necessary to maintain life. Monitoring of daily food intake and ingestive behavior is an important area that has direct implications on human health, as inadequate or excessive energy intake may lead to development of medical conditions such as malnutrition and underweight, or overweight and obesity, respectively. Understanding ingestive behavior is also a key in diagnosis and treatment of eating disorders such as anorexia, bulimia, and binge eating.

Food provides the chemical energy needed for functioning of the vital organs and performing physical activity, with the excess energy being stored in glycogen and adipose tissue for future use. The balance between energy intake from food and energy expended on basal metabolism and physical activity is an essential factor for maintaining a steady body weight in humans. A persistent imbalance between these two components is the cause of long-term changes in body weight, potentially leading to abnormal weight loss or gain. While hunger and malnourishment still remain an issue for a large part of the world's population, obesity has recently overtaken hunger as a global health threat.

Obesity, defined as the excessive accumulation of body fat, is the result of a chronic weight gain produced when the energy obtained from foods overcomes the energy expended. Excessive food intake (especially intake of calorie-rich foods now widely available on a global scale) may be a major contributor to the obesity epidemic. For example, in the United States, the prevalence of obesity reached a total of 35.5% among adults and 16.9% among adolescents in 2009–2010 [1]. Individuals suffering from obesity may potentially face a number of health issues ranging from cardiovascular problems to diabetes, and can expect a reduction in their life expectancy [2].

Eating disorders are serious mental disorders that cause disturbances in eating habits or weight-control behavior of individuals [3]. Anorexia nervosa, bulimia nervosa, and binge eating are the most common eating disorders with lifetime prevalence ranging from 0.6 to 4.5% in the United States [4]. Individuals with anorexia nervosa restrict their food intake by dieting, fasting, or excessive physical activity from the fear of gaining weight and distorted perceptions of their body shape and size. People with bulimia nervosa have periods of excessive food intake (binging) followed by a feeling of guilt that leads to extreme ways to compensate for binging (deliberate vomiting, crash dieting, and strenuous exercise). Binge eating is similar in terms of symptoms to bulimia nervosa, but does not include the extreme compensatory reactions typical to bulimia.

Both obesity and eating disorders are medical conditions highly resistant to treatment and can have severe physical and physiological health consequences [5]. The monitoring of food intake is considered to be the basis for behavioral treatment of obesity and eating disorders, which can be managed with dietary modification and control. Thus, monitoring of food intake is extremely important for identifying, understanding, and correcting food-intake patterns of individuals.

Traditionally, the ingestive behavior in humans is assessed by the means of self-monitoring. Various self-reporting methods were developed to estimate the timing and duration of intake as well as to characterize the intake in terms of amount of food consumed, energy, and nutrient intake [6]. Methods such as dietary records, 24-hr dietary recall, food frequency questionnaires, and diet history are widely used. All of these methods rely on a person's own declaration of what was eaten, when, where, and how much food was consumed. However, they suffer from underreporting, which is considered to be about 20% on average but may be as high as 50% [7,8]. The low accuracy of self-reporting food intake is mainly determined by two factors. First, there is a change in the eating behavior of individuals when they know they are being observed (the observation effect). Second, there is the tendency to either underestimate portion sizes or avoid reporting certain foods (reporting effect). For example, it has been shown that people tend to misreport or not report snacking at all, which may contribute significantly to the daily energy intake [9].

Thus, there is a critical need for developing methods for objective assessment of food intake, especially under free-living conditions. Such methods must provide accurate detection and characterization of food intake with minimal or no conscious effort from the subjects. The fundamental assumption is that the objectivity of monitoring will minimize or eliminate the reporting effect, while removing the conscious effort will minimize the observation effect.

Wearable sensors present a compelling possibility for monitoring of food intake. The dramatic technological advances in the past few decades allow building of miniature devices that can potentially detect the process of food ingestion and further characterize the ingested foods. A wearable ingestion sensor can potentially be very objective and capture all ingestion events, regardless of how short or insignificant they may seem. A wearable sensor could potentially capture timing, duration, and microstructure of food-intake episodes, characterize rate of ingestion, ingested mass, and nutritional and energy contents of food, without creating a reporting burden for the user. The individuals still need to comply with the requirement of wearing the sensor, but with an ergonomic design the wearer's burden can be minimal. The ingestion sensor can potentially recognize when it is being worn and when it is not, thus measuring the compliance of the individual.

Use of wearable sensors for monitoring of ingestion faces a number of challenges, with the main challenge being the great variety of foods that humans consume. Food is made from the most creative combinations of multiple ingredients, each with its own energetic and nutritional contents. Modifying just one ingredient may substantially change the nutritional properties of food (e.g., excluding butter or other forms of fat from a recipe may dramatically reduce the energy content). Foods have different physical properties and can be solid, liquid, semi-liquid, dry, moist, crispy, soft, and chewy, just to name a few possibilities. These physical properties are not necessarily correlated to the energy and nutritional content of food. The physical properties may also depend on whether the food is served raw or cooked in a certain way. Such variability of foods and food properties complicates and inevitably introduces errors even into the most sophisticated types of analyses.

The second challenge is the diversity of ingestive behaviors characteristic of humans. During an ingestion episode, we may eat a single food item or feast on a dozen or more of different food items. We may eat a little bit or a whole lot. We eat with hands, spoons, forks, chopsticks, or drink our food. We may have fairly stable meal times or customarily skip meals or eat at different times every day. Some individuals will eat fast, while others will eat slowly. Some will take breaks between different parts of a meal, while others will immediately start the next food item. We may eat during the waking hours or during the night. While most of us will consume a meal while seated, eating may also take place on the go or even lying down. If anything, the ingestive behavior is as variable and unpredictable as humans.

These two fundamental challenges create a variety of technological requirements for implementing wearable devices for food-intake monitoring. What sensor modality should be used? What are the capabilities and inherent limitations of such sensors? Can a given technology be made convenient, miniature, and lightweight to be worn for prolonged periods of time? What is the battery life of such a device? These and other questions need to be answered for any potential solution for food-intake monitoring.

This chapter presents an overview of the wearable sensors and accompanying methodologies proposed for food-intake monitoring. The overview is based on the recent research literature, with the focus on devices and methods developed in the Computer Laboratory of Ambient and Wearable Systems at The University of Alabama. Here, the task of monitoring ingestive behavior is considered as consisting of two subtasks: detection of food intake and characterization of food intake. Specifically, the task of detection of food intake includes:

- Detection and timing of each food-intake episode
- Measuring duration and microstructure of each episode

The task of characterization of food intake includes:

- Estimating the number and type of food items in a meal
- Estimating the mass and volume of ingestion
- Estimating caloric and nutritional content of a meal
- Measuring the rate of ingestion of each episode

This chapter is organized as follows. First, a description of the different sensor modalities developed for monitoring food intake is presented. Second, a description of the signal processing and pattern-recognition algorithms for automatic food-intake detection is presented. Third, methods of characterization of food intake are described. Fourth, an example of applying a wearable sensor for monitoring of food intake in free living is presented together with the challenges. Finally, future directions along with limitations and open issues concerning food intake are summarized at the end of the chapter.

2. WEARABLE SENSORS

Our society is increasingly dependent on wearable and mobile products that can offer effective ways to process information and interact with people. Many new approaches for monitoring of food intake are based on wearable sensors. These approaches can be divided in two main groups depending on how they are worn by individuals: handheld devices (such as mobile phones) and body-attached sensors.

2.1 Handheld Devices

Handheld devices are widely used to report and self-monitor daily intake by incorporating dietary software programs (electronic diaries). The use of mobile phones and personal digital assistants (PDAs) has significantly increased over the last decade, offering a new alternative for recording food intake. These computationally powerful devices can connect to the Internet anywhere and anytime, allowing Web-based interventions that can improve clinical management [10,11]. The advantage is that cell phones are carried by individuals most of the time, allowing them to make immediate food-intake annotations, which may help to improve cooperation and accuracy.

Dietary software programs have been developed and implemented into these electronic devices for a simpler and less burdensome food-intake monitoring (Figure 1). Such programs include a source database with thousands of food items with the corresponding nutritional information. For example, many diaries use the U.S. Department of Agriculture database, which has approximately 6,000 food items [12]. The integration of a food database into a user-friendly mobile application reduces the amount of time and labor required to report the intake when compared to paper-and-pencil diaries. Subjects simply log the foods consumed without the burden of searching and calculating the nutritional content of the meal. The software embedded in the electronic device automatically computes the total energy consumed in the meal from the amount and type of food reported. Electronic food diaries can also store the date and time of each entry, which can be used by clinicians or researchers to validate the reported data.

Another advantage of electronic diaries is that they can provide real-time feedback to the users about their nutrient intake such that individuals can make adjustments to their food intake to meet daily intake goals (e.g., calories, fat, carbohydrates, etc.). Additionally, some programs can connect to the Internet and upload the nutritional information for

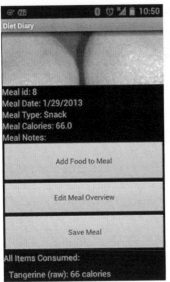

FIGURE 1 Example of an electronic diary implemented as a mobile phone application. Critical information about the consumed meals are recorded and stored in the phone.

further analysis by dietitians and clinicians. Finally, electronic diaries eliminate the issue of illegible handwriting present in paper-based methods [13].

Currently, most mobile phones are equipped with digital cameras that allow individuals to take high quality pictures of their meals. This technological advance has helped to improve the electronic diaries by providing additional information about a meal. Food-intake detection and characterization through food imagery is accomplished by taking photographs of the serving plate with the foods selected by an individual and of the plate's waste. Portion sizes of the food selection are estimated by a trained dietitian in the laboratory who compares the photographs taken by individuals against reference portion photographs of known quantities [14,15]. Estimation of the type and amount of foods consumed can also be done automatically by using methodologies based on image processing (see section 4.2). These estimates are entered into a food analysis program to derive the total mass and energy consumed along with macro- and micronutrients of food selection based on a source database. Accurate results are obtained when the pictures are taken always at the same angle (usually 45 degrees) with the serving plate occupying the entire field of view. This methodology has provided reliable results when used to measure energy intake of adults and children under different settings [16,17]. However, it may suffer from the limitations of self-reported intake as individuals need to remember to take pictures of the foods before and after consumption.

2.2 Body-Attached Sensors

Body-attached sensors monitor physiological processes in the body and focus on one or more stages of the food consumption process: hand-to-mouth gestures, bites, chewing, or swallowing.

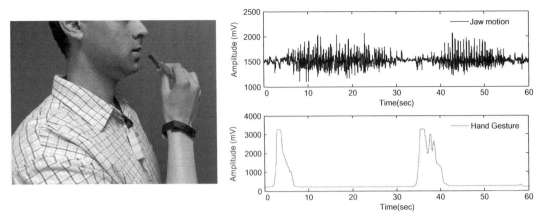

FIGURE 2 Left: Typical hand-to-mouth gesture occurring during food intake. Right: Hand gesture signal (bottom plot) collected during a short episode of food intake. The increase in signal amplitude occurs during a hand-to-mouth gesture related to a bite. The top plot illustrates a chewing sequence (captured by a jaw motion sensor) that follows the detected hand-to-mouth gesture.

2.2.1 Monitoring of Hand Gestures

Quite often, ingestion of food begins with a hand-to-mouth gesture followed by a bite. Consequently, ingestion events can potentially be detected by monitoring hand gestures occurring in daily living. Researchers have proposed a number of approaches integrating different sensor modalities into a wearable device [18–20].

In one approach, a proximity sensor operating in the radio frequency identification band was proposed to monitor hand-to-mouth gestures. This sensor consisted of an RF transmitter module worn on the inner side of the dominant arm and an RF receiver, which is a flat-loop antenna located in a package worn as a pendant around the neck (Figure 2, left) [21]. The location of the antennas was strategically selected such that the highest magnitude of the signal (when the antennas are aligned in parallel) is obtained for a typical hand-to-mouth gesture observed during food intake (Figure 2, left). The behavior of the proximity sensor is illustrated in Figure 2 (right), where a segment of the RF signal received by the antenna (bottom plot) shows two hand-to-mouth gestures. To help visualize these gestures within a food-intake episode, the top plot of Figure 2 (right) shows the chewing sequences occurring immediately after the hand-to-mouth gestures and captured by a jaw motion sensor. While the hand gesture sensor signal may not be sufficient by itself to discriminate hand-to-mouth gestures on various origins, it provides important features that can be used in a pattern-recognition algorithm [22].

In another approach, a watch-like device (Figure 3) was developed for measuring intake through an automatic tracking of wrist motions during hand-to-mouth gestures or "bites" [20]. This device incorporated a micro-electromechanical gyroscope to monitor the radial velocity of the wrist during food intake. The main advantages of this device are the low cost of the sensor and need for only a single wearable item (as compared with the RF sensor).

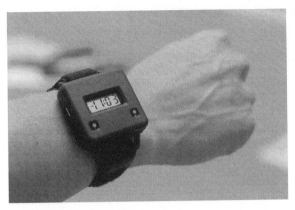

FIGURE 3 A wearable sensor for monitoring hand-to-mouth gestures associated with food intake incorporates a miniature gyroscope for capturing the motion of the wrist. *(Image courtesy of Dr. Adam Hoover [20].)*

Finally, accelerometers have been used for recognition of hand gestures [18,23]. Typically, the systems implementing this type of sensor use 3-axes accelerometers located on the arm to measure acceleration. The resulting signals are transmitted to a PC or smartphone via bluetooth or other wireless protocol for automatic detection of hand gestures.

A common limitation of hand gesture monitoring is that not only food intake, but many other activities can produce hand gestures that may be indistinguishable from gestures during food intake. This limitation stipulates either the use of additional sensors to capture independent indicators of food intake (such as in [24]) or the use of self-report in the form of turning the sensor on/off for each ingestion episode (such as in [20]). Another limitation is that people may use both hands during solid and liquid intake, which would require the use of two devices to reliably detect hand gestures.

2.2.2 Monitoring of Chewing

In a typical ingestion episode, a bite is followed by a sequence of chews and swallows, and this process is repeated throughout an entire meal. Thus, the ingestion of solid foods can be detected by monitoring the chewing process. Several sensing options have been evaluated with this purpose. Surface electromyography (EMG) can sense the activation of jaw muscles during mastication by placing electrodes over the skin surface [25]. Multi-point sheet-type sensors [26] and strain-gauge abutments [27] can also be used to measure bite and chewing forces. However, these sensors are likely to produce variations in an individual's normal mastication patterns because they are placed between the teeth.

Another sensing option is based on the monitoring of sounds produced in the chewing process. During food breakdown, portions of foods are crushed between the teeth, generating vibrations that are conducted through the teeth, the mandible, and the skull [28]. These vibrations are also propagated through the ear canal, which forms a natural cavity where they are audible. Therefore, an acoustic transducer can be placed in the proximity of the ear canal to capture the sounds of chewing. Several studies have been conducted based on this approach [29—32]. To reduce effects of environmental noise, a miniature in-ear microphone is used along with a reference microphone for noise cancellation (Figure 4). The acoustic signal is then processed in order to automatically detect and characterize food intake.

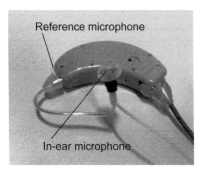

Reference microphone

In-ear microphone

FIGURE 4 A wearable sensor system consisting of: 1) an in-ear microphone to monitor the chewing and swallowing sounds, and 2) a reference microphone to capture environmental sounds. *(Image courtesy of Dr. Sebastian Päßler [32].)*

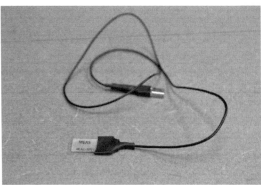 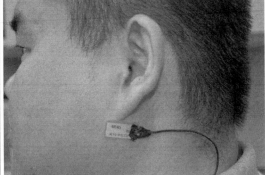

FIGURE 5 Left: Jaw motion sensor consisting of a piezoelectric film element that generates a voltage signal when flexed. Right: Sensor is placed below the earlobe and attached to the skin's surface using medical adhesive or medical tape.

Monitoring of the motion of the mandible (jaw) occurring during chewing is another reliable approach for capturing food intake. Monitoring of jaw motion, rather than monitoring chewing sounds, may result in a simpler and more accurate sensor system, as the sensor is less sensitive to environmental noise. The jaw motion can be detected by monitoring changes in skin curvature due to changes in distance between the mandible and temporal/occipital bones of the skull during chewing. Foil-strain gauges have been tested providing reasonable results but with unacceptably high energy consumption. A good option was found to be an off-the-shelf piezoelectric film strain gauge sensor (Figure 5, left), which can monitor dynamic skin strain with low power consumption [33]. The changes in skin curvature are detected by placing the strain sensor immediately below the earlobe (Figure 5, right). The simple structure of the chewing strain sensor may result in a less intrusive and simpler way to detect food intake. Additionally, this sensor is simply attached to the skin by medical adhesive or medical tape, and may last for a 24-hour period with a low risk of sensor detachment. Finally, the strain sensor is easily available in the market at a low cost.

Chewing has also been used in conjunction with digital imagery through the integration of miniature, lightweight cameras into wearable sensors to facilitate the assessment of dietary intake [34]. These cameras can potentially be worn by individuals throughout the day and can be triggered to automatically take pictures without requiring user input, thus reducing the misreporting bias [35]. An example of this trend is a wearable sensor platform [36] designed as a headset worn on the ear. The sensor consists of a microphone for detection of the chewing sound and a miniature camera directed toward the table where foods are consumed. The chewing activity triggers the camera to capture a video sequence of the food container. This action creates a food-intake log having a series of snapshots and time-stamps that are saved into a file to keep record of the food consumption history.

2.2.3 *Monitoring of Swallowing*

The swallowing process is a sequential, semi-automatic contraction and relaxation of muscles of the tongue, pharynx, and esophagus. Spontaneous swallowing is a functional act involving automatic and involuntary transport of saliva from the mouth to the stomach. When a person is awake, spontaneous swallowing of saliva occurs at a rate that varies from 1 to 2 swallows per minute [37−39]. During sleep, the production of saliva virtually ceases and spontaneous swallows are occasional, occurring generally in association with movement arousals [40,41]. During food intake, the swallowing process involves the passage of the bolus of food (or liquid) from the mouth to the stomach. Studies have demonstrated that the swallowing rate significantly increases during eating, suggesting that monitoring of swallowing episodes may provide suitable information to detect ingestion of food.

Videofluoroscopy and electromyography are considered the gold standard methods for swallowing studies [42,43]. However, the development of a wearable monitor based on any of these methods may not be feasible due to their invasiveness (subcutaneous EMG) and their dependency on bulky, expensive, and possibly unsafe equipment (videofluoroscopy). To overcome this issue, different sensor modalities have been proposed to monitor a particular phenomenon occurring during the swallowing process. For example, neck mounted accelerometers were used to capture throat vibrations, microphones were used to record sounds produced during swallowing, an electroglottograph device was evaluated to monitor changes in the electrical impedance of the neck, and magnetic sensors were implemented to measure the movement of the thyroid cartilage. This section describes some of these sensors in more detail and highlights their advantages and disadvantages.

The use of accelerometers has been proposed as a non-invasive, low-cost alternative for studying swallowing. The acceleration signal is produced during laryngeal displacement and the magnitude of such signal is a measure of the elevation of the larynx during swallowing [44]. Typically, miniature accelerometers are placed on the skin, over the throat, and at the level of the thyroid cartilage of the subject. However, this location may not be suitable for obese individuals due to accumulation of body fat under the chin. Additionally, the sensitivity of accelerometers to body motion and to orientation in the gravity field may also jeopardize the implementation of this sensor modality for freely moving individuals.

Monitoring of swallowing by acoustical means has been relying on miniature microphones that are able to record the sounds generated when the bolus of food passes through the pharynx in the pharyngeal phase of the swallowing process. This characteristic sound can be captured from different locations including inside the ear canal, on the mastoid bone, or over the laryngopharynx. Research has shown that microphones placed over the

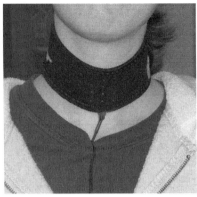

FIGURE 6   Left: Miniature microphone used to monitor swallowing events. Right: Acoustic sensor is placed over the laryngopharynx using a collar.

laryngopharynx (Figure 6) offer the advantage of stronger signals as the sensors are located closer to the origin of the swallowing sound [45].

A novel approach for swallowing detection is based on an electroglottograph (EGG) device. The EGG sensor measures the variation in the transverse electrical impedance across the neck at larynx level. The EGG signal is recorded by applying a high frequency signal (3 MHz) to guard-ring electrodes placed on the surface of the neck. EGG has been widely used for speech and swallowing analysis [46,47], but only recently has been applied to food-intake detection [48]. The passage of the bolus of food during the swallowing process causes significant variations in the electrical impedance across the larynx from the baseline, allowing detection of swallowing. The advantage of this sensor modality is that the EGG signal is virtually insensitive to external acoustic noise because of the physical principles used in EGG measurement. Disadvantages of this approach are related to motion artifacts that may arise during free living and may affect the EGG signal.

Recently, a magnetic sensor system was proposed to detect swallowing events by monitoring the movement of the thyroid cartilage [49]. This system also contained a piezoelectric microphone to detect swallowing sounds. The microphone and the magnetic coils were mounted in a holding unit that was positioned at the neck level. The distance between two coils (oscillation and detection coils) placed on each side of the thyroid cartilage was used for detection of swallowing. The level of voltage induced in the detection coil by a 20 kHz magnetic field produced in the oscillation coil was proportional to the length between coils.

3. SIGNAL PROCESSING AND PATTERN-RECOGNITION METHODS FOR AUTOMATIC DETECTION OF FOOD INTAKE

Wearable sensors convert a physiological event associated with ingestion into an electrical signal. Most often, the sensor signal carries not only information about ingestion but also information about other actions of the individual. For example, an acoustic swallowing sensor will register not only spontaneous and food-related swallowing but will also respond to speech, walking, and other activities. Thus, specialized signal processing and pattern-recognition algorithms are required for discriminating swallowing from other activities, and for detection of food intake from the time sequence of swallows. The following is a description of commonly used signal processing and pattern-recognition methods for analysis of the signals collected by different sensor modalities and automatic detection of food intake.

3.1 Food-Intake Detection from Imagery

Capturing an image of the foods being consumed and writing an entry in a multimedia food diary is, by itself, a method for food-intake detection that captures information about event timing, and in some cases, event duration. Using a digital diary requires no specialized processing to detect food intake. This simplicity carries a major limitation: individuals need to always carry the device and remember to take pictures of every food item they eat, subjecting the method to all of the relevant limitations of self-report. To alleviate this problem, a potential solution is to utilize automatic image recognition to detect food intake from continuous video monitoring. In [50] a wearable device was developed for food-intake detection from imagery. The principal assumption of this approach is that different features from the video footage such as motion patterns, presence of circular objects, surrounding sounds, and time of the day can provide valuable information to detect ingestion episodes. The food-intake detection faces many challenges, including privacy concerns for continuously captured video and the ability to automatically recognize the great variety of foods and make judgment whether the food is being consumed or just viewed by the user.

3.2 Detection of Hand Gestures

The process of ingestion begins when individuals take foods and put them into their mouth. This simple event requires a hand-to-mouth gesture. Automatic detection of hand-to-mouth gestures is thus a fundamental component of any system including a gesture sensor.

Automatic detection of gestures in the RF-based proximity system [21] can be reliably achieved by a simple algorithm detecting when the strength of the proximity signal is above a predefined threshold. Such simplicity of the gesture detection is an advantage of the RF sensor.

Use of inertial sensors in [20] demonstrated a signal pattern associated with a bite that can be detected by more sophisticated signal analysis. The comparison of orientation of the wrist when picking up the food with the orientation of the wrist when putting the food into the mouth found that a roll of the wrist occurred in most cases independently of the utensils or fingers used. Based on this motion pattern, a pattern-recognition algorithm was developed and implemented in a microcontroller to automatically detect bites. The algorithm looked at the velocity of the rolling motion measured by the sensor and at the minimum time between the two rolls present in a single bite. The methodology was tested under laboratory settings to determine the accuracy of bite detection.

Results showed a sensitivity of 94% for "bite" counting during controlled meals and of 86% during uncontrolled meals.

Neither of the described hand gesture detection algorithms can automatically differentiate between food-related and non-food-related hand gestures (e.g., using a napkin, head scratching, etc.) and are thus incapable of directly detecting food intake. Although food-intake episodes may potentially have a relatively high density of hand-to-mouth gestures, the experimental data show a high incidence of hand gestures during the waking day that does not allow simple differentiation based on the rate of gesture occurrence. Nevertheless, hand gestures carry important information that can be used in multi-sensor food-intake detection.

3.3 Food-Intake Detection from Chewing

With a few exceptions, chewing is a reliable indicator of solid food intake. Therefore, many methods attempt to detect food intake through computerized recognition of chewing.

Automatic recognition of chewing was achieved by processing the signal from a piezoelectric strain sensor by a pattern-recognition algorithm [33]. Analysis of the jaw motion signals in the frequency domain showed a strong signal in the frequency range between 1.25 Hz to 2.5 Hz, which results from the rhythmic up-and-down and side-to-side movements of the lower jaw during chewing. Such characteristic jaw movements are absent or less marked during inactivity, talking, or walking. Frequency and time domain features extracted from the strain sensor signal were used to train a subject-independent classifier based on support vector machines (SVM). Results indicated that the models were able to discriminate between chewing and other activities with an averaged accuracy of around 81% and a time resolution of 30 s. Later analysis [51] further improved the recognition results by incorporating features extracted from the voice frequency band in the range of 100 to 250 Hz.

Similar attempts have been performed in automatic recognition of the chewing sounds. Signal processing and pattern-recognition algorithms have been developed to discriminate between chewing sounds and sounds not related to eating or drinking (i.e., speech, ambient noise, etc.). One of the earlier studies implemented a naïve Bayes classifier trained with features from the frequency domain [31]. The pattern-recognition system achieved an overall food-intake recognition rate of 86.6% averaged across two subjects that completed a total of 375 chewing sequences, most of them in laboratory settings. A later analysis of the chewing sound revealed that most of energy is found in the frequency range below 4 kHz [32] and that no characteristic frequency with high energy content can be used to distinguish the chewing process from other activities, thus requiring noise cancelation techniques to reliably detect chewing sounds. Using the signals from the sensor shown in [52], chewing detection was performed by a signal processing algorithm developed to compute the energy of the acoustic signals from each microphone. The ratio between energy values was compared to an adaptive threshold value to discriminate food-intake sounds with an overall accuracy of 85%.

Overall, chewing is a fairly reliable indicator of food intake that can be accurately detected by non-intrusive wearable sensors. This factor makes recognition of chewing one of the more popular approaches to detection of food intake. At the same time, chewing is not a perfect predictor of food intake. Most of liquids are not being chewed during ingestion and thus cannot be detected through chewing. Many semi-liquids (e.g., yogurt) and even some of the solid foods (e.g., fufu) are also not being chewed during consumption. At the same time, there are several non-food items that may be chewed continuously but not swallowed (e.g., chewing gum, chewing tobacco, betel, etc.) thus potentially triggering a false detection. Therefore, food-intake detection from chewing may need additional indicators of food intake (such as swallowing or hand gestures) for reliable detection.

3.4 Food Intake Detection from Swallowing

Swallowing is potentially the most reliable indicator of food intake as all of the ingested food is eventually swallowed. Since spontaneous swallowing happens naturally during the waking hours, detection of food intake from swallowing can be a two-step process: first, individual swallows are recognized and, second, food intake is detected from the time series of swallows. This section presents methods developed for both swallowing recognition and food-intake detection from swallowing.

Several studies performed automatic recognition of swallowing events using accelerometers [53,54]. However, they were mostly focused on establishing objective criteria for detecting swallowing disorders rather than on detection of food intake. Pattern-recognition algorithms were implemented to discriminate normal swallows, dysphagic swallows, and artifacts showing promising results that can potentially be extended to detect food-intake swallows [55–57].

A two-step methodology for automatic detection of food intake based on the acoustical swallowing sensor data was recently presented. The first step used Mel-scale Fourier spectrum features and SVM signal classification to recognize individual swallowing events [58] with average accuracy of 84.7% on a dataset containing more than 64 hours of acoustic data and about 10,000 swallows from 20 subjects. Results indicated that the highest accuracy of identifying individual swallowing events was observed during quiet periods of no food intake (88%) and the lowest accuracy was observed during food-intake periods, where talking and/or background noise was present (82.9%). The swallowing recognition model was further improved by incorporating principal component analysis (PCA) and a smoothing algorithm to improve recognition rates for intra- and inter-subject models [59].

The second step performed the detection of food intake based on the frequency of swallowing. The instantaneous swallowing frequency (*ISF*) averaged over a sliding window of 30 s duration (*EISF*) was used as the key predictor for discriminating between the "intake" and "no intake" classes. The best subject-independent model used a decision threshold T_{FL}^{INGEST} obtained by multiplying the floating average of the swallowing frequency over several epochs with a scaling factor α. The resulting model had the advantage of self-adjusting to individual variations in the swallowing rates. Training and validation of this model demonstrated an average detection accuracy of 87% [39].

Another study used the time sequence of swallows as the main predictor of food-intake activity [60]. The absolute difference in time between a swallow occurring at time t_i and d neighboring swallows was used as features to create food-intake detection models.

A supervised machine learning technique (SVM) was used to create a subject-independent model whereas an unsupervised clustering technique (K-means) was used to create individual models. The swallowing data used to train the models was collected from 18 subjects and contained 4,045 spontaneous saliva swallows and 5,811 food-intake swallows. Results showed that unsupervised individual models presented a better performance than the subject-independent model (93.9% vs. 89.9%), most likely due to ability of unsupervised methods to adapt to individual traits in the data.

Food-intake detection from swallowing has also been studied in conjunction with an electroglottograph sensor and a one-step approach that does rely on recognition of individual swallowing events [48]. A pattern-recognition algorithm was developed to detect food-intake based on the changes in electrical impedance across the larynx. Signals captured by the EGG device were divided into non-overlapping windows of 30 seconds and features were extracted by performing wavelet decomposition. Subject-independent food-intake detection models were trained with these features and artificial neural networks. The models were created using data from a study involving 30 subjects performing unrestricted consumption of four meals each under laboratory settings. The study also included the monitoring of swallowing sounds through a miniature microphone. The performance of the EGG method was compared to an acoustic method using the same one-step algorithm. Results showed a statistically significant difference between the average food-intake detection accuracy for the EGG-based method (90.1%) and the acoustic-based method (83.1%).

In summary, food-intake recognition from swallowing achieves results similar to those from chewing. Also, similarly to chewing, the fundamental challenge of the signal processing and pattern-recognition is differentiation of swallowing and/or food intake from other activities of daily living. Swallowing is present in consumption of all foods: liquids, semi-liquids, and solids, and thus deserves to be one of the prime predictors of food intake.

4. METHODS FOR CHARACTERIZATION OF FOOD INTAKE

Successful detection of food intake is the first step for understanding the ingestive behavior. Once the fact of food intake is established, the process of ingestion and the consumed meal need to be characterized. Particularly, the number and type of food items in a meal, the mass and volume of ingestion, caloric and nutritional content of a meal, and the rate of ingestion can be estimated by analysis of the sensor data. This section presents a review of several approaches aimed at food-intake characterization.

4.1 Recognition of Number of Foods in a Meal

A methodology for recognizing the number of foods consumed is an important first step in characterization of food intake. As with other aspects of technology for food-intake monitoring, the available methods either rely on analysis of imagery or on analysis of physiological signals collected by wearable sensors.

Image processing methods attempt to automatically segment the image into regions that may be attributed to different foods, and potentially identify the food based on image features [61,62]. The recognition procedure starts by determining the region of interest, where the food may be located on an image. This is typically done by detecting round objects such as plates, bowls, or glasses and analyzing the regions inside of the enclosing circle. The next step of the procedure is the multi-scale image segmentation (such as normalized cut [63]) that uses both coarse and fine details to identify image segments that may contain food items. After the segmentation, color and texture features extracted from the image segment are used to recognize a particular food item that may be represented by this segment. The classification may be performed by an SVM or some other robust classifier. Since an image segment may not contain the entire food item, the next step of the procedure merges the segments that are in spatial proximity to each other and have the same class label (belong to the same food item). The outcome processing is the number of food items found in the image and the food type of each food item. While image segmentation is capable of estimating the number of foods in a meal, the authors of [61,62] quantified their approach in terms of accuracy in recognizing specific food types, which varied between 56% for 19 food items [62] to 44% for 32 food items [61].

Use of physiological signals to quantify the number of foods in a meal can be illustrated by a method based on monitoring of chewing and swallowing [64]. The fundamental assumption for the proposed method is that food of different physical properties result in varying chewing and swallowing patterns, allowing differentiating food items without prior knowledge of quantity of the food items consumed. To demonstrate feasibility of this approach, chewing and swallowing data was collected from 17 subjects participating in experiments involving the consumption of five different foods: cheese pizza, yogurt, apple, peanut butter sandwich, and water. Three different types of features were used to segment a meal into different food items. The first feature was the location in time of each swallow, which was important for grouping swallows associated to a certain food type. The second feature was the time to preceding swallow (TPS), which indicated the time difference between a swallow occurring at time t_i and the previous swallow occurring at time t_{i-1}. The last feature was the number of chews preceding a swallow (CPS), which was an indication of the number of chews observed between two consecutive swallows.

Since chewing and swallowing patterns for a given food may vary from individual to individual, the collected data was analyzed with two different unsupervised clustering techniques to identify groups of food within a meal: affinity propagation (AP) and agglomerative hierarchical clustering (AHC). The use of unsupervised learning enabled adaptation to individual traits while preserving the generality of the approach. Results showed that an overall accuracy of 95% was obtained when estimating the number of foods using a model created with AHC technique. On the other hand, an accuracy of 90% was obtained for a model created with AP technique. A limitation of the proposed clustering approach was that the food items were consumed in a predefined sequence and unmixed to eliminate the uncertainty caused by inter-food variation. Further studies are needed to test the clustering algorithm's performance on unrestricted meals with larger variability of food items. However, this methodology for food clustering should be directly applicable to unrestricted meals as food intake is cumulative over time and the consumption normally happens sequentially in a bite-by-bite manner.

Non-invasive monitoring of chewing sounds is another alternative for recognizing the number of foods in a meal. Recognition of the different food consumed in a meal may also be possible as the chewing of foods with different properties (i.e., moisture, crispiness, crunchiness, etc.) may produce characteristic sounds. By means of pattern-recognition techniques, foods with similar chewing sound patterns can potentially be grouped into categories containing foods with similar physical properties. This approach has been studied in [32], where 51 participants consumed seven different types of solid foods and a drink. The chewing sounds were monitored by a miniature microphone inserted into the ear canal, with hidden Markov models recognizing chewing and swallowing events. The chewing sound recognition achieved 79% classification accuracy on the food items tested. These results indicate that analysis of chewing sound is likely to be able to detect the number of foods in the meal, as long as these foods possess distinct physical properties and are consumed with chewing.

4.2 Estimation of Ingested Mass and Energy Intake from Imagery

The human brain can rapidly discern the energy content of foods by a simple visual inspection of different items in a process that involves object categorization, reward assessment, and decision making [65]. In an attempt to mimic this process, image-processing techniques have been used to estimate the food's portion size from imagery, which is then used to calculate the amount of energy intake by using a food database.

The estimation of the portion size from a picture is a difficult task due to the lack of distance information in a typical image. The same object may look bigger or smaller depending on the distance to camera, camera angle, and focal length of the camera. A typical solution to this problem is use of dimensional referents (fiducial marks) in each picture that allow the estimation of the food volume by computer algorithms. Several approaches have been investigated that included different types of referents, such as a checkered tablecloth, calibration cards, checkerboards, and circular plates of known dimensions [15,66,67]. Another option is to use lights from LEDs or laser diodes to produce a spotlight in the field of view when the picture is taken [68]. In most cases, a single picture was used to estimate food's volume and volumes computed from before and after the meal pictures were used to estimate amount of food consumed.

Different approaches have been proposed to estimate food's volume from a single picture [15,68,69]. One approach evaluated different algorithms to determine the location and orientation of an object plane based on the intrinsic parameters of the camera [68]. The results obtained were used to measure dimensional variables, such as length and thickness, by selecting various feature points in the food image. The values obtained for these variables provided an estimate of the food volume. Another approach evaluated different image-processing algorithms to segment (separate) the food items from the background of the picture [15]. The segmented image was used as the input of a process involving camera calibration and 3-D volume reconstruction that automatically estimated the volume of foods. One of the recent studies [70] used image processing techniques to automatically estimate portion weights from the mobile phone images. For the set of 19 different foods, the ratio of predicted weight to known weight varied from 0.89 to 4.61, while for a smaller subset of nine foods the ratio was in the range of 0.8 to 1.2. The largest errors were observed for foods such as lettuce, French fries, and garlic bread, while the smallest errors were obtained for strawberry jam, milk, orange juice, and cheeseburger sandwich. Use of 3-D models that fit the shape of the 2-D food image to a pre-constructed shape such as a cylinder, etc., has been reported to produce more accurate results (Figure 7). In [71], the error of estimating volume for 17 different foods was reported as 3.7%, which may be considered as highly accurate. Comparable results (7.2% average error for five food items) were reported in [72] for another 3-D-model-based method.

Although these methods show promising results towards the automatic estimation of portion sizes, food portion estimation from imagery needs further development. Computer algorithms can use images to estimate the volume of food consumed; however, the research combining automatic weight estimation with food item recognition for automatic energy content estimation is yet to emerge. Indeed, there are many challenges to overcome both in improving the accuracy of volume and mass estimation and in accurate identification and energy density estimation for the great variety of foods that we may find on the table. As a potential solution, food-intake characterization from imagery can be implemented in a semi-automatic process, where a human nutritionist visually identifies each food consumed and the computer algorithm automatically estimates the volume of food and calculates calories and nutrient values from the food database. A promising new direction is use of 3-D cameras or camera motion to assess the volume of food.

4.3 Estimation of Ingested Mass and Energy Intake from Counts of Chews, Swallows, and Hand Gestures

Characterization of food intake has also been studied using metrics derived from chewing, swallowing, and hand-to-mouth gestures. Mathematical models were proposed to estimate mass and energy consumed in a meal using counts of chews and swallows captured by wearable sensors.

In the approach presented in [39], individualized (subject-dependent) linear models were developed based on the hypothesis that chews and swallows can estimate the total mass of food ingested with acceptable accuracy. A model for estimation of the total

FIGURE 7   Example of food's volume estimation using 3-D models. *(Image courtesy of Dr. Mingui Sun [71].)*

ingested mass of solid food used the total number of chews and swallows within a period of ingestion as predictors was

$$M_S = \frac{1}{2}(\overline{M}_{sw}^S \times N_{sw}^S + \overline{M}_{chew} \times N_{chew})$$

where $\overline{M}_{sw}^S$ was subject's average mass per swallow of solid food, N_{sw}^S was the total number of swallows of solid foods, $\overline{M}_{chew}$ was the subject's average mass per chew, and N_{chew} was the total number of chews. The model to estimate the mass of liquids ingested used only the number of swallows as liquid consumption does not involve chewing. Thus, the mass of liquids was calculated as

$$M_L = \overline{M}_{sw}^L \times N_{sw}^L$$

where $\overline{M}_{sw}^L$ was the subject's average mass per swallow of liquid and N_{sw}^L was the total number of swallows of liquid. The values of the parameters $\overline{M}_{sw}^S$, $\overline{M}_{sw}^L$, and $\overline{M}_{chew}$ were statistical estimates of the average mass consumed by an individual in a swallow or a chew. Using individualized instead of population-based models is required due to the high inter-subject variability of the mass-per-chew and mass-per-swallow parameters [39]. These models were implemented and evaluated on data from 16 subjects consuming five different food items in a sequential manner. Results showed that the mass estimation model achieved an average accuracy of 91.8% for solid-food intake and an average accuracy of 83.8% for liquid intake.

In further development of this approach by the same investigator team, a similar method was developed to estimate the total amount of energy consumed in an unrestricted meal [24]. This new method involved two steps. First, the amount of mass ingested was estimated for each food item using the individual models based on chews and swallows. Second, the energy intake was estimated by multiplying the estimated mass by the caloric density (CD) of each consumed food, which was extracted from a nutritional analysis of food imagery by a qualified nutritionist. The total energy consumed was calculated by summation of energy estimates for each food ingested. Prediction models were developed based on training and validation meals. The training meals consisted of three meals of identical size and content that were consumed in three separate visits. Counts of chews and swallows for the training meals were used to estimate the model parameters $\overline{M}_{sw}^S$, $\overline{M}_{sw}^L$, and $\overline{M}_{chew}$. The validation meal was a new meal selection that included different solid foods and drinks from the training meal. The prediction models in this experiment were evaluated in a more realistic scenario of food intake where the dataset included information from a wide variety of foods (a total of 45 different foods) and assumed no restriction in the way the food was consumed. The performance of the models was compared to diet diaries and photographic food records to determine how well the estimation of energy intake matched the estimations obtained with weighed food records (gold-standard method).

Results showed that the models estimated the energy consumed in the training meals with a reporting error of 15.83%, which was significantly lower than the reporting errors obtained for the other methods (27.86% for diet diaries and 19.95% for photographic food records). Additionally, the proposed models presented the lowest reporting bias

(−8.6 kCal compared to −60 kCal for diet diary and 83.6 for photographic records). On the other hand, estimation of energy intake for the validation meals was not significantly different from either diet diaries or photographic methods. This was caused by an increase in the reporting error for the validation meal probably due to the different physical properties of the food items consumed during the training and the validation meals. Additionally, in the validation stage, it was observed that the accuracy of the self-report methods was very high given that this was just a single meal in a controlled setting. As the accuracy of self-report tends to decrease with the duration of the recordings [7] (due to either underestimation or subjects forgetting to take pictures of food), it is reasonable to think that, over several days of observation, energy estimation models based on chews and swallows may produce better results than self-report as they do not rely on subject participation.

The use of chewing and swallowing for mass estimation has been also explored in the sensors relying on acoustical detection of chewing sound. In [73], bite weight prediction was attempted based on four features extracted from the chewing sounds. On a test with three foods (potato chips, lettuce, and apple), the reported error varied between 19% to 31% for individual, subject-calibrated models, and between 41% to 62% for subject-independent models. Even though these results were obtained on a small set of foods and smaller population (eight subjects), they largely confirm the individuality of mass ingestion as expressed in the counts of chews.

5. APPLICATIONS

5.1 Laboratory Vs. Free-Living Monitoring

Under laboratory conditions, monitoring of food intake through paper-based methods, electronic diaries, or wearable sensors may provide reasonable agreement between the food records and the actual amounts eaten. However, laboratory research presents significant deficiencies when compared to free-living research [74]. Most of these deficiencies are caused by real-world variability that directly affects the eating behavior of individuals and is usually missing in laboratory settings. Food selection, timing of meals (i.e., determined by external schedules, work, or social commitments), food-intake environments, and food-intake behavior may vary dramatically day-to-day and between individuals in free living, while lab experiments typically are severely constrained. Additionally, important changes in the eating behavior of an individual during the course of a day and over different days (weekdays vs. weekends) occurring in free living may be controlled or eliminated in laboratory experiments [75]. Finally, the importance of variables can be overestimated and important effects can be missed as laboratory experiments are usually short in duration. Consequently, monitoring of eating behavior of individuals under free-living conditions is far more challenging than laboratory experiments.

Monitoring of food intake in the real world and over extended periods of time generates more complex datasets than monitoring in the laboratory. Methodologies developed for automating the process of food-intake detection should be able to handle the inherent intra- and inter-subject variability. The intra-subject variability is increased by the different activities performed by an individual over the course of a day (i.e., walking, talking, eating, sleeping, working, etc.). The inter-subject variability is reinforced by the diversity of the population, which, in some cases, may have different eating patterns and lifestyles. These issues make it difficult to generalize food-intake models created with laboratory data to free-living data while maintaining an acceptable performance. Ideally, methods of food-intake detection and characterization should be created using innovative methodologies based on free-living data.

5.2 Wearable Devices for Free-Living Monitoring

A novel wearable device, the Automatic Ingestion Monitor (AIM), has been developed and evaluated for objective monitoring of food intake under free-living conditions (Figure 8) [24]. AIM presented three major benefits over self-reported intake. First, AIM is a wearable device that has the ability to monitor 24 hours of ingestive behavior without relying on self-report or any other actions from subjects. AIM wirelessly integrated three different sensor modalities for an accurate monitoring: a jaw motion sensor to monitor chewing, a proximity sensor to monitor hand-to-mouth gestures, and an accelerometer to

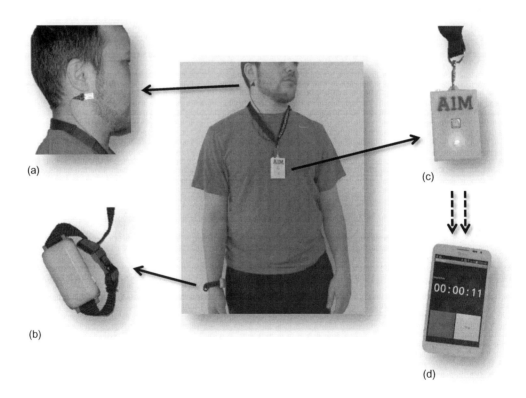

FIGURE 8   The Automatic Ingestion Monitor (AIM) consisting of four main parts: (a) the jaw motion sensor, (b) the wireless module, (c) the proximity sensor, and (d) the smartphone.

monitor body motion. Second, AIM is able to reliably detect food-intake episodes in the presence of real-life artifacts using a robust pattern-recognition methodology for detection of food intake. The detection methodology contains several steps, such as sensor information fusion, feature extraction, and classification. The sensor fusion step removes portions of the signal that cannot be food intake based on statistically derived rules. For example, it is highly uncommon to eat solid foods during moderate to vigorous exercise, or during sleep. Both of these activities (exercise and sleeping) can be reliably detected from the accelerometer signal and corresponding signal intervals not included into further consideration for food-intake detection. The feature extraction step computes a number of time, frequency, and time-frequency domain features from the sensor signals. Food-intake detection is based on an artificial neural network implementing a subject-independent classification model that requires no individual calibration. Third, the AIM device and food-intake detection methodology were validated in an objective study where an average food-intake recognition rate of 89.8% was achieved. Individuals with origins from five different countries and having different lifestyles and ingestive behaviors participated in the validation study. They wore AIM in free living during 24 hours without any restrictions on their eating behavior and activities.

The results of the validation study revealed that AIM can potentially provide an accurate prediction of the food-intake episodes occurring over the course of a day in a free-living population. However, several questions remain to be answered. One question is related to the capability of AIM for detecting liquid intake. In the validation study, the results showed the recognition rate for solid food intake only. Previous studies suggest that certain intake of liquids (such as gulping large quantities of a drink) may be detected through the monitoring of jaw motion [33] while others, such as sipping, may be undetected. Another question is related to the acceptance of the device by subjects. AIM was designed as a pendant device worn on a lanyard around the neck, which intended to satisfy the need for a socially acceptable device; further miniaturization of the device is needed to make it less obtrusive. Finally, although the food-intake detection was performed offline, the ultimate goal of AIM is to perform real-time recognition and characterization of food intake and to deliver feedback about an individual's intake behavior.

6. SUMMARY AND CONCLUSIONS

Detection and categorization of food intake is a difficult task given the variability and complexity of human food-intake behavior. New methodologies are constantly being developed attempting to achieve objective and accurate measurements of ingestion. The current trend is to replace inaccurate methods relying on self-report by new, less burdensome methods that require minimal participation of subjects.

Historically, portable electronic food diaries first facilitated the reporting task by allowing individuals to simply search and log the consumed foods on mobile phones or PDAs. The electronic diaries evolved to include image capture to further improve assessment of food intake. Complex image processing algorithms are being developed to facilitate image analysis. However, even modern food diaries and food imagery require active participation of the subjects in reporting their intake, thus misreporting and underreporting may still affect the measurements.

Development of body-worn sensors for food-intake detection and characterization has gained substantial interest in recent years. From the functional perspective, the approaches presented in this chapter show that wearable sensors such as microphones, accelerometers, and piezoelectric and magnetic sensors are capable of capturing information about events occurring during the eating process (hand-to-mouth gestures, bites, chewing, and swallowing). Most of the sensor modalities presented in this chapter also respond to excitations from the sources other than the physiological processes of interest, thus requiring sophisticated signal processing and pattern-recognition algorithms to separate artifacts from food intake. The future in development of wearable sensors for monitoring of ingestion lies in further miniaturization and advancements in reliability and functionality, especially with the focus on characterization of food intake.

The social acceptability of wearable devices is another open issue that needs to be carefully addressed in future research. Most of the current research focuses mainly on assessing the functionality of the device and does not present comprehensive analysis on the sensor's burden. Wearable sensors for monitoring ingestive behavior should be non-invasive, unobtrusive, comfortable, and should minimally impact the way people eat. If some of these requirements are missing, people would tend to be non-compliant and abandon the device. Moreover, if the ingestive behavior is significantly altered by the presence of the sensors, then the benefits of using such sensors may be greatly reduced.

Computer algorithms process signals captured by the sensors to detect and discriminate ingestion episodes from other activities. In the majority of the approaches food intake is detected from recognition of chewing or swallowing, potentially in combination with each other or with hand gestures. To date, no published research has reported on the use of hand gestures alone for detection of food intake. Automatic recognition of jaw motion during chewing or chewing sounds translates into 80% to 90% food-intake detection accuracy in most of the reported studies, typically performed under laboratory conditions. Similar accuracies have been reported for swallowing-based approaches. The variability of free-living behaviors may negatively impact the accuracy of these methods, and thus most of them need a thorough testing in the real world. Another major consideration is use of subject-dependent (individual) models vs. subject-independent (group) models. For example, acoustical recognition of swallowing demands an individual model as the swallowing sound varies from individual to individual. On the other hand, jaw motion during chewing exhibits lower variability and thus is easier to use in a group model. Also, use of unsupervised machine learning methods may help with adapting to individual traits in the sensor signals. Finally, the computational demands of the algorithms should also be taken into consideration. Many of the reported algorithms rely on computationally intensive time-frequency decompositions (e.g., wavelets) and classifiers (e.g., support vector machines) that may be unsuitable for limited computational resources of a wearable sensor. Future research should strive to develop robust, subject-independent, and computationally efficient methods of food-intake detection and characterization.

Food-intake characterization is still one of the biggest challenges facing eating behavior research. The diversity of foods makes it problematic to automatically and accurately capture energy intake through wearable sensors. Although many complex algorithms have been investigated, obtaining accurate and reliable algorithms for estimating energy intake is an active topic of research. Multi-modal fusion and merging of information acquired from different sources, such as physiological sensor and food imagery, is a promising approach in improving the accuracy of the mass, caloric, and nutritional intake estimates. New sensor modalities that directly or indirectly measure the nutritional content of food, possibly through intra-oral sensing, may need to be developed to improve reliability of such estimates. Similarly to food-intake detection, use of group, rather than subject-dependent models, is an important consideration in the algorithm development.

Given the early age of wearable sensors, most of the studies to date have relied on laboratory experiments to test the feasibility of suggested approaches. Food-intake detection and characterization becomes substantially more complex in free living due to the presence of artifacts originated by real-world activities that are controlled or missing in a laboratory. Food intake in free living is also more variable and complex than any laboratory or cafeteria study. Therefore, it is critical to test any proposed methodologies in realistic conditions of free living, and the expectation is that future research will reveal more and more of such experiments.

In summary, ongoing research on use of wearable sensor technology for study of ingestive behavior is addressing the need for innovative methodologies for dietary assessment. The development of wearable devices that integrate different non-invasive sensor modalities seems to be the main route selected by researchers towards an objective and accurate monitoring of food intake. However, a substantial number of challenges have to be resolved before a practical, accurate, and non-invasive monitoring of food intake becomes reality. A successful implementation of such devices and methodologies would have tremendous impact on the population as it may help to study and correct ingestive behaviors associated with obesity and eating disorders.

References

[1] K.M. Flegal, M.D. Carroll, B.K. Kit, C.L. Ogden, Prevalence of Obesity and Trends in the Distribution of Body Mass Index Among US Adults, 1999-2010, JAMA J. Am. Med. Assoc. 307 (5) (2012) 491–497.

[2] S.J. Olshansky, D.J. Passaro, R.C. Hershow, J. Layden, B.A. Carnes, J. Brody, et al., A potential decline in life expectancy in the United States in the 21st century, N. Engl. J. Med. 352 (11) (2005) 1138–1145.

[3] C.G. Fairburn, P.J. Harrison, Eating disorders, Lancet 361 (9355) (2003) 407–416.

[4] J.I. Hudson, E. Hiripi, H.G. Pope Jr, R.C. Kessler, The prevalence and correlates of eating disorders in the National Comorbidity Survey Replication, Biol. Psychiatry 61 (3) (2007) 348–358.

[5] D. Sánchez-Carracedo, D. Neumark-Sztainer, G. López-Guimerà, Integrated prevention of obesity and eating disorders: barriers, developments and opportunities, Public Health Nutr. 15 (12) (2012) 2295–2309.

[6] F.E. Thompson, A.F. Subar, Dietary assessment methodology, Nutrition in the Prevention and Treatment of Disease, second ed., Academic Press, San Diego, CA, 2008.

[7] A.H.C. Goris, E.P. Meijer, K.R. Westerterp, Repeated measurement of habitual food intake increases under-reporting and induces selective under-reporting, Br. J. Nutr. 85 (05) (2001) 629–634.

[8] A.H. Goris, M.S. Westerterp-Plantenga, K.R. Westerterp, Undereating and underrecording of habitual food intake in obese men: selective underreporting of fat intake, Am. J. Clin. Nutr. 71 (1) (2000) 130–134.

[9] S.D. Poppitt, D. Swann, A.E. Black, A.M. Prentice, Assessment of selective under-reporting of food intake by both obese and non-obese women in a metabolic facility, Int. J. Obes. Relat. Metab. Disord. J. Int. Assoc. Study Obes 22 (4) (1998) 303–311.

[10] M.-J. Park, H.-S. Kim, K.-S. Kim, Cellular phone and Internet-based individual intervention on blood pressure and obesity in obese patients with hypertension, Int. J. Med. Inf. 78 (10) (2009) 704–710.

[11] S.-I. Kim, H.-S. Kim, Effectiveness of mobile and internet intervention in patients with obese type 2 diabetes, Int. J. Med. Inf. 77 (6) (2008) 399–404.

[12] NDL/FNIC Food Composition Database Home Page. [Online]. Available: < http://ndb.nal.usda.gov/ >. (Accessed: 02.07.14)

[13] C. Johannes, J. Woods, S. Crawford, H. Cochran, D. Tran, B. Schuth, Electronic versus paper instruments for daily data collection, Ann. Epidemiol. 10 (7) (2000) 457.

[14] C.K. Martin, H. Han, S.M. Coulon, H.R. Allen, C.M. Champagne, S.D. Anton, A novel method to remotely measure food intake of free-living individuals in real time: the remote food photography method, Br. J. Nutr. 101 (03) (2009) 446–456.

[15] F. Zhu, M. Bosch, I. Woo, S.Y. Kim, C.J. Boushey, D.S. Ebert, et al., The Use of Mobile Devices in Aiding Dietary Assessment and Evaluation, IEEE J. Sel. Top. Signal Process. 4 (4) (2010) 756–766.

[16] D.A. Williamson, H.R. Allen, P.D. Martin, A.J. Alfonso, B. Gerald, A. Hunt, Comparison of digital photography to weighed and visual estimation of portion sizes, J. Am. Diet. Assoc. 103 (9) (2003) 1139–1145.

[17] J.A. Higgins, A.L. LaSalle, P. Zhaoxing, M.Y. Kasten, K.N. Bing, S.E. Ridzon, et al., Validation of photographic food records in children: are pictures really worth a thousand words? Eur. J. Clin. Nutr. 63 (8) (2009) 1025–1033.

[18] M. Popa, Hand gesture recognition based on accelerometer sensors, 2011 7th Int. Conf. Netw. Comput. Adv. Info. Manage. (NCM) (2011) 115–120.

[19] H. Junker, O. Amft, P. Lukowicz, G. Tröster, Gesture spotting with body-worn inertial sensors to detect user activities, Pattern Recognit. 41 (6) (2008) 2010–2024.

[20] Y. Dong, A. Hoover, J. Scisco, E. Muth, A new method for measuring meal intake in humans via automated wrist motion tracking, Appl. Psychophysiol. Biofeedback 37 (3) (2012) 205–215.

[21] P. Lopez-Meyer, Y. Patil, T. Tiffany, E. Sazonov, Detection of Hand-to-Mouth Gestures Using a RF Operated Proximity Sensor for Monitoring Cigarette Smoking, Open Biomed. Eng. J. 9 (2013) 41–49.

[22] J.M. Fontana, M. Farooq, E. Sazonov, Estimation of Feature Importance for Food Intake Detection Based on Random Forests, 2013 Annu. Int. Conf. IEEE Eng. Med. Biol. Soc.(EMBC) (2013).

[23] R. Xu, S. Zhou, W.J. Li, MEMS Accelerometer Based Nonspecific-User Hand Gesture Recognition, IEEE Sens. J. 12 (5) (2012) 1166–1173.

[24] J.M. Fontana, M. Farooq, E. Sazonov, Automatic Ingestion Monitor: A Novel Wearable Device for Monitoring of Ingestive Behavior, IEEE Trans. Biomed. Eng. 61 (no. 6) (2014) 1772–1779 < http://ieeexplore.ieee.org/xpl/articleDetails.jsp?reload = true&arnumber = 6742586 >. (Accessed 02.07.14)

[25] K. Fueki, T. Sugiura, E. Yoshida, Y. Igarashi, Association between food mixing ability and electromyographic activity of jaw-closing muscles during chewing of a wax cube, J. Oral Rehabil. 35 (5) (2008) 345–352.

[26] K. Kohyama, E. Hatakeyama, T. Sasaki, T. Azuma, K. Karita, Effect of sample thickness on bite force studied with a multiple-point sheet sensor, J. Oral Rehabil. 31 (4) (2004) 327–334.

[27] V.A. Bousdras, J.L. Cunningham, M. Ferguson-Pell, M.A. Bamber, S. Sindet-Pedersen, G. Blunn, et al., A novel approach to bite force measurements in a porcine model in vivo, Int. J. Oral Maxillofac. Surg. 35 (7) (2006) 663–667.

[28] P.J. Lillford, The Materials Science of Eating and Food Breakdown, MRS Bull. 25 (12) (2000) 38–43.

[29] J. Nishimura and T. Kuroda, "Eating habits monitoring using wireless wearable in-ear microphone," in 3rd International Symposium on Wireless Pervasive Computing, 2008. ISWPC 2008, 2008, pp. 130–132.

[30] M. Shuzo, S. Komori, T. Takashima, G. Lopez, S. Tatsuta, S. Yanagimoto, et al., Wearable Eating Habit Sensing System Using Internal Body Sound, J. Adv. Mech. Des. Syst. Manuf 4 (1) (2010) 158–166.

[31] O. Amft, A wearable earpad sensor for chewing monitoring, 2010 IEEE Sens. (2010) 222–227.

[32] S. Päßler, M. Wolff, W.-J. Fischer, Food intake monitoring: an acoustical approach to automated food intake activity detection and classification of consumed food, Physiol. Meas. 33 (6) (2012) 1073–1093.

[33] E. Sazonov, J.M. Fontana, A Sensor System for Automatic Detection of Food Intake Through Non-Invasive Monitoring of Chewing, IEEE Sens. J. 12 (5) (2012) 1340–1348.

[34] A.R. Doherty, S.E. Hodges, A.C. King, A.F. Smeaton, E. Berry, C.J.A. Moulin, et al., Wearable Cameras in Health: The State of the Art and Future Possibilities, Am. J. Prev. Med. 44 (3) (2013) 320–323.

[35] G. O'Loughlin, S.J. Cullen, A. McGoldrick, S. O'Connor, R. Blain, S. O'Malley, et al., Using a Wearable Camera to Increase the Accuracy of Dietary Analysis, Am. J. Prev. Med. 44 (3) (2013) 297–301.

[36] J. Liu, E. Johns, L. Atallah, C. Pettitt, B. Lo, G. Frost, et al., An Intelligent Food-Intake Monitoring System Using Wearable Sensors, 2012 Ninth Int. Conf. Wearable Implantable Body Sens. Netw.(BSN) (2012) 154—160.

[37] C.S.C. Lear, J.B. Flanagan Jr., and C.F.A. Moorrees, The frequency of deglutition in man Arch. Oral Biol., 10, 1, pp. 83—99, IN13—IN15.

[38] M. Pehlivan, N. Yüceyar, C. Ertekin, G. Çelebi, M. Ertaş, T. Kalayci, et al., An electronic device measuring the frequency of spontaneous swallowing: Digital Phagometer, Dysphagia 11 (4) (1996) 259—264.

[39] E. Sazonov, S.A.C. Schuckers, P. Lopez-Meyer, O. Makeyev, E.L. Melanson, M.R. Neuman, et al., Toward Objective Monitoring of Ingestive Behavior in Free-living Population, Obesity 17 (10) (2009) 1971—1975.

[40] I. Lichter, R.C. Muir, The pattern of swallowing during sleep, Electroencephalogr. Clin. Neurophysiol. 38 (4) (1975) 427—432.

[41] W.J. Dodds, The physiology of swallowing, Dysphagia 3 (4) (1989) 171—178.

[42] T.A. Hughes, P. Liu, H. Griffiths, B.W. Lawrie, C.M. Wiles, Simultaneous electrical impedance tomography and videofluoroscopy in the assessment of swallowing, Physiol. Meas. 17 (2) (1996) 109—119.

[43] D.S. Cooper, A.L. Perlman, Electromyography in the functional and diagnostic testing of deglutition, Deglutition Its Disord. Anat. Physiol. Clin. Diagn. Manag. (1996) 255—285.

[44] N.P. Reddy, A. Katakam, V. Gupta, R. Unnikrishnan, J. Narayanan, E.P. Canilang, Measurements of acceleration during videofluorographic evaluation of dysphagic patients, Med. Eng. Phys. 22 (6) (2000) 405—412.

[45] E. Sazonov, S. Schuckers, P. Lopez-Meyer, O. Makeyev, N. Sazonova, E.L. Melanson, et al., Non-invasive monitoring of chewing and swallowing for objective quantification of ingestive behavior, Physiol. Meas. 29 (5) (2008) 525—541.

[46] J.L. Schultz, A.L. Perlman, D.J. VanDaele, Laryngeal movement, oropharyngeal pressure, and submental muscle contraction during swallowing, Arch. Phys. Med. Rehabil. 75 (2) (1994) 183—188.

[47] S. Nozaki, J. Kang, I. Miyai, T. Matsumura, Electroglottographic evaluation of swallowing in Parkinson's disease, Rinshō Shinkeigaku Clin. Neurol 34 (9) (1994) 922—924.

[48] M. Farooq, J.M. Fontana, E. Sazonov, A novel approach for food intake detection using electroglottography, Physiol. Meas. 35 (no. 5) (2014) 739 < http://iopscience.iop.org/0967-3334/35/5/739 >. (Accessed 02.07.14)

[49] A. Kandori, T. Yamamoto, Y. Sano, M. Oonuma, T. Miyashita, M. Murata, et al., Simple Magnetic Swallowing Detection System, IEEE Sens. J. 12 (4) (2012) 805—811.

[50] M. Sun, J.D. Fernstrom, W. Jia, S.A. Hackworth, N. Yao, Y. Li, et al., A Wearable Electronic System for Objective Dietary Assessment, J. Am. Diet. Assoc. 110 (2010) 45—47.

[51] J.M. Fontana, E.S. Sazonov, A robust classification scheme for detection of food intake through non-invasive monitoring of chewing, 2012 Annu. Int. Conf. IEEE Eng. Med. Biol. Soc.(EMBC) (2012) 4891—4894.

[52] S. Passler, W.-J. Fischer, Food Intake Activity Detection Using a Wearable Microphone System, 2011 7th Int. Conf. Intell. Environ. (IE) (2011) 298—301.

[53] K. Takahashi, M.E. Groher, K. Michi, Methodology for detecting swallowing sounds, Dysphagia 9 (1) (1994) 54—62.

[54] J. Lee, C.M. Steele, T. Chau, Time and time—frequency characterization of dual-axis swallowing accelerometry signals, Physiol. Meas. 29 (9) (2008) 1105—1120.

[55] A. Das, N.P. Reddy, J. Narayanan, Hybrid fuzzy logic committee neural networks for recognition of swallow acceleration signals, Comput. Methods Programs Biomed. 64 (2) (2001) 87—99.

[56] J. Lee, S. Blain, M. Casas, D. Kenny, G. Berall, T. Chau, A radial basis classifier for the automatic detection of aspiration in children with dysphagia, J. NeuroEng. Rehabil. 3 (1) (2006) 14.

[57] S. Damouras, E. Sejdic, C.M. Steele, T. Chau, An Online Swallow Detection Algorithm Based on the Quadratic Variation of Dual-Axis Accelerometry, IEEE Trans. Signal Process. 58 (6) (2010) 3352—3359.

[58] E. Sazonov, O. Makeyev, P. Lopez-Meyer, S. Schuckers, E. Melanson, M. Neuman, Automatic detection of swallowing events by acoustical means for applications of monitoring of ingestive behavior, IEEE Trans. Biomed. Eng. 57 (3) (2010) 626—633.

[59] O. Makeyev, P. Lopez-Meyer, S. Schuckers, W. Besio, E. Sazonov, Automatic food intake detection based on swallowing sounds, Biomed. Signal Process. Control 7 (6) (2012) 649—656.

[60] P. Lopez-Meyer, O. Makeyev, S. Schuckers, E. Melanson, M. Neuman, E. Sazonov, Detection of Food Intake from Swallowing Sequences by Supervised and Unsupervised Methods, Ann. Biomed. Eng. 38 (8) (2010) 2766−2774.

[61] F. Zhu, M. Bosch, N. Khanna, C.J. Boushey, E.J. Delp, Multilevel segmentation for food classification in dietary assessment, 2011 7th Int. Symp. Image Signal Process. Anal. (ISPA) (2011) 337−342.

[62] F. Zhu, M. Bosch, T. Schap, N. Khanna, D.S. Ebert, C.J. Boushey, et al., Segmentation Assisted Food Classification for Dietary Assessment, Proc. SPIE. 7873 (2011) 78730B.

[63] J. Shi, J. Malik, Normalized Cuts and Image Segmentation, IEEE Trans. Pattern Anal. Mach. Intell. 22 (1997) 888−905.

[64] P. Lopez-Meyer, S. Schuckers, O. Makeyev, J.M. Fontana, E. Sazonov, Automatic identification of the number of food items in a meal using clustering techniques based on the monitoring of swallowing and chewing, Biomed. Signal Process. Control 7 (5) (2012) 474−480.

[65] U. Toepel, J.-F. Knebel, J. Hudry, J. le Coutre, M.M. Murray, The brain tracks the energetic value in food images, NeuroImage 44 (3) (2009) 967−974.

[66] W. Jia, Y. Yue, J.D. Fernstrom, Z. Zhang, Y. Yang, M. Sun, 3D localization of circular feature in 2D image and application to food volume estimation, 2012 Annu. Int. Conf. IEEE Eng. Med. Biol. Soc. (EMBC) (2012) 4545−4548.

[67] M. Sun, Q. Liu, K. Schmidt, L. Yang, N. Yao, J. D. Fernstrom, M. H. Fernstrom, J. P. DeLany, and R. J. Sclabassi, Determination of food portion size by image processing in 30th Annual International Conference of the IEEE Engineering in Medicine and Biology Society, 2008. EMBS 2008, 2008, pp. 871−874. <http://ieeexplore.ieee.org/xpls/abs_all.jsp?arnumber=4649292&tag=1>. (Accessed 02-July-2014)

[68] W. Jia, Y. Yue, J.D. Fernstrom, N. Yao, R.J. Sclabassi, M.H. Fernstrom, et al., Imaged based estimation of food volume using circular referents in dietary assessment, J. Food Eng. 109 (1) (2012) 76−86.

[69] C. K. Martin, S. Kaya, and B. K. Gunturk, Quantification of food intake using food image analysis in Annual International Conference of the IEEE Engineering in Medicine and Biology Society, 2009. EMBC 2009, 2009, pp. 6869−6872. <http://ieeexplore.ieee.org/xpls/abs_all.jsp?arnumber=5333123>. (Accessed 02-July-2014)

[70] C.D. Lee, J. Chae, T.E. Schap, D.A. Kerr, E.J. Delp, D.S. Ebert, et al., Comparison of known food weights with image-based portion-size automated estimation and adolescents' self-reported portion size, J. Diabetes Sci. Technol. 6 (2) (2012) 428−434.

[71] H.-C. Chen, W. Jia, Y. Yue, Z. Li, Y.-N. Sun, J.D. Fernstrom, et al., Model-based measurement of food portion size for image-based dietary assessment using 3D/2D registration, Meas. Sci. Technol. 24 (10) (2013) 105701.

[72] C. Xu, Y. He, N. Khanna, C. Boushey, E. Delp, Model-based food volume estimation using 3D pose, Proc. 2013 IEEE Int. Conf. Image Process. (2013) 2534−2538September 15-18, Melbourne, Australia

[73] O. Amft, M. Kusserow, G. Troster, Bite Weight Prediction From Acoustic Recognition of Chewing, IEEE Trans. Biomed. Eng. 56 (6) (2009) 1663−1672.

[74] J.M. de Castro, Eating behavior: lessons from the real world of humans, Nutrition 16 (10) (2000) 800−813.

[75] J.M. de Castro, Seasonal rhythms of human nutrient intake and meal pattern, Physiol. Behav. 50 (1) (1991) 243−248.

Index

Note: Page numbers followed by "*f*," "*t*," and "*b*" refer to figures, tables, and boxes, respectively.

CPI Antony Rowe
Eastbourne, UK
October 26, 2014